Introduction to
Biotechnology

Introduction to
Biotechnology

Second Edition

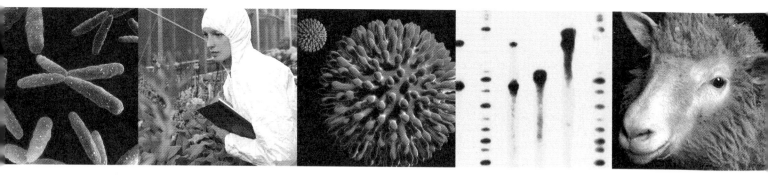

William J. Thieman

Ventura College, Emeritus

Michael A. Palladino

Monmouth University

PEARSON

Benjamin Cummings

San Francisco Boston New York
London Toronto Sydney Tokyo Singapore Madrid
Mexico City Munich Paris Cape Town Hong Kong Montreal

Editor-in-Chief: Beth Wilbur
Executive Director of Development: Deborah Gale
Acquisitions Editor: Becky Ruden
Project Editor: Leata Holloway
Managing Editor: Michael Early
Production Supervisor: Lori Newman
Production Management: Maria McColligan, Nesbitt Graphics, Inc.
Copy Editor: Anne Lesser
Compositor: Nesbitt Graphics, Inc.
Design Manager: Mark Ong
Interior Designer: Jerilyn Bockorick, Nesbitt Graphics, Inc.
Cover Designer: Rich Leeds, Big Wig Design
Illustrators: Nesbitt Graphics, Inc.
Photo Researcher: Maureen Spuhler
Director, Image Resource Center: Melinda Patelli
Image Rights and Permissions Manager: Zina Arabia
Image Permissions Coordinator: Elaine Soares
Manufacturing Buyer: Michael Penne
Executive Marketing Manager: Lauren Harp
Text printer: Edwards Brothers
Cover printer: Phoenix Color Corp.

Cover Photo Credits (from left to right): Chromosome, Sebastian Kaulitzki/Shutterstock; Scientist inspecting plants, Image Source/Veer; VIRUS, Sebastian Kaulitzki/Shutterstock; DNA image, Courtesy of Orchid Cellmark, Inc., Germantown, Maryland; Dolly, the cloned sheep, Najlah Feanny/CORBIS SABA; and Pipette background image, Caren Brinkema/Science Faction/Getty Images

Library of Congress Cataloging-in-Publication Data
Thieman, William J.
  Introduction to biotechnology / William J. Thieman, Michael A. Palladino. — 2nd ed.
    p.; cm.
Includes bibliographical references and index.
  ISBN 978-0-321-49145-9
  1. Biotechnology. I. Palladino, Michael Angelo. II. Title.
  [DNLM: 1. Biotechnology. 2. Genetic Techniques. 3. Molecular Biology. TP 248.2 T433i 2009]

  TP248.2.T49 2009
  660.6—dc22
2008019077

ISBN 0321491459/9780321491459

PEARSON

Benjamin
Cummings

2 3 4 5 6 7 8 9 10—EDB—13 12 11 10 09
www. pearsonhighered. com

To my wife Billye, the love of my life,
and to the hundreds of biotechnology graduates
who are now doing good science at biotechnology companies
and loving every minute of it.

W. J. T.

To Maria and Angelo, my parents,
for providing all the love and support
a family could ask for.

M. A. P.

About the Authors

Mr. Bill Thieman in the lab.

Dr. Palladino in the lab with undergraduate
biology student researchers.

William J. Thieman taught biology at Ventura College for 35 years and biotechnology for 11 years before retiring from full time teaching in 2005. He continues to teach the biotechnology course at Ventura parttime. He received his B.A. in biology from California State University at Northridge in 1966 and his M.A. degree in Zoology in 1969 at UCLA. In 1993, he started a biotechnician training program at Ventura College where he has been teaching since 1970. In 1995, he added laboratory skills components to the course and articulated it as a state-approved vocational program.

Mr. Thieman has taught a broad range of undergraduate courses including general, human, and cancer biology. He received the Outstanding Teaching Award from the National Biology Teachers Association in 1996 and the 1997 and 2000 Student Success Award from the California Community Colleges Chancellor Office. The Economic Development Association presented the biotechnology training program at Ventura College its 1998 Program for Economic Development Award for its work with local biotechnology companies. His success at acquiring grants to support the program was recognized at the National Center for Resource Development Conference in 2007.

Michael A. Palladino is Dean of the School of Science, Technology and Engineering, and an Associate Professor of Biology at Monmouth University in West Long Branch, New Jersey. He received his B.S. degree in Biology from Trenton State College (now known as The College of New Jersey) in 1987, and his Ph.D. in Anatomy and Cell Biology from the University of Virginia in 1994. From 1994 to 1999, he was a faculty member at Brookdale Community College in Lincroft, New Jersey. He joined the Monmouth faculty in 1999.

Dr. Palladino has taught a wide range of undergraduate courses. He has received several awards for research and teaching including the Distinguished Teacher Award from Monmouth University, Caring Heart Award from the New Jersey Association for Biomedical Research, the New Investigator Award of the American Society of Andrology, and the Outstanding Colleague Award from Brookdale Community College. At Monmouth, he has an active lab of undergraduates involved in research on the cell and molecular biology of male reproductive organs. He is founder and director for the New Jersey Biotechnology Educators Consortium, a statewide association for biotechnology teachers.

Dr. Palladino is author of *Understanding the Human Genome Project*, the first volume in the *Benjamin Cummings Special Topics in Biology Series* for which he also serves as series editor. Dr. Palladino recently joined the writing team of W.S. Klug, M.R. Cummings and C.A. Spencer on *Concepts of Genetics* and *Essentials of Genetics*, both published by Benjamin Cummings.

Preface

It is hard to imagine a more exciting time to be studying biotechnology. Advances are occurring at a dizzying pace, and biotechnology has made an impact on many aspects of our everyday lives. *Introduction to Biotechnology* is the first biotechnology textbook written specifically for the diverse backgrounds of undergraduate students. Appropriate for students at both two- and four-year schools and vocational technical schools, *Introduction to Biotechnology* provides students with the tools for practical success in the biotechnology industry through its balanced coverage of molecular biology, details on contemporary techniques and applications, integration of ethical issues, and career guidance.

Introduction to Biotechnology was designed with several major goals in mind. The text aims to provide:

- An engaging and easy-to-understand narrative that is appropriate for a diverse student audience with varying levels of scientific knowledge.

- Assistance to instructors teaching all major areas of biotechnology and help to students learning fundamental scientific concepts without overwhelming and excessive detail.

- An overview of historic applications while emphasizing modern, cutting-edge, and emerging areas of biotechnology.

- Insights on how biotechnology applications can provide some of the tools to solve important scientific and societal problems for the benefit of humankind and the environment.

- Inspiration for students to ponder the many ethical issues associated with biotechnology.

Introduction to Biotechnology provides broad coverage of topics including molecular biology, bioinformatics, genomics, and proteomics. We have strived to incorporate balanced coverage of basic molecular biology with practical and contemporary applications of biotechnology to provide students with the tools and knowledge to understand the field.

In our effort to introduce students to the cutting-edge techniques and applications of biotechnology, we dedicated specific chapters to such emerging areas as agricultural biotechnology (Chapter 6), forensic biotechnology (Chapter 8), bioremediation (Chapter 9), and aquatic biotechnology (Chapter 10). Consideration of the many regulatory agencies and issues that impact the biotechnology industry are discussed in Chapter 12. In addition to the ethical issues included in each chapter as *You Decide* boxes, a separate chapter (Chapter 13) is dedicated to ethics and biotechnology.

Features

Introduction to Biotechnology is specifically designed to provide several key elements that will help students enjoy learning about biotechnology and prepare them for a career in biotechnology.

Learning Objectives

Each chapter begins with a short list of learning objectives that present key concepts that students should understand after studying each chapter.

Abundant Illustrations

Approximately 200 figures and photographs provide comprehensive coverage to support chapter content. Illustrations, instructional diagrams, tables, and flow charts present step-by-step explanations that visually help students learn about laboratory techniques and complex processes that are important in biotechnology.

Career Profiles

A special box at the end of each chapter introduces students to different job options and career paths in the biotechnology industry and provides detailed information on job functions, salaries, and guidance for preparing to enter the workforce. Experts currently working in the biotechnology industry have contributed information to many of these **Career Profile** boxes. We strongly encourage students to refer to these profiles if they are interested in learning more about careers in the industry.

You Decide

From genetically modified foods to stem cell research, there are an endless number of topics in biotechnology that provoke strong ethical, legal, and social questions and dilemmas. **You Decide** boxes stimulate discussion in each chapter by presenting students with information that relates to the social and ethical implications of biotechnology followed by a set of questions for them to consider. The goal of these boxes is to help them understand *how* to consider ethical issues and formulate their own informed decisions.

Tools of the Trade

Biotechnology is based on the application of various laboratory techniques or tools in molecular biology, biochemistry, bioinformatics, genetics, mathematics, engineering, computer science, chemistry, and other disciplines. **Tools of the Trade** boxes in each chapter present modern techniques and technologies related to the content of each chapter to help students learn about the techniques and laboratory methods that are the essence of biotechnology.

Question & Answer (Q&A) Boxes

Q & A boxes in each chapter present students with questions they might be wondering about as they read the book. The answers provide background information and enrich student learning.

Questions & Activities

Questions are included at the conclusion of each chapter to reinforce student understanding of concepts. Activities frequently include internet assignments that ask students to explore a cutting-edge topic. Answers to these questions are provided at the end of the text.

References and Further Reading

A short list of references at the end of each chapter is provided as a starting point for students to learn more about a particular topic in biotechnology. We have carefully chosen articles that will help students and motivate them to learn more about a subject. Typically these references include primary research papers, review articles, and articles from the popular literature.

Keeping Current: Web Links

Because biotechnology is such a rapidly changing discipline, it is virtually impossible to keep a textbook updated on new and exciting discoveries. To help students access the most current information available, a rich complement of high-quality web links to some of the best sites in biotechnology and related disciplines are available on the Companion Website for this text, www.pearsonhighered.com/ biotechnology. Realizing that web links may frequently change, we tried to select substantive sites that are well established. We encourage you to let us know when you encounter address changes so we may update them.

Glossary

Like any technical discipline, biotechnology has a lexicon of terms and definitions that are routinely used when discussing processes, concepts, and applications. The most important terms are shown in **boldface type** throughout the book and are defined as they appear in the text. Definitions of these key terms are included in a glossary at the end of the book.

New Features for the Second Edition

The second edition of *Introduction to Biotechnology* includes several new instructor resources and exciting features:

- Increased end-of-chapter Questions and Activities, including more Internet-based exercises.
- A computerized test bank with multiple choice questions for each chapter; electronic files for all images in the textbook; and PowerPoint Lecture Outline slides, conveniently located on the Instructor's Resource Center, www.pearsonhighered.com/educator.
- New *You Decide* entries have been added to stimulate student interest in controversial areas of biotechnology.

In addition, each chapter has been thoroughly revised and updated to provide students with current information on emerging areas of biotechnology. Of special note are the following changes:

- **Chapter 1: The Biotechnology Century and Its Workforce** includes a new section on the "Business of Biotechnology" describing biotech company organization and structure, top biotechnology and pharmaceutical companies, and updated data on the biotechnology industry worldwide.
- **Chapter 2: An Introduction to Genes and Genomes** provides new content on transcription factors.
- **Chapter 3: Recombinant DNA Technology and Genomics** includes new sections on RNA interference (RNAi) and genomics and bioinformatics, updated content on the Human Genome Project, new content on comparative genomics, stone age genomics, and the "omics" revolution, and new content on high-throughput computer automated DNA sequencing, and real-time PCR.
- **Chapter 4: Proteins as Products** includes new information on mass spectrometry as the method of choice for protein sequencing and protein identification, and protein microarray studies.
- **Chapter 5: Microbial Biotechnology** includes new information on potential biological pathogens for bioweapons attacks on food sources, the use of gene microarrays for detecting host responses to pathogens and the use of protein microarrays for detecting bioweapons pathogens. New sections on metagenomics, assembling genomes to produce manufactured viruses, and new information on the human papillomavirus vaccine, Gardasil.
- **Chapter 6: Plant Biotechnology** provides updates on how genetically engineered plants have been used to replace other crops for better pest resistance, higher yields, and new applications in fiber, bioplastics, biofuels, and bioremediation. New information

on plant polyploids, and using plant material to make bioethanol production economical.
- **Chapter 7: Animal Biotechnology** provides updates on using mice with partial human immune systems for testing drug responses, the use of transgenic animals to produce vaccines, and using transgenic animal bioreactors for the production of complex therapeutic proteins.
- **Chapter 8: DNA Fingerprinting and Forensic Analysis** incorporates updates on uses of DNA sequence forensic information, identification of crime and catastrophe victims, establishing or excluding paternity relationships, authenticating food products, and other interesting aspects of forensic analysis.
- **Chapter 9: Bioremediation** includes a new section on the genomics of bioremediation and updated information on microbe-powered fuel cells, phytoremediation of explosives, and a new case study example.
- **Chapter 10: Aquatic Biotechnology** includes new information on using PCR to detect oyster pathogens, and new information on emerging medical products from aquatic organisms.
- **Chapter 11: Medical Biotechnology** provides updated information on gene therapy approaches including antisense methods and RNAi, updates and new information on regenerative medicine and tissue engineering, stem cell research and legislation and recent developments, and microarrays and pharmacogenomics.
- **Chapter 12: Biotechnology Regulations** incorporates new information on patenting gene sequences and world views on cloning and human stem cell use, compared to the United States.
- **Chapter 13: Ethics and Biotechnology** was rewritten to provide updated information on genetically modified foods, the genetic information nondiscrimination act, and stem cell regulations. It also includes a new section on patient rights and biological materials and three new *You Decide* entries.

Supplemental Learning Aids

Introduction to Biotechnology Companion Website (www.pearsonhighered.com/biotechnology)

The Companion Website is designed to help students study for their exams and deepen their understanding of the text's content. Each chapter contains learning objectives, content reviews, flashcards, glossary terms, jpegs of figures, and an extensive collection of *Keeping Current Web Links* which explore related topics in other areas.

Instructor Resource Center (IRC)

The Instructor Resource Center www.pearsonhighered .com/educator is designed to support instructors teaching biotechnology. The IRC is an online resource that supports and augments material in the textbook. New instructor supplements available for download include:

- **Computerized Test Bank:** 10-20 multiple choice test questions per chapter
- **Jpeg Art Files:** electronic files of all text tables, line drawings, and photos
- **PowerPoint Lecture Outlines:** a set of PowerPoint presentations consisting of lecture outlines for each chapter augmented by key text illustrations

Instructors using *Introduction to Biotechnology* can contact their Benjamin Cummings sales representative to access the Instructor Resource Center free of charge.

Benjamin Cummings Special Topics in Biology Series

The *Benjamin Cummings Special Topics in Biology Series* is a series of booklets designed for undergraduate students. These small booklets present the basic scientific facts and the social and ethical issues surrounding current hot topics. Booklets in the series include:

- *Alzheimer's Disease* (ISBN 0-1318-3834-2)
- *Biology of Cancer* (ISBN 0-8053-4867-0)
- *Biological Terrorism* (ISBN 0-8053-4868-9)
- *Emerging Infectious Diseases* (ISBN 0-8053-3955-8)
- *Genetic Testimony* (ISBN 0-1314-2338-X)
- *Gene Therapy* (ISBN 0-8053-3819-5)
- *HIV and AIDS* (ISBN 0-8053-3956-6)
- *Mad Cows and Cannibals* (ISBN 0-1314-2339-8)
- *Stem Cells and Cloning, 2e* (ISBN 0-3215-9002-3) NEW EDITION!
- *Understanding the Human Genome Project, 2e* (ISBN 0-8053-4877-8)

Contact your local Benjamin Cummings sales representative about bundling booklets in this series with *Introduction to Biotechnology* or visit www.pearson highered.com for more information.

Acknowledgments

A textbook is the collaborative result of hard work from many dedicated individuals including students, colleagues, editors and editorial staff, graphics experts, and many others. First, we thank our family and friends for their support and encouragement while we spent countless hours of our lives on this project. Without your understanding and patience, this book would not be possible.

We gratefully acknowledge the help of many talented people at Benjamin Cummings, particularly the editorial staff. Editorial duties changed several times during the preparation of this edition, but each editor provided creative talents and dedication to the book. We thank Mercedes Grandin for her guidance and help with reviews during the transition between editions. Acquistions Editor Becky Ruden has been the guiding force for the second edition and we are indebted to her for believing in the book's mission and for her vision to support and improve the text. Becky's leadership has been exactly what this project has needed, and she has been a great asset. We thank Project Editor Leata Holloway for keeping us on schedule and for her attention to detail, patience, enthusiasm, editorial suggestions, and great energy for the project.

We thank Production Supervisor Lori Newman for guiding us through the production process. Thanks go to Executive Marketing Manager Lauren Harp for her talents and creative contributions in developing promotional materials for this book. Maureen Spuhler, Photo Researcher, did a wonderful job of identifying and securing key art and suggesting alternatives. We thank Maria McColligan, Project Manager at Nesbitt Graphics, Inc., for her expert work on the first and second editions of the textbook. We'd also like to thank cover designer Rich Leeds of Big Wig Design who leant a fresh new look for the 2nd edition and Anne Lesser for expert copyediting.

Undergraduate students at Monmouth University read many drafts of the manuscript for the first and second edition and critiqued art scraps for clarity and content. We thank former students in BY 201 – Intro-

duction to Biotechnology for their honest reviews, error finding, and suggestions from a student's perspective. In particular we thank Monica Gionet who was especially skilled at noting points of clarification prior to completing the second edition. We also thank Monmouth University graduate Robert Sexton for contributing to the Career Profile section of Chapter 3. The 2002 Introductory Biotechnology class students at Ventura College provided many useful suggestions while they tolerated the use of the "sketchy figures" in PowerPoint programs, which were the result of our first draft. Our students inspire us to strive for better ways to help us teach and to help them understand the wonders of biotechnology. We applaud you for your help in creating what we hope future students will deem to be a student-friendly textbook.

We also thank Dr. Daniel Rudolph for contributing to the Career Profile in Chapter 5 (Microbial Biotechnology) and Gef Flimlin for contributing to the Career Profile in Chapter 10 (Aquatic Biotechnology).

Finally, *Introduction to Biotechnology* has greatly benefited from valued input of many colleagues and instructors who helped us with scientific accuracy, clarity, pedagogical aspects of the book, and suggestions for improving drafts of each chapter. The many instructors who have developed biotechnology courses and programs and enthusiastically teach majors and nonmajors about biotechnology provided reviews of the text and art that have been invaluable for helping shape this textbook. Your constructive criticism helped us to revise drafts of each chapter, and your words of praise helped inspire us to move ahead. All errors or omissions in the text are our responsibility. We thank you all and look forward to your continued feedback. Reviewers of *Introduction to Biotechnology*, include:

For the second edition:

Steve Benson   *California State University East Bay*
Ming-Mei Chang   *SUNY Genesco*
Jim Cheaney   *Iowa State University*
Peter Eden   *Marywood University*

Timothy S. Finco   *Agnes Scott College*
Mary A. Farwell   *East Carolina University*
James T. Hsu   *Lehigh University*
Lisa Johansen   *University of Colorado, Denver*
Michael Lawton   *Rutgers University*
Caroline Mackintosh   *University of Saint Mary*
Toby Mapes   *A B Tech Community College*
Patricia Phelps   *Austin Community College*
Ronald Raab   *James Madison University*
Melody Ricci   *Victor Valley College*
Bill Sciarrapa   *Rutgers University*
Salvatore A. Sparace   *Clemson University*
Janice Toyoshima   *Evergreen Valley College*
Dennis Walsh   *Massachusetts Bay Community College*
Lianna Wong   *Santa Clara University*

For the first edition:

D. Derek Aday   *Ohio State University*
Marcie Baer   *Shippensburg University*
Joan Barber   *Delaware Tech & Community College*
Theresa Beaty   *LeMoyne College*
Peta Bonham-Smith   *University of Saskatchewan*
Krista Broten   *University of Saskatchewan*
Heather Cavenagh   *Charles Sturt University*
Wesley Chun   *University of Idaho*
Mark Flood   *Fairmont State College*
Kathryn Paxton George   *University of Idaho*
Joseph Gindhart   *University of Massachusetts, Boston*
Jean Hardwick   *Ithaca College*
George Hegeman   *Indiana University, Bloomington*
Anne Helmsley   *Antelope Valley College*
David Hildebrand   *University of Kentucky*
Paul Horgen   *University of Toronto at Mississauga*
James Humphreys   *Seneca College of Applied Arts & Technology*
Tom Ingebritsen   *Iowa State University*
Ken Kubo   *American River College*
Theodore Lee   *SUNY Fredonia*
Edith Leonhardt   *San Francisco City College*

Lisa Lorenzen Dahl   *Iowa State University*
Keith McKenney   *George Mason University*
Timothy Metz   *Campbell University*
Robert Pinette   *University of Maine*
Lisa Rapp   *Springfield Technical Community College*
Stephen Rood   *Fairmont State College*
Alice Sessions   *Austin Community College*
Carl Sillman   *Pennsylvania State University*
Teresa Singleton   *Delaware State University*
Sharon Thoma   *Edgewood College*
Danielle Tilley   *Seattle Community College*
Jagan Valluri   *Marshall University*
Brooke Yool   *Ohlone College*
Mike Zeller   *Iowa State University*

Whether you are a student or instructor, we invite your comments and suggestions for improving the next edition of *Introduction to Biotechnology*. Please write to us at the following addresses below or contact us via e-mail at bc.feedback@pearson.com.

Bill Thieman
Ventura College
Department of Biology
4667 Telegraph
Ventura, CA 93003
BThieman@vcccd.edu

Michael Palladino
Monmouth University
School of Science, Technology and Engineering
Department of Biology
400 Cedar Avenue
West Long Branch, NJ 07764
www.monmouth.edu/mpalladi

As students ourselves, we too continue to learn about biotechnology everyday. We wish you great success in your explorations of biotechnology!

W. J. T.
M. A. P.

Contents

About the Authors   vii
Preface   viii

1 The Biotechnology Century and Its Workforce   1

1.1 What Is Biotechnology and What Does It Mean to You?   2

A Brief History of Biotechnology   2
Biotechnology: A Science of Many
. Disciplines   5
Products of Modern Biotechnology   6
Ethics and Biotechnology   8

1.2 Types of Biotechnology   8

Microbial Biotechnology   8
Agricultural Biotechnology   8
Animal Biotechnology   10
Forensic Biotechnology   10
Bioremediation   11
Aquatic Biotechnology   12
Medical Biotechnology   12
Regulatory Biotechnology   13
The Biotechnology "Big Picture"   13

1.3 Biological Challenges of the 21st Century   14

What Will the New Biotechnology Century
Look Like?   14
A Scenario in the Future: How Might We Benefit
from the Human Genome Project?   14

1.4 The Biotechnology Workforce   18

The Business of Biotechnology   19
Organization of a Biotechnology Company   19
Jobs in Biotechnology   20
Salaries in Biotechnology   23
Hiring Trends in the Biotechnology Industry   24
Questions & Activities   25
References and Further Reading   25

2 An Introduction to Genes and Genomes   26

2.1 A Review of Cell Structure   27

Prokaryotic Cells   27
Eukaryotic Cells   28

2.2 The Molecule of Life   30

Evidence That DNA Is the Inherited Genetic
Material   30
DNA Structure   32
What Is a Gene?   33

2.3 Chromosome Structure, DNA Replication, and Genomes   33

Chromosome Structure   34
DNA Replication   37
What Is a Genome?   38

2.4 RNA and Protein Synthesis   39

Copying the Code: Transcription   40
Translating the Code: Protein Synthesis   43
Basics of Gene Expression Control   46

2.5 Mutations: Causes and Consequences   51

Types of Mutations   51
Mutations Can Be Inherited or Acquired   53
Mutations Are the Basis of Variation in
Genomes and a Cause of Human Genetic
Diseases   54
Questions & Activities   55
References and Further Reading   56

3 Recombinant DNA Technology and Genomics   57

3.1 Introduction to Recombinant DNA Technology and DNA Cloning   58

Restriction Enzymes and Plasmid DNA
Vectors   58

Transformation of Bacterial Cells and Antibiotic Selection of Recombinant Bacteria   61
Introduction to Human Gene Cloning   63

3.2 What Makes a Good Vector?   65
Practical Features of DNA Cloning Vectors   65
Types of Vectors   66

3.3 How Do You Identify and Clone a Gene of Interest?   68
Creating DNA Libraries: Building a Collection of Cloned Genes   68
Polymerase Chain Reaction   71

3.4 What Can You Do with a Cloned Gene? Applications of Recombinant DNA Technology   75
Gel Electrophoresis and Mapping Gene Structure with Restriction Enzymes   75
DNA Sequencing   78
Chromosomal Location and Gene Copy Number   80
Studying Gene Expression   81
Northern Blot Analysis   82

3.5 Genomics and Bioinformatics: Hot New Areas of Biotechnology   87
Bioinformatics: Merging Molecular Biology with Computing Technology   87
Examples of Bioinformatics in Action   87
A Genome Cloning Effort of Epic Proportion: The Human Genome Project   89
What Have We Learned from the Human Genome?   90
The Human Genome Project Started an "Omics" Revolution   92
Comparative Genomics   92
Stone Age Genomics   93
Questions & Activities   95
References and Further Reading   95

4 Proteins as Products   97

4.1 Introduction to Proteins as Biotech Products   98

4.2 Proteins as Biotechnology Products   98
Making a Biotech Drug   99
Medical Applications   100
Food Processing   100
Textiles and Leather Goods   101
Detergents   101
Paper Manufacturing and Recycling   101

Adhesives: Natural Glues   101
Bioremediation: Treating Pollution with Proteins   102

4.3 Protein Structures   102
Structural Arrangement   102
Protein Folding   103
Glycosylation   104
Protein Engineering   104

4.4 Protein Production   106
Protein Expression: The First Phase in Protein Processing   107

4.5 Protein Purification Methods   109
Preparing the Extract for Purification   109
Stabilizing the Proteins in Solution   109
Separating the Components in the Extract   110

4.6 Verification   114

4.7 Preserving Proteins   115

4.8 Scale-up of Protein Purification   116

4.9 Postpurification Analysis Methods   116
Protein Sequencing   116
X-ray Crystallography   116

4.10 Proteomics   117
Questions & Activities   118
References and Further Reading   118

5 Microbial Biotechnology   119

5.1 The Structure of Microbes   120
Yeast Are Important Microbes Too   121

5.2 Microorganisms as Tools   122
Microbial Enzymes   123
Bacterial Transformation   123
Electroporation   124
Cloning and Expression Techniques   125

5.3 Using Microbes for a Variety of Everyday Applications   127
Food Products   127
Therapeutic Proteins   131
Using Microbes Against Other Microbes   131
Field Applications of Recombinant Microorganisms   134

5.4 Vaccines   135
A Primer on Antibodies   136
How Are Vaccines Made?   137
Bacterial and Viral Targets for Vaccines   139

5.5 Microbial Genomes   141

Why Sequence Microbial Genomes?   141
Microbial Genome Sequencing Strategies   142
Selected Genomes Sequenced to Date   142
Sorcerer II: Traversing the Globe to Sequence
    Microbial Genomes   143
Viral Genomics   144
Assembling Genomes to Produce Human-Made
    Viruses   145

5.6 Microbial Diagnostics   145

Bacterial Detection Strategies   145
Tracking Disease-Causing
    Microorganisms   147
Microarrays for Tracking Contagious
    Diseases   147

5.7 Combating Bioterrorism   148

Microbes as Bioweapons   149
Targets of Bioterrorism   150
Using Biotechnology Against Bioweapons   151
Questions & Activities   153
References and Further Reading   154

6 Plant Biotechnology   155

6.1 Agriculture: The Next Revolution   156

6.2 Methods Used in Plant Transgenesis   157

Conventional Selective Breeding and
    Hybridization   157
Cloning: Growing Plants from Single
    Cells   157
Protoplast Fusion   157
Leaf Fragment Technique   158
Gene Guns   158
Chloroplast Engineering   159
Antisense Technology   159

6.3 Practical Applications in the Field   161

Vaccines for Plants   161
Genetic Pesticides: A Safer Alternative?   162
Safe Storage   163
Herbicide Resistance   163
Stronger Fibers   163
Enhanced Nutrition   164
The Future: From Pharmaceuticals to Fuel   164
Metabolic Engineering   166

6.4 Health and Environmental Concerns   167

Concerns About Human Health   168
Concerns About the Environment   168
Regulations   169
Questions & Activities   170
References and Further Reading   170

7 Animal Biotechnology   171

**7.1 Introduction to Animal
Biotechnology**   172

7.2 Animals in Research   172

Animal Models   172
Alternatives to Animal Models   174
Regulation of Animal Research   175
Veterinary Medicine as Clinical Trials   176
Bioengineering Mosquitoes to Prevent
    Malaria   177

7.3 Clones   177

Creating Dolly: A Breakthrough in
    Cloning   177
The Limits to Cloning   178
The Future of Cloning   179

7.4 Transgenic Animals   180

Transgenic Techniques   180
Improving Agricultural Products with
    Transgenics   181
Transgenic Animals as Bioreactors   183
Knockouts: A Special Case of Transgenics   184

**7.5 Producing Human Antibodies in
Animals**   186

Monoclonal Antibodies   186
Eggs as Antibody Factories   188
Questions & Activities   188
References and Further Reading   189

**8 DNA Fingerprinting and Forensic
Analysis**   190

**8.1 Introduction to DNA Fingerprinting
and Forensics**   191

8.2 What Is a DNA Fingerprint?   191

How Is DNA Typing Performed?   191

8.3 Preparing a DNA Fingerprint   192

Specimen Collection   192
Extracting DNA for Analysis   193
Restriction Fragment Length
    Polymorphism (RFLP) Analysis   193
The Southern Blot Technique   193
PCR and DNA Amplification   196
Dot Blot (or Slot Blot) Analysis   196
STR Analysis   196

8.4 Putting DNA to Use   196

The Narborough Village Murders   197
The Forest Hills Rapist   197

Terrorism and Natural Disasters Force
Development of New Technologies   199

8.5 DNA and the Rules of Evidence   200

DNA Fingerprinting and the Simpson and
Goldman Murders   201
Human Error and Sources of Contamination   201
DNA and Juries   202

**8.6 Familial Relationships and
DNA Profiles**   202

Mitochondrial DNA Analysis   202
Y-Chromosome Analysis   203

8.7 Nonhuman DNA Analysis   204

DNA Tagging to Fight Fraud   206
Questions & Activities   206
References and Further Reading   207

9 Bioremediation   208

9.1 What Is Bioremediation?   209

Why Is Bioremediation Important?   209

9.2 Bioremediation Basics   210

What Needs to Be Cleaned Up?   210
Chemicals in the Environment   211
Fundamentals of Cleanup Reactions   211
The Players: Metabolizing Microbes   213
Bioremediation Genomics Programs   215

9.3 Cleanup Sites and Strategies   217

Soil Cleanup   217
Bioremediation of Water   218
Turning Wastes into Energy   220

**9.4 Applying Genetically Engineered Strains
to Clean Up the Environment**   222

Petroleum-Eating Bacteria   222
Engineering *E. coli* to Clean Up Heavy Metals   223
Biosensors   224
Genetically Modified Plants and
Phytoremediation   224

**9.5 Environmental Disasters: Case Studies in
Bioremediation**   225

Jet Fuel and Hanahan, South Carolina   225
The Exxon *Valdez* Oil Spill   225
Oil Fields of Kuwait   226

**9.6 Future Strategies and Challenges
for Bioremediation**   227

Recovering Valuable Metals   228
Bioremediation of Radioactive Wastes   229
Questions & Activities   230
References and Further Reading   230

10 Aquatic Biotechnology   231

**10.1 Introduction to Aquatic
Biotechnology**   232

**10.2 Aquaculture: Increasing
the World's Food Supply
Through Biotechnology**   232

The Economics of Aquaculture   232
Fish Farming Practices   235
Improving Strains for Aquaculture   238
Enhancing Seafood Quality and Safety   239
Barriers and Limitations to Aquaculture   239
The Future of Aquaculture   242

**10.3 Molecular Genetics of Aquatic
Organisms**   242

Discovery and Cloning of Novel Genes   242
Genetic Manipulations of Finfish and
Shellfish   247

**10.4 Medical Applications of Aquatic
Biotechnology**   251

Drugs and Medicines from the Sea   251
Monitoring Health and Human Disease   254

10.5 Nonmedical Products   254

A Potpourri of Products   254
Biomass and Bioprocessing   255

**10.6 Environmental Applications of Aquatic
Biotechnology**   256

Antifouling Agents   256
Biosensors   257
Environmental Remediation   257
Questions & Activities   259
References and Further Reading   259

11 Medical Biotechnology   260

**11.1 The Power of Molecular Biology:
Detecting and Diagnosing Human Disease
Conditions**   261

Models of Human Disease   261
Biomarkers for Disease Detection   263
Detecting Genetic Diseases   263

**11.2 Medical Products and Applications of
Biotechnology**   268

The Search for New Medicines and Drugs   269
Artificial Blood   273
Vaccines and Therapeutic Antibodies   273

11.3 Gene Therapy   275

How Is It Done?   275

Curing Genetic Diseases: Targets for Gene
   Therapy 279
Challenges Facing Gene Therapy 281

11.4 The Potential of Regenerative Medicine 282

Cell and Tissue Transplantation 282
Tissue Engineering 285
Stem Cell Technologies 287
Cloning 293
Embryonic Stem Cell and Therapeutic Cloning
   Regulations in the United States 297

11.5 The Human Genome Project Has Revealed Disease Genes on All Human Chromosomes 298

Piecing Together the Human Genome
   Puzzle 298
Questions & Activities 303
References and Further Reading 303

12 Biotechnology Regulations 305

12.1 The Regulatory Framework 306

12.2 U.S. Department of Agriculture 308

Animal and Plant Health Inspection
   Service 308
Permitting Process 308
The APHIS Investigative Process 308
The Notification Process 309

12.3 The Environmental Protection Agency 309

Experimental Use Permits 309
The First Experimental Use Permit 309
Deregulation and Commercialization 310

12.4 Food and Drug Administration 311

Food and Food Additives 311
The Drug Approval Process 311
Good Laboratory (GLP), Clinical (GCP), and
   Manufacturing (GMP) Practices 312
Phase Testing of Drugs 312
Faster Drug Approval versus Public
   Safety 313

12.5 Legislation and Regulation: The Ongoing Role of Government 314

Labeling Biotechnology Products 316
The Fluvirin Failure 316

12.6 Introduction to Patents 317

The Value of Patents in the Biotechnology
   Industry 318
Patenting DNA Sequences 319

12.7 Biotechnology Products in the Global Marketplace 321

Questions & Activities 322
References and Further Reading 323

13 Ethics and Biotechnology 324

13.1 What Is Ethics? 325

Approaches to Ethical Decision Making 325
Ethical Exercise Warm-Up 326

13.2 Biotechnology and Nature 327

Cells and Products 328
GM Crops: Are You What You Eat? 328
Animal Husbandry or Animal Tinkering? 331
The Human Question 332
What Does It Mean to Be Human? 333
Spare Embryos for Research Versus Creating
   Embryos for Research 335
Cloning 336
Patient Rights and Biological Materials 337
Regulations in Flux 338
Your Genes, Your Self 338
More or Less Human? 339

13.3 Economics, the Role of Science, and Communication 340

Questions & Activities 342
References and Further Reading 343

Appendix 1 Answers to Questions &
   Activities A-1

Appendix 2 The 20 Amino Acids of
              Proteins A-9

Credits C-1

Glossary G-1

Index I-1

Chapter 1

The Biotechnology Century and Its Workforce

After completing this chapter you should be able to:

- Define biotechnology and understand the many scientific disciplines that contribute to biotechnology.

- Provide examples of historic and current applications of biotechnology and its products.

- List and describe different types of biotechnology and their applications.

- Provide examples of potential advances in biotechnology.

- Discuss how medical diagnosis will change as a result of biotechnology and provide examples of how data from the Human Genome Project will be used to diagnose and treat human disease conditions.

- Understand that there are pros and cons to biotechnology and many controversial issues in this field.

- Describe career categories and options in biotechnology.

- Develop an understanding of some important skills and training required to be part of the biotechnology workforce.

- Discuss hiring trends in the biotechnology industry.

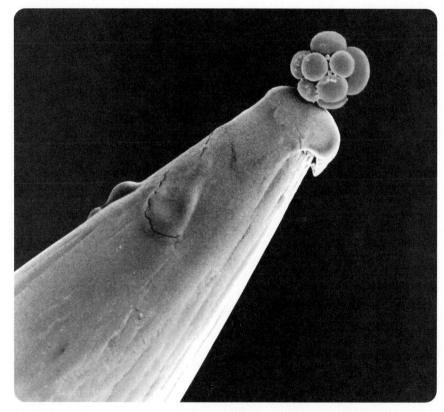

Miracle cells? This tiny cluster on the tip of a pin is a human embryo approximately three days after fertilization. Some scientists believe that stem cells contained within embryos may have the potential for treating and curing a range of diseases in humans through biotechnology. Use of these cells is also one of the most controversial topics in biotechnology.

If you have ever eaten a corn chip, you may have been impacted by biotechnology. Don't eat chips? How about sour cream, yogurt, cheese, or milk? In this century, more and more of the foods we eat will be produced by organisms that have been genetically altered through biotechnology. Such **genetically modified (GM) foods** have become a controversial topic over the last few years as have human embryos such as the one shown in the opening photo. This chapter was designed to provide you with a basic introduction to the incredible range of biotechnology topics that you will read about in this book. As you will see, biotechnology is a multidisciplinary science with many powerful applications and great potential for future discoveries.

The purpose of this chapter is not to provide a comprehensive review of the history of biotechnology and its current applications. Instead, we present a brief introduction and overview of many topics that we discuss in greater detail in future chapters. We begin by defining biotechnology and presenting an overview of the many scientific disciplines that contribute to this field. We highlight both historic and modern applications and define the different types of biotechnology that you will study in this book. At the end of the chapter, we discuss aspects of the biotechnology workforce and skills required to work in the industry. Be sure you are familiar with the different types of biotechnology and the key terms presented in this chapter because they will form the foundation for your future studies.

1.1 What Is Biotechnology and What Does It Mean to You?

Have you ever eaten a Flavr Savr™ tomato, been treated with a monoclonal antibody, received tissue grown from embryonic stem cells, or seen a "knock-out" mouse? Have you ever had a flu shot, known a person with diabetes who requires injections of insulin, taken a home pregnancy test, used an antibiotic to treat a bacterial infection, sipped a glass of wine, eaten cheese, or made bread? Although you may not have experienced any of the scenarios on the first list, at least one of the items on the second list must be familiar to you. If so, you have experienced the benefits of biotechnology.

Can you imagine a world free of serious diseases, where food is abundant for everyone and the environment is free of pollution? These scenarios are exactly what many people in the biotechnology industry envision as they dedicate their lives to this exciting science. Although you may not fully understand the range of disciplines and the scientific details of biotechnology, you have experienced biotechnology firsthand. **Bio-technology** is broadly defined as using living organisms, or the products of living organisms, for human benefit (or to benefit human surroundings) to make a product or solve a problem. Remember this definition. As you learn more about biotechnology, we expand and refine this definition with historical examples and modern applications from everyday life and look ahead to the biotechnology future.

You would be correct in thinking that biotechnology is a relatively new discipline that is only recently getting a lot of attention; however, it may surprise you to know that in many ways this science involves several ancient practices. As we discuss in the next section, old and new practices in biotechnology make this field one of the most rapidly changing and exciting areas of science. It affects our everyday lives and will become even more important during this century—what some have called the "century of biotechnology."

A Brief History of Biotechnology

If you ask your friends and family to define biotechnology, their answers may surprise you. They may have no idea about what biotechnology is. Perhaps they might tell you that biotechnology involves serious-looking scientists in white lab coats carrying out sophisticated and secretive gene-cloning experiments in expensive laboratories. When pressed for details, however, they probably will not be able to tell you how these "experiments" are done, what information is gained from such work, and how this knowledge is used. Although DNA cloning and the genetic manipulation of organisms are exciting modern-day techniques, biotechnology is not a new science. In fact, many applications represent old practices with new methodologies. Humans have been using organisms for their benefit in many processes for several thousand years. Historical accounts have shown that the Chinese, Greeks, Romans, Babylonians, and Egyptians, among many others, have been involved in biotechnology since nearly 2000 B.C.

Biotechnology does not mean hunting and gathering animals and plants for food; however, domesticating animals such as sheep and cattle for use as livestock is a classic example of biotechnology. Our early ancestors also took advantage of microorganisms and used **fermentation** to make breads, cheeses, yogurts, and alcoholic beverages such as beer and wine. During fermentation, some strains of yeast decompose sugars to derive energy, and in the process they produce ethanol (alcohol) as a waste product. When bread dough is being made, yeast (*Saccharomyces cerevisiae*, commonly called baker's yeast) is added to make the dough rise. This occurs because the yeast ferments sugar releasing carbon dioxide, which causes the dough to rise and creates holes in the bread. Alcohol produced by the yeast evaporates when the bread is cooked—but the remnants of alcohol remain in the semisweet taste of most bread. If you make bread or pizza dough at home, you have probably added store-bought *S. cerevisiae* from an envelope or jar to your dough mix. As you will discover when we discuss microbial biotechnology in Chapter 5, similar processes are very valuable for the production of yogurts, cheeses, and beverages.

For thousands of years, humans have used **selective breeding** as a biotechnology application to improve production of crops and livestock used for food purposes. In selective breeding, organisms with desirable features are purposely mated to produce offspring with the same desirable characteristics. For example, cross-breeding plants that produce the largest, sweetest, and most tender ears of corn is a good way for farmers to maximize their land to produce the most desirable crops (Figure 1.1a). Similar breeding techniques are used with farm animals including turkeys (to breed birds producing the largest and most tender breast meat), cows, chickens, and pigs. Other examples include breeding wild species of plants, such as lettuces and cabbage, over many generations to produce modern plants that are cultivated for human consumption. Many of these approaches are really genetic applications of biotechnology. Without realizing it—and without expensive labs, sophisticated equipment, Ph.D.-trained scientists, and well-planned experiments—humans have been manipulating genes for hundreds of years.

By selecting plants and animals with desirable characteristics, humans are choosing organisms with useful genes and taking advantage of their genetic potential for human benefit. Recently, scientists at the Children's Hospital of Boston produced a transparent zebrafish named Casper (Figure 1.1b). Casper was created by mating a zebrafish mutant that lacked reflective pigment with a zebrafish that lacked black pigment. Zebrafish are important experimental model organisms and scientists believe that Casper will be important for drug testing and *in vivo* studies of stem cells and cancer. Casper has already proven to be valuable for studying how cancer cells spread: scientists injected fluorescent tumor cells into the fish's abdominal cavity and were able to track migration of cancer cells to specific locations in the body.

One of the most widespread and commonly understood applications of biotechnology is the use of antibiotics. In 1928, Alexander Fleming discovered that the mold *Penicillium* inhibited the growth of a bacterium called *Staphylococcus aureus*, which causes skin disease in humans. Subsequent work by Fleming led to the discovery and purification of the **antibiotic** penicillin. Antibiotics are substances produced by microorganisms that will inhibit the growth of other microorganisms. In the 1940s, penicillin became widely available for medi-

(a)

(b)

Figure 1.1 Selective Breeding Is an Old Example of Biotechnology That Is Still Common Today
(a) Corn grown by selective breeding. From left to right is teosinte (*Zea canina*), selectively bred hybrids, and modern corn (*Zea mays*). (b) "Casper," a transparent zebrafish produced by selective breeding. See Figure 7.3b for a photo of a normal zebrafish.

cinal use to treat bacterial infections in humans. In the 1950s and 1960s, advances in biochemistry and cell biology made it possible to purify large amounts of antibiotics from many different strains of bacteria. **Batch (large-scale) processes**—in which scientists can grow bacteria and other cells in large amounts and harvest useful products in large batches—were developed to isolate commercially important molecules from microorganisms, as explained further in Chapter 4.

Since the 1960s, rapid development of our understanding of genetics and molecular biology has led to exciting new innovations and applications in biotechnology. As we have begun to unravel the secrets of DNA structure and function, new technologies have led to **gene cloning,** the ability to identify and reproduce a gene of interest, and **genetic engineering,** manipulating the DNA of an organism. Through genetic engineering, scientists are able to combine DNA from different sources. This process, called **recombinant DNA technology,** is used to produce many proteins of medical importance, including insulin, human growth hormone, and blood-clotting factors. From its inception, recombinant DNA technology has dominated many important areas of biotechnology, as we discuss in great detail beginning in Chapter 3. Throughout the book, you will see that recombinant DNA technology has led to hundreds of applications, including the development of disease-resistant plants, food crops that produce greater yields of fruits and vegetables, "golden rice" engineered to be more nutritious, and genetically engineered bacteria capable of degrading environmental pollutants.

Recombinant DNA technology has had a tremendous impact on human health through the identification of thousands of genes involved in human genetic disease conditions. The ultimate record of recombinant DNA technology on humans was produced by the **Human Genome Project,** an international effort that began in 1990. A primary goal of the Human Genome Project was to identify all genes—the **genome**—contained in the DNA of human cells and to map their locations to each of the 24 human chromosomes

(chromosomes 1 to 22 and the X and Y chromosomes). The Human Genome Project has provided unlimited potential for the development of new diagnostic approaches for detecting disease and molecular approaches for treating and curing human genetic disease conditions.

Just imagine the possibilities. The Human Genome Project can tell us the chromosomal location and code of *every* human gene from genes that control normal cellular processes and determine characteristics such as hair color, eye color, height, and weight to the myriad of genes that cause human genetic diseases (Figure 1.2). Over the next decade and beyond, we will witness many significant discoveries in human genetics. As a result of the Human Genome Project, new knowledge about human genetics will have tremendous and wide-ranging effects on basic science and medicine in the near future. In many ways, understanding the functions of all human genes is one of the great unknown and unsolved mysteries in biology. We explore the mysteries of the genome in several chapters of this book.

As you have just learned, biotechnology has a long and rich history. Future chapters are dedicated to exploring advances in biotechnology and looking ahead

Chromosome 13

114 million bases

Cholesterol-lowering factor	Cataract, zonular pulverulent
Deafness, autosomal dominant and recessive	Stem-cell leukemia/lymphoma syndrome
Vohwinkel syndrome	Spastic ataxia, Charlevoix-Saguenay type
Ectodermal dysplasia	
Muscular dystrophy, limb-girdle, type 2C	Pancreatic agenesis
Breast cancer, early onset	Maturity Onset Diabetes of the Young, type IV
Pancreatic cancer	Enuresis, nocturnal
Disrupted in B-cell neoplasia	Dementia, familial British
Leukemia, chronic lymphocytic, B-cell	Rieger syndrome, type 2
	X-ray sensitivity
MHC class II deficiency, group B	Rhabdomyosarcoma, alveolar
Hyperornithinemia Hyperammonemia Homocitrullinemia	Lung cancer, non small-cell
	Spinocerebellar ataxia
Serotonin receptor	Ceroid-lipofuscinosis, neuronal
Retinoblastoma	Microcoria, congenital
Osteosarcoma	Schizophrenia susceptibility
Bladder cancer	Xeroderma pigmentosum, group G
Pinealoma with bilateral Retinoblastoma	Coagulation Factor VIII deficiency
	Oguchi disease
Wilson disease	Stargardt disease, autosomal dominant
Postaxial polydactyly, type A2	Coagulation Factor X deficiency
	SRY (sex determining region Y)
Hirschsprung disease	Breast cancer, ductal
Propionicacidemia, types I or pccA	
Holoprosencephaly	
Bile acid malabsorption, primary	

Chromosome 21

50 million bases

Coxsackie and adenovirus receptor	Myeloproliferative syndrome, transient
Amyloidosis cerebroarterial, Dutch type	Leukemia, transient, of Down Syndrome
Alzheimer disease, APP-related	
Schizophrenia, chronic	Enterokinase deficiency
Usher syndrome, autosomal recessive	Multiple carboxylase deficiency
Amyotrophic lateral sclerosis	T-cell lymphoma invasion and metastasis
Oligomycin sensitivity	Mycobacterial infection, atypical
Jervell and Lange-Nielsen syndrome	Down syndrome (critical region)
Long QT syndrome	Autoimmune polyglandular disease, type I
Down syndrome cell adhesion molecule	Bethlem myopathy
Homocystinuria	Epilepsy, progressive myoclonic
Cataract, congenital, autosomal dominant	Holoprosencephaly, alobar
Deafness, autosomal recessive	Knobloch syndrome
Myxovirus (influenza) resistance	Hemolytic anemia
Leukemia, acute myeloid	Breast cancer
	Platelet disorder, with myeloid malignancy

Figure 1.2 Gene Maps of Chromosomes 13 and 21   The Human Genome Project has led to the identification of nearly all human genes and has mapped their location on each chromosome. The maps of chromosomes 13 and 21 indicate those genes known to be involved in human genetic disease conditions. Identifying these genes is an important first step toward developing treatments for many genetic diseases.

to what the future holds. As you study biotechnology, you will be introduced to what may seem to be an overwhelming number of terms and definitions. Be sure to use the index and glossary at the end of the book to help you find and define important terms.

Biotechnology: A Science of Many Disciplines

One of the many challenges you will encounter as you study biotechnology will be trying to piece together complex information from many different scientific disciplines. It is impossible to talk about biotechnology without considering the important contributions of the different fields of science. Although a primary focus of biotechnology involves the use of molecular biology to carry out genetic engineering applications, biotechnol-

ogy is not a single, narrow discipline of study. Instead, it is an expansive field that absolutely relies on contributions from many areas of biology, chemistry, mathematics, computer science, and engineering in addition to other disciplines such as philosophy and economics. Later in this chapter, we consider how biotechnology provides a wealth of employment opportunities for people who have been trained in diverse fields.

Figure 1.3 provides a diagrammatic view of the many disciplines that contribute to biotechnology. Notice that the "roots" are primarily formed by work in the **basic sciences**—research into fundamental processes of living organisms at the biochemical, molecular, and genetic levels. When pieced together, basic science research from many areas, with the help of computer science, can lead to genetic engineering approaches. At the top of the tree, applications of genetic engineering

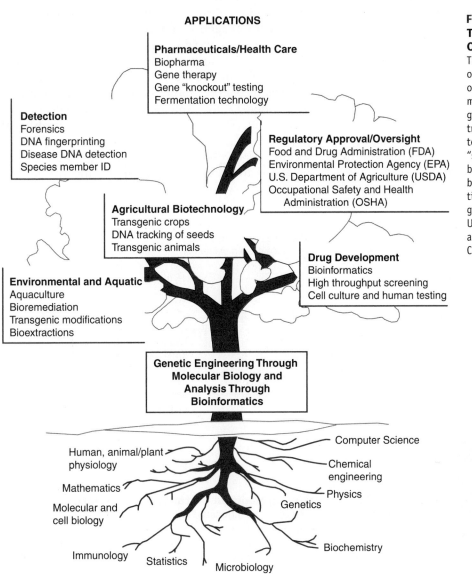

APPLICATIONS

Pharmaceuticals/Health Care
Biopharma
Gene therapy
Gene "knockout" testing
Fermentation technology

Detection
Forensics
DNA fingerprinting
Disease DNA detection
Species member ID

Regulatory Approval/Oversight
Food and Drug Administration (FDA)
Environmental Protection Agency (EPA)
U.S. Department of Agriculture (USDA)
Occupational Safety and Health
Administration (OSHA)

Agricultural Biotechnology
Transgenic crops
DNA tracking of seeds
Transgenic animals

Drug Development
Bioinformatics
High throughput screening
Cell culture and human testing

Environmental and Aquatic
Aquaculture
Bioremediation
Transgenic modifications
Bioextractions

Genetic Engineering Through Molecular Biology and Analysis Through Bioinformatics

Computer Science

Human, animal/plant physiology

Chemical engineering

Mathematics

Physics

Molecular and cell biology

Genetics

Immunology

Biochemistry

Statistics

Microbiology

Figure 1.3 The Biotechnology Tree: Different Disciplines Contribute to Biotechnology
The basic sciences are the foundation or "roots" of all aspects of biotechnology. The central focus or "trunk" for most biotechnological applications is genetic engineering. Branches of the tree represent different organisms, technologies, and applications that "stem" from genetic engineering and bioinformatics, central aspects of most biotechnological approaches. Regulation of biotechnology occurs through governmental agencies like the FDA, USDA, EPA, and OSHA, whose roles and responsibilities will be defined in Chapter 12.

can be put to work to create a product or process to help humans or our living environment.

A simplified example of the interdisciplinary nature of biotechnology can be summarized as follows. At the basic science level, scientists conducting research in microbiology at a college, university, government agency, or public or private company may discover a gene or gene product in bacteria that shows promise as an agent for treating a disease condition. Typically, biochemical, molecular biological, and genetic techniques would be used to better understand the role of this gene. This process also involves using computer science in sophisticated ways to study the sequence of a gene and to analyze the structure of the protein produced by the gene. Applying computer science to the study of DNA and protein data has created an exciting new field called **bioinformatics.**

Once basic research has provided a detailed understanding of this gene, the gene may then be used in a variety of ways, including drug development, agricultural biotechnology, and environmental and marine applications (Figure 1.3). The many applications of biotechnology will become much clearer as we cover each area. At this point keep in mind that biotechnology is a science that requires skills from many disciplines.

Products of Modern Biotechnology

Throughout the book, we consider many cutting-edge and innovative products and applications of biotechnology. Not only do we look at products for human use, but we also consider biotechnology applications of microbiology, marine biology, and plant biology, among other disciplines. The multitude of biotechnology products currently available are far too numerous to mention in this introductory chapter; however, many products reflect the current needs of humans—for example, pharmaceutical production, creating drugs for the treatment of human health conditions. In fact, more than 65% of biotechnology companies in the United States are involved in pharmaceutical production. In 1982, California biotechnology company Genentech received approval for recombinant insulin, used for the treatment of diabetes, as the first biotechnology product for human benefit (Figure 1.4). There are now several hundred drugs, vaccines, and diagnostics on the market with more than 300 biotechnology medicines in development targeting over 200 diseases. Nearly half of the new drugs in the development pipeline are designed to treat cancer. Table 1.1 provides a brief list of some of the top-selling biotechnology drugs and the companies that developed them. Diagnosis and/or treatment of a variety of human diseases and disorders, including AIDS, stroke, diabetes, and cancer, make up the bulk of biotechnology products on the market.

Many of the most widely used products of biotechnology are proteins created by gene cloning (Table 1.2). These proteins are called **recombinant proteins** because they are produced by gene-cloning techniques involving the transfer of genes from one organism to another. For example, the majority of

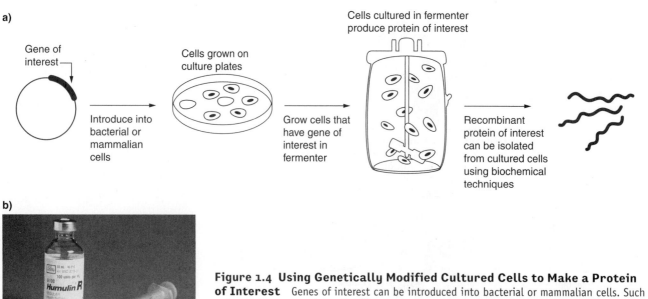

Figure 1.4 Using Genetically Modified Cultured Cells to Make a Protein of Interest   Genes of interest can be introduced into bacterial or mammalian cells. Such cells can be grown using cell culture techniques. Recombinant proteins isolated from these cells are used in hundreds of different biotechnology applications. In this example, mammalian cells are shown, but this process is also commonly carried out using bacteria.

Table 1.1 TOP-TEN BIOTECHNOLOGY DRUGS (WITH SALES OVER $1 BILLION)

Drug	Developer	Function (Treatment of Human Disease Conditions)
Procrit	Johnson & Johnson	Anemia
Epogen	Amgen	Anemia
Enbrel	Amgen & Wyeth	Rheumatoid arthritis
Aranesp	Amgen	Anemia
Remicade	Johnson & Johnson and Schering-Plough Corp.	Rheumatoid arthritis
Rituxan	Roche Holding Ltd.	Non-Hodgkin's Lymphoma
Neulasta	Amgen	Increases white blood cell count in cancer patients
Avonex	Biogen Idec Inc.	Multiple sclerosis
Neupogen	Amgen & Roche	Increases white blood cell count in cancer patients
Lantus	sanofi-aventis	Diabetes

these proteins are produced from human genes inserted into bacteria to produce recombinant proteins that are used to treat human disease conditions.

How genes are cloned and used to produce proteins of interest is discussed in great detail in Chapter 3. As an introduction to this idea, consider the diagram shown in Figure 1.4. As you will soon learn, scientists can identify a gene of interest and put it into bacterial cells or mammalian cells that are grown by a technique called **cell culture.** In cell culture, cells are grown in dishes or flasks within liquid culture media designed to provide the nutrients necessary for cell growth. Because only a limited number of cells can be grown in small culture dishes, the cells are transferred to large culturing containers called **fermenters** or **bioreactors,** in which cells containing the DNA of interest can be mass produced. Using techniques detailed in Chapter 4, scientists can harvest the protein produced by the gene of interest from these cells and use it in applications such as those described in Table 1.2.

If Tables 1.1 and 1.2 have not provided you with convincing examples of the importance of biotechnology on human health, consider that, in the near future, genes may be routinely introduced into humans as **gene therapy** approaches are employed to treat and cure human disease conditions. Genetics and tissue engineering may lead to the ability to grow organs for transplantation that would only rarely be rejected by their recipients. New biotechnology products from marine organisms will be used to treat cancers, strokes, and arthritis. Modern advances in medicine, driven by

new knowledge from the Human Genome Project, will likely result in healthier lives and potentially increase human lifespan.

Table 1.2 EXAMPLES OF PROTEINS MANUFACTURED FROM CLONED GENES

Product	Application
Blood factor VIII (clotting factor)	Used to treat hemophilia
Epidermal growth factor	Used to stimulate antibody production in patients with immune system disorders
Growth hormone	Used to correct pituitary deficiencies and short stature in humans; other forms used in cows to increase milk production
Insulin	Used to treat diabetes mellitus
Interferons	Used to treat cancer and viral infections
Interleukins	Used to treat cancer and stimulate antibody production
Monoclonal antibodies	Used to diagnose and treat a variety of diseases including cancers
Tissue plasminogen activator	Used to treat heart attacks and stroke

Q What are some products that could be produced by genetically engineering cultured cells?

A Human therapeutic proteins needed in bulk are one class of products produced using genetically engineered cultured cells. Insulin, blood-clotting factors, and hormones such as growth hormone used to treat dwarfism are among the many examples of therapeutic proteins produced this way. Many other enzymes involved in everyday life are also made using this technology. Rennin, an enzyme required to make cheese, is a good example. In the past, rennin was isolated from the stomachs of calves. By cloning and expressing rennin in bacteria, recombinant DNA technology now allows for rennin to be produced cheaply, in large amounts, and without having to sacrifice calves for this purpose.

Ethics and Biotechnology

Just like any other type of technology, the powerful applications and potential promise of biotechnology applications raises many ethical concerns, and it should be no surprise to you that not everyone is a fan of biotechnology. A wide range of ethical, legal, and social implications of biotechnology are a cause of great debate and discussion by scientists, the general public, clergy, politicians, lawyers, and many others around the world (Figure 1.5). Throughout this book, we present ethical, legal, and social issues for you to consider. Increasingly, you will be faced with ethical issues of biotechnology that may influence you directly. For

Figure 1.5   Biotechnology Is a Controversial Science That Presents Many Ethical Dilemmas

instance, now that organism cloning has been accomplished in mammals such as sheep, cows, and monkeys, some have suggested that human cloning be permitted. How do you feel about this idea? If in the future, you and your spouse were unable to have children by any other means, would you want the opportunity to create a baby by cloning? Look for "You Decide" boxes in each chapter where we present scenarios or ethical dilemmas for you to consider. Realize there are pros and cons and controversial issues associated with almost every application in biotechnology. Our goal is not to tell you *what* to think but to empower you with knowledge you can use to make your own decisions.

1.2 Types of Biotechnology

Now that you have learned about the many areas of science that contribute to biotechnology, you should recognize there are many different types of biotechnology. Consider this section an introduction to what you will learn in greater detail in the chapters that follow.

Microbial Biotechnology

In Chapter 5, we explore the many ways that microbial biotechnology impacts society. As we discussed previously, the use of yeast for making beer and wine is one of the oldest applications of biotechnology. By manipulating microorganisms such as bacteria and yeast, microbial biotechnology has created better enzymes and organisms for making many foods, simplifying manufacturing and production processes, and making decontamination processes for industrial waste product removal more efficient. Leaching of oil and minerals from the soil to increase mining efficiency is another example of microbial biotechnology in action. Microbes are also used to clone and produce batch amounts of important proteins used in human medicine including insulin and growth hormone.

Agricultural Biotechnology

Chapter 6 is dedicated to plant biotechnology and agricultural applications of biotechnology. In "ag-biotech," we examine a range of topics from genetically engineered, pest-resistant plants that do not need to be sprayed with pesticides to foods with higher protein or vitamin content and drugs developed and grown as plant products. Agricultural biotechnology is already a big business that is rapidly expanding. It has been estimated that agricultural biotechnology in the United States will be a $7 billion market by 2008.

YOU DECIDE

Genetically Modified Foods: To Eat or Not to Eat?

Many experts believe that genetically modified foods are safe and that they will provide significant benefits in the future. But public opinion on the use and safety of GM foods is mixed. About one third of Americans polled believe that using scientific methods, such as recombinant DNA technology, to enhance food flavor, color, nutrition, or freshness is wrong. Other polls indicate opposition to the use of GM foods may be as high as 50%. Skeptics frequently comment that "GM foods are against nature," and some people worry about potential health effects such as food allergies.

But it appears that Americans expect possible benefits in the future. Sixty-five percent of respondents in a 2000 Texas A&M poll indicated that GM foods will bring future benefits. If given a choice, many people have indicated they would look for another product rather than choose food labeled as genetically modified. This attitude raises another controversy we will consider later in the book, which is whether GM foods should be labeled as such.

Current U.S. regulations require labeling only if GM foods pose a health risk or if the product's nutritional value has changed. Although little evidence supports people's concerns about potential health risks of GM foods, not much research has been done either to support or to refute the claims of those concerned about the dangers of GM foods. What do you think about the use of GM foods? Would you be likely to buy GM foods if they were engineered to require fewer pesticide applica-

"THE LOWER-PRICED ITEMS CONTAIN GENETICALLY-MODIFIED FOODS NOT YET APPROVED FOR HUMAN USE."

tions than "natural" foods? What if GM foods stayed fresher longer? What if they were more nutritious and less expensive? How much risk should consumers be willing to take to reap the benefits of GM foods? Consider making a list of the questions you would want answered before you took your first bite of a GM food product. GM foods, to eat or not to eat, you decide.

Genetic manipulation of plants has been used for over 20 years to produce genetically engineered plants with altered growth characteristics such as drought resistance, tolerance to cold temperature, and greater food yields. Research conducted during the past 10 years clearly demonstrates that plants can be engineered to produce a wide range of pharmaceutical proteins in a broad array of crop species and tissues. Plants also offer certain advantages over bacteria for the production of recombinant proteins. For example, the cost of producing plant material with recombinant proteins is often significantly lower than producing recombinant proteins in bacteria.

The Presidential Advanced Energy Initiative of 2007 to allow biofuels to ease the "addiction" of the United States to foreign oil has been interpreted by advocates that 25% of U.S. energy would come from arable land by 2025. This goal will require significant advances in biotechnology to provide bioethanol sources other than corn, since this is not an efficient energy source. Agricultural waste, prairie grass, and other high cellulose sources will have to become efficient sources of energy through new decomposition and fermentation methods resulting from biotechnology. These challenges are well underway and are discussed in Chapter 6.

The use of plants as sources of pharmaceutical products is an application of agricultural biotechnology commonly called **molecular pharming.** For example, tobacco is a nonfood crop that has been the subject of many years of breeding and agronomic research. Tobacco plants have been engineered to produce recombinant proteins in their leaves, and these plants can be grown in large fields for molecular pharming. These and many other agricultural biotechnology applications are presented in Chapter 6.

Animal Biotechnology

In Chapter 7, we examine many areas of animal biotechnology, one of the most rapidly changing and exciting areas of biotechnology. Animals can be used as "bioreactors" to produce important products. For example goats, cattle, sheep, and chickens are being used as sources of medically valuable proteins such as **antibodies**—protective proteins that recognize and help body cells destroy foreign materials. Antibody treatments are being used to help improve immunity in patients with immune system disorders. Many other human therapeutic proteins produced from animals are in use, yet most of these proteins are needed in quantities that exceed hundreds of kilograms. To achieve this large-scale production, scientists can create female **transgenic animals** that express therapeutic proteins in their milk. Transgenic animals contain genes from another source. For instance, human genes for clotting proteins can be introduced into cows for the production of these proteins in their milk.

Animals are also very important in basic research. For instance, gene "knockout" experiments, in which one or more genes are disrupted, can be helpful for learning about the function of a gene. The idea behind a knockout is to disrupt a gene and then, by looking at what functions are affected in an animal as a result of the loss of a particular gene, determine the role and importance of that gene. Because many of the genes found in animals (including mice and rats) are also present in humans, learning about gene function in animals can lead to a greater understanding of gene function in humans. Similarly, the design and testing of drug and genetic therapies in animals often leads to novel treatment strategies in humans.

In 1997, scientists and the general public expressed surprise, excitement, and reservations about the announcement that scientists at the Roslin Institute in Scotland had cloned the now-famous sheep called Dolly (Figure 1.6). Dolly was the first mammal created by a cell nucleus transfer process, which we discuss in Chapter 7. Many other animals have been cloned since Dolly. Although animal cloning has elicited fears and concern about the potential for human cloning, scientists are generally excited about the techniques used to clone animals for a number of reasons. For instance, these techniques may lead to the cloning of animals that contain genetically engineered organs that can be transplanted into humans without fear of tissue rejection. Does a ready supply of donor organs of all types for all people who need an organ transplant sound like a good plan to you? If so, not everyone agrees. Animal cloning and the controversies surrounding organism cloning are important subjects discussed in Chapter 7.

Figure 1.6 Dolly, the First Mammal Produced by Nuclear Transfer Cloning   Dolly poses with her surrogate mother. Dolly was created by cloning technologies that may result in promising new techniques for improving livestock and cloning commercially valuable animals such as those containing organs for human transplantation. Unfortunately, Dolly developed early complications and was euthanized in February 2003.

Forensic Biotechnology

What do O. J. Simpson, Princess Anastasia, U.S. Army recruits, dinosaur bones, fecal bacteria, and 60,000-year-old Australian human fossils have in common? As you will learn in Chapter 8, all have been the subject of analysis by forensic biotechnology. **DNA fingerprinting**—a collection of methods for detecting an organism's unique DNA pattern—is a primary tool used in forensic biotechnology (Figure 1.7). Forensic biotechnology is a powerful tool for law enforcement that can lead to inclusion or exclusion of a person from suspicion, based on DNA evidence. DNA fingerprinting can be accomplished using trace amounts of tissue, hair, blood, or body fluids left behind at a crime scene. It was first used in 1987 to convict a rapist in England but is now routinely introduced as evidence in court cases throughout the world to convict criminals as well as free those wrongly implicated in a crime.

DNA fingerprinting has many other applications, including its use in paternity cases for pinpointing a child's father and for identifying human remains.

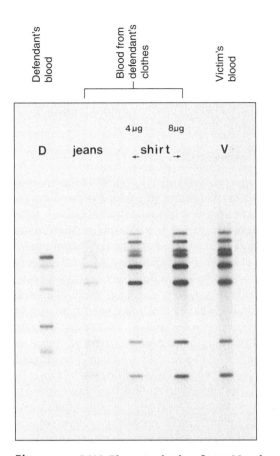

Figure 1.7 DNA Fingerprinting for a Murder Case
This photo shows the results of DNA fingerprinting techniques
(which you will learn about in Chapter 8) comparing the DNA
from bloodstains on a defendant's clothes to the DNA fingerprints
of a victim's blood. DNA fingerprinting cannot always be used to
determine definitely that an accused person has committed a
crime. In this case, DNA fingerprinting provides evidence that the
defendant can be linked to the crime scene, although it does not
mean the defendant is guilty of the murder.

Another application is the practice of DNA fingerprinting endangered species. This has already reduced
poaching and led to convictions of criminals by analyzing the DNA fingerprints of their "catch." Scientists
also use DNA fingerprinting to track and confirm the
spread of disease such as *Escherichia coli* in contaminated meat and to track diseases such as AIDS,
meningitis, tuberculosis, Lyme disease, and the West
Nile virus. Recently a French company even developed a gene expression test designed to determine if
expensive food products contain cheap, substitute,
mystery meats from species such as cats and eels.

Bioremediation

In Chapter 9, we discuss **bioremediation,** the use of
biotechnology to process and degrade a variety of natural

and manmade substances, particularly those that contribute to environmental pollution. Bioremediation is
being used to clean up many environmental hazards that
have been caused by industrial progress. Many processes
in bioremediation rely on applications of microbial
biotechnology. In the 1970s, the first U.S. patent for a
genetically modified microorganism was granted to
Ananda Chakrabarty. Chakrabarty and his colleagues
developed a strain of bacteria that was capable of degrading components in crude oil. One of the most publicized
examples of bioremediation in action occurred in 1989
following the *Exxon Valdez* oil spill in Prince William
Sound, Alaska (Figure 1.8). By stimulating the growth of
oil-degrading bacteria, which were already present in the
Alaskan soil, many miles of shoreline were cleaned up
nearly three times faster than they would have been had
chemical cleaning agents alone been used. Additionally,
the harsh treatment of the cleaning agents would have
further devastated the environment.

We also consider how domestic and industrial
sewage is treated and discuss how valuable metals such
as gold, silver, cobalt, nickel, and zinc can be recovered
from the environment through bioremediation. Once
again, microbial biotechnology plays important roles in
these processes.

Figure 1.8 Bioremediation in Action Strains of the
bacteria *Pseudomonas* were used to help clean Alaskan beaches
following the *Exxon Valdez* oil spill. Scientists on this Alaskan
beach are applying nutrients that will stimulate the growth of
Pseudomonas to help speed up the bioremediation process.

Aquatic Biotechnology

In Chapter 10, we explore the vast biotechnology possibilities offered by water—the medium that covers the majority of our planet. One of the oldest applications of aquatic biotechnology is **aquaculture,** raising finfish or shellfish in controlled conditions for use as food sources. Trout, salmon, and catfish are among many important aquaculture species in the United States. Aquaculture is growing in popularity throughout the world, especially in developing countries. It has recently been estimated that close to 30% of all fish consumed by humans worldwide are now produced by aquaculture.

In recent years, a wide range of fascinating new developments in aquatic biotechnology have emerged. These include the use of genetic engineering to produce disease-resistant strains of oysters and vaccines against viruses that infect salmon and other finfish. Transgenic salmon have been created that overproduce growth hormone leading to extraordinary growth rates over short growing periods, thus decreasing the time and expense required to grow salmon for market sale (Figure 1.9).

The uniqueness of many aquatic organisms is another attraction for biotechnologists. In our oceans, marine bacteria, algae, shellfish, finfish, and countless other organisms live under some of the harshest conditions in the world. Extreme cold, pressure from living at great depths, high salinity, and other environmental constraints are hardly a barrier because aquatic organisms have adapted to their difficult environments. As a result, such organisms are thought to be rich and valuable sources of new genes, proteins, and metabolic processes that may have important applications with human benefits. For instance, certain species of marine plankton and snails have been found to be rich sources of antitumor and anticancer molecules. Intensive research efforts are under way to better understand the wealth of potential biotechnology applications that our aquatic environments may harbor.

Medical Biotechnology

As you learned in Section 1.1, many biotechnology products, such as drugs and recombinant proteins, are being manufactured for human medical applications; however, these are just a few examples of **medical biotechnology.** Chapter 11 covers a range of different applications of medical biotechnology. Medical biotechnology is involved in the whole spectrum of human medicine. From preventative medicine to the diagnosis of health and illness to the treatment of human disease conditions, medical biotechnology has resulted in an amazing array of applications designed to improve human health. Over 325 million people worldwide have been helped by biotechnology drugs and vaccines. Although many powerful applications have already been designed and are currently being applied, the biotechnology century will see some of the greatest advances in medical biotechnology in our history.

It seems as though hardly a week goes by without news of a genetic breakthrough such as the discovery of a human gene involved in a disease process. Television, newspapers, and popular magazines all report important discoveries of new genes and other headlines involving

Figure 1.9 Aquatic Biotechnology Is an Emerging Science From using aquaculture to raise shellfish and finfish for human consumption to isolating biologically valuable molecules from marine organisms for medical applications, aquatic biotechnology has the potential for an incredible range of applications. Shown here is a genetically engineered salmon (top) bred to grow to adult size for market sale in half the time of a normal salmon (bottom).

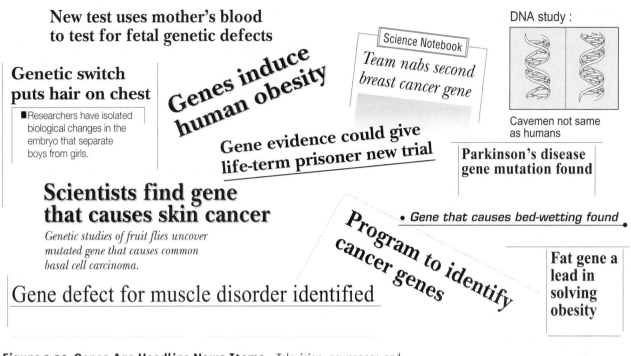

Figure 1.10 Genes Are Headline News Items Television, newspaper, and magazine headlines frequently report the discovery of genes involved in human disease conditions and many other everyday news items involving DNA.

DNA (Figure 1.10). Every day, new information from the Human Genome Project is helping scientists identify defective genes and decipher the details of genetic diseases such as sickle-cell anemia, Tay-Sachs disease, cystic fibrosis, causes of cancer, and forms of infertility, to give just a few examples. The Human Genome Project has already resulted in new techniques for genetic testing to identify defective genes and genetic disorders, and we explore many of these techniques in this book.

Gene therapy approaches, in which genetic disease conditions can be treated by inserting normal genes into a patient or replacing diseased genes with normal genes are being pioneered. In the near future, these technologies are expected to become increasingly more common. **Stem cell** technologies are some of the newest, most promising aspects of medical biotechnology, but they are also among the most controversial topics in all of science. Stem cells are immature cells that have the potential to develop and specialize into nerve cells, blood cells, muscle cells, and virtually any other cell type in the body. Stem cells can be grown in a laboratory and, when treated with different types of chemicals, they can be coaxed to develop into different types of human tissue that might be used in transplantations to replace damaged tissue. There are many exciting potential applications for stem cells, but, as we discuss in the next section and in Chapters 11 and 13, many complex scientific, ethical, and legal issues surround their use.

Regulatory Biotechnology

A very important aspect of the biotechnology business involves the regulatory processes that govern the industry. In much the same way that pharmaceutical companies must evaluate their drugs based on specific guidelines designed to maximize the safety and effectiveness of a product, most products of biotechnology must also be carefully examined before they are available for use. Two important aspects of the regulatory process include **quality assurance (QA)** and **quality control (QC).** QA measures include all activities involved in regulating the final quality of a product, whereas QC procedures are the part of the QA process that involves lab testing and monitoring of processes and applications to ensure consistent product standards. From QA and QC procedures designed to ensure that biotechnology products meet strict standards for purity and performance to issues associated with granting patents, resolving legal issues, and abiding by the regulatory processes required for clinical trials of biotechnology products in human patients, we consider these and other important biotechnology regulatory issues in Chapter 12.

The Biotechnology "Big Picture"

Although we have described the different types of biotechnology as distinct disciplines, do not think about biotechnology as a field with separate and unrelated disciplines. It is important to remember that almost all areas of biotechnology are closely interrelated. For example, applications of bioremediation are heavily based on using microbes (microbial biotechnology) to clean up environmental conditions. Even medical biotechnology relies on the use of microbes to produce recombinant proteins. A true appreciation of biotechnology involves understanding the biotechnology "big picture"—how biotechnology involves many different areas of science and how different types of biotechnology depend on each other. This interdependence of many areas of science will be put to the test in solving important problems in the 21st century.

1.3 Biological Challenges of the 21st Century

Numerous problems and challenges have the potential to be solved by biotechnology. For many of the greatest challenges—such as curing life-threatening human diseases—the barriers to overcoming these challenges are not insurmountable. Answers lie in our ability to better understand biological processes and design and adapt biotechnological solutions. For some applications that have been used for a few years, the biotechnology future is now. Rather than speculate about all of the ways that biotechnology may affect society in this century (an impossible task!), in this section we entice you with a few ideas on how medical biotechnology in particular will change our lives in the years ahead. In future chapters, we explore these and other ideas in much more detail.

What Will the New Biotechnology Century Look Like?

History will show that 2001 was a landmark in the biotechnology timeline. In February of that year, some of the world's most well-known molecular biologists gathered at a press conference to announce the publication of the rough draft of the human genome, a major accomplishment of the Human Genome Project. The DNA sequence—read as the letters A, G, C, and T—of human chromosomes was almost complete. One great surprise from this gathering was the announcement that the human genome consists of far less than 100,000 genes as had been expected. As you will learn in other chapters, the Genome Project was completed in 2003 and it has led to exciting new advances in biotechnology.

Identifying the chromosomal location and sequencing all genes in the human genome has greatly increased our understanding of the complexity of human genetics. Basic research on the molecular biology and functions of human genes and controlling factors that regulate genes is providing immeasurable insight into how genes direct the activities of living cells, how normal genes function, and how defective genes are the molecular basis of many human disease conditions. An understanding of human genes will also allow us to study human evolution, and by comparing our genes to other species such as chimpanzees, bacteria, flies, and even worms, we will develop a greater understanding of how humans are related to other species.

An advanced understanding of human genetic disease conditions will also transform medicine as it is currently practiced. A new era of medicine is on the horizon. But is the Human Genome Project a quick and simple way to find defective genes so that we can quickly and simply cure human disease? If you think so, then you are overlooking the complexity of biology. The human genome is not the "biological crystal ball" that will immediately solve all of our medical problems. Even when we have full knowledge of how genes work to allow a brain to assemble from a fertilized egg, we still won't know how the brain reasons or stores information as our memory. Identifying all human genes is just the tip of the iceberg for understanding how genes determine our health and susceptibility to disease. One benefit of this project will be in using it to decipher the **proteome,** the collection of proteins responsible for activity in a human cell. Even now that the Human Genome Project has been completed, scientists will continue to work on unlocking the secrets of how all human genes function, and how genes and proteins cause disease. A better understanding of human disease will require that we understand the structures and functions of the proteins that genes encode. But neither the genome nor the proteome is a software program that predetermines our health and our lives. Unlocking the mysteries of the human genome and human proteome alone makes the 21st century a most exciting time for scientific discovery.

A Scenario in the Future: How Might We Benefit from the Human Genome Project?

Imagine the following scene in the year 2015 or so. A man seeks advice at a local pharmacy. He recently switched from one major drug to another, and the current drug is not working any better than the first one for his arthritis. He tells his pharmacist, "This drug is so expensive and doesn't work any better for me than the last one, but I don't want to waste it or throw it out." "Well, sometimes the drugs don't work for everyone," says the pharmacist. This exchange represents one difficulty inherent in current health care strategies. Some drugs only work for some patients. How will the biotechnology century help this patient? The Human Genome Project might change medicine as we now know it and help patients such as this.

Many people currently experience the same problem that the man at the pharmacy encountered. The standard over-the-counter or routinely prescribed treatments available for arthritis and a host of other medical problems rarely work the same for everyone. Genome information has and will continue to result in the rapid, sensitive, and early detection and diagnosis of genetic disease conditions in humans of all ages from unborn children to the elderly. In the case of arthritis, we know there are different forms of arthritis with similar symptoms. Recent genetic studies have revealed that these different forms of the disease are caused by different genes. Increased knowledge about genetic disease conditions such as arthritis will lead to preventative medicine approaches designed to foster healthier lifestyles and new, safer, and more effective treatment strategies to cure disease.

Let us consider how identifying the genes causing arthritis in our imaginary patient can help him. From its inception, the Human Genome Project yielded immediate dividends in our ability to identify and diagnose disease conditions. The identification of disease genes has enabled scientists and physicians to screen for a wide range of genetic diseases. This screening ability will continue to grow in the future. One area expected to be a great aid in the diagnosis of genetic disease conditions will be applications involving **single nucleotide polymorphisms** (**SNPs;** pronounced "snips"). SNPs are single nucleotide changes or **mutations** in DNA sequences that vary from individual to individual (Figure 1.11). These subtle changes represent one of the most common examples of genetic variation in humans.

SNPs represent variations in DNA sequences that influence how we respond to stress and disease, and SNPs are the cause of genetic diseases such as sickle-cell anemia. Most scientists believe that SNPs will help them identify some of the genes involved in medical conditions such as arthritis, stroke, cancer, heart disease, diabetes, and behavioral and emotional illnesses, as well as a host of other disorders. One of the goals of the Human Genome Project was to identify SNPs and develop SNP maps of the human genome, which is being accomplished. An international group is developing HapMap, which will identify and verify millions of SNPs to compile a universal guide to human genetic variation. One company has already genotyped more than 4 million SNPs in 270 people. The significance of this can be seen in the drugs that have been developed for human breast cancer. The two well-known breast cancer genes,

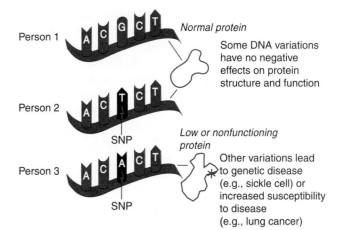

Figure 1.11 Single Nucleotide Polymorphisms A small piece of a gene sequence for three different individuals is represented. For simplicity, only one strand of a DNA molecule is shown. Notice how person 2 has a SNP in this gene, which has no effect on protein structure and function. Person 3 however, has a different SNP in the same gene. This subtle genetic change may affect how this person responds to a medical drug or influence the likelihood that person 3 will develop a genetic disease.

BRCA1 and *BRCA2*, have about 1,700 variants worldwide. Before these genes were known, all breast cancer was treated with the same cell toxin chemotherapy drugs. Now Herceptin, Rituxan, Gleevec, and Tarceva are available drug treatments linked to the testing for these two genes.

Testing one's DNA for different SNPs is one way to identify disease genes that a person may have. One way to do this is to isolate DNA from a small amount of a patient's blood and then apply this sample to a **DNA microarray,** also called a **gene chip.** As you will learn in Chapter 3, microarrays contain thousands of DNA sequences. Using sophisticated computer analysis, scientists can compare patterns of DNA binding between patient DNA and DNA on the microarray to reveal a patient's SNP patterns. For instance, researchers can use microarrays to screen a patient's DNA for a pattern of genes that might be expressed in a disease condition such as arthritis.

The discovery of SNPs is partially responsible for the emergence of a field called **pharmacogenomics,** a new type of biotechnology and a field in its infancy. Pharmacogenomics is really customized medicine. It involves tailor-designing drug therapy and treatment strategies based on the genetic profile of a patient—using our genetic information to determine the most effective and specific treatment approach (refer to Figure 11.7). Can pharmacogenomics solve some of our medical problems? Right now, doctors and physicians can only make guesses. Some day, however, with the right tools and human gene information, this will change.

Pharmacogenomics might help our arthritis patient in the pharmacy in 2015. We know that arthritis is a disease that shows familial inheritance for some individuals, and as mentioned earlier, a number of different genes are involved in different forms of arthritis. In many other cases of arthritis, a clear mode of inheritance is not seen. Perhaps there may be additional genes or nongenetic factors at work in these cases. A simple blood test from our patient could be used to prepare DNA for SNP and microarray analysis. SNP and microarray data could be used to determine which genes are involved in the form of arthritis that this man has. Armed with this genetic information, a physician could design a drug treatment strategy—based on the genes involved—that would be *specific* and *most effective* against this man's type of arthritis. A second man with a different genetic profile for his particular type of arthritis might undergo a different treatment than the first. This is the power of pharmacogenomics in action. It is predicted that eventually everyone will have a whole-genome scan to provide information for useful and specific treatment. Of course, such a screening for genes that are related to medical conditions must be done in an ethical fashion, with proper security and integration into the health care delivery system (Figure 1.12).

Figure 1.12 Secrets of the Human Genome In the future, we will have unprecedented knowledge of our genetic makeup including SNPs and other markers of genetic diseases. Can you think of possible ethical, legal, and social implications of such information?

The same principles of pharmacogenomics will also be applied to a host of other human diseases such as cancer. As you probably already know, many drugs currently used to treat different types of cancer through **chemotherapy** may be effective against cancerous cells but may also affect normal cells. Hair loss, dry skin, changes in blood cell counts, and nausea are all conditions related to the effects of chemotherapy on normal cells. But what if drugs that are effective against cancer cells could be designed so they had no effect on normal cells in other tissues? This may be possible as the genetic basis of cancer is better understood and drugs can be designed based on the genetics of different types of cancer. In addition, SNP and microarray information could also be used to figure out a person's risk of developing a particular type of cancer, long before he or she would otherwise begin to show the disease, especially when someone has a family history. Such information might be used to help that person develop changes in lifestyle such as diet and exercise habits that might be important for preventing disease.

Another example of the benefits of studying differences in human genotypes has been in the area of **metabolomics,** a biochemical snapshot of the small molecules produced during cellular metabolism, such as glucose, cholesterol, ATP, and signaling molecules that result from a cellular change. This snapshot directly reflects physiologic status and can be used to monitor drug effects on disease states. The exact number of human metabolites is unknown, but estimates of between 2,000 and 10,000 have been published. The use of this tool can distinguish between disease process and physiologic adaptation, and it can save time and money when incorporated into early-stage drug discovery. For example, a major drug company recently funded a study in which groups of mice were fed a diet designed to increase cholesterol, and then characterized their lipids in plasma, adipose tissue, and liver at intervals over a number of weeks. Over 500 "unusual" lipids were identified as the response. One group of mice was atherosclerotic susceptible, thus providing a good measure of differential response due to this disease physiology.

In Chapter 11 we also discuss examples of **nanotechnology,** applications that incorporate extremely small devices (a "nano" scale). Nanotechnology is an entirely new field that is rapidly emerging as a major research area. One promising application of nanotechnology has been the development of small particles that can be used to deliver drugs to cells (Figure 1.13).

In addition to advances in drug treatment, gene therapy represents one of the ultimate strategies for combating genetic disease. Gene therapy technologies involve replacing or augmenting defective genes

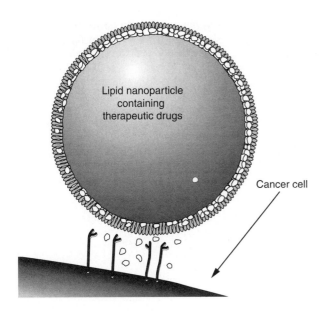

Lipid nanoparticle
containing
therapeutic drugs

Cancer cell

Figure 1.13 Nanobiotechnology in Action Nanoparticles containing chemotherapy agents can be specifically directed to target cancer cells by coating them with tumor specific proteins that bind to the target cells. In this way, chemotherapy agents that cannot pass through the cell membrane can be released specifically inside these target cancer cells.

with normal copies of a gene. Think about the potential power of this approach. Scientists are working on a variety of ways to deliver healthy genes into humans, such as using viruses to carry healthy genes into human cells. Promising techniques have been developed for treating some blood disorders and diseases of the nervous system such as Parkinson's disease (Figure 1.13). However, many barriers must be overcome before gene therapy becomes a safe, practical, effective, and well-established approach to treating disease.

Obstacles currently prevent gene therapy from being widely used in humans. For example, how can normal genes be delivered to virtually all cells in the body? What are the long-term effects of introducing extra genes into humans? What must be done to be sure that the normal protein is properly made after the extra genes are delivered into the body? As you will discover in Chapter 11, gene therapy applications are under increased scrutiny following complications in several patients including the tragic death of a gene therapy patient, Jesse Gelsinger, who died after a controversial gene therapy trial in 1999. Another exciting new technology with the potential for modification of a genetic defect by silencing a gene is being aggressively pursued using **small interfering RNA (siRNA;** see Chapter 3 for details of this process).

In the future, stem cell technologies are expected to provide powerful tools for treating and curing disease. As we briefly discussed in Section 1.2, stem cells are immature cells that can grow and divide to produce different types of cells such as skin, muscle, liver, kidney, and blood cells. Most stem cells are obtained from embryos **(embryonic stem cells,** or **ESCs).**

Some ESCs can be isolated from the cord blood of newborn infants. Recently, scientists have successfully isolated stem cells from adult tissues **(adult-derived stem cells,** or **ASCs).**

In the laboratory, stem cells can be coaxed to form almost any tissue of interest depending on how they are treated. Imagine growing skin cells, blood cells, neurons, and even tissues and whole organs in the lab and using these to replace damaged tissue or failing tissues and organs such as the liver, pancreas, and retina (Figure 1.14). **Regenerative medicine** is the phrase used to describe this approach. In the future, scientists may be able to collect stem cells from patients with genetic disorders, genetically manipulate these cells by gene therapy, and reinsert them into the patient from whom they were collected to help treat genetic disease conditions. Some of this work is already possible, and these technologies will be optimized in the near future.

We hope that the examples in this section demonstrated how the future is indeed bright for marvelous advances in medical biotechnology. Pharmacogenomics, gene therapy, and stem cell technologies are not the answers to all our genetic problems, but with continued rapid advances in genetic technology, many seemingly impossible problems may not be so insurmountable in the future. Here we presented basic examples of medical applications in the biotechnology century, but in future chapters you will learn about exciting applications from other areas of biotechnology that will potentially change our lives for the better. We conclude our introduction to the world of biotechnology by discussing career opportunities in the industry.

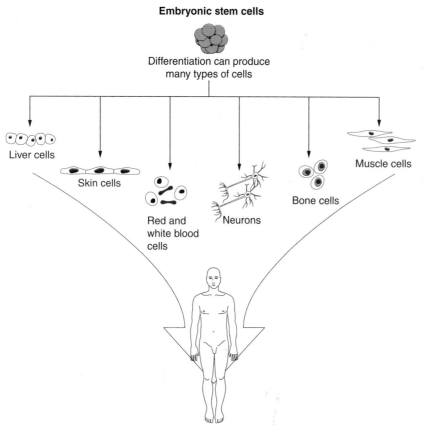

Embryonic stem cells

Differentiation can produce
many types of cells

Liver cells

Skin cells

Red and
white blood
cells

Neurons

Bone cells

Muscle cells

Transplantation to replace damaged or defective tissue

Figure 1.14 Embryonic Stem Cells Can Give Rise to Many Types of Differentiated Cells Embryonic stem cells are derived from embryos or early-stage fetuses. ES cells are immature cells that can be stimulated to develop (differentiate) into a variety of different cell types. Many scientists are excited about potential medical applications of ES cells, such as transplantation to replace damaged or defective tissues, but their use is very controversial.

1.4 The Biotechnology Workforce

How will the world prepare for the biotechnology century? Recent achievements of the Human Genome Project have created a range of new opportunities for biotechnology companies and individuals seeking employment in the biotechnology industry (Figure 1.15). One challenge will be to train people who can decipher growing mountains of genetic information and draw relevant conclusions about complex relationships among genes, health, and disease.

Figure 1.15 The Biotechnology Industry Provides Exciting Opportunities for Many Types of Scientists From biologists and chemists to engineers, information technologists, and salespeople, the biotechnology industry offers a great range of high-tech employment opportunities. Shown here is a senior undergraduate student working on a biotechnology research project. Gaining research experience as an undergraduate is an excellent way to prepare for a career in biotechnology.

Biotechnology scientists will need to be comfortable analyzing information from many sources, such as DNA sequence data, gene expression data from microarrays and SNPs, computer modeling data from DNA and protein structure analysis, and chemical data used to study molecular structures. Ultimately, biotechnology companies are looking for people who are comfortable analyzing complex data and sharing their expertise with others in team-oriented, problem-solving working environments. The biotechnology workforce depends on important contributions from talented people in many different disciplines of science.

The Business of Biotechnology

In 1976, Genentech Inc., a small company near San Francisco, California, was founded. Genentech is generally recognized as the first biotechnology company, and its success ushered in the birth of this exciting industry. Today, biotechnology is a global industry with hundreds of products on the market generating more than $63 billion in worldwide revenues, including $40 billion in sales of biological drugs (such as enzymes, antibodies, growth factors, vaccines, and hormones) in the United States. Many biotechnology companies are working on cures for cancer, in part because in the United States alone, nearly 40% of Americans will receive a diagnosis of cancer in their lifetime. Cancer is the second leading cause of death in the United States behind heart disease. Over 350 biotechnology products are currently in development targeting cancers, diabetes, heart disease, Alzheimer's and Parkinson's diseases, arthritis, AIDS, and many other diseases.

North America, Europe, and Japan account for approximately 95% of biotechnology companies, but biotechnology firms are found throughout the world with over 4,900 companies in 54 countries. Countries without a traditional history in research and development worldwide are turning to biotechnology for high-tech innovations. For example, biotechnology is a rapidly developing industry in India and China. Still many of the world's leading biotechnology companies are located in the United States (see Figure 1.16). There are currently around 1,500 biotechnology companies in the United States, many of which are often closely associated with colleges and universities or located near major universities where basic science ideas for biotechnological applications are generated. Visit the Biotechnology Industry Organization Website (www.bio.org) for information on biotechnology centers around the nation. These centers are excellent resources for biotechnology career information in your state. At this site, you can find biotechnology companies located near you and learn about their current products.

Organization of a Biotechnology Company

By now you may be wondering "what is the difference between a biotechnology company and a pharmaceutical company?" Most people can name pharmaceutical companies such as Merck, Johnson & Johnson, or Pfizer because they or a family member may have used one or more of their products, but most people cannot name a biotechnology company (Table 1.3) or explain why a biotech company is different than a pharmaceutical company. The large pharmaceutical companies are commonly referred to as "big pharma." Generally speaking, **pharmaceutical companies** are involved in drug development by chemically synthesizing or purifying compounds used to make the drug, products such as aspirin, antacids, and cold medicines. Pharmaceutical companies are typically not using living organisms to grow or produce a product (such as a recombinant protein) as is the focus of biotechnology companies. But these days the distinctions between the two are blurring because many large pharmaceutical companies are often involved in biotechnology-related research and product development either directly or indirectly by partnering with a biotechnology

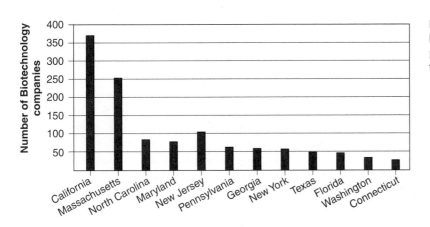

Figure 1.16 Distribution of U.S. Biotechnology Companies Public and private biotechnology companies are located throughout the United States.

Table 1.3 TOP-FIVE BIOTECHNOLOGY COMPANIES AND TOP-FIVE PHARMACEUTICAL COMPANIES BY REVENUE	
Biotech Companies	**Revenue ($Millions)**
Amgen	$12,022
Genentech	$6,633
Genzyme	$2,597
Biogen Idec	$2,377
Chiron	$1,921
Pharma	
Pfizer	$51,298
Johnson & Johnson	$50,656
Merck & Co.	$22,000
Bristol-Myers Squibb	$19,207
Eli Lilly & Co.	$14,645

Adapted from: Ernst & Young, *Beyond Borders: Global Biotechnology Report 2006* (www.ey.com/beyondborders). Revenue based on preliminary results reported by companies.

company. Also remember that biotechnology involves much more than drug development. There are many different companies of varying sizes dedicated to working on specific areas of biotechnology.

Biotechnology companies vary in size from small companies of less than 50 employees to large companies with over 300 employees. Historically many biotechnology companies begin as a small **startup company** formed by a small team of scientists who believe they may have a promising product to make (such as a recombinant protein to treat disease). The team must typically then seek investors to fund their company so they can buy or rent lab facilities, buy equipment and supplies, and continue the research and development necessary to make their product. If they are then successful in bringing a product to the market (a process that takes around 10 years on average), many startups are often bought by larger well-established companies.

There are similarities between how pharmaceutical companies and biotechnology companies are organized (Figure 1.17). We discuss many aspects of this organization in the next section where we describe different job opportunities in each area of a biotechnology company.

Jobs in Biotechnology

The biotechnology industry in the United States employs over 200,000 people. Biotechnology offers numerous employment choices such as laboratory technicians involved in basic research and development, computer programmers, laboratory directors, and sales and marketing personnel. All are essential to the biotechnology industry. In this section, we consider some of the job categories available in biotechnology.

Research and development

Development of a new biotech product is a long and expensive process. Individuals in **research and development (R&D)** are directly involved in the process of developing ideas and running experiments to determine if a promising idea (for example, using a recombinant protein from a recently cloned gene to treat a disease condition) can actually be developed into a product. It requires a great deal of trial and error. From the largest to the smallest biotechnology companies, all have some staff dedicated to R&D. On average, biotechnology companies invest at least four times more on R&D than any other high-tech industry. For some companies, the R&D budget is close to 50% of the operating budget. R&D is the lifeblood of most companies—without new discoveries, companies cannot make new products. The majority of positions in R&D usually require a bachelor's or associate's degree in chemistry, biology, or biochemistry (Figure 1.18). **Laboratory technicians** are responsible for duties such as cleaning and maintaining equipment used by scientists and keeping labs stocked with supplies. Technician positions usually require a B.A. in science or a B.S in biology or chemistry. **Research assistants** or **research associates** carry out experiments under the direct supervision of established and experienced scientists. These positions require a B.S. or M.S. in biology or chemistry. Research assistants and associates are considered "bench" scientists, carrying out research experiments under the direction of one or more principal or senior scientists. Assistants and associates perform research in collaboration with others. Involved in the design, execution, and interpretation of experiments and results, they may also be required to review scientific literature and prepare technical reports, lab protocols, and data summaries.

Principal or **senior scientists** usually have a Ph.D. with considerable practical experience in research and management skills for directing other scientists. These individuals are considered the scientific leaders of a company. Responsibilities include planning and executing research priorities of the company, acting as a spokesperson on company research and development at conferences, participating in patent applications, writing progress reports, applying for grants, and serving as an adviser to the top financial managers of the company. The job titles and descriptions we described can vary from company to company; how-

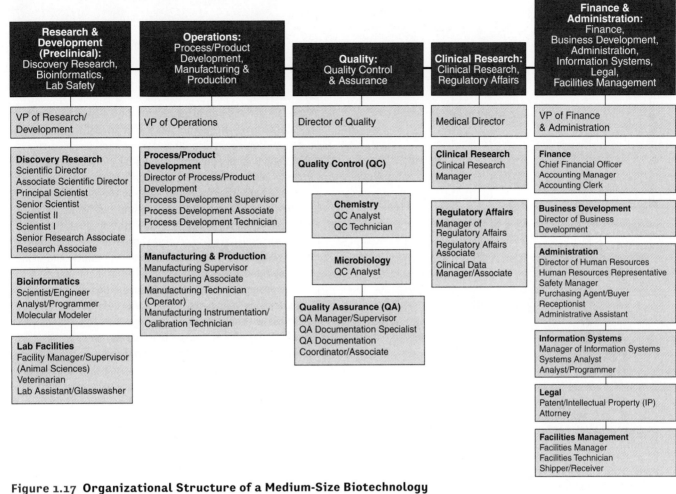

Figure 1.17 Organizational Structure of a Medium-Size Biotechnology Company Biotech companies range in size, but most companies have an organizational structure similar to those depicted in this figure. Notice the range of different aspects of the company from R&D through sales, marketing, and legal aspects of a product.

ever, if you are interested in making new scientific discoveries, then R&D might be an exciting career option for you!

The rapidly expanding field of bioinformatics, the use of computers to analyze and store DNA and protein data, requires an understanding of computer programming, statistics, and biology. Until recently, many experts in bioinformatics were computer scientists who had trained themselves in molecular biology or molecular biologists self-trained in computer science, database analysis, and mathematics. Today, people with computer science interests are being encouraged to take classes in biotechnology, and biotechnology students are being encouraged to take computer science classes. In addition, specific programs in bioinformatics are beginning to appear at major universities, four-year colleges, technical colleges, and community colleges to train people to become **bioinformaticists.**

Many speculate that the massive amount of data from the Human Genome Project will result in a merger of biotechnology and information technology. Bioinformaticists are needed to analyze, organize, and share DNA and protein sequence information. The human genome contains over 3 billion base pairs alone, and hundreds of thousands of bases of sequence data from other species are added to databases around the world each day. Sophisticated programs are required to analyze this information. How will biotechnology companies keep from drowning in an ever-increasing sea of data that has inundated biology and chemistry? Robust data-mining and data-warehousing systems are just beginning to enter the bioinformatics market. To put this in perspective, a financial database for a major bank might have 100 columns representing different customers, with 1 million rows of data. A major pharmaceutical database, in contrast, may

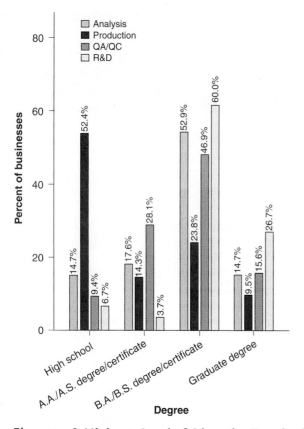

Figure 1.18 Minimum Level of Education Required of Entry-Level Technicians The Resource Group performed a survey of entry-level education requirements for 69 biotechnology companies in the three-county area of central coast California. Results are comparable with other areas of the United States and indicate that R&D generally requires a greater amount of training than other job areas.

YOU DECIDE

Generic Biotech Drugs?

As the industry of biotechnology has aged, many of the earlier biotech products that received patent approval are set to lose patent protection in the next few years. Patents for about 20 products with annual sales of over $10 billion expired in 2006. Patents are designed to provide a monopoly right for the developer of an invention, and in the United States patents can last for up to 20 years. There is controversy in the industry over whether many of these products will receive approval to become **generic drugs**. You may already know that generic drugs are copies of brand-name products that generally have the same effectiveness, safety, and quality as the original but are produced at a cheaper cost to the consumer than the brand-name drugs. One way that generics can cut costs is that they are often approved for use without having to undergo the same expensive safety and effects studies required for name drugs. Many biotechnology companies are fighting the production of generic biotech drugs, claiming that the higher costs of making a biological product such as an antibody compared with a pharmaceutically-produced drug earns them a right to manufacture named drugs and make a profit without competition from generics. Some also doubt whether a generic could be made at a greatly reduced price given that it is generally still more expensive to produce a biotech product than a pharmaceutical product. Similar questions about profit can be raised regarding drug costs in developing countries where many of the people who need a drug cannot afford it, although in this case many companies sell drugs in the developed world at a higher price and use some of these profits to provide drugs at low cost or free to developing nations. Should biotechnology companies be forced to produce generic drugs? You decide.

contain 30,000 columns for genes and only about 40 rows of patients. Computer modeling tools, such as neural networks and decision trees, are also widely used by bioinformaticists to identify patterns between SNP markers and a disease status. If you are interested in merging an understanding of biology with computer science skills, then bioinformatics may be a good career option for you.

Operations, biomanufacturing, and production

Operations, biomanufacturing, and *production* are terms that describe the division of a biotechnology company that oversees specific details of product development such as the equipment and laboratory processes involved in producing a product. This often includes **scale-up** processes in which cultured cells making a product must be grown in large scale. This is not a trivial task. As a simple analogy, scaling-up is the difference between cooking a meal for yourself versus preparing a full course Thanksgiving dinner for 50 people. Bio-

manufacturing and production units maintain and monitor the large-scale and large-volume equipment used during production and they ensure that the company is following proper procedures and maintaining appropriate records for the product. Biomanufacturing job details are specific to the particular product a company is manufacturing. Entry-level jobs include material handlers, manufacturing assistants, and manufacturing associates. Supervisory and management-level jobs usually require a bachelor's or master's degree in biology or chemistry and several years of experience in manufacturing the products or type of product being produced by that company. Manufacturing and production also involves many different types of engineers including those trained in chemical, electrical, environmental, or industrial engineering. Engineering positions usually require a B.A. degree in engineering or a M.S. degree in biology or an area of engineering.

Quality assurance and quality control

As we discuss in Chapter 12, most products from biotechnology are highly regulated by such federal agencies as the Food and Drug Administration (FDA), Environmental Protection Agency (EPA), and U.S. Department of Agriculture (USDA). These federal agencies require that manufacturing follow exact methods approved by regulatory officials. As discussed previously, the overall purpose of quality assurance is to guarantee the final quality of all products. Quality control efforts are designed to ensure that products meet stringent regulations mandated by federal agencies. In addition to guaranteeing that components of the product manufacturing process meet the proper specifications, QA and QC workers are also responsible for monitoring equipment, facilities and personnel, maintaining correct documentation, testing product samples, and addressing customer inquiries and complaints along with other responsibilities. Entry-level jobs in QC and QA include validation technician, documentation clerk, and QC inspector. Jobs usually require at least a B.S. degree in biology, and managerial or supervisory positions require more education. **Customer relation specialists** sometimes also work under QA divisions of a company. One function of customer relations is to investigate consumer complaints about a problem with a product and follow-up with the consumer to provide an appropriate response or solution to the problem encountered.

Clinical research and regulatory affairs

As we discuss in Chapter 12, in the United States, developing a drug product is a long and expensive process of testing the new drug candidate in volunteer subjects to ultimately receive new drug approval from the FDA. The clinical trial process, along with many other clinical and nonclinical areas of biotechnology, is regulated by a number of different agencies. As a result, every biotechnology company has staff monitoring regulatory compliance to ensure proper regulatory procedures are in place and being followed. Biotechnology companies involved in developing drugs for humans often have very large clinical research divisions with science and non-science personnel that conduct and oversee clinical trials.

Marketing, sales, finance, and legal divisions

Marketing and selling a variety of biotechnology products, from medical instruments to drugs, is a critical area of biotechnology. Most people employed in biotechnology marketing and sales have a B.S. degree in the sciences and familiarity with scientific processes in biotechnology, perhaps combined with coursework in business or even a B.A. degree in marketing. **Sales representatives** work with medical doctors, hospitals, and medical institutions to promote a company's products. **Marketing specialists** devise advertising campaigns and promotional materials to target customer needs for the products a company sells. Representatives and specialists frequently attend trade shows and conferences. An understanding of science is important because the ability to answer end-user questions is an essential skill in marketing and sales. **Finance divisions** of a biotechnology company are typically run by vice presidents or chief financial officers who oversee company finances and also often are involved in raising funds from partners or venture capitalists seeking investments in technology companies. **Legal specialists** in biotechnology companies typically work on legal issues associated with product development and marketing such as copyrights, naming rights, and obtaining patents. Staff in this area will also address legal circumstances that may arise if there are problems with a product or litigation from a user of a product.

Salaries in Biotechnology

People working in the biotechnology industry are making groundbreaking discoveries that fight disease, improve food production, clean up the environment, and make manufacturing more efficient and profitable. Although the process of using living organisms to improve life is an ancient practice, the biotechnology industry has only been around for about 25 years. As an emerging industry, biotechnology offers competitive salaries and benefits, and employees at almost all levels report high job satisfaction.

Salaries for life scientists who work in the commercial sector are generally higher than those paid to scientists in academia (colleges and universities). Scientists working in the biotechnology industry are among the most highly paid of those in the professional sciences. In 2006, in California alone, the biotechnology industry generated about $20 billion in personal wages and salary. In this same year, the top five biotechnology companies in the world spent an average of $93,400 on each employee.

According to a survey of more than 400 biotechnology companies conducted recently by the Radford Division of AON Consulting, Ph.D.s in biology, chemistry, and molecular biology with no work experience were starting at an average annual salary of $55,700 with senior scientists earning in excess of $120,000 a year. For individuals with a Master's degree in the same fields, the average salary was $40,600 annually, with a range from $60,000 to $70,000 per year for research associates, and $32,500 annually for those with a B.A. degree, with a range of $52,000 to $62,000 per year for research associates. Visit the Radford Biotechnology Compensation Report, the Commission on Professionals in Science and Technology, and the U.S. Office of Personnel Management on the Web for updates on the surveys used for the salary figures described in this section. Biotechnology salary reports websites are listed in the Keeping Current: Web Links at the Companion Website.

Based on a national survey, 56% of the college students entering biotechnology training programs had little or no science background. If you have the proper background in biology and good lab skills, many good positions are available at many different levels, but increasingly educational training at the community college, technical college, or four-year college or university level is becoming a requirement for employment in biotechnology.

Hiring Trends in the Biotechnology Industry

Career prospects in biotechnology are excellent. The industry has more than tripled in size since 1992, and worldwide company revenues have increased from approximately $8 billion in 1992 to nearly $28 billion in 2001. As a result of this prosperity, in the four years between 1995 and 1999, the most recent span for which such data are available, the U.S. biotechnology industry increased its employee workforce by 48.5%. This trend is expected to continue. Biotechnology firms and research labs have found that they are better off filling skilled technician-level jobs with people who have more specialized training than pursuing their more traditional practice of attempting to find people with master's and doctoral degrees. Many human resource hiring staff and recruitment firms indicate that there is a tremendous increase in the number of open positions in bioinformatics, proteomics, and genome studies.

There is currently also a hot job market for scientists in drug discovery. Larger biotechnology companies, no matter in which region, report that they are growing rapidly and find that almost every career choice is in demand. In particular, the most sought after jobs are more often roles that require team interaction, both inside and outside the company. Partnering has become the landscape of drug development, and skills in this area are required for any person's career in the industry.

Another trend that has reached a stage of critical importance is the need for people with multiple skill

CAREER PROFILE

Finding a Biotechnology Job That Appeals to You

Throughout the book we will use this career profile feature to highlight potential career options including educational requirements, job descriptions, salary, and related information. A number of websites are outstanding resources for biotechnology career information:

- Visit the Biotech Career Center (www.biotechcareercenter.com), a very good site for career materials, links to job resources, a wealth of information on over 600 biotechnology companies, and much more.

- Visit the Biotechnology Industry Organization website (www.bio.org) and access one of the biotechnology company sites.

- Visit the Access Excellence Careers in Biotechnology website (www.accessexcellence.org/RC/CC/) for job descriptions and excellent links to resources for careers in biotechnology.

- Visit the Bio-Link website (www.bio-link.org) to find useful biotechnology workforce resources. It has several sections of career information, job descrip-

tions and educational requirements, job posting sites, and state-by-state listings of biotechnology companies, among many other resources.

- The California State University Program for Education and Research in Biotechnology (CSUPERB) (www.csuchico.edu/csuperb/) is a great resource for educational and career materials in biotechnology. In particular, visit the "career site" and "job links" pages.

- Visit the Massachusetts Biotechnology Industry Organization (www.massbio.org/), and follow the "careers" link to one of the most comprehensive listing of job descriptions in the biotechnology industry, from vice president of research and development to glasswasher positions (yes, this person does what the title says!).

Search these sites for a biotechnology job that appeals to you. Next rewrite your résumé to fit this job description, or identify the coursework or experience you would need to apply successfully for this position. Print your results, and keep it for reference.

areas. For instance, an individual with a degree in biology, a minor in information technology, and coursework in mathematics can potentially have a great advantage in the job market, especially with companies seeking people with unique skill combinations. Employment prospects in biotechnology are exciting indeed. Opportunities are excellent for individuals with solid scientific training and good verbal and written communication skills coupled with a strong ability to work as part of a team in a collaborative environment.

QUESTIONS & ACTIVITIES

Answers can be found in Appendix 1.

1. Provide two examples of historical and current applications of biotechnology.

2. Pick an example of a biotechnology application, and describe how it has affected your everyday life.

3. Which area of biotechnology involves using living organisms to clean up the environment?

4. Describe how pharmacogenomics will influence the treatment of human diseases.

5. Distinguish between QC and QA, and explain why both are important for biotechnology companies.

6. Access the library of the National Center for Biological Information (www.ncbinlm.nih.org), and search for new information on adult-derived stem cells. This free source will provide abstracts and titles for full-text articles that can be obtained at other libraries.

7. Visit the "About Biotech" section of the Access Excellence website at www.accessexcellence.org/AB. This section provides an outstanding overview of historical and current applications of biotechnology. Survey the different topics presented at this site. Many interesting aspects and examples of biotechnology are described in student-friendly terms. Find a biotechnology topic that fascinates you, print out the information you are interested in, and share your newly discovered knowledge with a friend or family member who is unfamiliar with biotechnology.

8. Interacting with others in a group setting is an essential skill in most areas of science. As you learned in this chapter, biotechnology involves groups of scientists, mathematicians, and computing experts with different backgrounds collaborating to solve a problem or achieve a common goal. Discussing biology with other people is fun and beneficial to everyone working on the same problem in a company, and working with other students is an excellent way to help you learn and enjoy your studies in biotechnology. Teaching someone else is a great way to test your knowledge. Analyze your ability to work in a group by forming a study group for the next test. Assign a group leader who is responsible for organizing meetings and keeping the study group focused on helping each other learn. Make sure that everyone has a topic to present to the group and that all of you offer constructive criticism to their suggestions. If you cannot get together with your classmates in one room, share your thoughts via e-mail or ask your professor to set up an electronic bulletin board for discussion purposes.

9. The National Library of Medicine is a worldwide database of biology research publications in scientific journals and it can be accessed for free at http://www.ncbi.nlm.nih.gov/entrez/query.fcgi?DB=pubmed. Access PubMed and conduct a search on any topic of biotechnology that may be of interest to you to find recent papers on the latest new research developments in biotechnology.

10. Organism cloning has led to concerns that humans may be cloned. At least one group has even claimed to have already done so. Run a Google search with the terms *Raelians* and *cloning*. At the time this book was published, no claims of human cloning have been proven to be valid and most scientists are adamantly against human cloning. What do you think? Should humans be cloned?

References and Further Reading

Aggarwal, S. (2007). What's Fueling the Biotech Engine? *Nature Biotech.*, 25: 1097–1104.

Beyond Borders: The Global Biotechnology Report 2006. Ernst & Young, 2006. www.ey.com/beyond borders.

Flanagan, N. (2005). Bioresearch Highlights: Significance of SNPs. *GEN*, 25:1.

Lizeswski, K. (2006). Metabolomics Plays Crucial Discovery Role. *GEN*, 26:1.

Palladino, M. A. (2006). *Understanding The Human Genome Project*, 2e. San Francisco: Benjamin Cummings.

Robbins-Roth, C. (2001). *From Alchemy to IPO: The Business of Biotechnology*. Cambridge, MA: Perseus.

Visit www.pearsonhighered.com/biotechnology to download learning objectives, chapter summary, "Keeping Current" web links, glossary, flashcards, and jpegs of figures from this chapter.

Chapter 2

An Introduction to Genes and Genomes

After completing this chapter you should be able to:

■ Compare and contrast the structures of prokaryotic and eukaryotic cells.

■ Discuss important experiments that led scientists to determine that DNA is the inherited genetic material of living organisms.

■ Describe the structure of a nucleotide and explain how nucleotides join together to form a double-helical DNA molecule.

■ Describe the process of DNA replication and discuss the role of different enzymes in this process.

■ Understand what genomes are and appreciate why biologists are interested in studying genomes.

■ Describe the process of transcription and understand the importance of mRNA processing in creating a mature mRNA molecule.

■ Describe the process of translation including the roles of mRNA, tRNA, and rRNA.

■ Define gene expression and understand why gene expression regulation is important.

■ Discuss the role of operons in regulating gene expression in bacteria.

■ Name different types of mutations and provide examples of potential consequences of mutations.

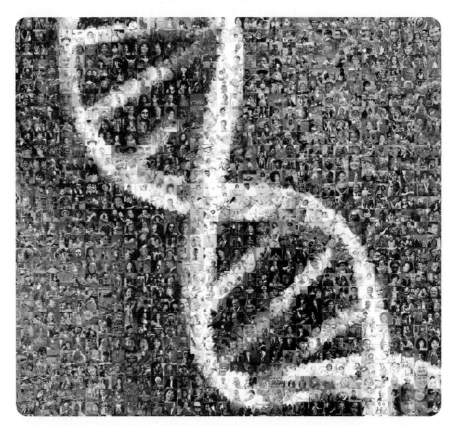

Encoded within DNA are genes that provide instructions controlling the activities of all cells. Genes influence our behavior; determine our physical appearances such as skin, hair, and eye color; and affect our susceptibility to genetic disease conditions.

Central to the study of biotechnology is an understanding of the structure of DNA as the molecule of life—the inherited genetic material. Later, in Chapter 3, we consider how extraordinary techniques in molecular biology enable biologists to clone and engineer DNA, manipulations that are essential for many applications in biotechnology. In this chapter, we review DNA structure and replication, discuss how genes code for proteins, and provide an introduction to the causes and consequences of mutations.

2.1 A Review of Cell Structure

Cells are the structural and functional units of all life-forms. Organisms such as bacteria consist of a single cell, whereas humans have approximately 75 trillion, including over 200 different types that vary in appearance and function. Cells vary greatly in size and complexity, from tiny bacterial cells to human neurons that may stretch for more than 3 feet from the spinal cord to muscles in the toes. But virtually all cells of an organism share a common component, genetic information in the form of **deoxyribonucleic acid (DNA).** Genes contained within DNA control numerous activities in cells by directing the synthesis of proteins. Genes influence our behavior; determine our physical appearance such as skin, hair, and eye color; and affect our susceptibility to genetic disease conditions. Before we begin our study of genes and genomes, we review basic aspects of cell structure and function and briefly compare different types of cells.

Prokaryotic Cells

Cells are complex entities with specialized structures that determine cell function. Generally, any cell can be divided into the **plasma (cell) membrane,** a double-

Table 2.1 PROKARYOTIC AND EUKARYOTIC CELLS

	Prokaryotic Cells	Eukaryotic Cells
Cell Types	True bacteria (eubacteria), Archaebacteria	Protists, fungi, plant, animal cells
Size	100 nm–10 μm	10–100 μm
Structure	No nucleus; DNA located in the cytoplasm. Lack organelles.	DNA enclosed in a membrane-bound nucleus. Many organelles.

layered structure of primarily lipids and proteins that surrounds the outer surface of cells; the **cytoplasm,** the inner contents of a cell between the nucleus and the plasma membrane; and **organelles** (a term that means "little organs"), structures in the cell that perform specific functions. Throughout this book, we not only consider how plant and animal cells play important roles in biotechnology but we also cover many biotechnology applications involving bacteria, yeast, and other microorganisms. Bacteria are referred to as **prokaryotic cells** or simply prokaryotes, named from the Greek words meaning "before nucleus" because they do not have a **nucleus,** an organelle that contains DNA in animal and plant cells. Prokaryotes include true bacteria (eubacteria), and cyanobacteria, a type of blue-green algae (Table 2.1) and members of Domain Archaea (ancient bacteria with some eukaryotic characteristics) that you will learn about in Chapter 5.

As shown in Figure 2.1, bacteria have a relatively simple structure. The outer boundary of a bacterium is defined by the plasma membrane, which is surrounded by a rigid cell wall that protects the cell. Except for ribosomes that are used for protein synthesis, bacteria have few organelles. The cytoplasm contains DNA,

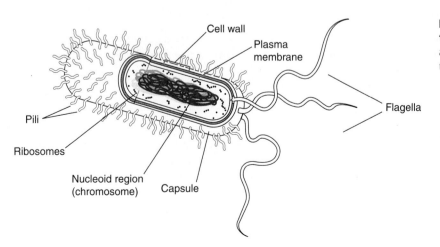

Figure 2.1 Prokaryotic Cell Structure Bacteria are prokaryotes. Shown here is a drawing of structures contained in a typical rod-shaped bacterium.

usually in the form of a single circular molecule, which is attached to the plasma membrane and located in an area called the nucleoid region of the cell (Figure 2.1). Some bacteria also have a tail-like structure called a flagellum that they use for locomotion.

Eukaryotic Cells

Plant and animal cells are considered **eukaryotic cells,** named from the Greek words meaning "true nucleus," because they contain a membrane-enclosed nucleus and many organelles. Eukaryotes also include fungi and single-celled organisms called protists, which include most algae. Diagrams of plant and animal cells are shown in Figure 2.2. The plasma membrane is a fluid, highly dynamic, and complex double-layered barrier composed of lipids, proteins, and carbohydrates. The membrane performs essential roles in cell adhesion (sticking cells to one another), cell-to-cell communication, and cell shape, and it is very important for transporting molecules into and out of the cell. The membrane also serves an important role as a selectively permeable barrier because it contains many proteins involved in complex transport processes that control which molecules can enter and leave the cell. For example, certain proteins such as insulin are released from the cell in a process called secretion; other molecules, such as glucose, can be taken into the cell and within mitochondria be converted into energy in the form of a molecule called **adenosine triphosphate**

(ATP). Membranes also enclose or comprise important parts of many organelles.

The cytoplasm of eukaryotes consists of **cytosol,** a nutrient-rich, gel-like fluid, and many organelles. The cytoplasm of prokaryotes also contains cytosol but few organelles. Think of each organelle as the compartment in which chemical reactions and cellular processes occur. Organelles allow cells to carry out thousands of different complex reactions simultaneously. Each organelle is responsible for specific biochemical reactions. For instance, lysosomes break down foreign materials and old organelles; organelles like the endoplasmic reticulum and Golgi apparatus synthesize proteins, lipids, and carbohydrates (sugars). By compartmentalizing reactions, cells can carry out a multitude of reactions in a highly coordinated fashion simultaneously without interference. Be sure to familiarize yourself with the functions of organelles presented in Figure 2.2 and Table 2.2.

In eukaryotic cells, the nucleus contains DNA. This organelle is a spherical structure enclosed by a double-layered membrane, the **nuclear envelope,** and is typically the largest structure in an animal cell. Nearly 6 feet of DNA is coiled into the nucleus of every human cell, and if the DNA in all human cells were connected end to end there would be enough to stretch to the sun and back about 500 times. Although the majority of DNA in a eukaryotic cell is contained within the nucleus, mitochondria and chloroplasts also contain small circular DNA molecules.

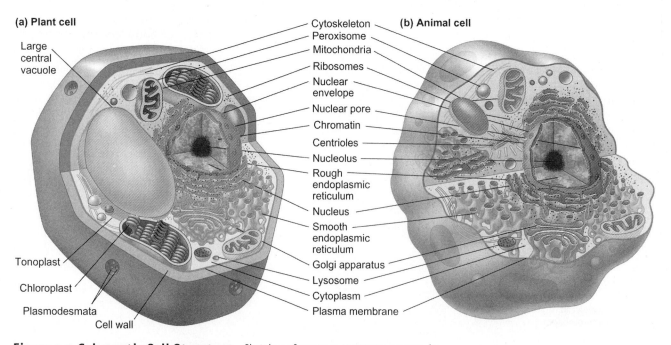

(a) Plant cell

Large central vacuole

Tonoplast

Chloroplast

Plasmodesmata

Cell wall

Cytoskeleton
Peroxisome
Mitochondria
Ribosomes
Nuclear envelope
Nuclear pore
Chromatin
Centrioles
Nucleolus
Rough endoplasmic reticulum
Nucleus
Smooth endoplasmic reticulum
Golgi apparatus
Lysosome
Cytoplasm
Plasma membrane

(b) Animal cell

Figure 2.2 Eukaryotic Cell Structure Sketches of common structures present in plant (a) and animal cells (b).

Table 2.2 EUKARYOTIC CELL STRUCTURE AND FUNCTION

Cell Part	Structure	Functions
Plasma Membrane	Membrane made of a double layer of lipids (primarily phospholipids, cholesterol) within which proteins are embedded; proteins may extend entirely through the lipid bilayer or protrude on only one face; externally facing proteins and some lipids have attached sugar groups	Serves as an external cell barrier; acts in transport of substances into or out of the cell; maintains a resting potential that is essential for functioning of excitable cells; externally facing proteins act as receptors (for hormones, neurotransmitters, and so on) and in cell-to-cell recognition
Cytoplasm	Cellular region between the nuclear and plasma membranes; consists of fluid cytosol (containing dissolved solutes), inclusions (stored nutrients, secretory products, pigment granules), and organelles (the metabolic machinery of the cytoplasm)	
Cytoplasmic organelles		
• Mitochondria	Rodlike, double-membrane structures; inner membrane folded into projections called cristae	Site of ATP synthesis; powerhouse of the cell
• Ribosomes	Dense particles consisting of two subunits, each composed of ribosomal RNA and protein; free or attached to rough endoplasmic reticulum	The sites of protein synthesis
• Rough endoplasmic reticulum	Membrane system enclosing a cavity (the cisterna) and coiling through the cytoplasm; externally studded with ribosomes	Sugar groups are attached to proteins within the cisternae; proteins are bound in vesicles for transport to the Golgi apparatus and other sites; external face synthesizes phospholipids and cholesterol
• Smooth endoplasmic reticulum	Membranous system of sacs and tubules; free of ribosomes	Site of lipid and steroid synthesis, lipid metabolism, and drug detoxification
• Golgi apparatus	A stack of smooth membrane sacs and associated vesicles close to the nucleus	Packages, modifies, and segregates proteins for secretion from the cell, inclusion in lysosomes, and incorporation into the plasma membrane
• Lysosomes	Membranous sacs containing hydrolases (digestive enzymes)	Sites of intracellular digestion
• Peroxisomes	Membranous sacs of oxidase enzymes	The enzymes detoxify a number of toxic substances; the most important enzyme, catalase, breaks down hydrogen peroxide
• Microtubules	Cylindrical structures made of tubulin proteins	Support cell and give it shape; involved in intracellular and cellular movements; form centrioles
• Microfilaments	Fine filaments of the contractile protein actin	Involved in muscle contraction and other types of intracellular movement; help form the cell's cytoskeleton
• Intermediate filaments	Protein fibers; composition varies	The stable cytoskeletal elements; resist mechanical forces acting on the cell
• Centrioles	Paired cylindrical bodies, each composed of nine triplets of microtubules	Organize a microtubule network during mitosis to form the spindle and asters; form the bases of cilia and flagella
• Cilia	Short, cell surface projections; each cilium composed of nine pairs of microtubules surrounding a central pair	Move in unison, creating a unidirectional current that propels substances across cell surfaces

(Continued)

Table 2.2 EUKARYOTIC CELL STRUCTURE AND FUNCTION (*CONTINUED*)

Cell Part	Structure	Functions
• Flagella	Like cilium, but longer; only example in humans is the sperm tail	Propels the cell
Nucleus	Largest organelle; surrounded by the nuclear envelope; contains fluid nucleoplasm, nucleoli, and chromatin	Control center of the cell; responsible for transmitting genetic information and providing the instructions for protein synthesis
• Nuclear envelope	Double-membrane structure; pierced by the pores; outer membrane continuous with the cytoplasmic endoplasmic reticulum	Separates the nucleoplasm from the cytoplasm and regulates passage of substances to and from the nucleus
• Nucleoli	Dense spherical (nonmembrane-bound) bodies, composed of ribosomal RNA and proteins	Site of ribosome subunit manufacture
• Chromatin	Granular, threadlike material composed of DNA and histone proteins	DNA contains genes
Central vacuole (plant cells)	Large membrane-enclosed compartment	Used to store ions, waste products, pigments, protective compounds
Chloroplasts (plant cells)	Membrane enclosed organelle containing stacked structures (grana) of chlorophyll-containing membrane sacs called thylakoids surrounded by an inner fluid (stroma)	Site of photosynthesis

2.2 The Molecule of Life

Virtually every course in biology involves some discussion of DNA, and DNA is routinely manipulated by students in college biology laboratories and in many high school classes. With the wealth of information available about many detailed aspects of DNA and genes, studying biology in the 21st century might give you the impression that the structure of DNA was always understood. However, the structure of the molecule of life and its function as genetic material were not always well known. Many extraordinary researchers and incredible discoveries have contributed to our modern-day understanding of DNA structure and function. We begin this section with a brief overview highlighting evidence for DNA as the genetic material, and then we discuss DNA structure.

Evidence That DNA Is the Inherited Genetic Material

In 1869, Swiss biologist Friedrich Miescher identified a cellular substance from the nucleus that he called "nuclein." Miescher purified nuclein from white blood cells and found that nuclein could not be broken down (degraded) by protein-digesting enzymes called proteases. This discovery suggested that nuclein was not solely made of proteins. Subsequent studies determined that this material had acidic properties, which led nuclein to be renamed "nucleic acids." DNA and **ribonucleic acid (RNA)** are the two major types of **nucleic acids.** While biochemists worked to identify the different components of nucleic acids, evidence that DNA is the inherited genetic material was first provided by the British microbiologist Frederick Griffith in 1928.

Griffith was studying two strains of the bacterium *Streptococcus pneumoniae*, a microbe that causes pneumonia. At the time Griffith carried out his studies, this strain was called *Diplococcus pneumoniae*. Griffith worked with a virulent (disease-causing) variety called the smooth strain (S cells) along with a harmless strain called rough cells (R cells). S cells are surrounded by a capsule (smooth coat) of proteins and sugars, whereas R cells lack this coat. When Griffith injected mice with living S cells, the mice died, and Griffith found live S cells in the blood of the dead mice (Figure 2.3). When live R cells were injected into mice, the mice lived and showed no living R cells in their blood (Figure 2.3). These experiments suggest that the protein coat was responsible for the death of the mice. To test this idea, Griffith then killed S cells by heating them, which destroys proteins in the coat. Not surprisingly, mice injected with heat-killed S cells lived, with no signs of live S cells in their blood.

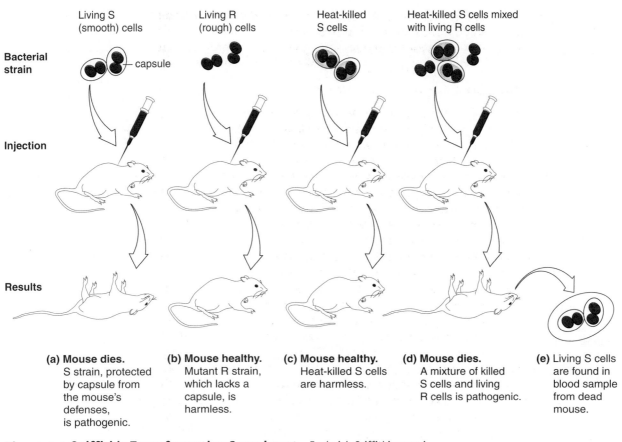

| Living S (smooth) cells | Living R (rough) cells | Heat-killed S cells | Heat-killed S cells mixed with living R cells |

Bacterial strain — capsule

Injection

Results

(a) Mouse dies.
S strain, protected by capsule from the mouse's defenses, is pathogenic.

(b) Mouse healthy.
Mutant R strain, which lacks a capsule, is harmless.

(c) Mouse healthy.
Heat-killed S cells are harmless.

(d) Mouse dies.
A mixture of killed S cells and living R cells is pathogenic.

(e) Living S cells are found in blood sample from dead mouse.

Figure 2.3 Griffith's Transformation Experiment Frederick Griffith's experiments with two strains of *Streptococcus pneumoniae* provided evidence that DNA is the genetic material of cells. The S strain of *S. pneumoniae* kills mice (a); the R strain is harmless (b). Heat-killed S cells are harmless (c). Mice injected with heat-killed S cells mixed together with live R cells died (d), and live S cells could be detected in the blood of dead mice (e). This result is a demonstration of transformation. Living R cells took in DNA from dead S cells, transforming the R cells into S cells.

But when Griffith mixed heat-killed S cells with living R cells in a tube and injected this mixture into mice, the mice died and living S cells were found in the blood of the dead mice. Where did the live S cells come from?

This experiment provided evidence that the genetic material from heat-killed S cells had transformed (changed) or converted R cells into S cells. Griffith's experiments demonstrated **transformation,** which is the uptake of DNA by bacterial cells. Heat treatment broke open some of the S cells, which released their DNA into the tube. Living R cells took up this S cell DNA, which transformed the properties of the R cells so they became virulent, resembling S cells. As you will learn later, transformation is a very powerful technique in molecular biology and routinely used to introduce genes into bacteria for DNA cloning, protein production, and other valuable purposes. Although Griffith hypothesized that some genetic factor was responsible for the transformation

results he saw, he didn't actually identify DNA as the "transforming factor." However, his experiments were instrumental in leading others in search of this factor.

In 1944, Oswald Avery, Colin MacLeod, and Maclyn McCarty purified DNA from large batches of *S. pneumoniae* grown in liquid culture. The experiments they performed provided definitive evidence that DNA is the genetic material and proved that DNA was the transforming factor in the Griffith experiments. In the now-famous Avery, MacLeod, McCarty experiment, they ground up (homogenized) mixtures of bacterial cells from *S. pneumoniae* and treated these extracts with proteases, RNA-degrading enzymes (RNase), or DNA-degrading enzymes (DNase). They subsequently carried out transformation experiments using these treated extracts. Extracts from killed S cells were mixed with living R cells and injected into mice. They demonstrated that DNase-treated extracts could not transform R cells to S cells because DNA in these mixtures was degraded by DNase. Extracts treated with protease or RNase still maintained their transforming ability because DNA in these mixtures remained intact. Although other studies with viruses were essential for determining the role of DNA, the work of Avery, MacLeod, and McCarty provided definitive evidence for DNA as the genetic material causing transformation.

DNA Structure

While evidence supporting DNA as hereditary material was building, a significant question still remained: What is the structure of DNA? Erwin Chargaff provided some insight to this question by isolating DNA from a variety of different species. Chemical analysis of DNA from different species revealed that the percentage of DNA bases called adenine was proportional to the percentage of bases called thymines, and that the percentage of cytosine bases in an organism's DNA were roughly proportional to the percentage of guanine. This valuable observation suggested that the bases adenine, thymine, cytosine, and guanine were somehow intricately related components of DNA structure—an important principle to remember because, as we explore next, these bases are essential components of DNA.

The building block of DNA is the **nucleotide** (Figure 2.4). Each nucleotide is composed of a (five-carbon) **pentose sugar** called deoxyribose, a phosphate molecule, and a **nitrogenous base.** The bases are interchangeable components of a nucleotide. Each nucleotide contains one base, either **adenine (A), thymine (T), guanine (G),** or **cytosine (C)**—the so-called As, Ts, Gs, and Cs of DNA.

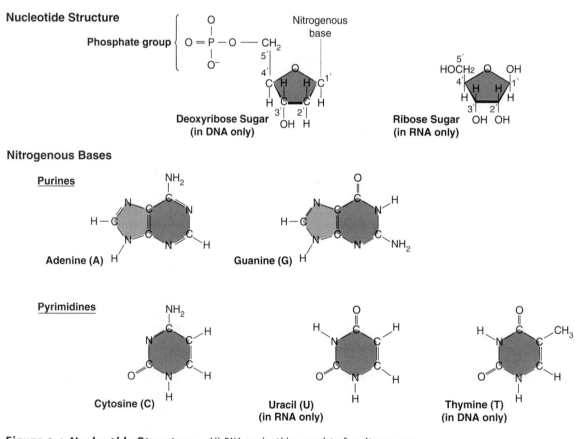

Figure 2.4 Nucleotide Structure All DNA nucleotides consist of a nitrogenous base, either adenine (A), cytosine (C), guanine (G), or thymine (T); a sugar; and a phosphate group. The sugar in DNA, deoxyribose, is called a pentose sugar because it contains five carbon atoms. RNA molecules contain a pentose sugar called ribose. The pentose sugar in DNA is called "deoxyribose" because it lacks an oxygen at carbon number 2 (2') of the sugar compared with the ribose sugar in RNA. A base is attached to carbon number 1 (1') of the sugar; the phosphate group is attached to carbon number 5 (5') of the sugar. Because of their structure, adenine and guanine are a type of base called purines, whereas cytosine, thymine, and uracil are called pyrimidines.

Nucleotides are the building blocks of DNA, but how are these structures arranged to form a DNA molecule? Many scientists have contributed to the answer to this question, but the definitive structure of DNA was finally revealed by James Watson and Francis Crick working at the Cavendish Laboratories in Cambridge, England. Chemists Rosalind Franklin and Maurice Wilkins, of University College, London, used X-ray crystallography to provide Watson and Crick with invaluable data on the structure of DNA. By firing an X-ray beam onto crystals of DNA, Franklin and Wilkins revealed a model of DNA indicating that its structure could be helical. From these data, Chargaff's findings, and other studies, Watson and Crick assembled a wire model of DNA.

Watson and Crick published "The Molecular Structure of Nucleic Acids: A Structure for Deoxyribose Nucleic Acid" in the prestigious journal *Nature* on April 25, 1953. The first paragraph of this paper reads, "We wish to suggest a structure for the salt of deoxyribose nucleic acid (D.N.A.). This structure has novel features which are of considerable biological interest." Given the importance of DNA and what we have learned about DNA structure over the last 50 years, this description might be one of the greatest understatements made in a published scientific paper. The significance of this discovery was appropriately recognized in 1962 when Watson, Crick, and Wilkins received the Nobel Prize in Medicine.

Watson and Crick determined that nucleotides are joined together to form long strands of DNA and that each DNA molecule consists of two strands that join together and wrap around each other to form a double helix (Figure 2.5). A strand of DNA is a string of nucleotides held together by **phosphodiester bonds** that connect the sugar of one nucleotide to the phosphate group of an adjacent nucleotide (Figure 2.5). The sequence of bases in a strand can vary. For instance, a nucleotide containing a C can be connected to a nucleotide containing an A, T, G, or another nucleotide containing a C.

Each strand of nucleotides has a **polarity** to it; there is a *5′ end* and a *3′ end* to the strand (Figure 2.5). This polarity refers to the carbons of the deoxyribose sugar. At the 5′ end of a strand, the phosphate at carbon 5 is not bonded to another nucleotide, but carbon 3 is involved in a phosphodiester bond. At the 3′ end, the phosphate at carbon 5 is bonded to another nucleotide, but carbon 3 is not joined to another nucleotide. Although this aspect of nucleotide structure may seem trivial, the polarity of DNA is important for replication and for routine manipulation of DNA in the laboratory.

Watson and Crick determined that each DNA molecule consists of two interconnecting strands that wrap around each other to form a right-handed double helix, perhaps the most famous molecular model in all of biology (Figure 2.5). The two strands of a DNA molecule are joined together by hydrogen bonds between **complementary base pairs** in opposite strands (Figure 2.5). Adenine only base pairs with thymine, and guanine only base pairs with cytosine. From this model, Chargaff's observations are easily understood. The proportions of As and Ts are equivalent in an organism's DNA as are the proportions of Gs and Cs because they pair with each other in a DNA molecule.

The two strands of nucleotides in a double helix are considered **antiparallel** because the polarity of each strand is reversed relative to each other (Figure 2.5). This orientation is necessary for the complementary base pairs to align and form hydrogen bonds with one another. The double helix resembles a twisted ladder of sorts. The rungs of this ladder consist of the complementary base pairs, and the sides of the ladder consist of sugar and phosphate molecules, creating the "backbone" of DNA.

What Is a Gene?

Genes are often described as units of inheritance, but what exactly is a gene? A **gene** is a sequence of nucleotides that provides cells with the instructions for the synthesis of a specific protein or a particular type of RNA. Most genes are approximately 1,000 to 4,000 nucleotides (nt) long, although many smaller and larger genes have been identified. By controlling the proteins produced by a cell, genes influence how cells, tissues, and organs appear, both through the microscope and with the naked eye. These inherited appearances are called **traits.** Through the DNA contained in your cells, you have inherited traits from your parents such as eye color and skin color. Genes not only influence cell metabolism and behavioral and cognitive abilities such as intelligence but also affect our susceptibility to certain types of genetic diseases.

Some traits are controlled by a single gene; others are determined by multiple protein-producing genes that interact in complex ways. In Section 2.4, we explore how genes direct protein synthesis in cells. Throughout this book, we consider examples of genes, their functions, and their many applications in different areas of biotechnology.

2.3 Chromosome Structure, DNA Replication, and Genomes

Before we consider how genes function, it is important that you understand how and why DNA is organized into chromosomes and how DNA is replicated in cells.

(a)

(b)

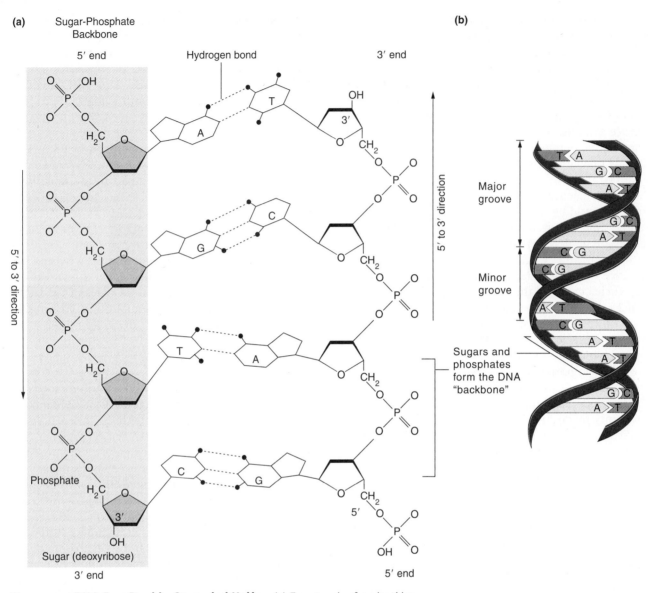

Figure 2.5 DNA Is a Double-Stranded Helix (a) Two strands of nucleotides are joined together by hydrogen bonds between complementary base pairs. Adenine bases (A) always base pair with thymine bases (T), and cytosine (C) base pairs with guanine (G). (b) The two strands wrap around each other so the overall structure of DNA is a double-stranded helix with a sugar-phosphate "backbone" where the bases are aligned in the center of the helix.

Chromosome Structure

Suppose you are presented with a challenge. If you solve it, you will earn free tuition for the rest of your undergraduate courses. You are given a basket containing 46 packages of different colored yarn all unraveled and intertwined. Your challenge is to sort the yarn into 46 even balls. How would you solve this challenge? If you start to cut the tangled pile of yarn randomly, you probably will not succeed. Of course, if you painstakingly unravel the yarn and wind each different colored yarn into a ball, you will eventually sort it into 46 even

piles. This analogy provides a highly simplified view of the challenge presented to a human cell when it has to divide and sort its DNA into even packages.

The 3 billion base pairs (bp) of DNA in every human cell would stretch to around 6 feet if unraveled—an amazing amount of material packed into the tiny nucleus of each cell. This DNA must be separated evenly when a cell divides; otherwise, the loss of DNA can have devastating consequences. Fortunately, such mistakes in DNA separation are rare, in part because cells can effectively separate and package DNA into chromosomes.

Inside the nucleus, DNA exists in a relatively unraveled state. This does not mean that the DNA is *uncoiled* from its double-helical structure, but rather the DNA is somewhat loosely arranged and not fully compacted into tightly coiled **chromosomes,** although all of the DNA in a chromosome remains together within the nucleus. When a cell is not dividing, DNA in the nucleus exists as an intricate combination of DNA and DNA-binding proteins called **histones** to form strings called **chromatin.** During cell division, chromatin is coiled into tight fibers that eventually wrap around each other to form a chromosome—a highly coiled and tightly condensed package of DNA and histone proteins (Figure 2.6).

The size and number of chromosomes vary from species to species. Most bacteria have a single circular chromosome that contains a few thousand genes. Eukaryotes typically contain one or more sets of chromosomes, which have a linear shape. Most human cells have two sets (pairs) of 23 chromosomes each, for a total of 46 chromosomes. Through the process of fertilization, you inherited 23 chromosomes from your mother **(maternal chromosomes)** and 23 chromosomes from your father **(paternal chromosomes).** These chromosome pairs are called **homologous pairs,** or **homologues.** Chromosomes 1 through 22 are known as the **autosomes;** the 23rd pair are called the **sex chromosomes**—consisting of X and Y chromosomes.

Human egg and sperm cells, called the sex cells or **gametes,** contain a single set of 23 chromosomes, called the **haploid number (*n*)** of chromosomes. All other cells of the body, such as skin cells, muscle cells, and liver cells, are known as **somatic cells.** Somatic cells from many organisms have two sets of chromosomes, called the **diploid number (2*n*)** of chromosomes. Human somatic cells contain 46 chromosomes. Somatic cells of a normal human male have 22 pairs of autosomes and an X and Y chromosome; cells of a normal female have 22 pairs of autosomes and two X chromosomes.

Sex chromosomes were so named because they contain genes that influence sex traits and the development of reproductive organs, whereas the autosomes were originally thought primarily to contain the genes that affect other body features unrelated to sex such as skin color and eye color. Although there are genes involved in sex organ determination that are present on the Y chromosome, there are other genes involved in sex determination that are present on autosomes, and the majority of genes on the X chromosome are not required for development of the reproductive organs.

Several characteristics are common to most eukaryotic chromosomes. Prokaryotes typically contain

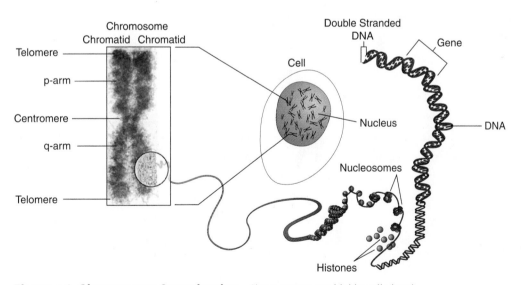

Figure 2.6 Chromosome Organization Chromosomes are highly coiled and condensed packages of DNA. In a nondividing cell, DNA exists in an unraveled state called *chromatin*. Histone proteins serve as particles around which DNA becomes tightly wound to give a "beads on a string" appearance when viewed with an electron microscope. During chromosome formation, which occurs when cells divide, chromatin is further compacted into tight fibers and supercoiled looped structures. Ultimately these supercoiled loops are tightly packed together with the assistance of other proteins to create an entire chromosome, a highly compact assembly of DNA. Each chromosome consists of two sister chromatids attached by a centromere. Chromosome arms are the portions of the chromatid on one side of the centromere, labeled as the *p* and *q arms*. The ends of a chromosome are called *telomeres*.

a circular chromosome with slightly different structures (see Chapter 5). Each eukaryotic chromosome consists of two thin rodlike structures of DNA called **sister chromatids** (Figure 2.6). The sister chromatids are exact replicas of each other, copied during DNA synthesis, which occurs just prior to chromosome formation. During cell division, each sister chromatid is separated so that newly forming cells receive the same amount of DNA as the original cell they arose from. Each eukaryotic chromosome has a single **centromere,** a constricted region of the chromosome consisting of intertwined DNA and proteins that join the two sister chromatids to each other. This region of a chromosome also contains proteins that attach chromosomes to organelles called microtubules, which play an essential role in moving chromosomes and separating sister chromatids during cell division.

The centromere delineates each sister chromatid into two arms—the short arm, called the **p arm,** and the long arm, or **q arm.** Each arm of a chromosome ends with a segment called a **telomere** (Figure 2.6). Telomeres are highly conserved repetitive sequences of nucleotides that are important for attaching chromo-

somes to the nuclear envelope. Telomeres are a subject of intense research. As we discuss in Chapter 11, changes in telomere length are believed to play a role in the aging process and in the development of certain types of cancers.

Karyotype analysis for studying chromosomes

One of the most common ways to study chromosome number and basic aspects of chromosome structure is to prepare a **karyotype.** In karyotype analysis, cells are spread on a microscope slide and then treated with chemicals to release and stain the chromosomes. For example, G-banding, in which chromosomes are treated with a DNA-binding dye called Giemsa stain, creates a series of alternating light and dark bands in stained chromosomes. Each stained chromosome shows a unique and reproducible banding pattern that can be used to identify different chromosomes. Chromosomes can be aligned and paired based on their staining pattern and their size (Figure 2.7). In humans, chromosome 1 is the largest chromosome and chromosome 21 is the smallest. Karyotypes are very valuable for studying and comparing chromo-

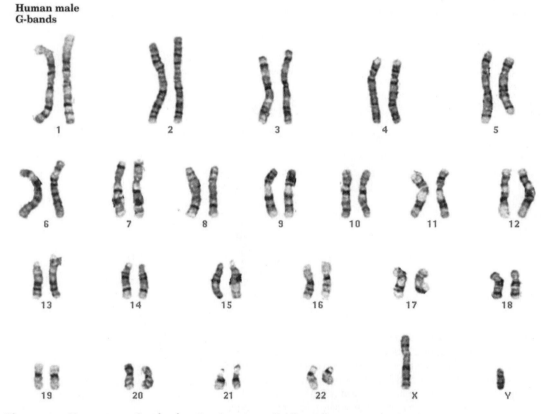

Figure 2.7 Karyotype Analysis   In a karyotype, dividing cells are spread out onto a glass microscope slide to release their chromosomes. Chromosomes are stained and aligned based on their overall size, position of the centromere, and their staining pattern to create a karyotype.

some structure. In Chapter 11, we consider how karyotype analysis can be used to identify human genetic
disease conditions associated with abnormalities in
chromosome structure and number.

DNA Replication

When a cell divides, it is essential that the newly
created cells contain equal copies of replicated DNA.
Somatic cells divide by a process called **mitosis**
wherein one cell divides to produce daughter cells
each of which contains an identical copy of the DNA of
the original (parent) cell. For instance, a human skin
cell divides to produce two daughter cells, each containing 23 pairs of chromosomes. Gametes are formed
by a process called **meiosis** wherein a parent cell
divides to create up to four daughter cells, which can
be either sperm or egg cells. During meiosis, the chromosome number in daughter cells is cut in half to the
haploid number. Sperm and egg cells contain a single
set of 23 chromosomes. Through sexual reproduction,
a fertilized egg called the **zygote** is formed. The
zygote, which divides by mitosis to form an embryo
and eventually a complete human, contains 46 chromosomes: 23 paternal chromosomes and 23 maternal
chromosomes.

> **Q** Do all species have the same
> number of chromosomes?
>
> **A** No. Chromosome number is almost
> as diverse as the number of different
> species. Human cells have a haploid number of 23. In
> parentheses are haploid numbers for other species:
> fruit flies (4), yeast (16), cats (19), and dogs (39).

Prior to cell division by either mitosis or meiosis,
DNA must be replicated in the cell. DNA replication
occurs by a process called **semiconservative replication.** Figure 2.8 shows an overview of this process.
Before replication begins, the two complementary
strands of the double helix must be pulled apart into
single strands. Once separated, the two strands serve
as templates for copying two new strands of DNA. At
the end of this process, two new double helices are
formed. Each helix contains one original DNA
(parental) strand and one newly synthesized strand,
thus the term "semiconservative."

DNA replication occurs in a series of stages involving a number of different proteins. Because prokaryotes contain circular chromosomes, DNA replication in

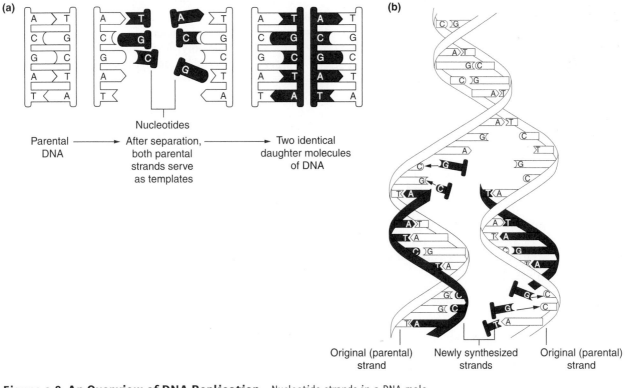

Figure 2.8 An Overview of DNA Replication   Nucleotide strands in a DNA molecule must first be separated (a). Each strand serves as a template for the synthesis of
new strands producing two DNA molecules, each containing one original strand and one
newly synthesized strand (b).

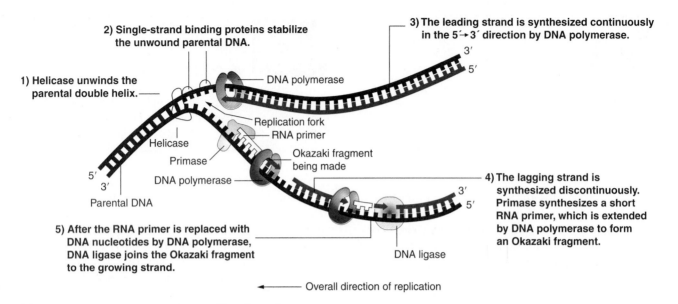

Figure 2.9 Semiconservative Replication of DNA

prokaryotes is slightly different from that in eukaryotes. Here we consider the main players and key concepts in DNA replication overall, but keep in mind there are subtle differences that distinguish the replication process in prokaryotes and eukaryotes. Replication is initiated by **DNA helicase,** an enzyme that separates the two strands of nucleotides, literally "unzipping" the DNA by breaking hydrogen bonds between complementary base pairs (Figure 2.9). The separated strands form a replication fork. As helicase unwinds the DNA, **single-strand binding proteins** attach to each strand and prevent them from base pairing and reforming a double helix. This step is important because the DNA strands must be held apart during DNA replication. The separation of complementary strands occurs in regions of the DNA called **origins of replication.** Bacterial chromosomes have a single origin. Because of their large size, eukaryotic chromosomes have multiple origins. Starting DNA replication at multiple origins allows eukaryotic chromosomes to be copied rapidly.

The next step in DNA replication involves the addition of short segments of RNA approximately 10 to 15 nucleotides long. These sequences, called RNA primers, are synthesized by an enzyme called **primase** (in eukaryotes, a form of DNA polymerase called α acts as the primase). Primers start the process of DNA replication because they serve as binding sites for **DNA polymerases,** the key enzymes that synthesize new strands of DNA. Several different forms of DNA polymerase are involved in copying DNA. In bacteria, the enzyme **DNA polymerase III** (called DNA polymerase δ in eukaryotes) binds to each single strand, moving along the strand and using it as a template to

copy a new strand of DNA. During this process, DNA polymerase uses nucleotides present in the cell to synthesize complementary strands of DNA. DNA polymerase always works in one direction, synthesizing new strands in a 5′ to 3′ orientation and adding nucleotides to the 3′ end of a newly synthesized strand (Figure 2.9) by forming phosphodiester bonds between the phosphate of one nucleotide and the sugar in the previous nucleotide.

Because DNA polymerase only proceeds in a 5′ to 3′ direction, replication along one strand, the **leading strand,** occurs in a continuous fashion (Figure 2.9). Synthesis on the opposite strand, the **lagging strand,** occurs in a discontinuous fashion because DNA polymerase must wait for the replication fork to open. On the lagging strand, short pieces of DNA called Okazaki fragments (named after Reiji and Tuneko Okazaki, the scientists who discovered these fragments) are synthesized as the DNA polymerase works it way out of the replication fork. Covalent bonds between Okazaki fragments in the lagging strand are formed by **DNA ligase** to ensure there are no gaps in the phosphodiester backbone. Finally, the RNA primers are removed, and these gaps are filled by DNA polymerase.

Remember the functions of enzymes involved in DNA synthesis: In the next chapter, we discuss how DNA polymerase and DNA ligase are routinely used in the lab during DNA cloning and analysis experiments.

What Is a Genome?

DNA contains the instructions for life—genes. All of the DNA in an organism's cells is called the **genome.** Contained in the human genome are approximately

A Absolutely not. Genome size varies greatly from organism to organism, but the size of an organism's genome does not relate to its complexity. Humans and mice share a similar number of base pairs (approximately 3 billion) and a similar number of genes, around 20,000 to 25,000 estimated genes, although some genome scientists believe the number of human genes may be closer to 18,000. Plants such as *Arabidopsis thaliana* contain approximately 25,000 genes in a 97 million base pair genome; fruit flies (*Drosophila melanogaster*) have around 13,000 genes in a genome of 165 million base pairs. Although nonscientists might not consider mice or plants to be as "complex" as humans, genome studies tell us that complexity is far more than just the number of genes an organism contains. It is incorrect to think about humans as being more complex than other life-forms. For instance, *A. thaliana*, a weed that has proven valuable for many studies in genetics, contains genes that allow it to derive energy from sunlight by photosynthesis. Human cells cannot convert energy by photosynthesis. All living organisms are complex with unique capabilities dictated by genes and the way proteins produced by genes interact with one another.

20,000 genes scattered among 3 billion base pairs of DNA. The study of genomes, a discipline called **genomics,** is currently one of the most active and rapidly advancing areas of biological science. Throughout this book, we discuss aspects of the **Human Genome Project,** a worldwide effort to identify all human genes on each chromosome along with several other goals. The Human Genome Project is an enormous undertaking in genomics that is providing scientists with exciting insight into human genes, their locations, and functions.

2.4 RNA and Protein Synthesis

Genes govern the activities and functions within a cell by directing the synthesis of proteins. Some of the myriad functions of these essential molecules follow:

- Proteins are necessary for cell structure as important components of membranes and the cytoplasm.
- Proteins carry out essential reactions in the cell as enzymes.

- Proteins perform critical roles as hormones and other "signaling" molecules that cells use to communicate with one another.
- Receptor proteins bind to other molecules such as hormones and transport proteins, enabling molecules to enter and leave cells.
- Proteins in the form of antibodies recognize and destroy foreign materials in the body.

Quite simply, cells cannot function without proteins. How does DNA make proteins? Actually, DNA does not make proteins directly. To synthesize proteins, genes are first copied into molecules called **messenger RNA (mRNA)** (Figure 2.10). RNA synthesis is called **transcription** because genes are literally transcribed (copied) from a DNA code into an RNA code. In turn, mRNA molecules, which are exact copies of genes, contain information that is deciphered into instructions for making a protein through a process known as **translation.**

Other than the fact that RNA molecules are single stranded, the chemical composition of RNA is very similar to that of DNA. Its bases are also very similar to DNA. One key difference is that RNA contains a base called uracil (U) instead of thymine (T) (see Figure 2.4).

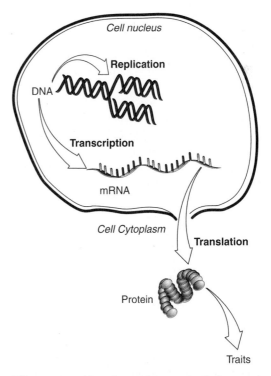

Figure 2.10 The Flow of Genetic Information in Cells DNA is copied into RNA during the process of transcription. RNA directs the synthesis of proteins during translation. Through proteins, genes control the metabolic and physical properties or traits of an organism.

TOOLS OF THE TRADE

Enzymes Involved in DNA Replication Have Valuable and Essential Roles in Molecular Biology Research

This chapter introduces the important principles of DNA structure, replication, transcription, and translation and provides an essential background on genes and genomes. The principles we discussed here are important for understanding not just what genes are and how they function but also many of the components involved in processes such as DNA replication that have become essential tools for research in molecular biology.

In the next chapter, you will learn how scientists can identify, clone, and analyze genes, a fascinating aspect of molecular biology research and biotechnology. The technology for cloning and studying genes only became possible as scientists learned more about the enzymes involved in processes such as DNA replication. For instance, DNA polymerase, the key enzyme that synthe-sizes DNA during semiconservative replication, is widely used in molecular biology labs to copy DNA. Similarly, RNA polymerase is also widely used in molecular research. Also, many DNA cloning procedures rely on the use of DNA ligase to cut and paste together DNA fragments from different sources—a process called recombinant DNA technology.

Because most of these enzymes are relatively inexpensive and readily available from supply companies, their use has become commonplace in most molecular biology labs. Without our understanding of enzymes that act on DNA and RNA and their functions, many modern applications of biotechnology and a majority of techniques routinely used in molecular biology research would be impossible.

The other primary difference is that RNA contains a pentose sugar called ribose, which has a slightly different structure than the deoxyribose sugar contained in DNA.

An easy way to remember the difference between transcription and translation is to remember that translation involves a change in code from RNA to protein, much like translating one language to another. Through production of mRNA and protein synthesis, DNA controls the properties of a cell and its traits (Figure 2.11). This process of transcription and translation directs the flow of genetic information in cells, controlling a cell's activities and properties. Here we study basic principles of transcription and translation and aspects of how gene expression can be controlled by cells.

Copying the Code: Transcription

How is DNA used as a template to make RNA? **RNA polymerase** is the key enzyme of transcription. Inside the nucleus, RNA polymerase unwinds the DNA helix and then copies one strand of DNA into RNA. Unlike DNA replication where the entire DNA molecule is copied, transcription occurs only in segments of a chromosome that contain genes. How does RNA polymerase know where to begin transcription? Adjacent to most genes is a **promoter,** specific sequences of nucleotides that allow RNA polymerase to bind at specific locations next to genes. As we discuss in more detail later in this chapter, proteins called **transcription factors** help RNA polymerase find the promoter and bind to DNA, and sequences called **enhancers** can also play important roles in transcription.

After RNA polymerase binds to a promoter, it unwinds a region of the DNA to separate the two strands. Only one of the strands, called the **template strand** (the opposite strand is called the *coding strand*), is copied by RNA polymerase. The presence of a promoter determines which strand of DNA will be transcribed. Once properly oriented, RNA polymerase proceeds in a 5' to 3' direction along the DNA template strand to copy a complementary strand of RNA by forming phosphodiester bonds between ribonucleotides in much the same way that DNA copies DNA polymerase (Figure 2.11). When RNA polymerase reaches the end of a gene, it encounters a termination sequence. These sequences either bind specific proteins or base pair to create loops at the end of the RNA. As a result, the RNA polymerase and newly formed RNA are released from the DNA molecule. Unlike DNA replication where the DNA is only copied once each time a cell divides, multiple copies of mRNA are transcribed from each gene during transcription. Sometimes a cell transcribes thousands of copies of mRNA from a gene. Later in this section, you will see that cells with a high requirement for a particular protein generally produce large amounts of mRNA to encode that protein.

Transcription produces different types of RNA

We have already seen that mRNA is produced when many genes are copied into RNA. Two other types of RNA, **transfer RNA (tRNA)** and **ribosomal RNA**

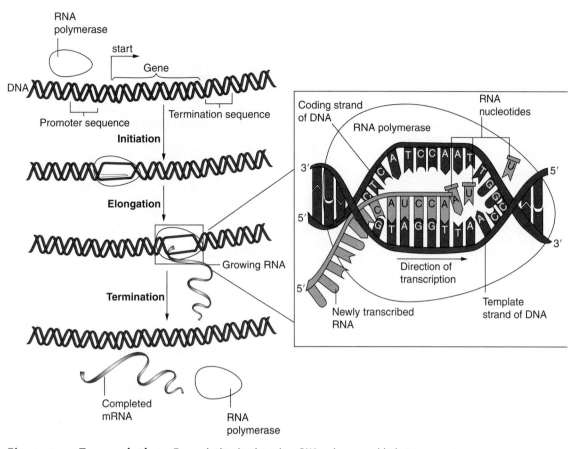

Figure 2.11 Transcription Transcription begins when RNA polymerase binds to DNA at a promoter region adjacent to a gene sequence and unwinds the DNA. RNA polymerase moves along the DNA template copying one strand into a molecule of RNA. When RNA polymerase reaches a termination sequence, it releases from the DNA and transcription ends.

(rRNA), are also produced by transcription. Different RNA polymerases produce each type of RNA. As we will soon learn, only mRNA carries information that codes for the synthesis of a protein, but tRNA and rRNA are also essential for protein synthesis.

Scientists recently discovered a new class of non-protein coding RNA molecules called **microRNAs (miRNAs).** MicroRNAs are part of a rapidly growing family of small RNA molecules about 20 to 25 nucleotides in size that play novel roles in regulating gene expression. Later in the chapter we briefly discuss the role of miRNAs.

mRNA processing

In eukaryotic cells, the initial mRNA copied from a gene is called a **primary transcript (pre-mRNA).** This mRNA is immature and not fully functional. Primary transcripts undergo a series of modifications, collectively called mRNA processing, before they are ready for protein synthesis. One modification involves **RNA splicing** (Figure 2.12). When the details of transcription were first worked out, scientists

were surprised to learn that genes are interrupted by stretches of DNA that do not contain protein coding information, called **introns.** Introns are interspersed between **exons,** protein coding sequences of a gene. Introns and exons are copied during transcription of mRNA. Before mRNA can be used to make a protein, the exons must be spliced together. As a simple analogy, think of introns as randomly inserted letters in a sentence that must be removed before the sentence can make sense. In the splicing process, introns are cut out of the primary transcript and adjacent exons are spliced to form a fully functional mRNA molecule with no introns.

Splicing provides flexibility in the types of proteins that can ultimately be produced from a single gene. When genes were first discovered, scientists thought that a single gene could produce only one protein. But as the process of splicing was revealed, it became clear that when a gene contains several exons, splicing doesn't always occur in the same way. As a result, multiple proteins can be produced from a single gene. In a complex process called **alternative splicing,**

(a)

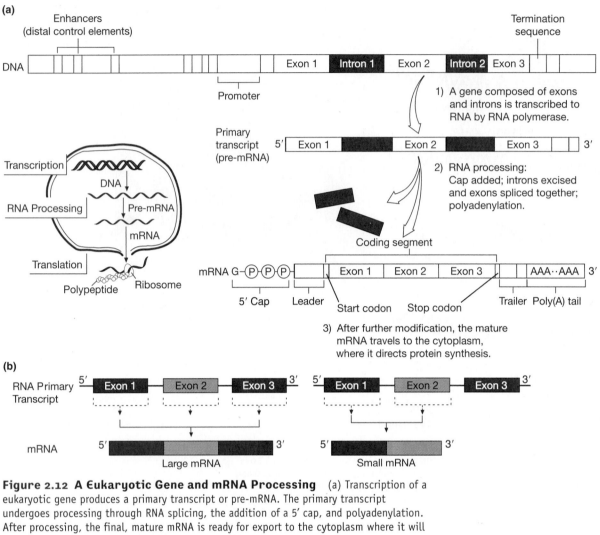

Figure 2.12 A Eukaryotic Gene and mRNA Processing (a) Transcription of a eukaryotic gene produces a primary transcript or pre-mRNA. The primary transcript undergoes processing through RNA splicing, the addition of a 5′ cap, and polyadenylation. After processing, the final, mature mRNA is ready for export to the cytoplasm where it will be translated into a protein. (b) Alternative splicing can produce different mRNAs and protein products from the same gene. Notice that the larger mRNA on the left contains three exons spliced together but that the shorter mRNA on the right contains only two exons spliced together.

splicing sometimes can join together certain exons and *cut out* other exons, essentially treating them as introns (Figure 2.12b). This process creates multiple mRNAs of different sizes from the same gene. Each mRNA can then be used to produce different proteins with different, sometimes unique, functions. Alternative splicing allows several different protein products to be produced from the same gene sequence. For instance, certain genes used to produce antibodies are alternatively spliced to produce some antibody proteins that attach to the surface of cells as well as other antibodies with different structures causing them to be secreted into the bloodstream. Similar splicing occurs with neurotransmitter genes among many others. In fact, scientists originally believed that the human genome contained approximately 100,000 genes based on a

predicted number of proteins made by human cells. As you will learn in Chapter 3, genome scientists were very surprised to find that the human genome contains only about 20,000 genes. Much of the reason for this discrepancy is that many human genes can be spliced in different ways, and it has been estimated that around 60% of human genes may use alternative splicing.

Another type of processing occurs at the 5′ end of mRNA where a guanine base containing a methyl group is added (Figure 2.12). Known as a 5′ cap, this structure plays a role in ribosome recognition of the 5′ end of the mRNA molecule during translation. Lastly, in a process called **polyadenylation,** a string of adenine nucleotides around 100 to 300 nucleotides in length is added to the 3′ end of the mRNA creating a poly(A) "tail" (Figure 2.12). This tail protects mRNA from RNA-degrading enzymes in the cytoplasm, increasing its stability and availability for translation. Following processing, a mature mRNA leaves the nucleus and enters the cytoplasm where it is now ready for translation (Figure 2.12).

Translating the Code: Protein Synthesis

The ultimate function of a gene is to produce a protein. We have seen how RNA is made through transcription. Here we look at a brief overview of translation, using information in mRNA to synthesize a protein from amino acids. Translation occurs in the cytoplasm of cells as a multistep process that involves several different types of RNA molecules. It will be much easier for you to understand the details of translation if you are familiar with important functions of each type of RNA. These are the major components of translation:

- Messenger RNA (mRNA)—an exact copy of a gene. Acts as a "messenger" of sorts by carrying the genetic code, encoded by DNA, from the nucleus to the cytoplasm where this information can be read to produce a protein. Messenger RNAs vary in size from approximately 1,000 nucleotides to several thousand nucleotides in length.

- Ribosomal RNA (rRNA)—short single-stranded molecules around 1,500 to 4,700 nucleotides long. Ribosomal RNAs are important components of **ribosomes,** organelles that are essential for protein synthesis. Ribosomes recognize and bind to mRNA and "read" the mRNA during translation.

- Transfer RNA (tRNA)—molecules that transport amino acids to the ribosome during protein synthesis. Transfer RNA molecules are approximately 75 to 90 nucleotides in size.

We have already discussed the details of mRNA structure, but before we explore the details of translation, you need to be familiar with the genetic code of mRNA and the specific structures of ribosomes and tRNA.

The genetic code

What is the **"genetic code"** contained within mRNA? As you will learn shortly, ribosomes read the code and then produce proteins, which are formed by joining the building blocks called **amino acids** (see Appendix 2). A chain of amino acids linked together by covalent bonds is a **polypeptide.** Some proteins consist of a single polypeptide chain; others contain several polypeptide chains that must wrap and fold around each other to form complicated three-dimensional structures. We discuss protein structure in more detail in Chapter 4. Proteins can contain combinations of up to 20 different amino acids, yet there are only four bases in mRNA molecules, so how does this code work? What information is the ribosome decoding to tell a cell what amino acids belong in a protein? If there are only four nucleotides in mRNA, how can mRNA provide information coding for 20 different amino acids?

The answers to these questions lie in the genetic code, a fascinating aspect of biology because it is a universal language of genetics used by virtually all living organisms. The code works in three-nucleotide units called **codons,** which are contained within mRNA molecules. Each codon codes for a single amino acid (Table 2.3). For instance, notice that the codon UAC codes for the amino acid tyrosine, and the codon UGC codes for the amino acid cysteine. Although each codon codes for one amino acid, there is flexibility in the genetic code. There are 64 different potential codons corresponding to all possible combinations of the four possible bases assembled into three nucleotide codons (4^3). But because there are only 20 amino acids, most amino acids may be coded for by more than one codon. For example, notice in Table 2.3 that the amino acid lysine may be coded for by AAA and AAG. Having a redundancy of codons increases the efficiency of translation. Some codons are present in mRNAs with greater frequency than others, just as some words in the English language are preferred over others with identical meanings.

Also contained in the genetic code are codons that tell ribosomes where to begin translation and end translation. The start codon, AUG, codes for the amino acid methionine and signals the starting point for mRNA translation. As a result, the first amino acid in many proteins is methionine, although this amino acid is removed shortly after translation in some proteins. Stop codons terminate translation. UGA is a commonly used stop codon in many mRNA, but UAA and UAG are other stop codons (Table 2.3). Stop codons do not code for amino acids; they simply signal the end of translation.

Because the genetic code is universal, it is used by cells in humans, bacteria, plants, earthworms, fruit flies, and all other species. There are some subtle differences to the code in certain species, but at the basic level it operates the same way throughout biology. Because the code is universal, biologists can use techniques called recombinant DNA technology to clone a human gene such as the insulin gene and insert it into a bacteria so that bacterial cells transcribe and translate insulin—a protein they normally do not produce. Another helpful aspect of the universal genetic code is that it enables scientists to clone a gene in one species such as a mouse and then use sequence information

Table 2.3 THE GENETIC CODE

		Second Position				
		U	C	A	G	
First Position (5' End)	U	UUU ⎤ Phenylalanine (Phe) UUC ⎦ UUA ⎤ Leucine (Leu) UUG ⎦	UCU ⎤ UCC ⎥ Serine (Ser) UCA ⎥ UCG ⎦	UAU ⎤ Tyrosine (Tyr) UAC ⎦ UAA *Stop* UAG *Stop*	UGU ⎤ Cysteine (Cys) UGC ⎦ UGA *Stop* UGG Tryptophan	U C A G
	C	CUU ⎤ CUC ⎥ Leucine (Leu) CUA ⎥ CUG ⎦	CCU ⎤ CCC ⎥ Proline (Pro) CCA ⎥ CCG ⎦	CAU ⎤ Histidine (His) CAC ⎦ CAA ⎤ Glutamine (Gln) CAG ⎦	CGU ⎤ CGC ⎥ Arginine (Arg) CGA ⎥ CGG ⎦	U C A G
	A	AUU ⎤ AUC ⎥ Isoleucine (Ile) AUA ⎦ AUG Methionine (Met) START	ACU ⎤ ACC ⎥ Threonine (Thr) ACA ⎥ ACG ⎦	AAU ⎤ Asparagine (Asn) AAC ⎦ AAA ⎤ Lysine (Lys) AAG ⎦	AGU ⎤ Serine (Ser) AGC ⎦ AGA ⎤ Arginine (Arg) AGG ⎦	U C A G
	G	GUU ⎤ GUC ⎥ Valine (Val) GUA ⎥ GUG ⎦	GCU ⎤ GCC ⎥ Alanine (Ala) GCA ⎥ GCG ⎦	GAU ⎤ Aspartic Acid (Asp) GAC ⎦ GAA ⎤ Glutamic Acid (Glu) GAG ⎦	GGU ⎤ GGC ⎥ Glycine (Gly) GGA ⎥ GGG ⎦	U C A G

Third Position (3' End)

from the mouse gene to identify a similar gene in humans. Because different species share a common genetic code, this approach is a very common strategy for identifying human genes, including many involved in disease processes.

Ribosomes and tRNA molecules

Ribosomes are complex structures consisting of aggregates of rRNA and proteins that form structures called subunits. Each ribosome contains two subunits, the large and small subunits. These subunits associate to form two grooves, called the **A site** and the **P site,** into which tRNA molecules can bind, and an **E site** through which tRNA molecules leave the ribosome (Figure 2.13).

Transfer RNAs are small molecules less than 100 nucleotides long. Transfer RNA molecules fold in intricate ways, and several nucleotides in a tRNA base pair with each other. As a result, a tRNA assumes a structure called a cloverleaf because as regions of the molecule base pair, other unpaired segments create loops. At one end of each tRNA is an amino acid attachment site (Figure 2.13b). Enzymes in the cytoplasm called aminoacyl tRNA synthetases attach a single amino acid to each tRNA molecule, creating what is known as an **aminoacyl** or "charged" **tRNA.** Charged tRNA molecules carry their amino acids to the ribosome and bind within grooves of the ribosomes at the A site. At the opposite end of each tRNA molecule is a three-

nucleotide sequence called an **anticodon.** Different amino acids have different anticodon sequences. As you will learn shortly, anticodons are designed to complementary base pair with codons in mRNA. Now that you know the "players" of translation—mRNA, ribosomes, and tRNA—we examine how these components come together to produce a protein.

Stages of translation

There are some fundamental differences between translation in prokaryotes and eukaryotes. Here we provide an overview of basic aspects of the three major stages of translation in eukaryotes: initiation, elongation, and termination. The beginning of translation is called initiation. During initiation, the small ribosomal subunit binds to the 5' end of the mRNA molecule by recognizing the 5' cap of the mRNA. Other proteins called initiation factors are also involved in guiding the small subunit to the mRNA. The small subunit moves along the mRNA until it encounters the start codon, AUG. Pausing at the start codon, the small subunit waits for the correct tRNA, called the initiator tRNA, to come along (Figure 2.13). This tRNA has the amino acid methionine (met) attached to it (remember that most proteins begin with this amino acid) and contains the anticodon UAC. The UAC anticodon binds to the start codon by complementary base pairing (Figure 2.13); then the large ribosomal subunit binds to this complex containing the small subunit, initiation factors, mRNA,

(a)

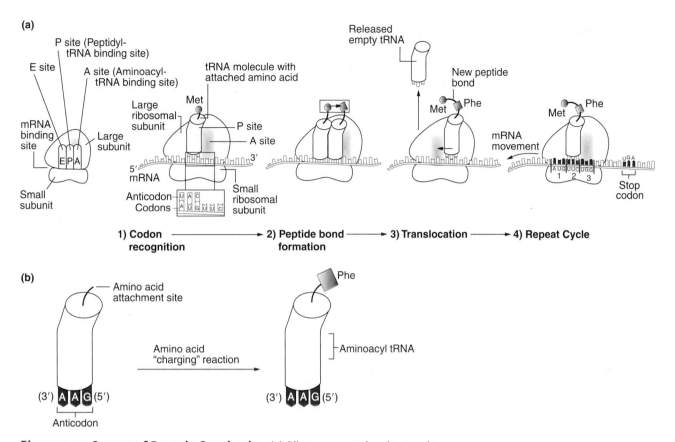

1) **Codon** ⟶ 2) **Peptide bond** ⟶ 3) **Translocation** ⟶ 4) **Repeat Cycle**
recognition **formation**

(b)

Figure 2.13 Stages of Protein Synthesis (a) Ribosomes contain a large and a small subunit. Shown here is a ribosome attached to mRNA. Ribosomes contain two binding sites for tRNA molecules, called the A site and the P site. Abbreviated steps of translation are shown 1 to 4. (b) Shown is a diagrammatic example of the tRNA symbol used in this book. At one end of each tRNA is an amino acid binding site and at the opposite end is a 3-nucleotide anticodon sequence.

and initiator tRNA. After all these components are in place, the ribosome can start translating a protein.

The next cycle of translation is called elongation because during this phase additional tRNAs enter the ribosome, one at a time, and a growing polypeptide chain is elongated. The ribosome, paused at the second codon, waits for the (second) tRNA to enter the A site. In Figure 2.13, notice that the second codon is UUC, which codes for the amino acid phenylalanine (phe). The phe-tRNA enters the A site of the ribosome and the anticodon (AAG) base pairs with the codon. After two tRNAs are attached to the ribosome, an enzyme in the ribosome called **peptidyl transferase** catalyzes the formation of a peptide bond between the amino acids (attached to their tRNAs). Peptide bonds join together amino acids to form a polypeptide chain.

After the amino acids are attached to each other, the initiator tRNA, without methionine attached, pauses briefly in the E site and then is released from the ribosome. Released "empty" tRNAs are recycled by the cell. A new amino acid is attached to the tRNA so it can be used again for translation. The newly forming polypeptide remains attached to the tRNA in the A

site. During a phase called translocation, the ribosome shifts so the tRNA and growing protein move into the P site of the ribosome. The tRNA with a growing polypeptide chain attached is called a peptidyl tRNA. The A site of the ribosome is now aligned with the third codon in sequence (UGG, which codes for tryptophan), and the ribosome waits for the proper aminoacyl tRNA to enter the A site. The cycle continues as described to attach the next amino acid (tryptophan) to the growing protein and repeats itself as the ribosome moves along the mRNA.

Elongation cycles continue to form a new protein until the ribosome encounters a stop codon (for instance, UGA). This signals the third stage of translation called *termination*. Remember that stop codons do not code for an amino acid. Proteins called *releasing factors* interact with the stop codon to terminate translation. The ribosomal subunits come apart and release from the mRNA, and the newly synthesized protein is released into the cell. Ribosomes do recycle and subsequently can bind to any other mRNA molecule (not just the mRNA for one particular gene) and start the process of translation again.

YOU DECIDE

Access to Biotechnology Products for Everyone?

Now that you have studied what genes are and how they are used to create proteins, in Chapter 3 you will learn how genes can be identified, cloned, and studied in great detail. One benefit of gene cloning has been the identification of genes involved in human disease conditions. As a result, it is possible to make many gene products in the laboratory and use them for medical purposes. For instance, when the gene for insulin was cloned in bacteria, it became possible to produce large amounts of insulin for treating people with diabetes. Similarly, cloning of the gene for human growth hormone (hGH), which stimulates growth of bones and muscles during childhood, provided a readily available source of this hormone. Legally available by prescription only, hGH is used widely and effectively to treat children with certain forms of short stature or dwarfism. Illegal use of hGH by professional athletes has received much attention in recent years. Dwarfism is generally defined as a condition that results in an adult height of 4'10" or shorter. The availability of hGH and other products of biotechnology raises an ethical question. Should hGH be available to everyone who wants taller children or only those children with dwarfism? Suppose parents wanted their average-size son to be taller so he would have a better chance of making his high school varsity basketball team. Should these parents be able to give their son hGH simply to enhance his height? You decide.

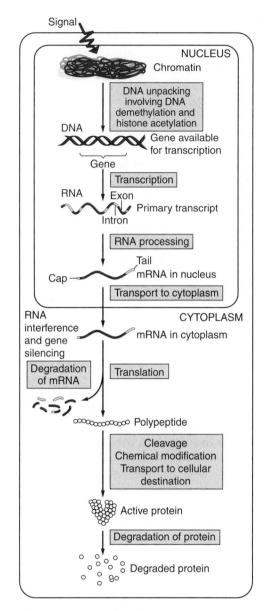

Figure 2.14 Levels of Gene Expression Regulation   Prokaryotic and eukaryotic cells can regulate gene expression in a variety of complex ways. This figure summarizes the primary ways gene expression can be regulated in eukaryotic cells. Notice that gene expression regulation can occur at many different "levels" beginning with how chromatin is folded or chemically modified to controlling degradation or turnover of a protein once it is made. Regulating transcription is a commonly used control mechanism in both prokaryotic and eukaryotic cells.

Basics of Gene Expression Control

Biologists use the term **gene expression** to refer to the production of mRNA (and sometimes protein) by a cell. Cells are exquisitely effective at controlling gene expression and translation to accommodate their needs. All genes are not transcribed and translated at the same rate in all cells. All cells of an organism contain the same genome, so how and why are skin cells different from brain cells or liver cells? Different cell types have different properties and carry out different functions because cells can *regulate* or control the genes they express. At any given time in a cell, only certain genes are "turned on" or expressed to produce proteins while many other genes are silenced or repressed. These genes may only be expressed by cells at certain times, in response to specific cues from inside or outside of the cell, to make proteins as needed. These cues can be environmental signals such as temperature changes,

nutrients in the external environment, hormones, or other complex chemical signals exchanged by cells.

How can genes be turned on and off in response to different signals? Biologists call this process **gene regulation.** Prokaryotic cells and eukaryotic cells regulate gene expression in a variety of complex ways (Figure 2.14). One common mechanism used by both types of cells is called **transcriptional regulation—**

controlling the amount of mRNA transcribed from a particular gene as a way to turn genes on or off. Here we provide an introduction to transcriptional regulation and consider basic examples of this process in eukaryotes and prokaryotes.

Transcriptional regulation of gene expression

Because the amount of protein translated by a cell is often directly related to the amount of mRNA in the cell, cells can regulate the amount of mRNA produced for any given gene to control indirectly the amount of protein a cell produces. How do cells know which genes to turn on and which to shut off? To understand transcriptional regulation we need to look at the role of promoter sequences more closely.

Promoters are found "upstream" of gene sequences, meaning they are found at the 5' end of a gene. Genes in prokaryotes and eukaryotes do not all use the same promoter sequences. In eukaryotes, common promoter sequences found upstream of many genes include a **TATA box** (TATAAAA), located about 30 nucleotides (–30) upstream of the start site of a gene and a **CAAT box** (GGCCAATCT) located

about 80 nucleotides (–80) upstream of a gene (Figure 2.15).

Earlier in this chapter we discussed how RNA polymerase initiates transcription by binding to promoter sequences adjacent to genes. For most eukaryotic genes, RNA polymerase cannot properly recognize and bind to a promoter unless transcription factors are also present at the promoter. Transcription factors are DNA binding proteins that can bind promoters and interact with RNA polymerase to stimulate transcription of a gene (Figure 2.15). In eukaryotes and prokaryotes, common transcription factors interact with promoters for many genes; however, both types of cells also use specific transcription factors that only interact with certain promoters. Transcription of some genes also depends on the binding of specific transcription factors to regulatory sequences adjacent to the promoter. In addition, many genes that are tightly regulated by cells also contain regulatory sequences called **enhancers.**

Enhancer sequences are usually located around 50 or more base pairs upstream of the promoter, but they can also be located downstream of a gene. Enhancer sequences bind regulatory proteins, generally referred

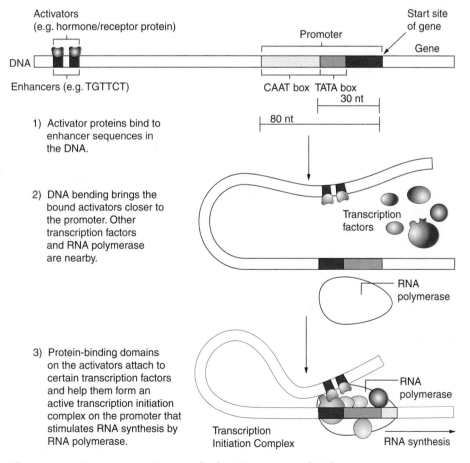

Figure 2.15 Promoters, Transcription Factors, and Enhancers

to as *activators*. Activator molecules interact with transcription factors and RNA polymerase forming a complex that stimulates (activates) transcription of a gene. Cells use a wide variety of different activator molecules. Each activator binds to a particular enhancer sequence. Some activator molecules can be hormones. For instance, the male sex steroid testosterone can act as an activator to stimulate gene expression. You may know that testosterone stimulates cellular activities such as muscle and hair growth in developing boys, but how does testosterone work? This hormone binds to a receptor protein inside cells. The testosterone-receptor protein complex acts as an activator to bind to a specific enhancer element in DNA called an *androgen-response element* (5′-TGTTCT-3′). These elements are usually found close to a promoter. In turn, testosterone and its receptor stimulate gene expression. The female sex steroid estrogen works in a similar manner.

But testosterone and other activators don't stimulate expression of all genes in all cells. Activators act only on those genes that contain enhancer sequences they can bind to. For instance, testosterone stimulates expression of genes involved in muscle growth and hair growth because these genes contain androgen-response elements. Transcription of other genes without androgen-response elements are not directly affected by the hormone. Incidentally, steroid abuse by bodybuilders and athletes looking to increase muscle mass and tone can cause serious long-term health effects in part because steroids abnormally stimulate gene expression for prolonged periods of time.

Through activators and enhancers, cells can use transcriptional control to regulate gene expression and control cellular activities. Some genes even contain repressor sequences that decrease transcription. Because different cells produce different transcription factors and activator molecules, genes can be turned on in some tissues and not others. Skin cells turn on different genes than muscle cells do, so each cell type produces different proteins giving each cell type different functions. Consequently, tissue- and cell-specific gene expression is one way for cells to control the proteins they express even though all body cells contain the same genome. These important control mechanisms are part of why different cells have different functions.

In addition, identifying the promoters, enhancer sequences, and transcription factors that bind these sections of a gene is important for making many biotechnology products. For instance, identifying transcription factors that stimulate the expression of proteins needed for bone growth and development is helping scientists develop new drugs that can be used to stimulate bone growth in people suffering from forms of arthritis when their cells no longer produce bone-growth stimulating factors.

Transcription factors are a large category of DNA-binding proteins

A wide variety of transcription factors can be involved in transcription and in the control of gene expression. Some transcription factors bind to the promoters of most genes and are generally required for RNA polymerase to begin transcription. However many transcription factors have true regulatory roles in controlling gene expression as described in the previous section. These regulatory transcription factors contain regions that allow them to bind to specific DNA sequences to control expression of certain genes. These regions, called **DNA-binding domains,** have folded structural arrangements of amino acids called **motifs** that interact directly with DNA.

A number of different DNA-binding motifs are used by transcription factors, and they typically bind to DNA by fitting within the major groove of the DNA molecule (Figure 2.16). These motifs are typically named according to the shape of the amino acids involved in creating the motif, which can include alpha helices and beta sheets (see Figure 4.3 for a review of protein structure). For example, the *lac* repressor regulatory protein described in Figure 2.17 is an example of a helix-turn-helix containing protein.

Bacteria use operons to regulate gene expression

Bacteria are very important organisms for many applications of biotechnology such as producing human proteins. In several sections of this book, we discuss how gene expression in bacteria can be regulated for a particular purpose. Many initial studies on gene regulation were carried out in bacteria. Scientists discovered that bacteria use a variety of mechanisms to regulate gene expression. Bacteria and other microorganisms can and must rapidly control gene expression in response to environmental conditions such as growth nutrients, temperature, and light intensity. One interesting aspect of gene expression and regulation in bacteria is that many bacterial genes are organized in arrangements called **operons.** Operons are essentially clusters of several related genes located together and controlled by a single promoter. The genes of an operon can be regulated in response to changes within the cell, and many genes controlling nutrient metabolism by bacteria are organized as operons. Bacteria can use operons to tightly regulate gene expression in response to their nutrient requirements. Here we present a well-studied, classic example of gene regulation in bacteria by describing the ***lac* operon** (Figure 2.17).

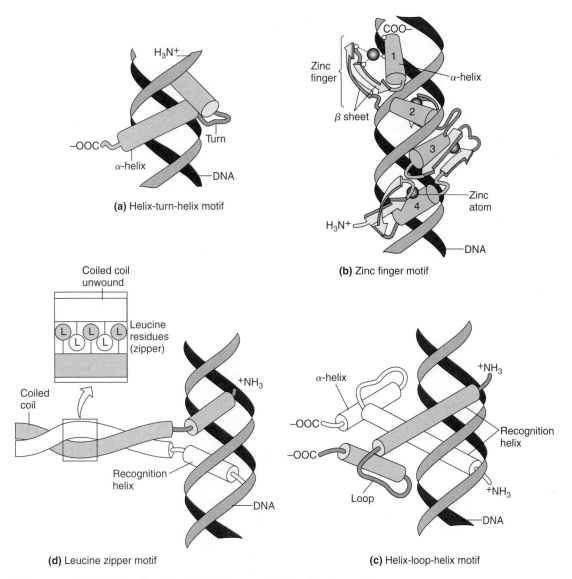

(a) Helix-turn-helix motif

(b) Zinc finger motif

(d) Leucine zipper motif

(c) Helix-loop-helix motif

Figure 2.16 DNA-Binding Motifs in Transcription Factors Common DNA-binding motifs in many transcription factors are shown here. (a) The helix-turn-helix motif consists of α-helices joined by a short turn with the helices fitting into the major groove of DNA. (b) Zinc finger motifs involve groups of two cysteine and two histidine amino acids that bind a zinc atom and fold the amino acids into a "fingerlike" loop that binds to DNA. (c) Leucine zipper motifs consist of parallel arrangements of the amino acid leucine bound to each other with adjacent α-helices that bind to DNA. (d) Helix-loop-helix motifs have a shorter α-helix connected to a longer α-helix through an interconnecting loop of amino acids.

The *lac* operon consists of the following three genes:

- *lacZ*, encoding the enzyme β-*galactosidase*
- *lacY*, encoding the enzyme *permease*
- *lacA*, encoding the enzyme *acetylase*

Together, these three enzymes are necessary for the transport and breakdown of lactose by bacterial cells. Lactose, a sugar present in milk, is an important energy source for many bacteria. For bacteria to metabolize lactose, the sugar must be transported into cells by permease and then degraded into glucose and galactose by β-galactosidase. The function of acetylase is not clear, although it may play a role in protecting cells against toxic products of lactose degradation. The *lac* operon is regulated by a protein called the **lac repressor,** which is encoded by a separate gene called the *lacI* gene. When bacteria are grown in the absence of lactose, the repressor protein uses helix-turn-helix motifs

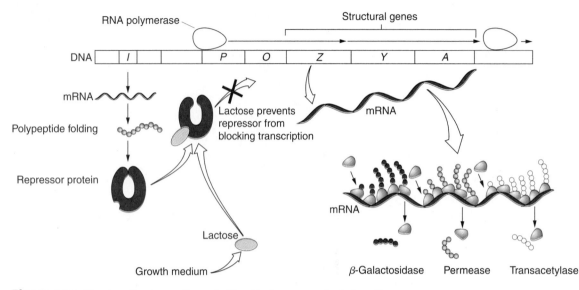

Figure 2.17 The *lac* Operon   By controlling the *lac* operon, bacteria cells can regulate gene expression in response to availability of the sugar lactose. In the absence of lactose, the *lac* repressor binds to the operator blocking transcription of the operon. In the presence of lactose, lactose binds and inactivates the repressor allowing transcription of the operon to occur.

to bind a sequence within the *lac* operon promoter (p) called the **operator** (o). By binding to the operator, the repressor blocks RNA polymerase from binding to the promoter and blocks transcription of the *Z*, *Y*, and *A* genes in the operon (Figure 2.17). This is a nice way for bacteria to control their metabolism. Why expend energy transcribing genes and translating proteins if there is no lactose available for producing energy?

Conversely, in the presence of lactose, the sugars act as inducer molecules that stimulate transcription of the *lac* operon. Lactose binds to the *lac* repressor, changing the shape of the repressor protein and preventing it from binding to the operator (Figure 2.17). With no repressor in its way, RNA polymerase can bind to the *lac* promoter and stimulate transcription of the operon. Transcribed mRNA from the operon is translated to produce the enzymes required by the cell to metabolize lactose.

MicroRNAs "silence genes" revealing a recent discovery of gene expression regulation

In 1998, scientists studying the roundworm *Caenorhabditis elegans* discovered small (21 or 22 nt) double-stranded pieces of non-protein coding RNA called **short interfering RNA (siRNA),** so named because they were shown to bind to mRNA and subsequently block or interfere with translation of bound mRNAs. For a short time it appeared that perhaps siRNAs were restricted to *C. elegans*, but while looking for siRNAs in other species, scientists recently discovered that microRNAs (miRNAs) are another category of small RNA molecules that do not encode proteins. Genes for miRNAs have been identified in all multicellular organisms studied to date.

Similar to siRNAs, miRNAs are regulatory molecules that regulate gene expression by "silencing" gene expression through blocking translation of mRNA or by causing degradation of mRNA. MicroRNA genes produce RNA transcripts that are processed by enzymes (Drosha and Dicer) into short single-stranded pieces of 21 to 22 nucleotides in size (Figure 2.18). The processed miRNAs then bind to a complex of proteins (called RISC), which allows them to bind to mRNAs that are complementary to the miRNA sequence. Binding of miRNA to an mRNA sequence can either block translation by ribosomes or trigger enzymatic degradation of an mRNA. Both mechanisms inhibit or silence expression by preventing the mRNA from being translated into a protein.

RNA-based mechanisms of gene silencing are generally referred to as **RNA interference (RNAi).** Identifying human genes that are silenced by miRNAs is currently a very active area of research. In Chapters 3 and 11, we discuss how scientists are working on exploiting RNAi as a potentially promising way to use small RNA molecules to selectively turn off or silence genes involved in human disease conditions. As evidence of how valuable RNAi is as a research technique, the 2006 Nobel Prize in Physiology or Medicine was awarded to Andrew Fire of Stanford University and Craig Mello of the University of the Massachusetts School of Medicine for their discovery of this method for switching genes off.

We conclude this chapter with a brief discussion of how genes can be affected by mutation.

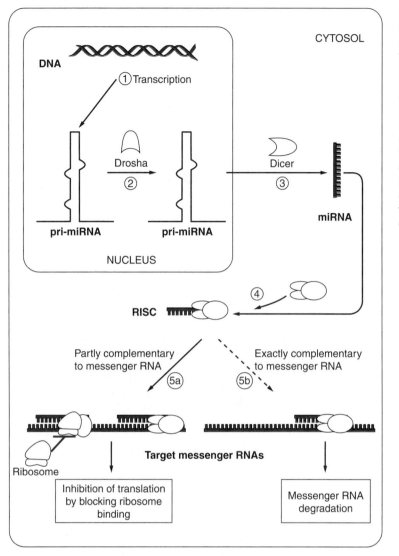

Figure 2.18 MicroRNAs Silence Gene Expression Gene silencing by miRNAs is a multi-step process. (1) MicroRNA genes transcribe a primary transcript called pri-RNA that (2) folds into a hairpin loop similar to the folding patterns of tRNA molecules. (3) The enzyme Dicer cleaves pri-mRNAs into single-stranded miRNAs that are 21 to 22 nucleotides long. (4) MicroRNAs bind to a cluster of proteins to form a ribonucleoprotein complex. (5) This miRNA complex can then bind to mRNA molecules by complementary base pairing with sequences that are an exact match or a close match to complementary sequences in the mRNA and miRNA. (5a) miRNA binding then silences gene expression by blocking translation or by targeting the bound mRNA for degradation by enzymes in the cell.

2.5 Mutations: Causes and Consequences

A **mutation** is a change in the nucleotide sequence of DNA. Mutations are a major cause of genetic diversity. For instance, the underlying basis of the evolution of species to develop and acquire new characteristics is governed by mutations of genes over time. Mutations can also be detrimental. Mutation of a gene can result in the production of an altered protein that functions poorly or in some cases no longer encodes a functional protein. Such mutations can cause genetic diseases. In this section, we provide an overview of different types of mutations and their consequences.

Types of Mutations

There are many different causes of mutation. Sometimes mutations can occur through spontaneous events such as errors during DNA replication. For instance, DNA polymerase can insert the wrong nucleotide into a newly synthesized strand of DNA, say, inserting a T where a C belongs. Even though enzymes in cells work to detect and correct mistakes, errors occasionally occur during DNA replication. Mutations can also be induced by environmental causes. For example, chemicals called **mutagens,** many of which mimic the structure of nucleotides, can mistakenly be introduced into DNA and change DNA structure. Exposure to X-rays or ultraviolet light from the sun can also mutate DNA (That glowing tan you may enjoy during the summer is not as healthy as you think!).

Regardless of how mutations arise, depending on the type of mutation, they may have no effect on protein production or they can dramatically change protein production and protein structure and function. Mutations can involve large changes in genetic information or single nucleotide changes in a gene, such as changing an A to a C or a G to a T. The most common

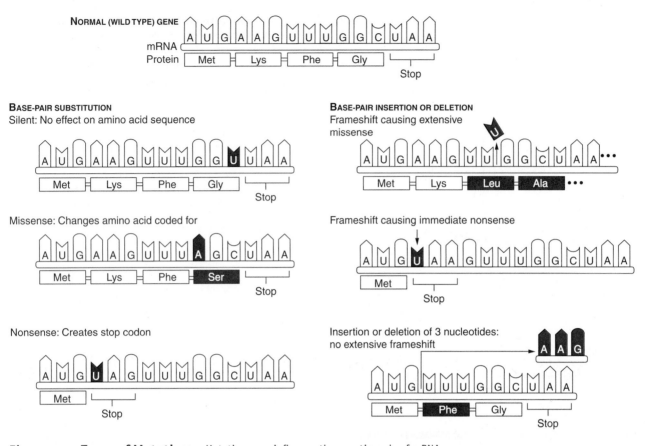

Figure 2.19 Types of Mutations   Mutations can influence the genetic code of mRNA and resulting proteins translated from a gene. Shown here is a portion of mRNA copied from a gene, but mutations generally occur within DNA. Mutations in a gene can have different consequences on the protein translated.

mutations in a genome are single nucleotide changes (or a few nucleotide mutations) called **point mutations.** Point mutations often involve base pair substitutions, in which a base pair is replaced by a different base pair; insertions, in which a nucleotide is inserted into a gene sequence; or deletions, the removal of a base pair (Figure 2.19).

Mutations ultimately exert their effects on a cell by changing the properties of a protein, which in turn can affect traits.

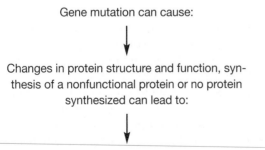

Gene mutation can cause:

↓

Changes in protein structure and function, synthesis of a nonfunctional protein or no protein synthesized can lead to:

↓

Change or loss of a trait

Proteins are large, complicated molecules. To work correctly, most proteins must be folded into complex three-dimensional shapes. Changing one or two amino acids in a protein can alter the overall shape of a protein, dramatically disrupting its function or in some cases preventing the protein from functioning at all. Mutations may have no effect on a protein if the mutation changes the codon sequence of a gene to another codon that codes for the same amino acid (Figure 2.19). These are considered **silent mutations** because they have no effect on the structure and function of the protein. Similarly, a mutation can change a codon so a different amino acid is coded for. These **missense mutations** can also be considered "silent" if the new amino acid coded for doesn't change protein structure and function. However, if the newly coded amino acid changes the structure of the protein, then its function may be significantly changed. Later, we consider a dramatic example of how a single-nucleotide mutation is responsible for the human genetic condition called sickle-cell disease.

Sometimes mutations called **nonsense muta-tions** change a codon for an amino acid into a stop codon, which causes an abnormally shortened pro-tein to be translated, usually creating a nonfunc-tional protein. Insertions or deletions can also dramatically affect the protein produced from a gene by creating **"frameshifts."** As shown in Figure 2.19, inserting a nucleotide (U in this example) causes the reading frame of the codons to be shifted to the right of the insertion, changing the protein encoded by the mRNA. Frameshifts often create nonfunctional proteins.

Mutations Can Be Inherited or Acquired

It is important to realize that not all mutations have the same effect on body cells. The effects of a mutation depend not only on the type of mutation that occurs but also on the cell type in which a mutation occurs.

Gene mutations can be inherited or acquired. **Inherited mutations** are those mutations passed to offspring through gametes—sperm or egg cells. As a result, the mutation is present in the genome of all of the offspring's cells. Inherited mutations can cause birth defects or inherited genetic diseases. We will study a number of different inherited genetic diseases throughout this book.

Acquired mutations occur in the genome of somatic cells (remember that these include all other cell types except for the gametes) and are not passed along to offspring. Although not inherited, acquired mutations can cause abnormalities in cell growth leading to cancerous tumor formation, metabolic dis-orders, and other conditions. For instance, prolonged exposure to ultraviolet light can cause acquired muta-tions in skin cells leading to skin cancer. Understand-ing the genetic basis of cancer and many other human diseases is a major area of biotechnology research.

CAREER PROFILE

Careers in Genomics

It is an incredibly exciting time to consider a career in genomics. Never before have there been greater opportuni-ties or a wider variety of career options for anyone inter-ested in genomes. In addition to studying the human genome, scientists are actively involved in studying the genomes of many other species including model organ-isms such as mice, fruit flies, zebrafish, agriculturally important crop plants and plant pests, disease-causing microbes, and marine organisms. As enormous amounts of genome information become available, decades of work will be necessary to study what different genes do. Deciphering the secrets contained in genomes will involve the combined efforts of many people.

A recent publication from the National Institutes of Health (*Genetic Basics*, NIH Publication No. 01–662, www.nigms.nih.gov) proclaimed that "Help Wanted" signs are up all over the world to recruit thousands and thousands of human brains to contribute to the study of genomics. Career opportunities in genomics primarily fall into four major categories: laboratory scientists, clinical doctors, genetic counselors, and bioinformatics experts. Laboratory scientists are the so-called bench scientists because they conduct experiments at the lab bench daily. Lab scientists are often involved in carrying out experiments to discover and clone genes and study their functions. Entry-level lab scientist opportunities exist as lab technicians for those with associate's and bachelor's degrees, and higher-level scientist positions

such as laboratory director positions require a master's degree or a doctor of philosophy (Ph.D.) degree. Clinical doctors are M.D.-trained physicians who conduct research and interact with patients as part of research teams. Clinical doctors may be involved in research studies to treat patients with new genetics-based treat-ments such as gene therapy protocols. Physicians are also important because genomics is having and will have profound effects on medicine and the treatment of human diseases. Genetic counselors help people under-stand genomics information such as how genes affect disease susceptibility. Counselor positions usually require a M.S. degree with training in psychology, biol-ogy, and genetics.

Bioinformatics is a discipline that merges biology with computer science. Bioinformatics involves storing, analyzing, and sharing gene and protein data. The tremendous amount of DNA sequence data being generated by genome projects has made bioinformatics a rapidly emerging area that is absolutely essential for genomics. Bioinformaticians generally have solid training in both biology and computer science, and they work together with bench scientists to analyze genome data. Most positions in bioinformatics require a B.S., M.S., or Ph.D. degree.

Visit the Human Genome Program Information page in Keeping Current: Web Links at the Companion Website for an outstanding site on career possibilities in genetics that also includes links to many other valuable resources.

Mutations Are the Basis of Variation in Genomes and a Cause of Human Genetic Diseases

Mutations form the molecular basis of many human genetic diseases. Sickle-cell disease was the first genetic disease discovered whose cause was pinpointed to a particular mutation (Figure 2.20). Sickle-cell disease is created by a single nucleotide change, a base pair substitution, in the gene coding for one of the polypeptides in the protein **hemoglobin,** the oxygen-binding protein in red blood cells. Red blood cells contain millions of hemoglobin molecules, each of which consists of four polypeptide chains (Figure 2.20). A point mutation in this gene, technically called the β-globin

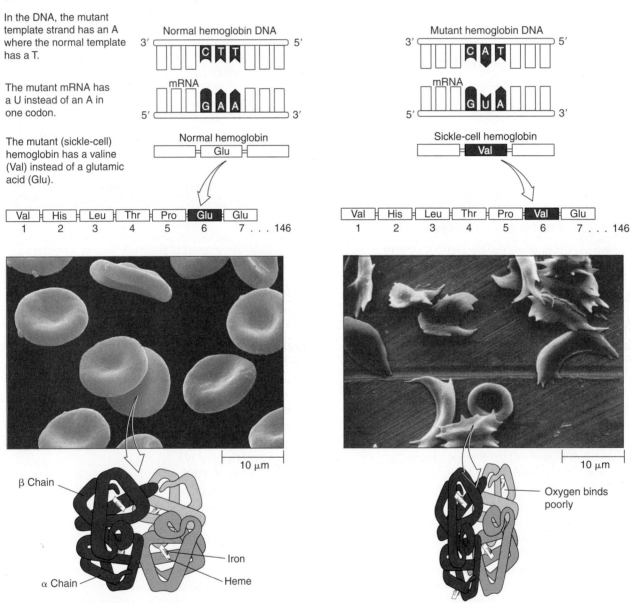

(a) Normal hemoglobin and normal red blood cells

In the DNA, the mutant template strand has an A where the normal template has a T.

The mutant mRNA has a U instead of an A in one codon.

The mutant (sickle-cell) hemoglobin has a valine (Val) instead of a glutamic acid (Glu).

(b) Sickle-cell hemoglobin and sickled red blood cells

Figure 2.20 Molecular Basis of Sickle-Cell Disease   (a) Hemoglobin, the oxygen-binding protein of red blood cells, consists of four polypeptide chains. A portion of the normal hemoglobin gene is shown here along with its transcribed mRNA and the first seven amino acids for one of the hemoglobin polypeptides, which contains a total of 146 amino acids. (b) The defective gene that causes sickle-cell disease contains a single base pair change in its sequence that alters the hemoglobin protein translated. This subtle change alters the shape of red blood cells causing them to sickle. Sickled cells are fragile, they can clump and block blood vessels, and they do not transport oxygen well.

gene, changes the genetic code so the gene codes for a different amino acid at position 6 in one of the hemoglobin polypeptides. As a result, the amino acid valine is inserted into sickle-cell hemoglobin instead of glutamic acid. Individuals with two defective copies of the hemoglobin gene suffer the effects of sickle-cell disease. This subtle mutation alters the oxygen-transporting ability of hemoglobin and dramatically changes the shape of red blood cells to an abnormal sickled shape. Sickled cells block blood vessels, and patients suffer from poor oxygen delivery to tissues, causing joint pain and other symptoms. Sickle-cell disease is one of the most well understood inherited genetic disorders.

In Chapter 3, we explore how scientists working on the Human Genome Project have identified and are analyzing the roughly 3 billion base pairs that comprise human DNA.

As scientists have learned more about the human genome, they have discovered that DNA sequences from people of different backgrounds around the world are very similar. Regardless of ethnicity, human genomes are approximately 99.9% identical. In other words, you have about 99.9% the same DNA sequences as President George W. Bush, Shaquille O'Neal, Britney Spears, Saddam Hussein, Michael Jackson—or virtually any other human.

But because there are about 0.1% differences in DNA between individuals, or around one base out of every thousand, this means there are roughly 3 million differences between different individuals. Many of these variations are created by mutation, and most of these have no obvious effects; however, other mutations strongly influence cell functions, behavior, and susceptibility to genetic diseases. These variations are important and are the basis for differences in all inherited traits among people, from height and eye color to personality, intelligence, and lifespan.

Most genetic variation between human genomes is created by substitutions of individual nucleotides called

"You're lucky nobody was hurt. Your base pairs are out of alignment and that has your reading frames all messed up."

Figure 2.21 Biotechnology Is Being Used to Detect and Correct Mutations

single nucleotide polymorphisms (SNPs). For instance, at a particular sequence, a certain base in a given region of DNA sequence from President Bush may read "A," Shaquille O'Neal's sequence may read "T," Britney Spears's sequence may read "C," and the same site in your sequence may read "G." Most SNPs are harmless because they occur in intron regions of DNA; however, when they do occur in exons, they can affect the structure and function of a protein, which can influence cell function and result in disease. Sickle-cell disease is caused by an SNP. Refer to Figure 1.11, which shows a comparison of a gene sequence from three different people. An SNP in this gene in person 2 may have no effect on protein structure and function if it is a silent mutation. Other SNPs (person 3) cause disease if they change protein structure and function.

In Chapter 11, we consider several genetic disease conditions, discuss how defective genes can be detected, and examine how scientists are working on gene therapy approaches to cure these diseases (Figure 2.21).

QUESTIONS & ACTIVITIES

Answers can be found in Appendix 1.

1. Compare and contrast genes and chromosomes, and describe their roles in the cell.

2. If the sequence of one strand of a DNA molecule is 5'-AGCCCCGACTCTATTC-3', what is the sequence of the complementary strand?

3. What does the phrase "gene expression" mean?

4. Suppose you identified a new strain of bacteria. If the DNA content of this organism's cells is 13% adenine, approximately what percentage of this organism's genome consists of guanine? Explain your answer.

5. Provide at least three important differences between DNA and RNA.

6. Consider the following sequence of mRNA: 5'-AGCACCAUGCCCCGAACCUCAAAGUGAAA-CAAAAA-3'. How many codons are included in this mRNA? How many amino acids are coded for by this sequence? Use Table 2.3 to determine the amino acid sequence encoded by this mRNA.

Note: Remember that mRNA molecules are actually much larger than the very short sequence shown here.

7. Consider the following sequence of DNA:
5'-TTTATGGG TTGGCCCGGGTCATGATT- 3'
3'-AAATACCCAACCGGGCCCAGT ACTAA- 5'

   a. Transcribe each of these sequences into mRNA. Which DNA sequence (top or bottom strand) produces a functional mRNA containing a start codon? What is the amino acid sequence of the polypeptide produced from this mRNA?

   b. For the DNA strand producing a functional mRNA, number each base from left to right with the first base numbered "1." Insert a "T" between bases 10 and 11, representing a base insertion mutation. Transcribe an mRNA from this new strand and translate it into a protein. Compare the amino acid sequence of this protein to the one translated in (a). What happened? Explain.

8. Name the three types of RNA involved in protein synthesis and describe the function of each.

9. What is gene expression regulation? Why is it important?

10. Diagram the process of semiconservative replication of DNA by drawing a replication fork and indicating important enzymes, proteins, and other components involved in this process. Provide a *one-sentence* description of the function of each component.

11. Why do some mutations affect protein structure and function that can result in disease whereas other mutations have no significant effects on protein structure and function?

12. In a nucleotide of DNA, which carbon in the deoxyribose sugar is bonded to the nitrogenous base?

References and Further Reading

Ast, G. (2005). Alternative Genome. *Scientific American*, 292: 58–65.

Becker, W. M., Kleinsmith, L. J., and Hardin, J. (2006). *The World of the Cell*, 6/e. San Francisco: Benjamin Cummings.

Dennis, C., and Gallagher, R., eds. (2001). *The Human Genome*. New York: Palgrave.

Gibbs, W. W. (2003). The Unseen Genome: Gems Among the Junk. *Scientific American*, 289: 46–53.

Mattick, J. S. (2004). The Hidden Genetic Program of Complex Organisms. *Scientific American*, 291: 60–67.

Ridley, M. (2000). *Genome: The Autobiography of a Species in 23 Chapters*. New York: HarperCollins.

Watson, J. D., Baker, T. A., Bell, S. P., Gann, A., Levine, M., and Losick, R. (2004). *Molecular Biology of the Gene*, 5/e. San Francisco: Benjamin Cummings.

Visit www.pearsonhighered.com/biotechnology to download learning objectives, chapter summary, "Keeping Current" web links, glossary, flashcards, and jpegs of figures from this chapter.

Chapter **3**

Recombinant DNA Technology and Genomics

After completing this chapter you should be able to:

- Define recombinant DNA technology and explain how it is used to clone and manipulate genes.

- Compare and contrast different types of vectors; describe their practical features and applications in molecular biology.

- Discuss how DNA libraries are created and screened to clone a gene of interest.

- Describe how agarose gel electrophoresis, restriction enzyme mapping, and DNA sequencing can be used to study gene structure.

- Explain common techniques used to study gene expression.

- Be familiar with RNA interference (RNAi) as a powerful new technique for silencing gene expression.

- Understand potential scientific and medical consequences of the Human Genome Project, and discuss its ethical, legal, and social issues.

- Define genomics and bioinformatics and explain why these disciplines are rapidly developing major areas of research.

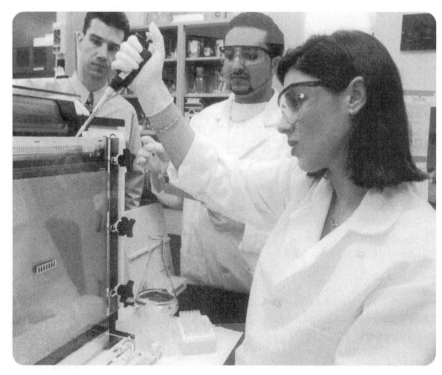

Undergraduate biology students working on a recombinant DNA experiment.

As you learned in Chapter 1, biotechnology is not a new science. We have used crop and livestock improvement practices for a long time, and have used microbes to make food and beverages through fermentation for many years. However, the modern era of biotechnology began when DNA cloning techniques were developed. Beginning in the 1970s and continuing over the next three decades, amazing and rapidly developing laboratory methods in **recombinant DNA technology** and **genetic engineering** have changed molecular biology, basic science, and medical research forever. In this chapter, we present an overview of the history of genetic manipulation and discuss landmark discoveries in recombinant DNA technology that have revolutionized many areas of science and medicine. We then take a look at an amazing range of modern techniques that scientists use to clone genes and study gene structure and function. The chapter concludes with an introduction to bioinformatics, a relatively new and powerful field.

3.1 Introduction to Recombinant DNA Technology and DNA Cloning

When scientists James Watson and Francis Crick discovered that the structure of DNA is a double-helical molecule, they hinted about the potential importance and impact of this discovery. However, not even these two Nobel Prize winners could have imagined the astonishing pace at which molecular biology would advance over the next half century.

As you learned in Chapter 2, in the years before and after Watson and Crick's discovery, many other scientists contributed to our modern understanding of DNA as the genetic material of living cells. A number of researchers studied DNA structure and replication in bacteria and in **bacteriophages.** Bacteriophages, often simply called *phages*, are viruses that infect bacterial cells. Much of what we know about DNA replication and DNA-synthesizing enzymes has been learned from studying bacteria and phages. For example, a key enzyme involved in DNA replication is called DNA ligase. Recall from Chapter 2 that ligase joins together adjacent DNA fragments (Okazaki fragments) during DNA replication. DNA ligase is an important enzyme in recombinant DNA technology.

Bacteria such as *Escherichia coli*, which are present naturally in the intestines of animals, including humans, have served an important role as experimental **model organisms** for studies in genetics and molecular biology. *E. coli* continues to be a favorite lab organism for many experiments in biotechnology. Important roles of bacteria and viruses in biotechnology and applications of microbial biotechnology are discussed in more detail in Chapter 5.

In the late 1960s, many scientists were interested in gene cloning and they speculated that it might be possible to clone DNA by cutting and pasting DNA from different sources (recombinant DNA technology). As you begin to learn about biotechnology, it may seem that the terms *gene cloning, recombinant DNA technology,* and *genetic engineering* describe the same process. In fact, these techniques are slightly different methodologies that are interrelated. As you will see in this chapter, recombinant DNA technology is commonly used to make gene cloning possible, whereas genetic engineering often relies on recombinant DNA technology and gene cloning to modify an organism's genome. However, the terms *recombinant DNA technology* and *genetic engineering* are frequently used interchangeably. *Clone* is derived from a Greek word that describes a cutting (of a twig) that is used to propagate or copy a plant. A modern biological definition of a clone is a molecule, cell, or organism that was produced from another single entity. The laboratory methods required for gene cloning as described in this chapter are different from the techniques used to clone whole organisms such as Dolly the sheep. Organism cloning is discussed in Chapter 7.

Restriction Enzymes and Plasmid DNA Vectors

In the early 1970s, gene cloning became a reality. Many near simultaneous discoveries and collaborative efforts among several researchers led to the discovery of two essential components that made gene cloning and recombinant DNA techniques possible—**restriction enzymes** and **plasmid DNA.** Restriction enzymes are DNA-cutting enzymes, and plasmid DNA is a circular form of self-replicating DNA that scientists can manipulate to carry and clone other pieces of DNA.

Microbiologists in the 1960s discovered that some bacteria are protected from destruction by bacteriophages because they can *restrict* phage replication. Swiss scientist Werner Arber proposed that restricted growth of phages occurred because some bacteria contained enzymes that could cut viral DNA into small pieces, thus preventing viral replication. Because of this ability, these enzymes were called *restriction enzymes*. Bacteria do not have immune systems to fend off phages, but restriction enzymes do provide a type of protective mechanism for some bacteria.

In 1970, working with the bacterium *Haemophilus influenzae*, Johns Hopkins University researcher Hamilton Smith isolated *Hin*dIII, the first restriction enzyme to be well characterized and used for DNA cloning. Restriction enzymes are also called restriction endonucleases (*endo* = "within," *nuclease* = "nucleic acid cutting enzyme") because they cut *within* DNA sequences as opposed to enzymes that cut from the ends of DNA sequences (exonucleases). Smith demonstrated that *Hin*dIII could be used to cut or digest DNA into small fragments. In 1978, Smith shared a Nobel Prize with Werner Arber and Daniel Nathans for their discoveries on restriction enzymes and their applications.

Restriction enzymes are primarily found in bacteria, and they are given abbreviated names based on the genus and species names of the bacteria from which they are isolated. For example, one of the first restriction enzymes to be isolated, *Eco*RI, is so named because it was discovered in the *E. coli* strain called RY13. Restriction enzymes cut DNA by cleaving the phosphodiester bond (in the sugar-phosphate backbone) that joins adjacent nucleotides in a DNA strand. However, restriction enzymes do not just randomly cut DNA, nor do all restriction enzymes cut DNA at the same locations. Like other enzymes, restriction enzymes show specificity for certain **substrates.** For these enzymes, the substrate is DNA. As shown in Figure 3.1a, restriction enzymes bind to, recognize, and cut (digest) DNA within specific sequences of bases called a **recognition sequence or restriction site.**

Restriction enzymes are commonly referred to as four-base pair or six-base pair cutters because they typically recognize restriction sites with a sequence of four or six nucleotides. Eight-base pair cutters have also been identified. These recognition sequences are palindromes—the arrangement of nucleotides reads the same forward and backward on opposite strands of the DNA molecule. (Remember the word "madam" or the phrase "a toyota" as examples of palindromes.) Some restriction enzymes such as *Eco*RI cut DNA to create DNA fragments with overhanging single-stranded ends called "sticky" or **cohesive ends** (see Figure 3.1a); other enzymes generate fragments with double-stranded, nonoverhanging ends called **blunt ends.** Table 3.1 shows some common restriction enzymes, their source microorganisms, and their recognition sequences. Notice that the first three enzymes in the table are six-base pair cutters that produce DNA molecules with cohesive ends. The fourth enzyme (*Taq*I) is a four-base pair cutter that produces cohesive ends, and the lower three enzymes produce blunt-ended DNA fragments. Enzymes that produce cohesive ends are often favored over blunt-end cutters for many cloning experiments because DNA fragments with cohesive ends can easily be joined together. DNA from *any* source such as bacteria, humans, dogs, cats, frogs, dinosaurs, or ancient human remains can be digested by a particular restriction enzyme as long as the DNA has a restriction site for that enzyme. Keep this in mind as you read the next few pages. In the simplest sense, the discovery of restriction enzymes provided molecular biologists with the "scissors" needed to carry out gene cloning.

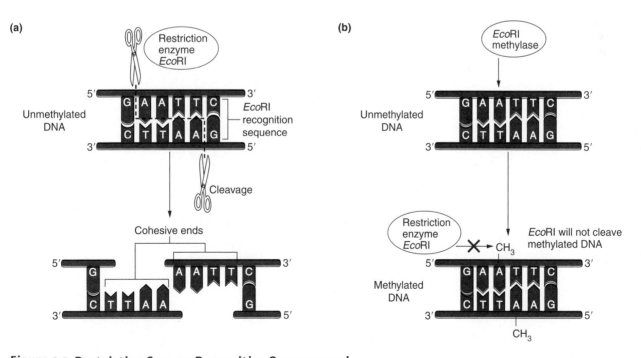

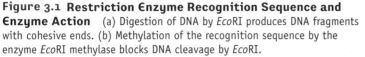

Figure 3.1 Restriction Enzyme Recognition Sequence and Enzyme Action   (a) Digestion of DNA by *Eco*RI produces DNA fragments with cohesive ends. (b) Methylation of the recognition sequence by the enzyme *Eco*RI methylase blocks DNA cleavage by *Eco*RI.

Table 3.1 COMMON RESTRICTION ENZYMES

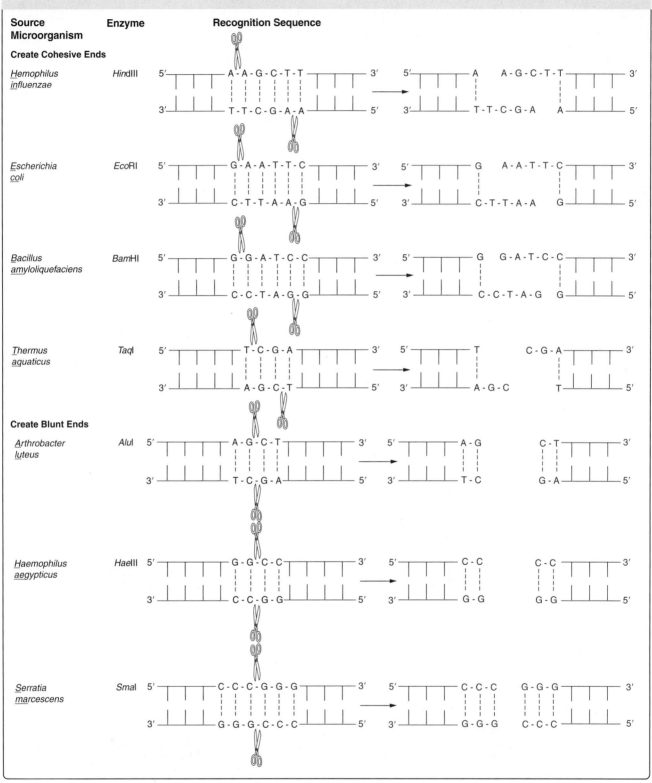

Source Microorganism	Enzyme	Recognition Sequence

Create Cohesive Ends

Hemophilus influenzae — HindIII

Escherichia coli — EcoRI

Bacillus amyloliquefaciens — BamHI

Thermus aquaticus — TaqI

Create Blunt Ends

Arthrobacter luteus — AluI

Haemophilus aegypticus — HaeIII

Serratia marcescens — SmaI

In the early 1970s, Paul Berg, Herbert Boyer, Stanley Cohen, and their colleagues at Stanford University used gene cloning to change molecular biology forever. Berg founded recombinant DNA technology when he created a piece of recombinant DNA by joining together (splicing) DNA from the *E. coli* chromosome and DNA from a primate virus called SV40 (simian virus 40). Berg isolated chromosomal DNA

Q Why don't restriction enzymes digest chromosomal DNA in bacterial cells?

A Bacteria are protected from enzymatic digestion of their DNA because some of the nucleotides in their DNA contain methyl groups that block restriction enzymes from digestion (see Figure 3.1b). Many phages do not normally methylate their DNA, so they are susceptible to DNA degradation by restriction enzymes; however, certain bacteriophages have evolved to use methylation as a way to avoid digestion by restriction enzymes. By attaching methyl groups to their DNA, phages use methylation to protect their DNA from being destroyed by restriction enzymes.

from *E. coli* and DNA from SV40, and then he cut both DNA samples with *Eco*RI. He then added *E. coli* DNA and viral DNA fragments to a reaction tube with the enzyme DNA ligase and succeeded in creating a hybrid molecule of SV40 and *E. coli* DNA. The importance of this discovery was fully recognized when Paul Berg won the 1980 Nobel Prize in Chemistry for this experiment, which demonstrated that DNA could be cut from different sources with the same enzyme and that the restriction fragments could be joined to create a recombinant DNA molecule. Berg shared this prize with Walter Gilbert and Frederick Sanger, who independently developed methods for sequencing DNA—a process we discuss in Section 3.4.

Berg worked with chromosomal DNA; Cohen was interested in the molecular biology of small circular pieces of DNA known as *plasmids*. Plasmid DNA is primarily found in bacteria. Plasmids are considered **extrachromosomal** DNA because they are present in the bacterial cytoplasm in addition to the bacterial chromosome. Plasmids are small (most average approximately 1,000 to 4,000 bp in size) and self-replicating (they duplicate independently of the bacterial chromosome). Cohen studied plasmid replication, the transfer of plasmids between bacteria, and mechanisms of bacterial resistance to antibiotics.

Cohen postulated that plasmids could be used as **vectors**—pieces of DNA that can accept, carry, and replicate (clone) other pieces of DNA. Cohen and Boyer worked with two bacterial plasmids to clone DNA successfully. They published results describing these experiments in 1973. Many consider this work the informal birth of recombinant DNA technology. Using *Eco*RI, a restriction enzyme previously isolated by Herbert Boyer, they cut both plasmids and then joined fragments from each plasmid together using DNA ligase to create new hybrid (recombinant) plasmids. Recall that DNA ligase catalyzes the formation of

phosphodiester bonds between nucleotides. Ligase can join together DNA with cohesive ends as well as blunt-ended fragments. As a result of these and other experiments, Cohen and Boyer produced the first plasmid vector for cloning purposes, called pSC101 and named "SC" for Stanley Cohen. pSC101 contained a gene for tetracycline resistance and restriction sites for several enzymes including *Eco*RI and *Hin*dIII. In subsequent work they used similar experiments to clone DNA from the South African claw-toed frog *Xenopus laevis* (another important model organism in genetics and developmental biology) into the *Eco*RI site of pSC101. Figure 3.2 illustrates how recombinant DNA can be formed in a process similar to that used in the Cohen and Boyer experiments.

Cohen and Boyer had created the first DNA cloning vector—a vehicle for the insertion and replication of DNA—and in 1980 they were awarded patents for pSC101 and for the gene splicing and cloning techniques they had developed. These experiments ushered in the birth of modern biotechnology because many of the current techniques used for gene cloning and gene manipulations are based on these fundamental methods of recombinant DNA technology.

In 1974, as a direct result of the Berg, Cohen, and Boyer experiments, gene cloning pioneers and cloning critics voiced concerns about the safety of genetically modified organisms. Scientists were concerned about what might happen if recombinant bacteria were to leave the lab or if such bacteria could transfer their genes to other cells or survive in other organisms including humans. In 1975, an invited group of well-known molecular biologists, virologists, microbiologists, lawyers, and journalists gathered at the Asilomar Conference Center in Pacific Grove, California, to discuss the benefits and potential hazards of recombinant DNA technology. As a result of the historic Asilomar meeting, the **National Institutes of Health (NIH)** formed the **Recombinant DNA Advisory Committee (RAC),** which was charged with evaluating risks of recombinant DNA technology and establishing guidelines for recombinant DNA research. In 1976, the RAC published a set of guidelines for working with recombinant organisms. The RAC continues to oversee gene cloning research, and compliance with RAC guidelines is mandatory for scientists working with recombinant organisms.

Transformation of Bacterial Cells and Antibiotic Selection of Recombinant Bacteria

Cohen also made another important contribution to gene cloning, which made the pSC101 cloning experiments possible. His laboratory demonstrated **transformation,** a process for inserting foreign DNA into

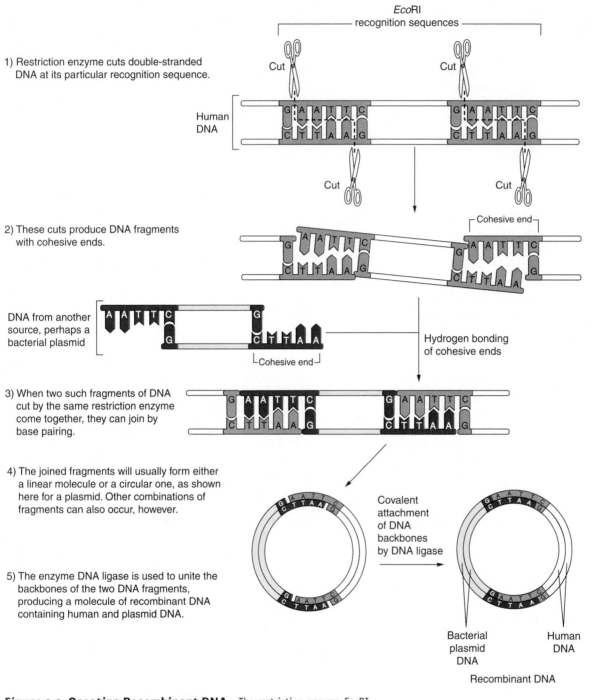

1) Restriction enzyme cuts double-stranded DNA at its particular recognition sequence.

2) These cuts produce DNA fragments with cohesive ends.

DNA from another source, perhaps a bacterial plasmid

3) When two such fragments of DNA cut by the same restriction enzyme come together, they can join by base pairing.

4) The joined fragments will usually form either a linear molecule or a circular one, as shown here for a plasmid. Other combinations of fragments can also occur, however.

5) The enzyme DNA ligase is used to unite the backbones of the two DNA fragments, producing a molecule of recombinant DNA containing human and plasmid DNA.

EcoRI recognition sequences

Cut

Cut

Human DNA

Cut

Cut

Cohesive end

Hydrogen bonding of cohesive ends

Cohesive end

Covalent attachment of DNA backbones by DNA ligase

Bacterial plasmid DNA

Human DNA

Recombinant DNA

Figure 3.2 Creating Recombinant DNA   The restriction enzyme *EcoRI* binds to a specific sequence (5'-GAATTC-3') and then cleaves the DNA backbone, producing DNA fragments. The single-stranded ends of the DNA fragments can hydrogen bond with each other because they have complementary base pairs. The enzyme DNA ligase can then catalyze the formation of covalent bonds in the DNA backbones of the fragments to create a piece of recombinant DNA.

bacteria. Cohen discovered that if he treated bacterial cells with calcium chloride solutions, added plasmid DNA to cells chilled on ice, and then briefly heated the cell and DNA mixture, plasmid DNA entered bacterial cells. Once inside bacteria, plasmids replicate and express their genes. (Bacterial transformation is explained in greater detail in Chapter 5.) A more modern transformation technique called **electroporation**

Q Why do bacteria have plasmids?

A Bacteria possessed plasmids long before molecular biologists dreamed of DNA cloning. Different types of plasmids exist naturally. A primary function of plasmids is that they can provide bacteria with resistance to antibiotics. This is possible because many plasmids contain antibiotic-resistance genes encoding proteins that can inactivate antibiotics or prevent antibiotics from entering bacterial cells. As you will learn in this chapter, antibiotic-resistance properties of plasmids are conveniently exploited by molecular biologists during gene-cloning experiments. Other types of plasmids (Col-plasmids) can produce molecules that destroy other bacteria. Plasmids called F-plasmids contain genes that produce proteins and can form a tube called a pilus that allows for the transfer of plasmid DNA between bacterial cells in a process called *conjugation*.

involves applying a brief (millisecond) pulse of high-voltage electricity to create tiny holes in the bacterial cell wall that allow DNA to enter. Electroporation can also be used to introduce DNA into mammalian cells and to transform plant cells.

Ligation of DNA fragments and transformation by any method are somewhat inefficient. During ligation, some of the digested plasmid ligates back to itself to create a recircularized plasmid that lacks foreign DNA. During transformation, a majority of cells do not take up DNA.

Now that you have learned how DNA can be inserted into a vector and introduced into bacterial cells, we consider how recombinant bacteria—those transformed with a recombinant plasmid—can be distinguished from a large number of nontransformed bacteria and bacterial cells that contain plasmid DNA without foreign DNA. This screening process is called **selection** because it is designed to facilitate the identification of (selecting *for*) recombinant bacteria while preventing the growth of (selecting *against*) nontransformed bacteria and bacteria that contain plasmid without foreign DNA. Cohen and Boyer used **antibiotic selection,** a technique in which transformed bacterial cells are plated on agar plates with different antibiotics, as a way to identify recombinant bacteria and nontransformed cells. For many years antibiotic selection was a widely used approach; however, more recently, cloning techniques often incorporate other more popular selection strategies such as "blue-white" selection (the reason for this name will soon be obvious). In blue-white selection, DNA is cloned into a restriction site in the *lacZ*

gene, as illustrated in Figure 3.3. Recall from Chapter 2 that the *lacZ* gene encodes β-galactosidase (β-gal), an enzyme that degrades the disaccharide lactose into the monosaccharides glucose and galactose. When interrupted by an inserted gene, the *lacZ* gene is incapable of producing functional β-gal.

Transformed bacteria are plated on agar plates that contain an antibiotic—ampicillin, in this example. Nontransformed bacteria cannot grow in the presence of ampicillin because they lack plasmid DNA containing an ampicillin resistance gene (amp^R). The agar also contains a chromogenic (color-producing) substrate for β-gal called X-gal (5-bromo-4-chloro-3-indolyl-β-D-galactopyranoside). X-gal is similar to lactose in structure and turns blue when cleaved by β-gal. As a result, nonrecombinant bacteria—those that contain plasmid that ligated back to itself without insert DNA—contain a functional *lacZ* gene, produce β-gal, and turn blue. Conversely, recombinant bacteria are identified as white colonies. Because these cells contain plasmid with foreign DNA inserted into the *lacZ* gene, β-gal is not produced, and these cells cannot metabolize X-gal. Therefore through blue-white selection, nontransformed and nonrecombinant bacteria are *selected against* and white colonies are identified or *selected for* as the desired colonies containing recombinant DNA. Colonies containing recombinant plasmid are **clones**—genetically identical bacterial cells each containing copies of the recombinant plasmid.

Introduction to Human Gene Cloning

Restriction enzymes, DNA ligase, and plasmids are the major tools of molecular biologists for manipulating and cloning genes from virtually any source. With the discovery of transformation, scientists had a way to introduce recombinant DNA into bacterial cells. Recombinant DNA technology made it possible to cut and join together virtually any DNA fragments, insert DNA into a plasmid (DNA cloned into a plasmid is commonly called "insert" DNA), and produce large amounts of the insert DNA by allowing bacteria to be the workhorses for replicating recombinant DNA.

If the cloned DNA fragment is a gene that encodes a protein product, bacterial cells could be used to synthesize the protein product of the cloned gene. We call this "expressing" a protein. Molecular biologists recognized that if human genes could be cloned and expressed, recombinant DNA technology would become an invaluable tool with powerful and exciting applications in research and medicine. Because bacteria can be grown in large-scale preparations (these processes are described in more detail in Chapters 4 and 5), scientists can produce large amounts of the cloned DNA and

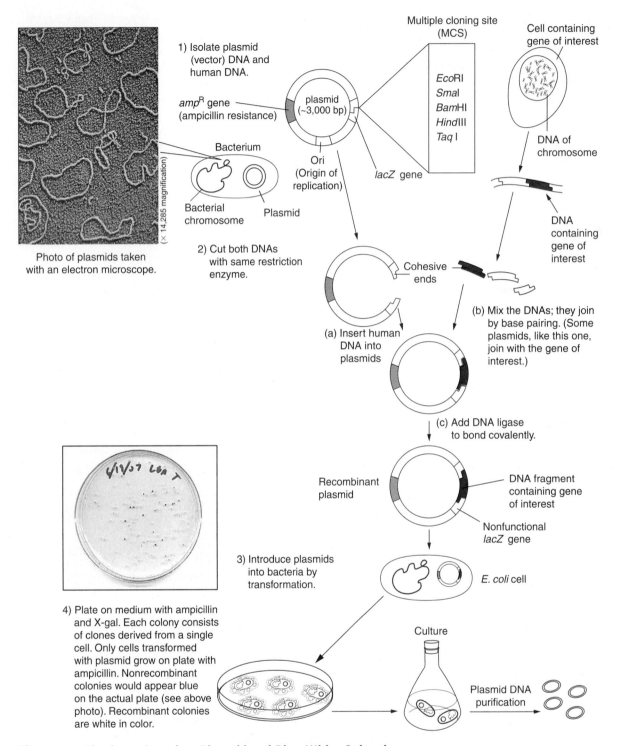

Photo of plasmids taken with an electron microscope.

(× 14,285 magnification)

1) Isolate plasmid (vector) DNA and human DNA.

amp^R gene (ampicillin resistance)

plasmid (~3,000 bp)

Multiple cloning site (MCS)

*Eco*RI
*Sma*I
*Bam*HI
*Hind*III
Taq I

Cell containing gene of interest

Bacterium

Ori (Origin of replication)

lacZ gene

Bacterial chromosome

Plasmid

DNA of chromosome

DNA containing gene of interest

2) Cut both DNAs with same restriction enzyme.

Cohesive ends

(a) Insert human DNA into plasmids

(b) Mix the DNAs; they join by base pairing. (Some plasmids, like this one, join with the gene of interest.)

(c) Add DNA ligase to bond covalently.

Recombinant plasmid

DNA fragment containing gene of interest

Nonfunctional *lacZ* gene

3) Introduce plasmids into bacteria by transformation.

E. coli cell

4) Plate on medium with ampicillin and X-gal. Each colony consists of clones derived from a single cell. Only cells transformed with plasmid grow on plate with ampicillin. Nonrecombinant colonies would appear blue on the actual plate (see above photo). Recombinant colonies are white in color.

Culture

Plasmid DNA purification

Figure 3.3 Cloning a Gene in a Plasmid and Blue-White Selection

isolate quantities of protein that would normally be very difficult or expensive to purify without cloning. As a result of recombinant DNA technology, a wide range of valuable proteins that are otherwise difficult to obtain can be produced from cloned genes.

Human genes can be cloned into plasmid DNA and replicated in bacteria using recombinant DNA technology. The first commercially available human gene product of recombinant DNA technology was human **insulin,** a peptide hormone produced by cells in the pancreas called beta cells. When blood glucose rises—for example, after eating a sugar-rich meal—insulin lowers blood glucose by stimulating glucose storage in liver and muscle cells as long chains of glucose called glycogen. Individuals with **type I,** or **insulin-dependent, diabetes mellitus** do not produce insulin on their own. As a result of this insulin deficiency, diabetics experience excessively high blood sugar levels (hyperglycemia), which, over time, can lead to serious damage to many bodily organs. In 1977, insulin was cloned into a bacterial plasmid, expressed in bacterial cells, and isolated by scientists at **Genentech,** the San Francisco, California, biotechnology company cofounded in 1976 by Herbert Boyer and Robert Swanson. Details of the techniques used for cloning insulin are examined in Chapter 5. Genentech, short for genetic engineering technology, is generally regarded as the first biotechnology company.

In 1982, the recombinant form of human insulin, called **Humulin,** became the first recombinant DNA product to be approved for human applications by the U.S. Food and Drug Administration. Shortly after insulin became available, growth hormone—used to treat children who suffer from a form of dwarfism—was cloned, and because of recombinant DNA technology, a wide variety of other medically important proteins that were once difficult to obtain in adequate amounts became readily available.

Prior to recombinant DNA technology, important hormones like insulin and growth hormone had to be isolated from tissues. Growth hormone was isolated from the pituitary glands of human cadavers. Not only was this process expensive and inefficient, but these isolations also carried with them the risk of unknowingly co-purifying viruses and other pathogens as contaminants that could be passed to people receiving the hormone. There are now several hundred products of recombinant DNA technology on the market with widespread applications in basic research, medicine, and agriculture.

With a basic understanding of the techniques involved in manipulating a piece of DNA, in the next section we go on to examine some important aspects of DNA vectors and how different vectors are chosen and used depending on what is to be accomplished.

3.2 What Makes a Good Vector?

The number of different DNA vectors, vector functions, and applications has increased substantially since Stanley Cohen constructed pSC101. Plasmids are still the most commonly used cloning vectors. Plasmids are popular because they allow for the routine cloning and manipulation of small pieces of DNA that form the foundation for many techniques used daily in a molecular biology laboratory. In addition, it is fairly simple to transform bacterial cells with plasmid DNA and relatively easy to isolate plasmid DNA from bacterial cells.

One of the first widely used plasmid DNA vectors, called pBR322, was designed to have genes for ampicillin and tetracycline resistance and several useful restriction sites. However, plasmid DNA cloning vectors have been engineered over the years to incorporate a number of other important features that have made pBR322 almost obsolete.

Practical Features of DNA Cloning Vectors

Modern plasmid DNA cloning vectors usually include most of the following desirable and practical features.

- *Size*—They should be small enough to be easily separated from the chromosomal DNA of the host bacteria.

- *Origin of replication* (ori)—The site for DNA replication that allows plasmids to replicate separately from the host cell's chromosome. The number of plasmids in a cell is called **copy number.** The normal copy number of plasmids in most bacterial cells is small (usually less than 12 plasmids per cell); however, many of the most desirable cloning plasmids are known as high-copy-number plasmids because they replicate to create hundreds or thousands of plasmid copies per cell.

- *Multiple cloning site* (MCS)—The MCS, also called a polylinker, is a stretch of DNA with recognition sequences for many different common restriction enzymes (see Figure 3.3a). These sites are engineered into the plasmid so that digestion of the plasmid with restriction enzymes does not result in the loss of a fragment of DNA. Rather, the circular plasmid simply becomes linearized when digested with a restriction enzyme. An MCS provides for great flexibility in the range of DNA fragments that can be cloned into a plasmid because it is possible to insert DNA fragments generated by cutting with many different enzymes.

- *Selectable marker genes*—These genes allow for the selection and identification of bacteria that have been transformed with a recombinant plasmid compared to nontransformed cells. Some of the most common selectable markers are genes for ampicillin resistance (amp^R) and tetracycline resistance (tet^R) and the *lacZ* gene used for blue-white selection.

- *RNA polymerase promoter sequences*—These sequences are used for transcription of RNA *in vivo* and *in vitro* by RNA polymerase. Recall from Chapter 2 that RNA polymerase copies DNA into RNA during transcription. *In vivo*, these sequences allow bacterial cells to make RNA from cloned genes, which in turn leads to protein synthesis. *In vitro* transcribed RNA can be used to synthesize

RNA "probes" that can be used to study gene expression, as described in Section 3.4.

- *DNA sequencing primer sequences*—These sequences permit nucleotide sequencing of cloned DNA fragments (as described in Section 3.4) that have been inserted into the plasmid.

Types of Vectors

Just as one screwdriver cannot be used for all sizes and types of screws, bacterial plasmid vectors cannot be used for all applications in biotechnology. There are limitations to how plasmids can be used in cloning. One primary limitation is the size of the DNA fragment that can be inserted into a plasmid. Insert size usually cannot exceed approximately 6 to 7 kilobases (1 kb = 1,000 bp). In addition, sometimes bacteria express proteins from eukaryotic genes poorly. As a result of these limitations, molecular biologists have worked to develop many other types of DNA vectors, each of which has particular benefits depending on the cloning application. Table 3.2 compares important features, sources, and applications of different types of cloning vectors.

Bacteriophage vectors

DNA from bacteriophage lambda (λ) was one of the first phage vectors used for cloning. The λ chromosome is a linear structure approximately 49 kb in size (Figure 3.4). Cloned DNA is inserted into restriction sites in the center of the λ chromosome. Recombinant chromosomes are then packaged into viral particles *in vitro*, and these phages are used to infect *E. coli* growing as a lawn (a continuous layer covering the plate). At each end of the λ chromosome are 12 nucleotide sequences called cohesive sites (COS) that can base pair with each other. When λ infects *E. coli* as a host, the λ chromosome uses these COS sites to circularize and then replicate. Bacteriophage λ replicates through a process known as a **lytic cycle.** As λ replicates to create more viral particles, infected *E. coli* are lysed (*lysed* or *lysis* means to split or rupture) by λ, creating zones of dead bacteria called **plaques** that appear as cleared spots on the bacterial lawn. Each plaque contains millions of recombinant phage particles. In Section 3.3, we discuss how plaques are screened to identify recombinant DNA. A primary advantage of these vectors is that they allow for the cloning of larger DNA fragments (up to approximately 25 kb) than

Table 3.2   A COMPARISON OF DNA VECTORS AND THEIR APPLICATIONS

Vector Type	Maximum Insert Size (kb)	Applications	Limitations
Bacterial plasmid vectors (circular)	~6–12 DNA	DNA cloning, protein expression, subcloning, direct sequencing of insert DNA	Restricted insert size; limited expression of proteins; copy number problems; replication restricted to bacteria
Bacteriophage vectors (linear)	~25	cDNA, genomic and expression libraries	Packaging limits DNA insert size; host replication problems
Cosmid (circular)	~35	cDNA and genomic libraries, cloning large DNA fragments	Phage packaging restrictions; not ideal for protein expression; cannot be replicated in mammalian cells
Bacterial artificial chromosome (BAC, circular)	~300	Genomic libraries, cloning large DNA fragments	Replication restricted to bacteria; cannot be used for protein expression
Yeast artificial chromosome (YAC, circular)	200–2,000	Genomic libraries, cloning large DNA fragments	Must be grown in yeast; cannot be used in bacteria
Ti vector (circular)	Varies depending on type of Ti vector used	Gene transfer in plants	Limited to use in plant cells only; number of restriction sites randomly distributed; large size of vector not easily manipulated

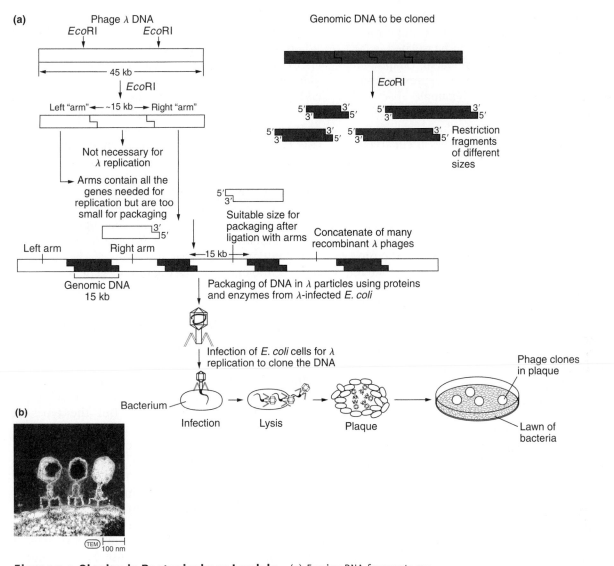

(a)

Phage λ DNA

*Eco*RI   *Eco*RI

Genomic DNA to be cloned

45 kb

*Eco*RI

Left "arm" ← ~15 kb → Right "arm"

*Eco*RI

5′ 3′   5′ 3′   Restriction
3′ 5′   3′ 5′   fragments
of different
5′ 3′   5′ 3′   sizes
3′ 5′   3′ 5′

Not necessary for
λ replication

Arms contain all the
genes needed for
replication but are too
small for packaging

5′
3′

Left arm   Right arm

3′
5′

Suitable size for
packaging after
ligation with arms   Concatenate of many
recombinant λ phages

←15 kb→

Genomic DNA
15 kb

Packaging of DNA in λ particles using proteins
and enzymes from λ-infected *E. coli*

Infection of *E. coli* cells for λ
replication to clone the DNA

Phage clones
in plaque

(b)

Bacterium

Infection   Lysis   Plaque   Lawn of
bacteria

TEM 100 nm

Figure 3.4 Cloning in Bacteriophage Lambda (a) Foreign DNA fragments are
cloned into the λ chromosome and then "packaged" into virus particles that are used to
infect *E. coli* plated as a lawn. Lysis of bacterial cells by virus creates plaques in the lawn of
E. coli. (b) Transmission electron micrograph of phage λ attached to the surface of *E. coli*.

plasmids. Many bacteriophage vectors can also be used
as protein expression vectors.

Cosmid vectors

Cosmid vectors contain COS ends of λ DNA, a plasmid origin of replication, and genes for antibiotic resistance, but most of the viral genes have been removed.
DNA is cloned into a restriction site, and the cosmid is
packaged into viral particles, as is done with bacteriophage vectors that are used to infect *E. coli* wherein
cosmids replicate as a low copy number plasmid.

Bacterial colonies are formed on a plate, and recombi-
nants can be screened by antibiotic selection. A primary advantage of cosmids is that they allow for the
cloning of DNA fragments in the 20- to 45-kb range.

Expression vectors

Protein **expression vectors** allow for the high-level
synthesis (expression) of eukaryotic proteins within
bacterial cells because they contain a prokaryotic
promoter sequence adjacent to the site where DNA is
inserted into the plasmid. Bacterial RNA polymerase

can bind to the promoter and synthesize large amounts of RNA (for the insert), which is then translated into protein. Protein may then be isolated using the biochemical techniques described in Chapter 4. However, it is not always possible to express a functional protein in bacteria. For example, bacterial ribosomes sometimes cannot translate eukaryotic mRNA sequences. If a protein is produced, it may not fold and be processed correctly, as occurs in eukaryotic cells that use organelles to fold and modify proteins. Also, making some recombinant products in bacteria can be a problem because *E. coli* often does not secrete proteins so expression vectors are often used in *Bacillus subtilis*, a strain more suitable for protein secretion.

In some cases, the host bacteria can recognize recombinant proteins as foreign and degrade the protein, whereas in others the expressed protein is lethal to the host bacterial cells. Certain viruses such as SV40 can be used to deliver expression vectors into mammalian cells. Typically SV40-derived vectors contain a strong (viral) promoter sequence for high-level transcription and a poly(A) addition signal for adding a poly(A) tail to the 3′ end of synthesized mRNAs. Variations of such vectors have been used for human gene therapy as described in Chapter 11.

Bacterial artificial chromosomes

Bacterial artificial chromosomes (BACs) are large low-copy-number plasmids, present as one to two copies in bacterial cells that contain genes encoding the F-factor (a unit of genes controlling bacterial replication). BACs can accept DNA inserts in the 100- to 300-kb range. It is still somewhat unclear why BACs can accept and replicate large pieces of DNA. BACs were widely used in the Human Genome Project to clone and sequence large pieces of human chromosomes.

Yeast artificial chromosomes

Yeast artificial chromosomes (YACs) are small plasmids grown in *E. coli* and introduced into yeast cells (such as *Saccharomyces cerevisiae*). A YAC is a miniature version of a eukaryotic chromosome. YACs contain an origin of replication, selectable markers, two telomeres, and a centromere that allows for replication of the YAC and segregation into daughter cells during cell division. Foreign DNA fragments are cloned into a restriction site in the center of the YAC. YACs are particularly useful for cloning large fragments of DNA from 200 kb to approximately 2 megabases (mb = 1 million bases) in size. Similar to BACs, YACs have also played an important role in the cloning efforts of the Human Genome Project.

Ti vectors

Ti vectors are naturally occurring plasmids (around 200 kb in size) isolated from the bacterium *Agrobacterium tumefaciens*, which is a soil-borne plant pathogen that causes a condition in plants called crown gall disease. When *A. tumefaciens* enters host plants, a piece of DNA (T-DNA) from the Ti plasmid (Ti stands for tumor-inducing) inserts into the host chromosome. T-DNA encodes for the synthesis of a hormone called auxin, which weakens the host cell wall. Infected plant cells divide and enlarge to form a tumor (gall). Plant geneticists recognized that if they could remove auxin and other detrimental genes from the Ti plasmid, the resulting vector could be used to deliver genes into plant cells. Ti vectors are widely used to transfer genes into plants, as you will learn in Chapter 6.

Now that we have examined different types of vectors and their applications, in the next section we turn our attention to how scientists can use recombinant DNA technology to identify and clone genes of interest.

3.3 How Do You Identify and Clone a Gene of Interest?

Cutting and pasting different pieces of DNA to produce a recombinant DNA molecule has become a routine technique in molecular biology. But the types of cloning experiments we have described so far allow for the *random* cloning of DNA fragments based on restriction enzyme cutting sites and not precise cloning of a single gene or particular piece of DNA of interest. For example, if you were interested in cloning the insulin gene and you simply took DNA from the pancreas, cut it with enzymes, and then ligated digested DNA into plasmids, you would create hundreds of thousands of recombinant plasmids and not just a recombinant plasmid with the insulin gene. Molecular biologists call this approach "shotgun" cloning because many fragments are randomly cloned at once, and no individual gene is specifically targeted for cloning. How would you know which recombinant plasmid contained the insulin gene? Moreover, if the insulin gene (or adjacent sequences) does not have recognition sites for the restriction enzyme you used, you may not have any recombinant plasmids containing the insulin gene. Even if you did create plasmids with the insulin gene, how would you separate these from the other recombinant plasmids? So how do you find a particular gene of interest and clone only the DNA sequence that you want to study? These questions can often be answered by a cloning approach known as DNA libraries.

Creating DNA Libraries: Building a Collection of Cloned Genes

Many cloning strategies begin by preparing a **DNA library.** Libraries are collections of cloned DNA fragments from a particular organism contained within bacteria or viruses as the host. Libraries can be saved for relatively long periods of time and "screened" to pick out different genes of interest. Two types of libraries are typically used for cloning, **genomic DNA libraries** and **complementary DNA libraries (cDNA libraries).** Figure 3.5 shows how genomic libraries and cDNA libraries are constructed.

Genomic versus cDNA libraries

In a genomic library, chromosomal DNA from the tissue of interest is isolated and then digested with a restriction enzyme (see Figure 3.5a). This process produces fragments of DNA that include the organism's entire genome. A plasmid, BAC, YAC, or bacteriophage vector is digested with the same enzyme, and DNA ligase is used to ligate genomic DNA pieces and vector DNA randomly. In theory, all DNA fragments in the genome will be cloned into a vector. Recombinant vectors are then used to transform bacteria, and each bacterial cell clone will contain a plasmid with a genomic DNA fragment. Consider each clone a "book" in this "library" of DNA fragments. One disadvantage of creating this type of library for eukaryotic genes is that nonprotein coding pieces of DNA called introns are cloned in addition to protein coding sequences (exons). Because a majority of DNA in any eukaryotic organism consists of introns, many of the clones in a genomic library will contain nonprotein coding pieces of DNA. Another limitation of genomic libraries is that many organisms, including humans, have such a large genome that searching for a gene of interest is like searching for a needle in a haystack.

In a cDNA library, mRNA from the tissue of interest is isolated and used for making the library. However, mRNA cannot be cut directly with restriction enzymes so it must be converted to a double-stranded DNA molecule. A viral enzyme called **reverse transcriptase (RT)** is used to catalyze the synthesis of single-stranded DNA from the mRNA (see Figure 3.5b). This enzyme is made by a class of viruses called **retroviruses**—so named because they are exceptions to the usual flow of genetic information. Instead of having a DNA genome that is used to make RNA, retroviruses have an RNA genome. After infecting host cells, they use RT to convert RNA into DNA so that they can replicate. Human immunodeficiency virus (HIV), the causative agent of acquired immunodeficiency syndrome (AIDS), is a retrovirus. Other retroviruses have important applications in biotechnology as gene therapy vectors. This is discussed in Chapter 11. Because this DNA has been synthesized as an exact copy of mRNA, it is called complementary DNA (cDNA). The mRNA is degraded by treatment with an alkaline solution or enzymatically digested; then DNA polymerase is used to synthesize a second strand to create double-stranded cDNA.

Because cDNA sequences do not necessarily have convenient restriction enzyme cutting sites at each end, short double-stranded DNA sequences called *linker sequences* are often enzymatically added to the ends of the cDNA. Linkers contain a restriction enzyme recognition site. Different linkers for different restriction enzyme sites are commercially available. By adding linkers, cDNA can now be ligated into a convenient restriction site in a vector of choice, often a plasmid. Recombinant plasmid is then used to transform bacteria.

A primary advantage of cDNA libraries over genomic libraries is that a cDNA library is a collection of actively *expressed* genes in the cells or tissue from which the mRNA was isolated. Introns are not cloned in a cDNA library, which greatly reduces the total amount of DNA that is cloned compared with a genomic library. For this reason, cDNA libraries are typically preferred over genomic libraries when attempting to clone a gene of interest. Another advantage of cDNA libraries is that they can be created and screened to isolate genes that are primarily expressed only under certain conditions in a tissue. For example, if a gene is only expressed in a tissue stimulated by a hormone, researchers make libraries from hormone-stimulated cells to increase the likelihood of cloning hormone-sensitive genes. Libraries have become such a routine aspect of molecular biology that many companies sell libraries prepared from a range of tissues from different species. One disadvantage is that cDNA libraries can be difficult to create and screen if a source tissue with an abundant amount of mRNA for the gene is not available. But as you will learn, a technique called the *polymerase chain reaction (PCR)* can frequently solve this problem.

Library screening

Once either a genomic library or a cDNA library is created, it must be *screened* to identify the genes of interest. One of the most common library screening techniques is called **colony hybridization** (see Figure 3.6). In colony hybridization, bacterial colonies from the library containing recombinant DNA are grown on an agar plate. A nylon or nitrocellulose filter is placed over the plate, and some of the bacterial cells attach to the filter at the same location where they are found on the plate. If bacteriophage vectors are used, phages are transferred onto the filter. The filter is treated with an alkaline solution to lyse bacteria and denature their DNA. The denatured DNA binds to the filter as single-stranded molecules. Typically, the filter is then incubated with a DNA **probe,** a single-stranded radioactive or nonradioactive DNA fragment that is complementary to the gene of interest because it can base pair by hydrogen bonding to the target DNA to be cloned. The probe binds to complementary sequences on the filter. This process of

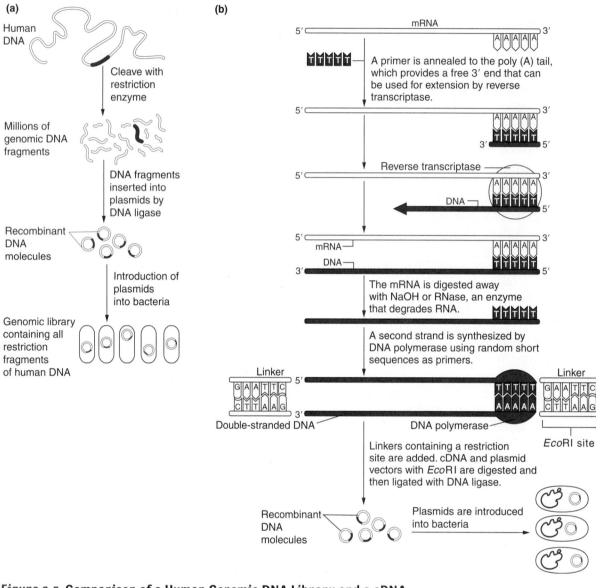

Figure 3.5 Comparison of a Human Genomic DNA Library and a cDNA Library   (a) Human DNA is cleaved with a restriction enzyme to create a series of smaller fragments that are cloned into plasmids. A (human) genomic library consists of a collection of bacteria each containing a different fragment of human DNA. In theory, all DNA fragments from the genome will be represented in the library. (b) In a cDNA library, mRNA is converted into cDNA by the enzyme reverse transcriptase. Linkers containing a restriction site are added to the cDNA to create cohesive ends. The cDNA can now be cloned into a plasmid vector for subsequent replication in bacteria.

probe binding is called **hybridization.** The filter is then washed to remove excess unbound probe and exposed to photographic film in a process called **autoradiography.** Anywhere the probe has bound to the filter, radioactivity from radioactive probes or released light (fluorescence or chemiluminescence) from nonradioactive probes exposes silver grains in the film. Depending on the abundance of the gene of

interest, there may only be a few colonies (or plaques) on the filter that hybridize to the probe. Film is developed to create a permanent record called an autoradiogram (or autoradiograph), which is then compared to the original plate of bacterial colonies to identify which colonies contained recombinant plasmid with the gene of interest. These colonies can now be grown in larger scale to isolate the cloned DNA.

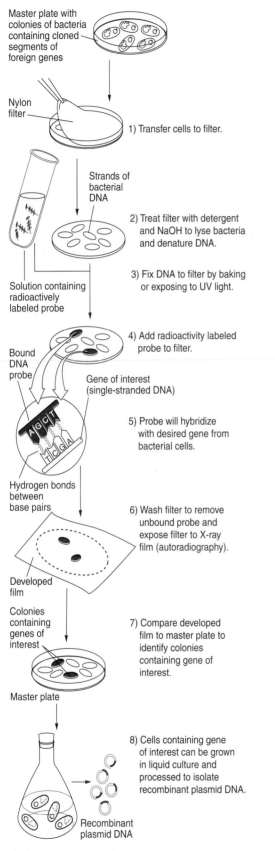

Master plate with colonies of bacteria containing cloned segments of foreign genes

Nylon filter

1) Transfer cells to filter.

Strands of bacterial DNA

2) Treat filter with detergent and NaOH to lyse bacteria and denature DNA.

3) Fix DNA to filter by baking or exposing to UV light.

Solution containing radioactively labeled probe

4) Add radioactivity labeled probe to filter.

Bound DNA probe

Gene of interest (single-stranded DNA)

5) Probe will hybridize with desired gene from bacterial cells.

Hydrogen bonds between base pairs

6) Wash filter to remove unbound probe and expose filter to X-ray film (autoradiography).

Developed film

Colonies containing genes of interest

7) Compare developed film to master plate to identify colonies containing gene of interest.

Master plate

8) Cells containing gene of interest can be grown in liquid culture and processed to isolate recombinant plasmid DNA.

Recombinant plasmid DNA

Figure 3.6   Colony Hybridization: Library Screening with a DNA Probe to Identify a Cloned Gene of Interest

The type of probe used for library screening often depends on what is already known about the gene of interest. For example, the screening probe is frequently a gene cloned from another species. A cDNA clone of a gene from rats or mice is often a very effective probe for screening a human genomic or cDNA library because many gene sequences in rats and mice are similar to those found in human genes. If the gene of interest has not been cloned in another species but some information is available about the protein sequence, a series of chemically synthesized **oligonucleotides** can be made based on a prediction of codons that can code for the known protein sequence. If some partial amino acid sequence is known for a protein encoded by a gene to be cloned, it is possible to "work backward" and design oligonucleotides based on the predicted nucleotides that coded for the amino acid sequence. In addition, if an antibody is available for the protein encoded by the gene of interest, an expression library, which results in protein expression in bacteria, can be used, and the library can be screened with the antibody to detect colonies expressing the recombinant protein.

Rarely does library screening result in the isolation of clones that contain full-length genes. It is more common to obtain clones with small pieces of the gene of interest (one reason why this occurs with cDNA libraries is because it may be difficult to isolate full-length mRNA or synthesize full-length cDNA for the gene of interest). When small pieces of a gene are cloned, scientists sequence these pieces and look for sequence overlaps. Overlapping fragments of DNA can then be pieced together like a puzzle in an attempt to reconstruct the full-length gene. This often requires screening the library several times with large numbers of bacteria being plated and used for colony hybridization. Looking for start and stop codons in the sequenced pieces is one way to predict if the entire gene has been pieced together. Through this process overlapping fragments can be pieced together to assemble an entire gene.

Later in this chapter we discuss how whole genome shotgun sequencing strategies can enable scientists to sequence entire genomes. Because of genomic studies, libraries are becoming a less common way to identify and clone genes. Instead of using a library to identify one or a few genes at a time, genomics enables scientists to identify sequences for all genes in a genome.

Polymerase Chain Reaction

Although libraries are very effective and commonly used for cloning and identifying a gene of interest, the **polymerase chain reaction (PCR)** is a much more rapid approach to cloning than building and screening a library. PCR is often the technique of choice. Developed in the mid-1980s by Kary Mullis, PCR turned out to be a revolutionary technique that has had an impact on many areas of molecular biology. In 1993, Mullis won a Nobel Prize in Chemistry for his invention. PCR is a technique for making copies or amplifying a *specific sequence* of DNA in a short period of time. The concept behind a PCR reaction is remarkably simple. Here, target DNA to be amplified is added to a thin-walled tube and mixed with deoxyribonucleotides (dATP, dCTP, dGTP, dTTP), buffer, and DNA polymerase. A paired set of **primers** is added to the mixture. Primers are short single-stranded DNA oligonucleotides usually around 20 to 30 nucleotides long. These primers are complementary to nucleotides flanking opposite ends of the target DNA to be amplified (see Figure 3.7).

The reaction tube is then placed in an instrument called a thermal cycler. In the simplest sense, a thermal cycler is a sophisticated heating block that is capable of rapidly changing temperature over very short time intervals. The thermal cycler takes the sample through a series of reactions called a PCR cycle (Figure 3.7). Each cycle consists of three stages. In the first stage, called *denaturation*, the reaction tube is heated to approximately 94°C to 96°C, causing separation of the target DNA into single strands. In the second stage, called *hybridization* (or *annealing*), the tube is then cooled slightly to around 50°C to 65°C, which allows the primers to hydrogen bond to complementary bases at opposite ends of the target sequence. During extension (or elongation), the last stage of a PCR cycle, the temperature is usually raised slightly (to about 70°C to 75°C) and DNA polymerase copies the target DNA by binding to the 3' ends of each primer and using the primers as templates. DNA polymerase adds nucleotides to the 3' end of each primer to synthesize a complementary strand. At the end of one cycle, the amount of target DNA has been doubled. The thermal cycler repeats these three stages again according to the total number of cycles determined by the researcher, usually 20 or 30 cycles.

One key to PCR is the type of DNA polymerase used in the reaction. Repeated heating and cooling required for PCR would denature and destroy most DNA polymerases after just a few cycles. Several sources of PCR-suitable DNA polymerases are available. One of the first and most popular enzymes for PCR is known as **_Taq_ DNA polymerase**. *Taq* is isolated from the *archaea* called *Thermus aquaticus*, a species that thrives in hot springs. Because *T. aquaticus* is adapted to live in hot water (it was first discovered

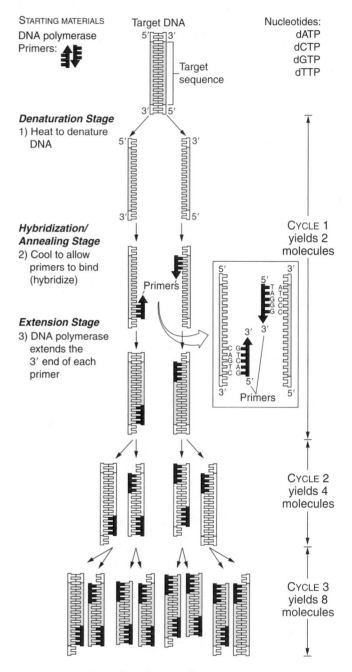

Figure 3.7 The Polymerase Chain Reaction

in the hot springs of Yellowstone National Park), it has evolved a DNA polymerase that can withstand high temperatures (such microbes are called *thermophiles* because of their ability to survive and thrive in extremely hot environments). Because *Taq* is stable at high temperatures, it can withstand the temperature changes necessary for PCR without being denatured. In 1989, the journal *Science* named *Taq* polymerase Molecule of the Year.

Q How do scientists determine what primer sequences and temperature conditions should be used for a PCR experiment?

A Designing primers and choosing the correct temperatures are critically important parameters. Software programs used for primer design make this process a lot easier, but even with such programs many aspects must be considered such as:

- Primers must bind only to specific sequences in the target DNA sequence of interest to avoid mispriming—primer binding to nontarget sequences.

- Complementary sequences for primer binding to the target DNA must be neither too far apart nor too close together.

- Primers should contain an approximately equal number of the four bases.

- Avoiding primers that can bind to each other forming "primer dimers" by making sure that primer pairs do not have guanine and cytosine nucleotides at their 3' ends.

Temperatures chosen for a PCR experiment are determined by the primer sequences and the requirements of the DNA polymerase being used for the experiment. The denaturation temperature is almost always around 94°C to 95°C for most experiments, but selecting the correct hybridization temperature is critical. If this temperature is too high, primers will not be able to bind to the target DNA. If the temperature is too low, primers may bind at nonspecific segments of DNA causing amplification of nontarget sequences. The hybridization temperature is determined largely by the A + T and G + C composition of the primers. Primers with a high content of G + C base pairs can hybridize at higher temperatures than primers with a high A + T content. Ideal temperatures for hybridization are calculated based on the G + C and A + T percentage in the primers.

The greatest advantage of PCR is its ability to amplify millions of copies of target DNA from a very small amount of starting material in a short period of time. Because the target DNA is doubled after every round of PCR, after 20 cycles of PCR approximately 1 million copies (2^{20}) of target DNA are produced from a reaction starting with one molecule of target DNA. New applications in PCR technology make it possible to determine the amount of PCR product made during an experiment through a technique called **quan-**

titative real-time PCR (qPCR), which uses primers made with fluorescent dyes and specialized thermal cyclers that enable researchers to quantify amplification reactions as they occur. We discuss qPCR later in this chapter. An excellent tutorial on PCR can be viewed at the Cold Spring Harbor DNA Learning Center website listed at the Companion Website.

PCR has widespread applications in research and medicine, such as making DNA probes, studying gene expression, amplifying minute amounts of DNA to detect viral pathogens and bacterial infections, amplifying DNA to diagnose genetic conditions, detecting trace amounts of DNA from tissues at a crime scene, and even amplifying ancient DNA from fossilized dinosaur tissue (Figure 3.8). Many of these applications are described in other chapters.

Cloning PCR products

PCR is often used instead of library screening approaches for cloning a gene because it is rapid and effective (Figure 3.9). A disadvantage of PCR cloning is that you need to know something about the DNA sequences that flank your gene of interest to design primers. Cloning by PCR is easiest if the gene has already been cloned in another species—for instance, using primers for a gene cloned previously from mice to clone the equivalent gene from humans.

There are many ways to clone a gene using PCR. One of the earliest approaches involved designing primers for a gene of interest that included restriction enzyme recognition sequences built into the primers. In this technique, the gene is amplified and the restriction sites are engineered into the primers to digest the PCR products with a restriction enzyme. These products are ligated into a vector that can be used for DNA sequencing. A more modern approach to PCR cloning takes advantage of an interesting quirk of thermostabile polymerases. As DNA is copied, *Taq* and other polymerases used for PCR normally add a single adenine nucleotide to the 3' end of all PCR products (see Figure 3.9). After amplifying a target gene, cloned PCR products can be ligated into plasmids called T vectors. T vectors contain a single-stranded thymine nucleotide at each end that can complementarily base pair with overhanging adenine nucleotides in PCR products. Once ligated into a T vector, the recombinant plasmid containing the cloned PCR product can be introduced into bacteria by transformation, and its nucleotide sequence can be determined.

Now that you have learned some of the most common strategies used to clone genes, in the next section we consider a wide range of different approaches that scientists use to study cloned genes.

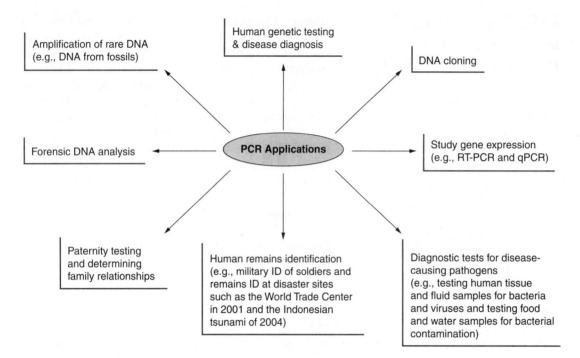

Figure 3.8 PCR Applications   Amplifying DNA by PCR has become an essential technique in molecular biology that has a wide range of different applications. Some of the more common biotechnology-related applications are represented in this figure.

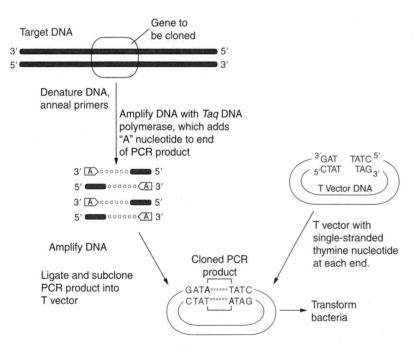

Figure 3.9 Cloning a Gene by PCR

3.4 What Can You Do with a Cloned Gene? Applications of Recombinant DNA Technology

Why clone DNA? What can you do with a cloned gene? There are numerous applications of gene cloning and recombinant DNA technology. Figure 3.10 summarizes common gene-cloning applications, many of which are discussed further in other chapters. In this section, we present some important introductory applications of gene cloning.

Gel Electrophoresis and Mapping Gene Structure with Restriction Enzymes

Typically, soon after a gene is cloned a type of physical map of the gene is created to determine which restriction enzymes cut the cloned gene and to pinpoint the location of these cutting sites. Knowing the **restriction map** of a gene is very useful for making clones of small pieces of

the gene (this is called *subcloning*) and for manipulating many relatively small pieces of DNA (for example, 100 to 1,000 bp) to sequence DNA and to prepare DNA probes to study gene expression.

To create a restriction map, cloned DNA is subjected to a series of single-digests with restriction enzymes as well as double-digests with combined enzymes. Researchers then use **agarose gel electrophoresis** (see Figure 3.11a), a common technique in molecular biology, to separate and visualize DNA fragments based on size. Finally the pattern of fragments created by the digests is analyzed to make the "map." Agarose is a material that is isolated from seaweed, melted in a buffer solution, and poured into a plastic tray. As the agarose cools, it solidifies to form a horizontal semisolid gel containing small holes or pores through which DNA fragments will travel. The percentage of agarose used to create the gel determines its ability to resolve DNA fragments of different sizes. Most applications generally involve gels that contain 0.5 to 2% agarose. A gel with a high percentage of agarose (say, 2%) is better suited for

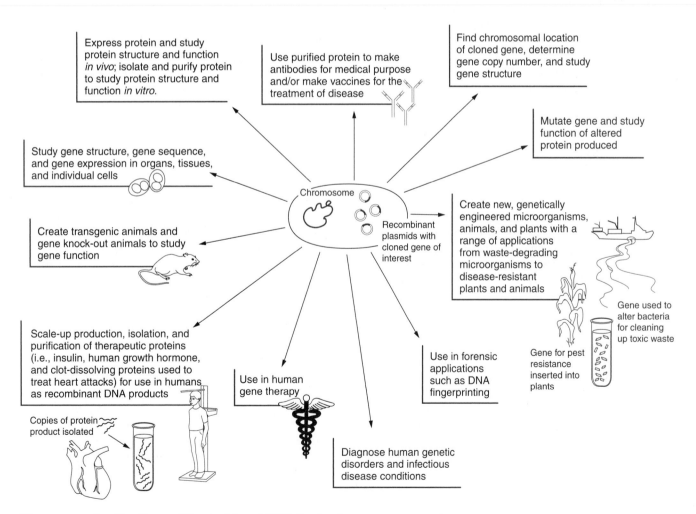

Figure 3.10 Applications of Recombinant DNA Technology

YOU DECIDE

To Patent or Not?

The Human Genome Project was completed ahead of schedule in part because of competition between publicly funded genome centers and privately funded companies such as Celera Genomics, originally directed by former NIH researcher Craig Venter. While at The Institute for Genomic Research (TIGR), Venter and colleagues were the first scientists to completely sequence the genome for a living organism, the bacterium *Haemophilus influenzae*. This group applied for patents on the nucleotide sequence for *H. influenzae* and on the bioinformatics technology used to analyze this genome.

Previously, Venter and his colleagues described a set of experiments in which they randomly cloned short pieces of cDNAs from human brain cells. These short sequences—called **expressed sequence tags (ESTs)**—could in theory be used as probes to identify full-length cDNAs. Some of Venter's ESTs were found to be identical to already-cloned genes, or a portion of a gene; others appeared to be novel gene sequences or junk DNA. Hoping to gain proprietary rights to full-length genes that could be identified from Venter's ESTs, TIGR applied for a patent. This request generated a great deal of controversy.

Should scientists be allowed to patent DNA sequences from naturally living organisms? What if a patent is awarded for only small *pieces* of a gene—even if no one knows what a DNA sequence does—just because some individuals or a company wants a patent to claim their stake at having cloned a piece of DNA first? What if there are no clear uses for the DNA sequences cloned? Can or should investigators who use a gene microarray or create a DNA library be allowed to patent the entire genome of any organism they have studied?

When granted a patent, scientists essentially have a monopoly on patented information for two decades from the patent filing date. Many believe that hoarding genome information is against the tradition of sharing information to advance science. Would awarding patents slow progress to clone genes if groups hoarded data and did not share information? Could or should a group stake a claim to a gene, thereby preventing others from working on it or developing a product from it?

Since 1980, the U.S. Patent and Trademark Office has granted patents on more than 20,000 genes or gene sequences and an estimated 20% of human genes have been patented. Some scientists are concerned that patents awarded for simply cloning a piece of DNA is awarding a patent for too little work. Given that computers do most of the routine work of genome sequencing, who should get the patent? What about individuals who figure out *what* to do with the gene? Should a genetically engineered living organism be patented? Engineered bacteria (for example, those used to clean up environmental pollution) and transgenic animals have been patented, as have clinically important genes such as beta-interferon. Can a group claim rights to anticipated future uses of the gene, even if there are no data to substantiate such claims? What if a gene sequence is involved in a disease for which a genetic therapy may be developed? What is the best way to use this information to advance medicine and cure disease?

Many scientists believe it is more appropriate to patent the novel technology used to discover and study genes and the applications of genetic technology such as gene therapy approaches than to patent the gene sequences themselves. Such technology patents have been awarded, although the guidelines for what constitutes novel technology are unclear at best.

From a commercial standpoint, one advantage of patenting is that it provides private companies with the incentive to get a medicine or technology to the marketplace. At the same time, this can slow progress on a cure by making the treatment expensive. To patent or not? You decide.

---

separating small DNA fragments because they will snake their way through the pores more easily than large fragments, which do not separate through the dense gel very well. A lower percentage of agarose is better suited for resolving large DNA fragments.

To run a gel, it is submerged in a buffer solution that will conduct electricity. DNA samples are loaded into small depressions in the gel called *wells*, and then an electric current is applied through electrodes at opposite ends of the gel. Separating DNA by electrophoresis is based on the fact that DNA migrates through a gel according to its charge and size (in base pairs). The sugar-phosphate backbone renders DNA negatively charged; therefore, when DNA is placed in an electrical field, it migrates toward the anode (positive pole) and is repelled by the cathode (negative pole). Because all DNA is negatively charged, regardless of the length or source, the rate of DNA migration and separation through an agarose gel depends on the *size* of a DNA molecule. Because migration distance is inversely proportional to the size of a DNA fragment, large DNA fragments migrate short distances through a gel and small fragments migrate faster through the gel.

Tracking dyes are added to monitor DNA migration during electrophoresis. After the desired time of electrophoresis, DNA in the gel is stained using dyes such as **ethidium bromide** that intercalate (penetrate) in between the base pairs of DNA. These dyes

(a)

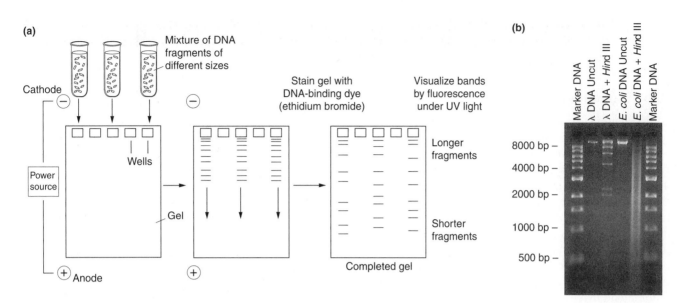

(b)

Figure 3.11 Agarose Gel Electrophoresis (a) DNA fragments can be separated and visualized by agarose gel electrophoresis. (b) Photograph of an agarose gel stained with ethidium bromide. Lanes labeled as "Marker DNA" were loaded with commercially prepared DNA size standards. These serve as a ladder of fragments of known size that are used to determine the size of experimental samples of DNA being analyzed. The lane labeled "λ DNA uncut" shows high molecular weight uncut chromosomal DNA from phage λ; "λ DNA + HindIII" shows a series of discrete fragments created when λ DNA is digested with the restriction enzyme HindIII. The lane labeled "E. coli DNA uncut" contains undigested chromosomal DNA and the adjacent lane shows E. coli chromosomal DNA digested with HindIII (E. coli DNA + HindIII). Notice how the HindIII-digested E. coli DNA produces a smear of bands unlike the set of discrete fragments visualized with HindIII-digested λ DNA. This smearing is due to the large size of the E. coli chromosome and the large number of cutting sites for HindIII; so many fragments are created that it is not possible to visualize discrete bands.

fluoresce when exposed to ultraviolet light. A permanent record of the gel is obtained by photographing the gel while it is exposed to ultraviolet light (see Figure 3.11b).

Once the DNA samples have been digested, separated, and visualized by gel electrophoresis, creating the actual restriction map is like assembling a puzzle. As illustrated in Figure 3.12, by comparing the single-digests with each double-digest, researchers can arrange the fragments in the correct order to create a map of restriction sites. Currently, because DNA sequencing of even relatively small pieces of DNA has become fairly common, restriction mapping can often be done using bioinformatics software (such as Web-cutter described in Questions & Activities problem 8 at

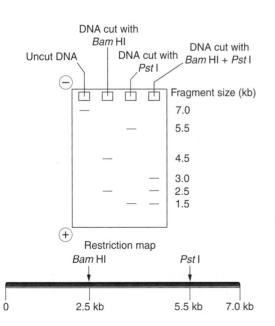

Figure 3.12 Restriction Mapping The location of restriction sites for *Bam*HI and *Pst*I is determined in a DNA fragment that is 7 kb long. Digestion with *Bam*HI cleaves the DNA into two fragments measuring 2.5 and 4.5 kb, indicating that the DNA was cut at a single site located 2.5 kb from one end. Digestion with *Pst*I cleaves the DNA into two fragments at 1.5 and 5.5 kb, indicating the DNA was cut at a single site located 1.5 kb from one end. A double-digest with both enzymes cleaves the DNA into three fragments, 3.0 kb, 2.5 kb, and 1.5 kb. Because the 3.0-kb fragment is not created by either *Bam*HI or *Pst*I digestion alone, it must represent the DNA located between the *Bam*HI and *Pst*I cutting sites. By arranging this "puzzle" of fragments, the restriction map at the bottom of the figure is the only map consistent with the pattern of fragments created by the digests in this experiment.

the end of this chapter) to identify restriction cutting sites in a DNA sequence without the need to actually digest DNA and create a map experimentally.

DNA Sequencing

After a gene is cloned, it is important to determine the nucleotide sequence of the gene, its exact order of As, Gs, Ts, and Cs. Knowing the DNA sequence of a gene can be helpful (1) to deduce the amino acid sequence of a protein encoded by a cloned gene, (2) to determine the exact structure of gene, (3) to identify regulatory elements such as promoter sequences, (4) to identify differences in genes created by gene splicing, and (5) to identify genetic mutations. Today many different methods for **DNA sequencing** are available including techniques for PCR "cycle" sequencing and computer-automated DNA sequencing. Initially, the most widely used sequencing approach was chain-termination sequencing, a manual method developed in 1977 by Frederick Sanger and often referred to as the Sanger method. In this technique, a DNA primer is hybridized to denatured template DNA, such as a plasmid containing cloned DNA to be sequenced, in a reaction tube containing deoxyribonucleotides and DNA polymerase. Because many modern plasmid vectors are designed with sequencing primer binding sites adjacent to the multiple cloning site, DNA polymerase can be used to extend a complementary strand from the 3' end of primers hybridized to the plasmid. The original approach utilized radioactively labeled primer sequences.

A small amount of a modified nucleotide called a **dideoxyribonucleotide (ddNTP)** is mixed in with the vector, primer, polymerase, and deoxyribonucleotides. A ddNTP differs from a normal deoxyribonucleotide (dNTP) because it has a hydrogen group attached to the 3' carbon of the deoxyribose sugar instead of a hydroxyl group-OH (see Figure 3.13a). When a ddNTP is incorporated into a chain of DNA, the chain cannot be extended because the absence of a 3'-OH prevents formation of a phosphodiester bond with a new nucleotide; hence, the chain is "terminated."

Four separate reaction tubes are set up. Each tube contains vector, primer, and all four dNTPs, but each tube also contains a small amount of one ddNTP. As synthesis of a new DNA strand from the primer begins, DNA polymerase randomly inserts a ddNTP into the sequence instead of a normal dNTP, preventing further synthesis of a complementary strand. Over time, there will be a ddNTP incorporated at all positions in the newly synthesized strands creating a series of fragments of varying lengths that are terminated at dideoxy residues. For the original Sanger technique, the DNA strands were separated on a thin polyacrylamide gel, which can separate sequences that differ in

length by a single nucleotide. Autoradiography was used to detect the radioactive sequencing fragments as shown in Figure 3.13c. The sequence determined from the autoradiogram is "read" from the bottom to the top as individual nucleotides. As shown in Figure 3.13c, the sequence determined from the autoradiogram is *complementary* to the sequence on the template strand in the vector.

Because of limitations in running a sequencing gel, the original Sanger method can only be used to sequence approximately 200 to 400 nucleotides in a single reaction; therefore, when sequencing a longer piece of DNA—for example, 1,000 base pairs—it was necessary to run multiple reactions to create overlapping sequences that could be pieced together to determine the entire, continuous sequence of 1,000 base pairs.

Because of this limitation, the original Sanger sequencing approach is a cumbersome method for the large-scale sequencing efforts such as those used for the Human Genome Project. The project was completed ahead of schedule in part because of the development of **computer-automated DNA sequencing methods** capable of sequencing long stretches of DNA (more than 500 bp in a single reaction). Computer-automated sequencing reactions use either ddNTPs, each labeled with a different colored nonradioactive fluorescent dye, or a sequencing primer labeled at its 5' end with a dye (Figure 3.13). A single reaction tube is used and the original manual style approach of using polyacrylamide gels followed by autoradiography has been replaced by separating sequencing reactions on a single lane of an ultrathin diameter tube gel called a *capillary gel*. As DNA fragments move through the gel, they are scanned with a laser beam. The laser stimulates the fluorescent dye on each DNA fragment, which emits different wavelengths of light for each ddNTP. The emitted light is collected by a detector that amplifies and then feeds this information to a computer to process and convert the light patterns to reveal the DNA sequence.

Automated DNA sequencers often contain multiple capillary gels that are several feet long, allowing many bases to be separated. As a result some instruments run as many as 96 capillary gels, each producing around 900 bases of sequence. With these instruments it is possible to generate approximately 2 million bases of sequence in a day.

A number of different groups around the world are working on next-generation sequencing methods designed to generate highly accurate and long stretches of DNA sequence, greater than 1 gigabase (billion bases) of DNA per reaction, at a low cost. These sequencers perform DNA amplification without cloning DNA in bacteria and utilize DNA sequencing reactions that do not involve the chain termination techniques pioneered

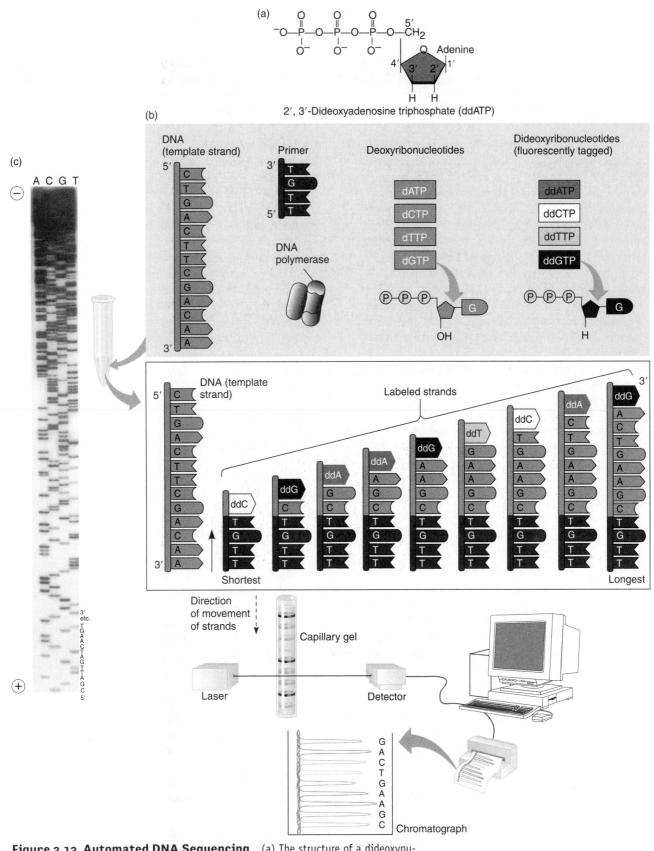

Figure 3.13 Automated DNA Sequencing (a) The structure of a dideoxynucleotide (ddNTP). Note that the 3′ group attached to the carbon is a hydrogen rather than a
hydroxyl group (OH). Because another nucleotide cannot be attached at the 3′ end of a
ddNTP, these nucleotides are the key to DNA sequencing by the Sanger method as shown
(b). Computer automated approaches separate DNA fragments using a capillary gel. A laser
and detector are used to detect fluorescence of each dideoxynucleotide. (c) Autoradiogram
of a dideoxy sequencing gel. The letters over the lanes (A, C, G, and T) correspond to the
particular dideoxynucleotide used in the sequencing reaction analyzed in the lane.

TOOLS OF THE TRADE

Restriction Enzymes

Restriction enzymes are little more than sophisticated "scissors" that molecular biologists use to manipulate DNA. Working with restriction enzymes has become easier over the 30 years since Hamilton Smith and others pioneered their use. Scientists no longer need to purify their own enzymes from bacteria and prepare their own buffers for working with restriction enzymes. Over 300 restriction enzymes are commercially available rather inexpensively. Many enzymes are readily available because they have been cloned using recombinant DNA technology and so they are made and isolated in large quantities. Commercially prepared enzymes come in conveniently sized prepackages with buffer solutions that provide all the components necessary for optimal enzyme activity. If researchers need to work with an enzyme with which they are unfamiliar, they can use the restriction enzyme database REBASE (http://rebase.neb.com/rebase/rebase.html), an outstanding tool for locating enzyme suppliers and enzyme specifics.

In addition, a variety of software packages and websites are available to assist scientists who work with restriction enzymes and DNA sequences. For example, imagine you are a molecular biologist who just cloned and sequenced a 7,200-bp piece of DNA and you want to see if there is an enzyme that will cut your gene to create a 250-bp piece of DNA for a probe you want to make. Not too long ago, if you had a lot of enzymes in your freezer, you could digest this DNA and run gels to see if you could get a 250-bp piece, but this imprecise approach took a lot of time and resources. If you had sequenced your gene, you could scan the sequence with your eyes looking for a restriction site of interest; a very time-consuming and eye-straining effort! The Internet makes this task much easier because many websites function as online tools for analyzing restriction enzyme cutting sites. For example, in Webcutter (http://rna.lundberg.gu.se/cutter2/), DNA sequences can be entered and searched to determine restriction enzyme cutting patterns (see Questions & Activities problem 8).

The widespread application of recombinant DNA techniques in many areas of biological and medical research has led to hundreds of technique books, websites, and journals. In Biotechniques (www.BioTechniques.com), a popular monthly journal, biologists publish and share information on molecular cloning techniques. Several sites that are commonly used for designing PCR primers and other applications are provided in Keeping Current: Web Links on the Companion Website.

---

by Sanger. Some of these techniques use DNA ligase to attach DNA to beads where it is amplified by PCR and then bound to dye labeled oligonucleotides. The sequence is then determined by computer analysis of binding patterns. Keep an eye out for the new generation of high-powered sequencers expected to reach the market in the next 2–3 years.

Chromosomal Location and Gene Copy Number

When a new gene is cloned, it is often helpful to identify the chromosomal location of the gene and to determine if the gene is present as a single copy in the genome or if multiple copies of the gene exist.

Fluorescence *in situ* hybridization

A technique called **fluorescence *in situ* hybridization** (**FISH;** *in situ* means "in place") can be used to identify which chromosome contains a gene of interest. For example, if you just cloned a human gene believed to be involved in intelligence, with the help of FISH you could determine on which chromosome this gene resides. In FISH, chromosomes are isolated from cells such as white blood cells and spread out on a glass microscope slide. A cDNA probe for the gene of interest is chemically labeled with fluorescent nucleotides and then incubated in solution with the slide. The probe hybridizes with complementary sequences on the slide. The slide is washed and then exposed to fluorescent light. Wherever the probe has bound to a chromosome, the fluorescently labeled probe is illuminated to indicate the presence of that gene (Figure 3.14).

To determine which of the 23 human chromosomes show fluorescence, they are aligned according to the length and staining patterns of their chromatids to create a karyotype. Fluorescence on more than one chromosome indicates either multiple copies of the gene or related sequences that may be part of a gene family. FISH is also used to analyze genetic disorders. For example, FISH analysis can be performed on a karyotype of fetal chromosomes from a pregnant woman to determine if a developing fetus has an abnormal number of chromosomes.

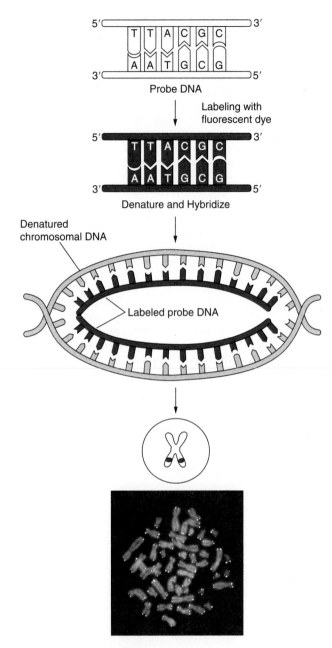

Figure 3.14 Fluorescence *in Situ* Hybridization White spots at the tips of each chromosome indicate fluorescence from a probe binding to telomeres.

Southern blotting

Another technique called **Southern blot analysis** (Southern blotting or hybridization) is frequently used to determine gene copy number. Developed by Ed Southern in 1975, Southern blotting begins by digesting chromosomal DNA into small fragments with restriction enzymes. DNA fragments are separated by agarose gel electrophoresis (Figure 3.15). However, the number of restriction fragments generated by digesting

chromosomal DNA is often so great that simply running a gel and staining the DNA does not resolve discrete fragments. Rather, digested DNA appears as a continuous smear of fragments in the gel. Southern blotting is used to visualize only *specific* fragments of interest. Following electrophoresis, the gel is treated with an alkaline solution to denature the DNA; then, the fragments are transferred onto a nylon or nitrocellulose membrane using a technique called *blotting*.

Blotting can be achieved by setting up a gel sandwich in which the gel is placed under the nylon membrane, filter paper, paper towels, and a weight to allow for wicking of a salt solution through the gel that will transfer DNA onto the nylon by capillary action (Figure 3.15). Alternatively, pressure or vacuum blotters can be used to transfer DNA onto nylon. The nylon blot is then baked or briefly exposed to UV light to attach the DNA permanently. Now that the DNA is bound to the nylon membrane as a solid support, the blot is incubated with a radioactive probe (or nonradioactively labeled probe) in much the same way that colony hybridizations are carried out. The blot is washed to remove extraneous probes and then exposed to film by autoradiography. Wherever the probe has bound to the blot, radioactivity in the probe develops silver grains on the film to expose bands on the blot, creating an autoradiogram (see Figure 3.15). By interpreting the number of bands on the autoradiogram, it can be possible to determine gene copy number.

The development of Southern blot analysis was an important technique that formed the basic principles for **Northern blotting** (the separation and blotting of RNA molecules as discussed in the next section) and **Western blotting** (the separation and blotting of proteins). (Northern and Western blotting techniques were not named after scientists named "Northern" and "Western"; rather they were named as tongue-in-cheek references to Ed Southern, the founder of Southern blots.) Southern blotting has many other applications including gene mapping, related gene family identification, genetic mutation detection, PCR product confirmation, and DNA fingerprinting (a topic we discuss in Chapter 8). An excellent animation of how Southern blot analysis is used in DNA fingerprinting can be found at the website for the Cold Spring Harbor DNA Learning Center, which can be found on the Companion Website.

Studying Gene Expression

Many molecular biologists are involved in research studying gene expression and the regulation of gene expression. A range of different molecular techniques are available for studying gene expression. Most methods involve analyzing mRNA produced by a tissue.

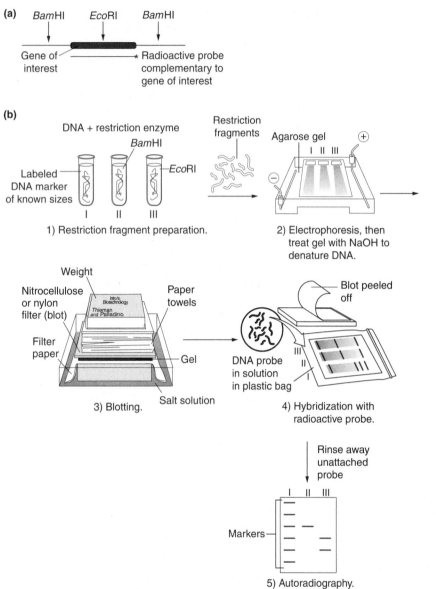

(a)

Gene of interest — Radioactive probe complementary to gene of interest

(b)

1) Restriction fragment preparation.

2) Electrophoresis, then treat gel with NaOH to denature DNA.

3) Blotting.

4) Hybridization with radioactive probe.

5) Autoradiography.

Figure 3.15 Southern Blot Analysis of DNA Fragments (a) Region of DNA for a gene of interest to be studied by Southern blot analysis (b). Steps involved in Southern blotting: (1) The DNA samples to be analyzed are digested with restriction enzymes. (2) Then the mixtures of the restriction fragments from each sample are separated by electrophoresis. (3) When the samples are blotted, capillary action pulls a salt solution upward from a filter paper wick through the gel, transferring the DNA to a nylon filter or blot. The single strands of DNA stick to the blot, positioned in bands exactly as on the gel. (4) The blot is exposed to a solution containing a radioactively labeled probe (a single-stranded DNA complementary to the DNA sequence of interest), which attaches by base pairing to restriction fragments of complementary sequence. (5) Film is laid over the blot. The radioactivity in the bound probe exposes the film to form an image corresponding to specific DNA bands on the blot that base pair with the probe.

This is often a good measure of gene expression because the amount of mRNA produced by a tissue is often equivalent to the amount of protein the tissue makes.

Northern Blot Analysis

A common technique for studying gene expression is Northern blot analysis. The basic methodology of a Northern blot is similar to Southern blot analysis. In a Northern blot, RNA is isolated from a tissue of interest and separated by gel electrophoresis (the RNA is not digested with enzymes). RNA is blotted onto a nylon membrane and then hybridized to a probe as described for Southern blots. Exposed bands on the autoradiogram show the presence of mRNA for the gene of interest and the size of the mRNA (Figure 3.16a). In addition, the amounts of mRNA produced by different tissues can be compared and quantified.

Reverse transcription PCR

Sometimes the amount of RNA produced by a tissue is below the level of detection by Northern blot analysis. PCR allows for detecting minute amounts of mRNA from even very small amounts of starting tissue. For instance, PCR has been a great tool for molecular biologists studying gene expression in embryos and developing tissues where the amount of tissue for analysis is very small. Because RNA cannot be directly amplified by PCR, a technique called **reverse transcription**

(a)

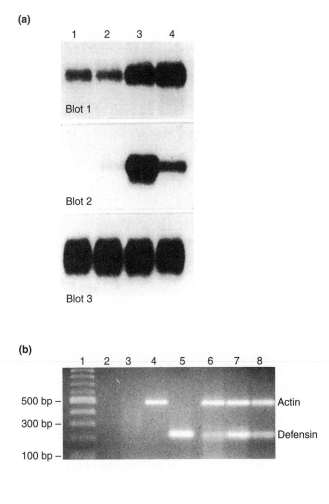

Figure 3.16 Analyzing Gene Expression by Northern Blot Analysis and RT-PCR Northern blot analysis and RT-PCR are two common techniques for analyzing the amount of mRNA produced by a tissue (gene expression). (a) Blot 1 is a portion of an autoradiogram from a Northern blotting experiment in which RNA from four rat tissues: 1, seminal vesicles; 2, kidney; 3 and 4, different segments of the epididymis (a male reproductive organ) was blotted onto nylon and then probed with a radioactive cDNA probe for a gene involved in protecting tissues from damage by harmful free radicals, atoms, or molecules with unpaired numbers of electrons. Notice how the amount of mRNA detected in lanes 3 and 4 (as indicated by the size and darkness of each band) is greater than the amount of mRNA in lanes 1 and 2. Blot 2 shows an autoradiogram from blot 1 that was stripped of bound probe and reprobed with a probe for a different gene. Blot 3, the same blot shown in the other two panels, was washed and stripped of bound probe and then reprobed with a radioactive cDNA for a gene (cyclophilin) that is expressed at nearly the same levels in virtually all tissues. (b) Agarose gel from an RT-PCR experiment in which RNA from rat tissues was reverse transcribed and amplified with primers for β-actin (an important component of the cytoplasm of cells) and/or β-defensin-1 (a gene that encodes a peptide that provides protection against bacterial infections in many tissues). Lane 1 contains DNA size standards of known size (often called a ladder) increasing in 100-bp increments. Lane 2 is a negative control sample in which primers were added to a PCR experiment without cDNA. Lane 3 is a negative control sample in which cDNA was added to a PCR experiment without primers. Notice that lanes 2 and 3 do not show any amplified PCR product because amplification will not occur without cDNA as target DNA (lane 2) or without primers (lane 3). Lane 4, kidney cDNA amplified with actin primers. Lane 5, kidney cDNA amplified with defensin primers. Lanes 6, 7, and 8 show PCR products from cDNA of three different rat reproductive tissues that were amplified with both actin and defensin primers. Notice how lanes 6 and 8 show relatively even amounts of actin and defensin PCR products; greater amounts of defensin PCR product are shown in lane 7. These differences reflect the different amounts of defensin mRNA made by these tissues.

PCR (RT-PCR) is carried out. In RT-PCR, isolated RNA is converted into double-stranded cDNA by the enzyme reverse transcriptase in a process similar to how cDNA for a library is made. The cDNA is then amplified with a set of primers specific for the gene of interest. Amplified DNA fragments are electrophoresed on an agarose gel and evaluated to determine expression patterns in a tissue (see Figure 3.16b).

Real-time PCR

New applications in PCR technology make it possible to determine the amount of PCR product made during an experiment through a technique called **real-time** or **quantitative PCR (qPCR),** which enables researchers to quantify amplification reactions as they occur in "real time" (Figure 3.17). There are several ways to run real-time PCR reactions, but the basic procedure involves the use of specialized (and expensive) thermal cyclers that use a laser to scan a beam of light through the top or bottom of each PCR tube. Each reaction tube contains either a dye-containing probe or DNA-binding dye that emits fluorescent light when illuminated by the laser. The light emitted by these dyes correlates to the amount of PCR product amplified. Light from each tube is captured by a detector that relays information to a computer to provide a readout on the amount of fluorescence after each cycle and can be plotted and analyzed to quantitate the number of PCR products produced after each cycle.

Two of the most widely used approaches for real-time PCR involve the use of a dye called SYBR Green and TaqMan probes. SYBR Green is a dye that binds double-stranded DNA. As more double-stranded DNA is copied with each round of real-time PCR, there are more DNA copies to bind SYBR Green, which increases the amount of fluorescent light emitted.

TaqMan probes are complementary to specific regions of the target DNA between where the forward and reverse primers for PCR bind (Figure 3.17). TaqMan probes contain two dyes. One dye, the reporter,

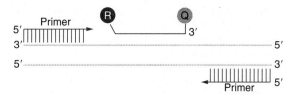

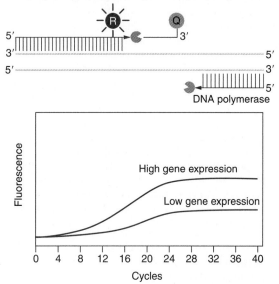

1. **Hybridization.** Forward and reverse PCR primers bind to denatured target DNA. TaqMan probe with reporter (R) and quencher (Q) dye binds to target DNA between the primers. When probe is intact, emission by the reporter dye is quenched.

2. **Extension.** As DNA polymerase extends the forward primer, it reaches the TaqMan probe and cleaves the reporter dye from the probe. Released from the quencher the reporter can now emit light when excited by a laser.

3. **Detection.** Emitted light from the reporter is detected and interpreted to produce a plot that quantitates the amount of PCR product produced with each cycle.

Figure 3.17 Real-Time PCR (a) The TaqMan method of real-time PCR involves a pair of PCR primers along with a probe sequence complementary to the target gene. The probe contains a reporter dye (R) at one end and a quencher dye (Q) at the other end. When the quencher dye is close to the reporter dye, it interferes with fluorescence released by the reporter dye. When *Taq* DNA polymerase extends a primer to synthesize a strand of DNA it cleaves the reporter dye off of the probe allowing the reporter to give off energy. (b) Each subsequent PCR cycle removes more reporter dyes so increased light emitted from the dye can be captured by a computer to produce a readout of fluorescence intensity with each cycle.

is located at the 5′ end of the probe and can release fluorescent light when excited by laser light from the thermal cycler. The other dye, called a quencher, is attached to the 3′ end of the probe. When these two dyes are close to each other, the quencher dye interferes with the fluorescent light released from the reporter dye. However, as *Taq* DNA polymerase extends each primer, it removes the reporter dye from

the end of the probe (and eventually removes the entire probe). Now that the reporter dye is separated from the quencher, the fluorescent light released by the reporter can be detected by the thermal cycler. Detected light is analyzed by a computer to produce a plot displaying the amount of fluorescence emitted with each cycle. Because real-time PCR does not involve running gels, it is a powerful and rapid technique for measuring and quantitating changes in gene expression, particularly when multiple samples and different genes are being analyzed.

In situ hybridization

When a gene is expressed in an organ with many different cell types, for example kidney or brain tissue, Northern blot analysis and RT-PCR can only tell you that the organ is expressing the mRNA of interest, but they cannot determine the cell type (that is, epithelial cells versus connective tissue cells) expressing the mRNA. A technique called ***in situ* hybridization** is commonly used to determine the cell type that is expressing a particular mRNA. In this technique, the tissue of interest is preserved in a fixative solution and then embedded in a wax-like material or resin. This allows researchers to slice the tissue into thin sections about 1 to 5 µm thick and to attach them to a microscope slide. Sometimes frozen sections of tissue are used for these experiments. The slide is incubated with a fluorescent dye or radioactively-tagged RNA or DNA probe for the gene of interest. Many investigators also use nonisotopic techniques to label probes for *in situ* hybridization. The probe hybridizes to mRNA within cells in their native place. Slides are covered with a photographic emulsion to develop silver grains as in autoradiography, or alternative detection methods are used for nonradioactive fluorescent probes. For some studies, PCR can even be performed directly on tissue sections *in situ* as a way to determine cell type expression for a given gene.

Gene microarrays

DNA microarray analysis is another technique for studying gene expression that has rapidly gained popularity in recent years because it enables researchers to test all the genes expressed in a tissue very quickly (see Figure 3.18). A microarray, also known as a gene chip, is created using a small glass microscope slide. Single-stranded DNA molecules are attached or "spotted" onto the slide using a computer-controlled high-speed robotic arm called an *arrayer*, fitted with a number of tiny pins. Each pin is immersed in a small amount of solution containing millions of copies of different DNA molecules (such as cDNAs for different genes), and the arrayer fixes this DNA onto the slide at specific locations (points or spots) recorded by a computer. A

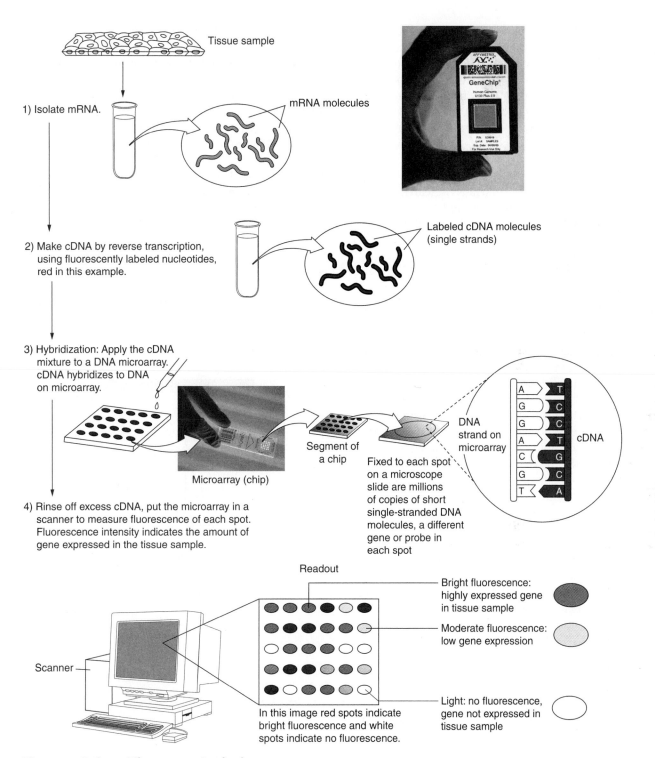

1) Isolate mRNA.

Tissue sample

mRNA molecules

2) Make cDNA by reverse transcription, using fluorescently labeled nucleotides, red in this example.

Labeled cDNA molecules (single strands)

3) Hybridization: Apply the cDNA mixture to a DNA microarray. cDNA hybridizes to DNA on microarray.

Microarray (chip)

Segment of a chip

DNA strand on microarray

cDNA

Fixed to each spot on a microscope slide are millions of copies of short single-stranded DNA molecules, a different gene or probe in each spot

4) Rinse off excess cDNA, put the microarray in a scanner to measure fluorescence of each spot. Fluorescence intensity indicates the amount of gene expressed in the tissue sample.

Readout

Scanner

Bright fluorescence: highly expressed gene in tissue sample

Moderate fluorescence: low gene expression

Light: no fluorescence, gene not expressed in tissue sample

In this image red spots indicate bright fluorescence and white spots indicate no fluorescence.

Figure 3.18 Gene Microarray Analysis

single microarray can have over 10,000 spots of DNA, each containing unique sequences of DNA for a different gene.

To use a microarray to study gene expression, scientists extract mRNA from a tissue of interest. The mRNA, or sometimes the cDNA, is then tagged with a fluorescent dye and incubated overnight with the microarray. During this incubation, mRNA hybridizes to spots on the microarray that contain complementary DNA sequences. The microarray is washed and then scanned by a laser that causes the mRNA hybridized to the microarray to fluoresce. These fluorescent spots reveal which genes are expressed in the tissue of interest, and the intensity of fluorescence indicates the relative amount of expression. The brighter the spot, the more mRNA is expressed in that tissue. Microarrays can also be run by labeling cDNAs from two or more tissues with different colored fluorescent dyes. Gene expression patterns from the different tissues are compared based on the color of spots that appear following hybridization and detection.

For many species, including humans, entire genomes are available on microarrays. Researchers are also using microarrays to compare patterns of expressed genes in tissues under different conditions. For example, cancer cells can be compared with normal cells to look for genes that may be involved in cancer formation. As you will learn in Chapter 11, results of such microarray studies can be used to develop new drug therapy strategies to combat cancer and other diseases.

Protein expression and purification

Bacteria that have been transformed with recombinant plasmids containing a gene of interest can often be used to produce the protein product of the isolated gene. Large quantities of bacteria can be grown in a fermenter, and the protein can be isolated using techniques that are described in Chapter 4. We conclude our study of recombinant DNA technology and its applications in the next section by taking a look at bioinformatics, one of the newest and most rapidly developing fields in biology.

Gene mutagenesis studies

Of the many different ways that scientists work with and study cloned genes, there are a wide-variety of techniques that can be used to study the structure and function of protein produced by a specific gene. One approach is called **site-directed mutagenesis.** In this technique mutations can be created in specific nucleotides of a cloned gene contained in a vector. The gene can then be expressed in cells, which results in translation of a mutated protein. This allows researchers to study the effects of particular mutations on protein structure and functions as a way to determine what nucleotides are important for specific functions of the protein. Site-directed mutagenesis can be a very valuable way to help scientists identify critical sequences in genes that produce proteins involved in human diseases.

RNA interference

In 1998, researchers Craig C. Mello of the University of Massachusetts Medical School and Andrew Z. Fire of Stanford University published groundbreaking work in which they used double-stranded pieces of RNA (dsRNA) to inhibit or silence expression of genes in the

nematode roundworm *Caenorhabditis elegans*. This naturally occurring mechanism for inhibiting gene expression is known as **RNA interference (RNAi).** As Mello, Fire, and other researchers discovered, dsRNA can be bound by an RNA-digesting enzyme called **dicer** that cuts dsRNA into 21- to 25-nucleotide-long snippets of RNA molecules called **small interfering RNAs (siRNAs).**

These siRNAs are bound by a protein-RNA complex called the **RNA-induced silencing complex (RISC).** RISC unwinds the double-stranded siRNAs, releasing single-stranded siRNAs that bind to complementary sequences in mRNA molecules. Binding of siRNAs to mRNA results in degradation of the mRNA (by the enzyme slicer) or blocks translation by interfering with ribosome binding (Figure 3.19). RNAi is similar in mechanism to the gene-silencing actions of miRNAs discussed in Chapter 2, but a primary difference is that siRNAs are generated from dsRNA.

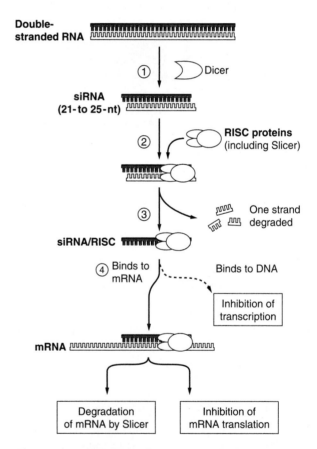

Figure 3.19 RNA Interference (1) Double-stranded RNA is cleaved by the enzyme dicer into siRNAs. (2) RISC proteins bind siRNAs and degrade one of the two strands (3) to produce single-stranded siRNAs. (4) Single-stranded siRNAs bind to complementary sequences on mRNA molecules in the cytoplasm and interfere with (silence) gene expression through triggering mRNA degradation by slicer or by inhibiting translation of the mRNA by ribosomes.

Techniques incorporating RNAi have developed rapidly as methods for regulating gene expression and as a potential way to target and inactivate specific genes with high efficiency. In Chapter 11, we consider examples of how biotechnology and pharmaceutical companies are working on RNAi techniques for silencing genes involved in human diseases. RNAi has become such a rapidly developing technology that recent estimates indicate the use of RNAi reagents will grow from about $400 million in 2005 to $850 million by 2010. The 2006 Nobel Prize in Physiology or Medicine was awarded to Andrew Fire and Craig Mello for their discovery of this natural method for switching genes off. Nobel Prizes are typically awarded decades after the honored work has been completed, so this unusually quick recognition is a clear indication of the value of RNAi as a powerful and promising research tool.

3.5 Genomics and Bioinformatics: Hot New Areas of Biotechnology

Cloning individual genes using libraries and other techniques described in this chapter will continue to be important approaches in recombinant DNA technology. However over the last 15 years or so, a number of advances in cloning and sequencing technologies have increasingly led to the use of strategies for cloning, sequencing, and analyzing entire genomes—an exciting and rapidly developing science called **genomics.** You will learn about many applications of genomics as you study biotechnology. Here we provide an introduction to how scientists can study whole genomes.

Whole genome "shotgun" sequencing strategies

As powerful as traditional recombinant DNA techniques are, it became increasing apparent that if scientists wanted to study complex biological processes that involve many genes—such as most cancers—cloning one or even a few genes at a time is a slow process yielding only incremental information about genes. As a result a number of scientists started working on strategies for cloning and sequencing entire genes—a strategy commonly called **shotgun sequencing** or **shotgun cloning.** The analogy is that cloning individual genes using libraries is equivalent to using a rifle to hit a specific spot on a target (e.g., cloning a specific gene), whereas a shotgun would randomly hit many spots on a target with little precision. In shotgun sequencing, the entire genome, introns and exons, is cloned and sequenced then individual genes are sorted out later through bioinformatics.

One shotgun strategy for constructing the sequence of whole chromosomes involves using restriction enzymes to digest pieces of entire chromosomes (Figure 3.20). This process can produce thousands of fragments that have to be sequenced. Each chromosome piece is sequenced separately, and then computer programs are used to align the fragments based on overlapping sequence pieces. This approach enables scientists to reconstruct the entire sequence of a whole chromosome, but as you can see in the figure, the use of computers to organize and compare the DNA sequences of these fragments is essential. This is **bioinformatics** in action—an interdisciplinary field that applies computer science and information technology to promote an understanding of biological processes.

In 1995, scientists used a shotgun strategy at The Institute for Genomic Research to sequence the 1.8 million base pairs in the genome of a strain of bacteria called *Haemophilus influenzae*. This was the first time an entire genome for any organism had been sequenced using this approach. Many doubted that shotgun-sequencing strategies could be used effectively to sequence larger genomes, but development of this technique and novel sequencing strategies rapidly accelerated progress of the Human Genome Project.

Bioinformatics: Merging Molecular Biology with Computing Technology

When scientists clone and sequence a newly identified gene or DNA sequence, they report their findings in scientific publications and submit the sequence data to databases so other scientists who may be interested in this sequence information can have access to it. Database manipulations of DNA sequence data were some of the first applications of bioinformatics, which involves the use of computer hardware and software to study, organize, share, and analyze data related to gene structure, gene sequence and expression, and protein structure and function.

As genome data has rapidly accumulated, resulting in an enormous amount of information being stored in public and private databases, bioinformatics has become an essential tool that allows scientists to share and compare data, especially DNA sequence data. As we discussed in Chapter 1, bioinformatics is one of the most rapidly developing career areas of biotechnology. We discuss basic applications of bioinformatics throughout the book; however, we introduce the field here with a few very common applications.

Examples of Bioinformatics in Action

Even before whole genome sequencing projects, scientists had been accumulating a tremendous amount of

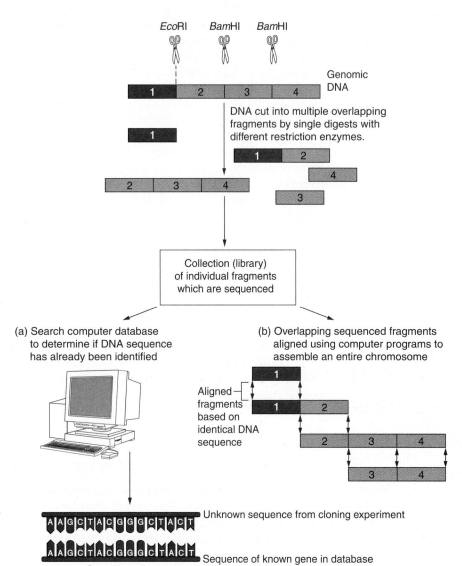

Figure 3.20 Whole Genome "Shotgun" Sequencing and Two Examples of Bioinformatics Applications   Shown is a simplified example of a piece of genomic DNA cut into smaller pieces (labeled 1, 2, 3, and 4) by *Eco*RI and *Bam*HI. Regardless of how a piece of DNA is cloned, DNA sequence analysis is an essential part of learning more about the cloned DNA. (a) Typically such analysis begins by searching the unknown sequence against a database of known sequences. In this example, sequence alignment reveals that the unknown sequence cloned is an exact match with a sequence of a "known" gene already in the database. Note: For simplicity, only one strand of sequenced DNA is being compared to the database. (b) Another application of bioinformatics involves the use of computer programs to align DNA fragments based on nucleotide sequence overlaps. The scheme shown here is similar to one approach used in the Human Genome Project to assemble completed sequences of entire chromosomes.

sequence information from a variety of different organisms, necessitating the development of sophisticated databases that could be used by researchers around the world to store, share, and obtain the maximum amount of information from protein and DNA sequences. Databases are essential tools for archiving and sharing data with researchers and the public. Because high-speed computer-automated DNA sequencing techniques were developed nearly simultaneously with expansion of the Internet as an information tool, many DNA sequence databases became available through the Internet.

When scientists who have cloned a gene enter their sequence data into a database, these databases search the new sequence against all other sequences in the database and create an alignment of similar nucleotide sequences if it finds a match (Figure 3.20a). This type of search is

often one of the first steps taken after a gene is cloned because it is important to determine if the sequence has already been cloned and studied or if the gene sequence is a novel one. Among other applications, these databases can also be used to predict the sequence of amino acids encoded by a nucleotide sequence and to provide information on the function of the cloned gene. In addition, as we discuss in more detail in Chapter 11, one strategy for constructing the sequence of whole chromosomes involves a shotgun approach in which restriction enzyme digested pieces of entire chromosomes are sequenced separately and then computer programs are used to align the fragments based on overlapping sequence pieces called **contigs** (contiguous sequences) (Figure 3.20b). This approach enables scientists to reconstruct the entire sequence of a whole chromosome.

Human *OB* gene
```
gtcaccaggatcaatgacatttcacacacg---tcagtctcctccaaacagaaagtcacc
|||||||||||||||||||||||||||||||   ||  ||  |||  ||||  ||||   |||||
gtcaccaggatcaatgacatttcacacacgcagtcggtatccgccaagcagagggtcact
```
Mouse *ob* gene

```
ggtttggacttcattcctgggctccaccccatcctgaccttatccaagatggaccagaca
||  |||||||||||||||||||||||  |||||||||  ||||   || ||||||||||||||||||
ggcttggacttcattcctgggcttcacccattctgagtttgtccaagatggaccagact
```

```
ctggcagtctaccaacagatcctcaccagtatgccttccagaaacgtgatccaaatatcc
||||||||||| |||||| |||||||||||   |||||||||  ||| ||| | || ||| ||
ctggcagtctatcaacaggtcctcaccagcctgccttcccaaaatgtgctgcagatagcc
```

Figure 3.21 Comparison of the Human and Mouse *ob* Genes Partial sequences for these two genes are shown with the human gene on top and the mouse gene sequence below it.

Many databases are maintained throughout the world. One of the most widely used gene databases is called **GenBank.** It is the largest publicly available database of DNA sequences and contains the National Institutes of Health (NIH) collection of DNA sequences. GenBank shares and acquires data from databases in Japan and Europe. Maintained by the National Center for Biotechnology Information (NCBI) in Washington, D.C., GenBank contains more than 65 billion bases of sequence data from over 100,000 species, and it doubles in size roughly every 14 months. The NCBI is a gold mine of bioinformatics resources that creates public-access databases and develops computing tools for analyzing and sharing genome data. The NCBI has also designed user-friendly ways to access and analyze an incredible amount of data on nucleotide sequences, protein sequences, molecular structures, genome data, and even scientific literature.

DNA database searching: Try it yourself

For example, an NCBI program called **Basic Local Alignment Search Tool** (**BLAST;** www.ncbi.nlm .nih.gov/BLAST) can be used to search GenBank for sequence matches between cloned genes. Go to the BLAST website and click "standard nucleotide-nucleotide BLAST [blastn]." In the search box type in the following sequence: AATAAAGAAC CAGGAGTGGA. Imagine that this sequence is from a piece of a gene that you just cloned and sequenced and you want to know if anyone cloned this gene before you. Click the "Blast!" button. Your results will be available in a minute or two. Click the "Format!" button to see the results of your search. A page will appear with the results of your search (you may need to scroll down the page to find the sequence alignment). What did you find?

Figure 3.21 shows a BLAST search alignment comparing the human and mouse gene sequences for an obesity gene (*ob*) that produces a hormone called

leptin. Leptin plays a role in fat metabolism, and mutations in the leptin gene can contribute to obesity (see also Figure 11.1). Notice how the nucleotide sequence for these two genes is very similar as indicated by the vertical lines between identical nucleotides.

The Human Genome Nomenclature Committee, supported by the NIH, establishes rules for assigning names and symbols to newly cloned human genes. Each entry into GenBank is provided with an **accession number** that scientists can use to refer back to that cloned sequence. For example, go to the GenBank* website listed at the Companion Website and then type in the accession number U14680 and click "GO." What gene is identified by this accession number? Click on the accession number link. Notice that GenBank provides the original journal reference that reported this sequence, the single letter amino acid code of the protein encoded by this gene, and the nucleotide sequence (cDNA) of this gene. Alternatively, you can type in the name of a gene or a potential gene that you are interested in to see if it has already been cloned and submitted to GenBank.

GenBank is only one example of an invaluable database that is an essential tool for bioinformatics. Many other specialized databases exist with information such as single-nucleotide polymorphism data, BAC and YAC library databases, and protein databases that catalog amino acid sequences and three-dimensional protein structures.

A Genome Cloning Effort of Epic Proportion: The Human Genome Project

It is a very exciting time to be studying biotechnology. We are fortunate to witness an unprecedented project

---

*The BLAST search will identify a match with a human gene for early-onset breast cancer, *BRCA1*. The GenBank search identifies the same gene by its accession number.

that involves many of the techniques described in this chapter, the **Human Genome Project (HGP).** Initiated in 1990 by the U.S. Department of Energy (DOE), the HGP was an international collaborative effort with a 15-year plan to identify all human genes, originally estimated at 80,000 to 100,000 genes, and to sequence the approximately 3 billion base pairs thought to comprise the 24 different human chromosomes (chromosomes 1 to 22, X, Y). The Human Genome Project was also designed to accomplish the following:

■ Analyze genetic variations among humans. This included the identification of single-nucleotide polymorphisms (SNPs; see Figure 1.11).

■ Map and sequence the genomes of **model organisms,** including bacteria, yeast, roundworms, fruit flies, and mice.

■ Develop new laboratory technologies such as high-powered automated sequencers and computing technologies (including databases of genome information) that can be used to advance our analysis and understanding of gene structure and function.

■ Disseminate genome information among scientists and the general public.

■ Consider ethical, legal, and social issues that accompany the Human Genome Project and genetic research.

In the United States, public research on the HGP was coordinated by the National Center of Human Genome Research, a division of the NIH and the DOE. The project funded seven major sequencing centers in the United States. Over time the project grew to become an international effort with contributions from scientists in 18 countries, but the work was primarily carried out by the International Human Genome Sequence Consortium, which involved nearly 3,000 scientists working at 20 centers in six countries: China, France, Germany, Great Britain, Japan, and the United States. The estimated budget for completing the genome was $3 billion, a cost of $1 per nucleotide. Driven in part by competition from private companies and the development of computer-automated DNA sequences such as the one described in Figure 3.14 and bioinformatics, the Human Genome Project turned out to be a rare government project that completed all of its initial goals, and several additional goals, more than 2 years ahead of schedule—and under budget.

One of the most aggressive competitors on the project was a private company called Celera Genomics (aptly named from a word meaning "swiftness") directed by Dr. J. Craig Venter, who we discussed ear-

lier in this section as the scientist directing The Institute for Genomic Research when it used shotgun cloning to sequence the genome for *H. influenza*.

Celera announced its intention to use novel technologies for shotgun cloning, such as the ones used to sequence the genome for *H. influenza*, and newly developed, high-powered computer-automated DNA sequencers to sequence the entire human genome in three years. Fearful of how private corporations might control the release of genome information, U.S. government groups involved in the Human Genome Project were effectively forced to keep pace with private groups to stay competitive in the race to complete the genome.

In 1998, as a result of accelerated progress, a revised target date of 2003 was set for completion of the project. On June 26, 2000, leaders of the Human Genome Project and Celera Genomics participated in a press conference with President Clinton to announce that a rough "working draft" of approximately 95% of the human genome had been assembled (nearly four years ahead of the initially projected timetable). At a joint press conference on February 12, 2001, Dr. Francis Collins, director of the NIH National Human Genome Research Institute and director of the Human Genome Project, and Dr. J. Craig Venter of Celera announced that a series of papers describing the initial analysis of the genome working draft sequence were published by their research groups in the prestigious journals *Nature* and *Science*, respectively. Scientists spent the next two years working to fill in thousands of gaps in the genome by completing the sequencing of pieces not yet finished, correcting misaligned pieces, and comparing sequences to ensure the accuracy of the genome. On April 14, 2003, the International Human Genome Sequencing Consortium announced that its work was done. A "map" of the human genome was essentially complete with virtually all bases identified and placed in their proper order.

What Have We Learned from the Human Genome?

One of the most surprising findings of the project was that the human genome consists of only 20,000 to 25,000 protein-coding genes, not 100,000 genes as predicted. The prediction was based primarily on estimates that human cells make approximately 100,000 to 150,000 proteins. One reason why the actual number of genes is so much lower than the predicted number is the discovery of large numbers of gene families with related functions. In addition, many genes that code for multiple proteins through alternative splicing have been found. It has been estimated that over half of all human genes may produce multiple proteins through alternative splicing.

Analysis of human genes by functional categories has provided genome scientists with a snapshot of the numbers of genes involved in different molecular functions. Figure 3.22 shows one proposed interpretation of assigned functions to human genes based on gene sequence similarity to genes of known function. Not surprisingly, many genes encode enyzmes while other large categories of genes encode proteins involved in signaling and communication within and between cells, and DNA and RNA binding proteins. Notice that this estimate also shows that approximately 42% of human genes have no known function, although recent evidence suggests that functions for over half of our genes remain unknown. Keep this in mind if you are interested in a career in genetics research because understanding what these genes do will provide exciting career opportunities for many years into the future.

Here is a summary of important highlights summarizing some of the basic concepts scientists have learned from the Human Genome Project:

- The human genome consists of approximately 3.1 billion base pairs; 2.85 billion base pairs have been fully sequenced.
- The genome is approximately 99.9% the same between individuals of all nationalities and backgrounds.
- Less than 2% of the genome codes for genes.

- The vast majority of our DNA is non–protein-coding, and repetitive DNA sequences account for at least 50% of the noncoding DNA.
- The genome contains approximately 20,000 to 25,000 protein-coding genes.
- Many human genes are capable of making more than one protein, allowing human cells to make perhaps 80,000 to 100,000 from only 20,000 to 25,000 genes.
- Functions for over half of all human genes are unknown.
- Chromosome 1 contains the highest number of genes. The Y chromosome contains the fewest genes.
- Much of the human genome shows a high degree of sequence similarity to genes in other organisms.
- Thousands of human disease genes have been identified and mapped to their chromosomal locations.

How will we benefit from the HGP? The complete sequence of the human genome has been described as the "blueprint" of humanity containing the scientific keys to understanding our biology and behaviors. Identifying all human genes is not as important as understanding *what* these genes do and *how* they function. We will not fully understand the function of all human genes for many years, if ever. One immediate impact of the HGP will be the identification of genes associated

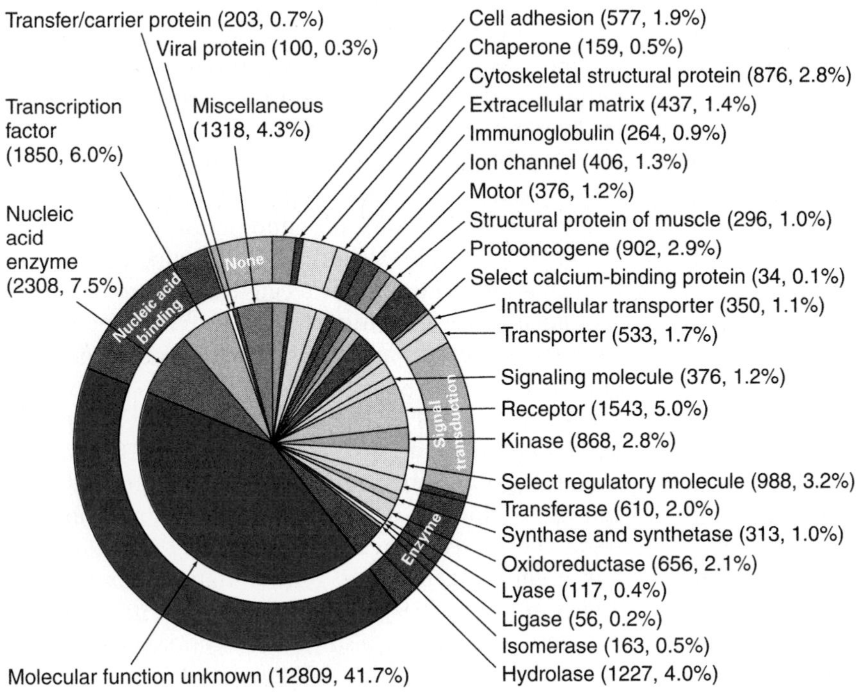

Transfer/carrier protein (203, 0.7%)
Viral protein (100, 0.3%)
Transcription factor (1850, 6.0%)
Miscellaneous (1318, 4.3%)
Nucleic acid enzyme (2308, 7.5%)
Nucleic acid binding
None
Signal transduction
Enzyme
Molecular function unknown (12809, 41.7%)

Cell adhesion (577, 1.9%)
Chaperone (159, 0.5%)
Cytoskeletal structural protein (876, 2.8%)
Extracellular matrix (437, 1.4%)
Immunoglobulin (264, 0.9%)
Ion channel (406, 1.3%)
Motor (376, 1.2%)
Structural protein of muscle (296, 1.0%)
Protooncogene (902, 2.9%)
Select calcium-binding protein (34, 0.1%)
Intracellular transporter (350, 1.1%)
Transporter (533, 1.7%)
Signaling molecule (376, 1.2%)
Receptor (1543, 5.0%)
Kinase (868, 2.8%)
Select regulatory molecule (988, 3.2%)
Transferase (610, 2.0%)
Synthase and synthetase (313, 1.0%)
Oxidoreductase (656, 2.1%)
Lyase (117, 0.4%)
Ligase (56, 0.2%)
Isomerase (163, 0.5%)
Hydrolase (1227, 4.0%)

Figure 3.22 Proposed Functions for the Numbers of Human Genes Assigned to Different Functional Categories Parentheses show percentages based on the 2001 published data by Venter et al. for 26,383 genes.

with human genetic diseases. In Chapter 11 we discuss some of the major findings of the human genome, particularly those that relate to human disease genes (see Figures 3.26 and 3.27). A better understanding of the genetic basis of many diseases will lead to the development of new strategies for disease detection and innovative therapies and cures. For updated information on the HGP and for an outstanding overview of goals, sequencing and mapping technologies, and the ethical, legal, and social issues of the HGP, visit the DOE Human Genome Project Information Site and other human genome sites listed at the Companion Website.

The Human Genome Project Started an "Omics" Revolution

The Human Genome Project and genomics are largely responsible for ushering in a new area of biological research—the "omics." It seems that every year, new or existing areas of biological research are being described as having an omics connection. For example,

- Proteomics—studying all of the proteins in a cell
- Metabolomics—studying proteins and enzymatic pathways involved in cell metabolism
- Glycomics—studying the carbohydrates of a cell
- Interactomics—studying the complex interactions of protein networks in a cell.
- Transcriptomics—studying all genes expressed (transcription) in a cell

As further evidence of the impact of genomics, a new field of nutritional science, called nutritional genomics or **nutrigenomics,** has emerged. Nutrigenomics is focused on understanding interactions between diet and genes. Several companies provide nutrigenomics tests in which they use microarrays or other genetic tests to analyze your genotype for genes thought to be associated with different medical conditions or aspects of nutrient metabolism. These companies then provide a customized report on nutrition, and claim that they can suggest diet changes you should make to improve your health and prevent illness based on your genes. Whether nutrigenomics is actually a valid scientific approach is a subject of debate by scientists.

Comparative Genomics

It might surprise you that in addition to studying the human genome, the HGP involved mapping and sequencing genomes from a number of model organisms, including *E. coli,* a model plant called *Arabidopsis thaliana,* the yeast *Saccharomyces cerevisiae,* the fruit fly *Drosophila melanogaster,* the nematode roundworm *Caenorhabditis elegans,* and the mouse *Mus musculus,*

among other species. Complete genome sequences of these model organisms have been incredibly useful for **comparative genomics** studies that allow researchers to study gene structure and function in these organisms in ways designed to understand gene structure and function in other species including humans. Because we share many of the same genes as flies, roundworms, and mice, such studies will also lead to a greater understanding of human evolution. The number of genes we share with other species is very high, ranging from about 30% of genes in yeast to about 80% of the genes in mice and about 95% of genes in chimpanzees (Table 3.3). Recently, the genome for "man's best friend" was completed and revealed that we share about 75% of our genes with dogs. Human DNA even contains around 100 genes that are also present in many bacteria. In 2006, researchers at Baylor College of Medicine completed the 814 million bp genome of the sea urchin (*Strongylocentrus purpuratus*), an invertebrate that has served as an important model organism particularly for developmental biologists. Of the 23,500 genes in the urchin genome, many include genes with important functions in humans.

The genomes of many model organisms have been completely sequenced, and there are literally hundreds of genomics projects underway worldwide. The NIH is working on a cancer genome project called the **Cancer Genome Atlas Project** to map important genes and genetic changes involved in cancer. Because of rapid advances in sequencing technologies, several companies have even proposed to sequence **personalized genomes** for individual people. In 2006, the X Prize Foundation announced the Archon X Prize for Genomics, a project to award $10 million to the first group to develop technology capable of sequencing 100 human genomes with a high degree of accuracy in 10 days for under $10,000 per genome. Other groups are working on sequencing a personalized genome for a mere $1,000.

In 2007, the Connecticut company 454 Life Sciences together with researchers at Baylor University sequenced James Watson's genome for approximately $1 million. 454 Life Sciences is developing an innovative DNA sequencer and decided that "Project Jim," sequencing the genome of the co-discoverer of the DNA structure, was a good high profile way to develop and promote their sequencing technology. James Watson provided 454 scientists with a blood sample in 2005 and by mid-2007 they presented Dr. Watson with two DVDs containing his genome sequence. Watson has allowed his sequence to be available to researchers except for the sequence of his apolipoprotein E gene (*ApoE*). *ApoE* gene mutations can indicate a disposition for Alzheimer's disease. Human genome pioneer Craig Venter also had his genome sequenced by scientists at The Craig J. Venter Institute.

Table 3.3 COMPARISON OF SELECTED GENOMES

Organism (scientific name)	Approximate Size of Genome (date completed)	Number of Genes	Approximate Percentage of Genes Shared with Humans	Web Access to Genome Databases
Bacterium (*Escherichia coli*)	4.1 million bp (1997)	4,403	Not determined	www.genome.wisc.edu/
Chicken (*Gallus gallus*)	1 billion bp (2004)	~20,000–23,000	60%	http://genomeold.wustl.edu/projects/chicken
Dog (*Canis familiaris*)	6.2 million bp (2003)	~18,400	75%	http://www.ncbi.gov/genome/guide/dog
Chimpanzee (*Pan troglodytes*)	~3 billion bp (initial draft, 2005)	~20,000–24,000	96%	http://www.nature.com/nature/focus/chimpgenome/index.html
Fruit fly (*Drosophila melanogaster*)	165 million bp (2000)	~13,600	50%	www.fruitfly.org
Humans (*Homo sapiens*)	~2.9 billion bp (2004)	~20,000–25,000	100%	www.doegenomes.org
Mouse (*Mus musculus*)	~2.5 billion bp (2002)	~30,000	~80%	www.informatics.jax.org
Plant (*Arabidopsis thaliana*)	119 million bp (2000)	~26,000	Not determined	www.arabidopsis.org
Rat (*Rattus norvegicus*)	~2.75 billion bp (2004)	~22,000	80%	www.hgsc.bcm.tmc.edu/projects/rat
Roundworm (*Caenorhabditis elegans*)	97 million bp (1998)	19,099	40%	genomeold.wustl.edu/projects/celegans
Yeast (*Saccharomyces cerevisiae*)	12 million bp (1996)	~5,700	30%	genomeold.wustl.edu/projects/yeast.index.php

A division of the NIH is also working on an Environment Genome Project designed to catalog variations in genes affected by environmental toxins, genes that degrade toxins, and genes that help cells repair DNA. Throughout this book you will learn about many applications of genomics. As another example, in Chapter 5 we discuss microbial genome sequencing projects that are sequencing genomes for literally hundreds of newly identified microbes.

Recently the first genome for a tree, the black cottonwood (a type of poplar) was sequenced. To date, the poplar's 45,555 genes are the highest number found in a genome. Scientists anticipate using data from this project to help the forestry industry make better products, including biofuels or even genetically engineered poplars to capture high levels of carbon dioxide from the atmosphere. The honeybee genome was also recently completed, and information from this project will be used to

understand bee genetics and behavior to help the honey-producing industry as well as advance an understanding of how bee toxins produce allergic responses.

Stone Age Genomics

As yet another example of how genomics has taken over areas of DNA analysis, a number of labs around the world are involved in analyzing "ancient" DNA. These studies are generating fascinating data from minuscule amounts of ancient DNA from bone and other tissues and fossil samples that are tens of thousands of years old. Analysis of DNA from a 2,400-year-old Egyptian mummy, mammoths, Pleistocene-age cave bears, and Neanderthals are some of the most prominent examples of **Stone Age genomics,** also called **paleogenomics.** In 2005, researchers from McMaster University in Canada and Pennsylvania

CAREER PROFILE

Career Options in the Biopharmaceutical Industry: Perspectives from a Recent Graduate

Biopharmaceutical companies use many biological disciplines to discover new treatments and to develop them from the research and development stages through to commercial manufacturing, marketing, and finally sales. What makes the biopharmaceutical industry different from the pharmaceutical industry? Pharmaceutical companies, in general, develop chemically based drug compounds that treat illness nonspecifically, whereas the biopharmaceutical industry uses knowledge of biological systems to develop biologically derived drug compounds that treat illness in a highly specific manner. Although the biopharmaceutical industry is currently much smaller than the pharmaceutical industry, its growth is triggered by the increasing awareness of the potential to treat illnesses effectively and possibly to find cures for illnesses that were once thought incurable.

Career paths within the biopharmaceutical industry are similar to those in the pharmaceutical industry, owing to the similarity in the path of drug development. The career possibilities are wide ranging; however, gaining employment with a biopharmaceutical company is not easy. Competition is fierce, and the truth is that no matter how well a person may fit a position, there is always someone who will be just as qualified. For that reason, start thinking about career options shortly after selecting a major. The best way to begin is by becoming familiar with the industry. Many of the websites cited in this book, particularly those in Chapter 1, are excellent resources for learning about different job opportunities. This knowledge will enable you to customize your education to better serve a particular job function—a sure way to get ahead of the competition.

Many entry-level positions are available to those with associate's and bachelor's degrees; however, having a bachelor's degree will guarantee a slightly higher starting salary. In fact, many of the higher-paying entry-level positions require an advanced degree, such as a master's degree or even a Ph.D. Knowing this may greatly influence your decision to continue your education before entering the workforce. However, as important as educational training is, employers are constantly looking for experienced individuals. Gaining experience through volunteer work, independent study research, or an internship provides the best opportunity to gain real-life experience while completing your studies. Internships in the biotechnology industry not only are an excellent way to learn about different roles in a company to see what you like but also may lead to employment after graduation.

If you cannot find the ideal job in the industry after graduation, temporary employment is another good way to begin with a biopharmaceutical company. Many placement services work with job seekers and biotechnology companies to find people who can fill a need for a short period of time; however, if you do a good job in a temporary position, the company may try to find a permanent position for you.

When considering a position, it is important to understand the job description as well as what is required of that position. Numerous companies hire people as lab technicians, but the job descriptions for the position are as numerous as the jobs. Because every company is different, picking a career based on a job title alone is difficult. For example, within a company, lab technicians can work in such different areas as quality control, research and development, and diagnostics. All these areas have very different functions, but the lab technicians use the same tools and knowledge to perform their daily work. For example, quality control lab technicians follow very controlled and highly precise procedures to test the quality of a product; their main function is to ensure that a good product is being produced. Technicians working in a diagnostic capacity are responsible for testing a new product or procedure to make sure it works correctly and then for recommending improvements. Research and development or discovery lab technicians are responsible for just that; their main function is to develop new products as well as to find new uses for old products. It is not required that you have prior experience in every experimental assay or lab procedure to become a lab technician. A good company will always teach you what you need to know to do the specific job; however, all lab technicians must have good lab skills and be organized, detail oriented, motivated, and, most importantly, enthusiastic.

A potential downside of working in the biopharmaceutical industry is that all work is tailored around developing, producing, and selling a product. Creativity and originality are not always incorporated into the daily routine of life in the laboratory. The work environment is usually very controlled because all operations are regulated by the FDA and must be documented rigorously. Also, most products are produced through biological processes, which means that work schedules are linked directly to the biological process such as waiting for cells to grow to the right density. In some cases, this can provide a shorter work week or less structured work hours.

In the biopharmaceutical industry, the fast-paced work environment is always changing and always demanding that its employees work hard and efficiently. Each day brings new obstacles and challenges, which makes working in this industry so exciting.

—Contributed by Robert Sexton (B.S., Biology, M.B.A., Monmouth University), Regulatory Affairs, Quality and Compliance, sanofi-aventis.

State University published partial sequence data (about 13 million bp) from a 27,000-year-old woolly mammoth. This study showed that there is approximately 98.5% sequence identity between mammoths and African elephants.

A team of scientists led by Svante Pääbo at the Max Planck Institute for Evolutionary Anthropology in Germany plans to produce a rough draft of the Neanderthal (*Homo neanderthalensis)* genome in about two years. In 1997, a team led by Pääbo sequenced portions of Neanderthal mitochondrial DNA from a fossil. In late 2006, Pääbo's group, along with a number of scientists in the United States, reported the first sequence of about 65,000 bp of nuclear DNA isolated from bone of a 38,000-year-old Neanderthal sample from Croatia. Because Neanderthals are close relatives of humans, sequencing the Neanderthal genome is expected to provide a tremendous opportunity to use comparative genomics to advance our understanding of evolutionary relationships between humans and humans.

QUESTIONS & ACTIVITIES

Answers can be found in Appendix 1.

1. Distinguish among gene cloning, recombinant DNA technology, and genetic engineering by describing each process and discussing how they are interrelated. Provide examples of each approach as described in this chapter.

2. Describe the importance of DNA ligase in a recombinant DNA experiment. What does this enzyme do and how does its action differ from the function of restriction enzymes?

3. Your lab just determined the sequence of a rat gene thought to be involved in controlling the fertilizing ability of rat sperm. You believe a similar gene may control fertility in human males. Briefly describe how you could use what you know about this rat gene combined with PCR to clone the complementary human gene. Be sure to explain your experimental approach and the necessary lab materials. Also, explain in detail any procedures necessary to confirm that you have a human gene that corresponds to your rat gene.

4. What features of plasmid cloning vectors make them useful for cloning DNA? Provide examples of different types of cloning vectors, and discuss their applications in biotechnology.

5. If you performed a PCR experiment starting with only one copy of double-stranded DNA, approximately how many molecules would be produced after 15 cycles of amplification?

6. Compare and contrast genomic libraries with cDNA libraries. Which type of library would be your first choice to use if you were attempting to clone a gene in adipocyte (fat) cells that encodes a protein thought to be involved in obesity? Explain your answer. What type of library would you choose if you were interested in cloning gene regulatory elements such as promotor and enhancer sequences?

7. Visit the Online Mendelian Inheritance in Man Site (OMIM) at the Companion Website then click on "Search the OMIM database." Type "diabetes" in the search box and then click "Submit Search." What did you find? Try typing "114480" in the search box. What happened this time? Alternatively, search for a gene you might be interested in and see what you can find in OMIM. If the results of your OMIM search are too technical, visit the "Genes & Disease" section of the NCBI site from the Companion Website then search for "diabetes."

8. Software analysis of DNA sequences has made it much easier for molecular biologists to study gene structure. This activity is designed to enable you to experience applications of DNA analysis software. Imagine that the following very short sequence of nucleotides, GGATCCGGCCGGAATT CGTA, represents one strand of an important gene that was just mailed to you for your research project. Before you can continue your research, you need to find out which restriction enzymes, if any, cut this piece of DNA.

Go to the Webcutter site from the Companion Website. Scroll down the page until you see a text box with the title, "Paste the DNA Sequence into the Box Below." Type the sequence of your DNA piece into this box. Scroll down the page, leaving all parameters at their default settings until you see "Please Indicate Which Enzymes to Include in the Analysis." Click on "Only the following enzymes:" and then use the drop-down menu and select *Bam*HI. Scroll to the bottom of the page and click the "Analyze sequence" button. What did you find? Is your sequence cut by *Bam*HI? Analyze this sequence for other cutting sites to answer the following questions. Is this sequence cut by *Eco*RI? How about *Sma*I? What happens if you do a search and scan for cutting sites with all enzymes in the database?

9. Go to the molecular biology section of The Biology Project website from the University of Arizona (http://www.biology.arizona.edu/ molecular_bio/problem_sets/Recombinant_ DNA_Technology/recombinant_dna.html).

Link to "Recombinant DNA Technology" and test your knowledge of recombinant DNA technology by working on the questions at this site.

10. Search the Web for the company Sciona, which markets the Cellf DNA evaluation kit for nutrigenomics. Do you think kits such as this should be used even though they are largely based on unproven information?

11. What is the structural difference between a deoxyribonucleotide (dNTP) and a dideoxyribonucleotide (ddNTP) used for DNA sequencing?

12. Describe several findings of the Human Genome Project.

References and Further Reading

Campbell, A. M., and Heyer, L. J. (2007). *Discovering Genomics, Proteomics, and Bioinformatics*, 2e. San Francisco: Benjamin Cummings.

Chaudhuri, J. D. (2005). Genes Arrayed Out for You: The Amazing World of Microarrays. *Medical Science Monitor*, 11: RA52–62.

Church, G. M. (2006). Genomes for All. *Scientific American*, 294: 47–54.

Hercher, L. (2007). Diet Advice from DNA? *Scientific American*, 297: 84–89.

IHGS Consortium. (2001). Initial Sequencing and Analysis of the Human Genome. *Nature*, 409: 860–891.

IHGS Consortium. (2004). "Finishing the Euchromatic Sequence of the Human Genome." *Nature*, 431: 931–945.

Krane, D. E., and Raymer, M. L. (2003). *Fundamental Concepts of Bioinformatics*. San Francisco: Benjamin Cummings.

Lau, N. C., and Bartel, D. P. (2003). Censors of the Genome. *Scientific American*, 289: 34–41.

Noonan, J. P., Coop, G., Kudaravalli, S., et al. (2006). Sequencing and Analysis of Neanderthal DNA. *Science*, 314: 1113–1118.

Palladino, M. A. (2006). *Understanding the Human Genome Project*, 2e. San Francisco: Benjamin Cummings.

Pennisi, E. (2006). The Dawn of Stone Age Genomics. *Science*, 314: 1068–1071.

Venter, J. C., Adams, M. D., Myers, E. W., et al. (2001). The Sequence of the Human Genome. *Science*, 291: 1304–1351.

Visit www.pearsonhighered.com/biotechnology to download learning objectives, chapter summary, "Keeping Current" web links, glossary, flashcards, and jpegs of figures from this chapter.

Chapter 4

Proteins as Products

After completing this chapter you should be able to:

■ Describe in general terms the molecular structure of proteins.

■ Provide three examples of the medical applications of proteins.

■ Explain the uses of some biotechnologically produced enzymes in industry.

■ List common household products that may include manufactured proteins as ingredients.

■ Discuss the advantages and disadvantages of microbial, fungi, plant, and animal sources for protein expression.

■ Explain why *E. coli* is frequently used for protein production.

■ Explain why protein glycosylation may determine the choice of a protein expression system.

■ Describe a general scheme for protein purification of hemoglobin.

■ Explain how the target protein is separated from other cell proteins given a specific purification sequence.

■ Explain proteomics and how it will affect future protein studies.

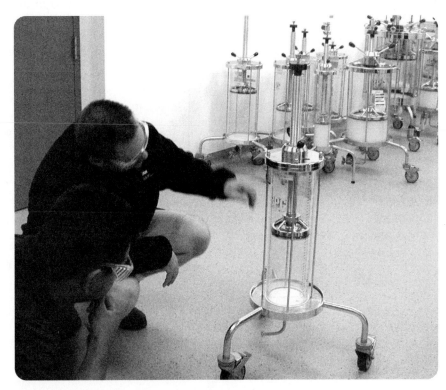

Lab Supervisor Preparing a Purification Column to Separate a Commercial Biological Product.

4.1 Introduction to Proteins as Biotech Products

Tropical rainforests, the deepest reaches of the ocean, boiling geysers in Yellowstone National Park, and whale skeletons—these are on the frontier of the scientific quest for proteins. **Proteins** are large molecules that are required for the structure, function, and regulation of living cells. Each protein molecule has a unique function in the biochemical reactions that sustain life. As researchers explore the proteins that occur in nature, they unlock secrets that govern growth, speed chemical decomposition, and protect us from disease.

The applications of proteins are as numerous as the proteins themselves. Consider whale skeletons, for example. During the natural decomposition process, the bones are often colonized by bacteria, some of which have evolved especially to digest the fatty residue on the bones. The proteins that the bacteria produce to break down the fats are adapted to the frigid waters of the deep sea. Researchers recognized that a substance with the ability to dissolve fats at cold temperatures would make a great additive for commercial laundry detergents.

Even after a protein is discovered in nature and an application is matched to its characteristics, a great deal of ingenuity is required to produce proteins in the necessary quality and quantity required for commercial use. For example, if we plan to mass-produce a great new cold-water detergent, we cannot rely on an unlimited supply of whale skeletons but rather we must find another source for those proteins. Fortunately, biotechnology can facilitate production of virtually any protein. We focus on those production processes in this chapter.

In 2000, the National Institutes of Health launched the Protein Structure Initiative, a 10-year $600 million effort to identify the structure of human proteins. This massive effort to further understand the structure of proteins will help researchers understand their function. More than 1,200 protein structures have been identified so far, and the knowledge of the relationship of a gene sequence to a protein's structure has been advanced. The public database that is part of the initiative currently holds more than 33,000 protein sequences, which means that only about 5% have been structured. The primary objective of the initiative is to be able to model unknown protein structures based on structural comparisons to those stored in the database.

We begin this chapter with a quick survey of the many applications of proteins in a variety of industries. Then we look at the nature of protein structures, paying special attention to the process of protein folding. With that as a foundation, we delve into some of the details of protein processing, beginning with the methods of expressing proteins. We then learn how expressed proteins are purified and examine the processes used to analyze and verify the final product. Although there is no one best method for processing proteins, several generally useful techniques, shown in Figure 4.1, are available. In this chapter, we look at those generally useful techniques, keeping in mind that the specifics of protein processing vary from case to case.

4.2 Proteins as Biotechnology Products

The use of proteins in manufacturing processes is a time-tested technology. For example, two of the oldest

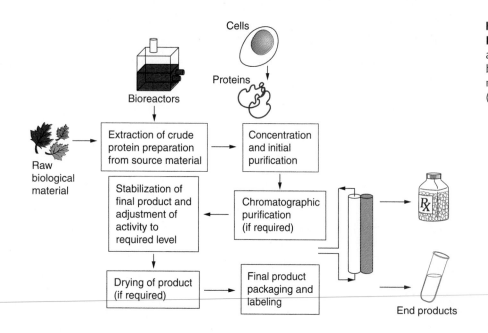

Figure 4.1 Basic Steps in Bioprocessing Purifications can be accomplished from raw materials or from bioreactors. The steps in the process must be devised and are often unique (and patentable).

food-processing endeavors depend on proteins: beer brewing and winemaking. Fermentation, used in beer and wine production, depends on the enzymes produced by the yeast and others added to the batch. Cheese-making is another industry that has always used proteins, and thanks to bioengineering, the protein source now used is from engineered bacteria (which substitute for the calves stomachs that originally provided the cheese-making enzyme). Even though the value of proteins in manufacturing had long been evident, we were not able to further this knowledge until the 1970s, when recombinant DNA technology was first developed and it became possible to produce specific proteins on demand. Since that time, the production of proteins has been the driving force behind the development of new products in a wide variety of industries.

Many of these applications depend on the power of a group of proteins called **enzymes** to speed up chemical reactions. Countless industries depend on enzymes to break down large molecules, a process called **depolymerization.** These enzymes include carbohydrases like **amylase,** which breaks down starch; **proteases,** which break down other proteins; and **lipases,** which break down fats. Such enzymes (like chymosin) are used in food and beverage production and in various bulk-processing industries, as shown in Figure 4.2.

Hormones that carry chemical messages and **antibodies** that protect the organism from disease are two other groups of proteins produced commercially, primarily for the medical industry. Hormones also have agricultural uses. For instance, hormones can stimulate the rooting of plant cuttings and encourage more rapid growth of meat animals. (We discuss hormones used in agriculture in more detail in Chapters 6 and 7.)

Making a Biotech Drug

Therapeutic proteins, such as monoclonal antibodies, blood proteins, and enzymes produced by living organisms to fight disease, can be thought of as biotech drugs. Unlike other medicines, biotech drugs are not synthetically produced (i.e., chemically synthesized by adding one compound at a time) but are usually produced through microbial fermentation or by mammalian cell culture. Today, there are nearly 400 new biotechnology medicines in the pipeline, and the majority are proteins. If all of these succeed in trials, it will significantly add to the approximately 40 biotech drugs currently in use.

Producing biotech drugs is a complicated and time-consuming process. Researchers can spend many years just identifying the relevant therapeutic

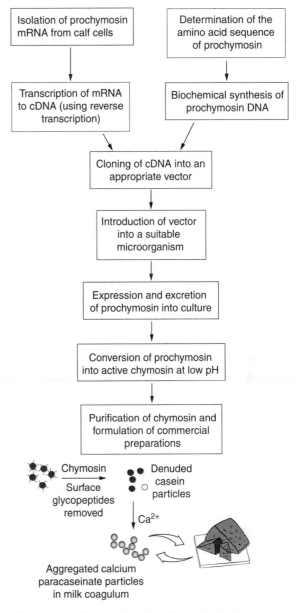

Figure 4.2 Cheese Production Casein, the primary ingredient in cheese, is the result of a chemical conversion that depends on chymosin. This enzyme has been obtained from the stomach walls of unweaned calves for centuries. Today 80% of the chymosin manufactured is produced (and purified) from genetically modified cells.

protein, determining its gene sequence, and working out a process to make the necessary molecules using biotechnology. Once the method for producing biotech drugs is determined, technicians can produce large batches of the protein products by growing host cells that have been transformed to contain the therapeutic gene in carefully controlled conditions in a bioreactor. (A bioreactor is any contained biological production system designed to promote enzymatic reactions.) The cells are stimulated to produce the target proteins through precise culture conditions that include a balance of temperature, oxygen, acidity, and other variables. At the appropriate time (whose duration varies depending on the protein produced and the nature of the organism), the proteins are isolated from the cultures, stringently tested at every step of purification (which we discuss later in the chapter), and formulated into pharmaceutically active products. Manufacturing technicians must strictly comply with Food and Drug Administration (FDA) regulations at all stages of the procedure (see Chapter 12 for examples of these regulations).

Medical Applications

Biotechnology proteins have revolutionized the health care and pharmaceutical industries in the past several decades. Many illnesses, from common conditions like diabetes to rare diseases like Gaucher's disease, can be treated by replacing missing proteins. In the case of diabetes, the missing protein is the hormone insulin. Not long ago, insulin had to be harvested from pigs or cows. This was less than ideal because human bodies often rejected this foreign protein. Researchers overcame the problem by turning to an unlikely source: the bacteria *Escherichia coli (E. coli)*. By inserting human genes into *E. coli*, they created microscopic insulin factories. (We look at the remarkable use of genetically engineered organisms as a source of proteins more closely later in this chapter.)

The FDA approved this new insulin in 1982, making it the first recombinant DNA drug. The ability to produce an abundant supply of human insulin has improved the health and lives of millions of people. (Table 4.1 lists some other protein-based pharmaceutical products.)

Another dramatic example of the potential use of proteins in health care is in the treatment of Gaucher's disease. In this rare disorder, a mutation results in the buildup of fats in the organs, including the brain. Left untreated, the disease can be fatal. The current treatment, lifelong replacement of a missing enzyme, is extremely costly. Human placentas are the only source of the enzyme, and between 400 and 2,000 placentas are required for a single dose. Thanks to biotechnology, it may be possible in a few years to harvest the essential enzyme from genetically altered tobacco plants.

In addition to the pharmaceutical industry, the manufacturing industries are also benefiting from the ready availability of bioengineered proteins. Enzyme (proteins) serve myriad purposes, such as making detergents work better, increasing the flow of oil in drilling operations, and cleaning contact lenses. (Table 4.2 lists some other protein-based industrial products.)

Food Processing

You can find plenty of examples of industrial uses of enzymes in your own kitchen cupboard. For decades

---

Table 4.1 SOME PROTEIN-BASED PHARMACEUTICAL PRODUCTS (MOST PRODUCED AS RECOMBINANT PROTEINS)

Protein	Application
Erythropoietins	Treatment of anemia
Interleukins 1, 2, 3, 4	Treatment of cancer, AIDS; radiation- or drug-induced bone marrow suppression
Monoclonal antibodies	Treatment of cancer, rheumatoid arthritis; used for diagnostic purposes
Interferons (α, β, γ, including consensus)	Treatment of cancer, allergies, asthma, arthritis, and infectious disease
Colony-stimulating factors	Treatment of cancer, low blood cell count; adjuvant chemotherapy; AIDS therapy
Blood clotting factors	Treatment of hemophilia and related clotting disorders
Human growth factor	Treatment of growth deficiency in children
Epidermal growth factor	Treatment of wounds, skin ulcers, cancer
Insulin	Treatment of diabetes mellitus
Insulin-like growth factor	Treatment of type II diabetes mellitus
Tissue plasminogen factor	Treatment after heart attack, stroke
Tumor necrosis factor	Cancer treatment
Vaccines	Vaccinate against hepatitis B, malaria, herpes

Table 4.2 SOME ENZYMES AND THEIR INDUSTRIAL APPLICATIONS

Enzyme	Application
Amylases	Digest starch in fermentation and processing
Proteases	Digest proteins for detergents, meat/leather, cheese, brewing/baking, animal/human digestive aids
Lipases	Digest lipids (fats) in dairy and vegetable oil products
Pectinases	Digest enzymes in fruit juice/pulp
Lactases	Digest milk sugar
Glucose isomerase	Produce high-fructose syrups
Cellulases/hemicellulases	Produce animal feeds, fruit juices, brewing converters
Penicillin acylase	Produce penicillin

the food processing industry has used proteins to improve baby food, canned fruit, cheeses, baked goods, beer, desserts, and dietetic foods. The enhancements are quite varied.

In bread, for example, proteins may be used to make starches easier for yeast to act on, allowing the dough to rise more quickly. As a result, the bread dough is easier to handle, and the final texture is more consistent. That same loaf of bread may owe its appetizing brown crust to enzyme additives that break down starches and encourage caramelization, which is the source of that golden brown color.

The fruit at the bottom of yogurt is firmer, soft drinks have a longer shelf life, and ice cream has a smoother texture and better freezing characteristics because protein products are used in their manufacture.

Textiles and Leather Goods

Other examples of industrial proteins are right in your closet. For more than a century, enzymes have been used in the textile industry to break down the starches used to "size" products during the manufacturing process. Enzymes are also replacing harsh chemicals and processes used to lighten and soften fabrics. The use of enzymatic bio-bleaches reduces the demand for ordinary bleaches and the time-consuming and costly environmental cleanup required after their use.

If you have wool items in your wardrobe, they also may have been improved by enzymes that have taken the place of harsh chemicals to clean the wool,

soften it, or shrink-proof it. If you have a leather jacket, shoes, or belt in your closet, they were also probably processed using enzymes. In fact, leather production has always depended on proteins to remove hair and fur from pelts and soften the tissue to make it workable. However, there was room for improvement in this technology, and today's leather is softer, cleaner, and stronger because of new biotechnology protein products.

The textile industry has many roles beyond filling your closet. Nonwoven products including geo-textiles used for mulch, erosion control, and road building all include flax, hemp, and jute fibers processed with enzymes. These fibers are also used in the noise-reducing panels in many cars. In the future, the air filter in your car may be made from the enzyme-processed fibers as well.

Detergents

As noted earlier, when enzymatic ingredients are added to detergents, they do a better job of cleaning and are more biodegradable. Laundry detergents take advantage of the specific roles of proteases, lipases, and amylases to dissolve stains in cooler water. Enzymes in detergents have also become more powerful so they can remove invisible residues that make it easier for fabric to become resoiled. If you look at the label of many laundry stain removers, you will see enzymes listed as the first, and sometimes only, active ingredient. Dishwasher detergents include enzymes that dissolve proteins and starches. Without enzymes, some films and spots cannot be removed from dishes.

Paper Manufacturing and Recycling

High-speed paper-manufacturing machines can be slowed or even halted when pitch accumulates on the moving parts, but protein-based products are able to remove it and keep the factories humming. Protein products have also reduced the negative environmental impacts of the paper industry. In the past, harmful chlorine bleaches were used to lighten paper products. Now enzymes can be used to do the same job, resulting in far less pollution. Even paper recycling benefits from enzymes. Office paper, for example, needs to be "de-inked," and enzymes make the process easier, which makes the recycled paper less expensive and therefore in greater demand.

Adhesives: Natural Glues

If you have ever tried to pull a barnacle or mussel from a rock in a tide pool, you know about the excellence of natural protein adhesives. The water-resistant glue that attaches mussels to the rocks is made of several types of proteins. Natural protein adhesives are strong and water insoluble, making them ideally suited for a wide variety of biotechnical applications. An important example occurs in the medical field. Because these natural adhesives are nontoxic, biodegradable, and rarely trigger an immune response, they can be used to reattach tendons and tissues, fill cavities in teeth, and repair broken bones.

Research into natural adhesives may have other benefits as well. The shipping industry is plagued by mussels and other mollusks attaching themselves to ship hulls and other underwater equipment. As scientists learn more about the protein glues produced by these creatures, they may discover a biodegradable inhibitor that can prevent this problem, which may in turn allow protein biotechnology to provide solutions for other problems.

Bioremediation: Treating Pollution with Proteins

In addition to reducing the amount of pollutants produced during industrial processes, proteins can also be used to clean up harmful wastes. Organic wastes from feedlots, homes, and businesses are a growing threat to the environment, especially aquatic ecosystems. Enzymes can be used to digest the organic wastes before they cause trouble.

Another promising new application for proteins is in neutralizing "heavy metal" pollutants like mercury and cadmium. These dangerous elements can persist in the environment, causing harm to organisms throughout the food chain. Because they are metals, they resist enzymatic breakdown, but that does not mean proteins cannot be used to neutralize them (see Chapter 9 for the details of this process). Through bioengineering, we can create microorganisms that have a sticky coat of **metallothioneins,** proteins that actually capture heavy metals. In this case, the pollutant is not dismantled or digested but simply made less dangerous. When toxic metals are bound to bacteria, they are less likely to be absorbed by plants and animals.

Researchers are currently using the power of genetic engineering to create new, better biological tools to attack and destroy toxic substances. Because the research process sometimes depends on randomly shuffling genes in the bacteria, the enzymes produced by the rearranged genes can be more or less reactive than those naturally produced. In a sense, scientists are accelerating the process of random mutation and evolution with the hope of discovering new, more efficient pollution-eating proteins, as described in greater detail in Chapter 9.

YOU DECIDE

Testing for the Best Product: Who Should Pay?

As noted elsewhere in this chapter, it takes about $500 million to bring a drug to market. Included in this cost is the price of drugs that are not approved owing to some adverse reaction or ineffectiveness. Usually, the purification process has been developed and the product is in human trials when this happens. Although drug costs are often high, many people do not realize these prices include the cost of research on products that did not make it to the drug stage. If we also add the ability to determine which drug is best for each patient (pharmacogenetics, Chapter 1) to the cost of bringing a drug to market, the cost climbs higher. Because biotechnology companies are supposed to make a profit for their stockholders, it is more profitable for the company to have everyone buy their drug, even if the drug is not entirely suited for them. What do you think is the best solution: higher prices and better drugs, or lower prices and drugs that do not always work?

4.3 Protein Structures

We saw in Chapter 2 that ribosomes are the factories that form proteins. To understand the processes of expressing and harvesting proteins, we need to look at the molecular structure of proteins more closely.

Proteins are complex molecules built of chains of amino acids. Like all molecules, proteins have specific molecular weights. They also have an electrical charge that causes them to interact with other atoms and molecules. This ability to interact is the key to the biological activity of proteins. Consider, for example, the way the chemical structure and electrical charge of an amino acid can influence its interactions with water: The molecules will be either **hydrophilic** (water loving, as if the amino acid were magnetically attracted to water molecules) or **hydrophobic** (water hating, as if the water molecules and amino acids magnetically repel one another).

Structural Arrangement

Proteins are capable of four levels of structural arrangement, all depending on the specific chemical sequences of their amino acid subunits. (See Figure 4.3.) These include the primary, secondary, tertiary, and quaternary structure.

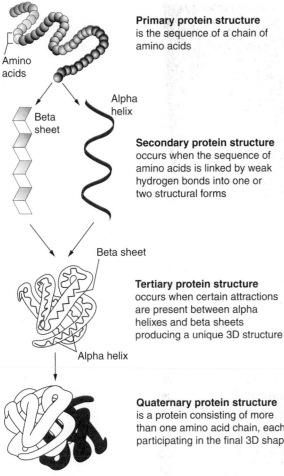

Primary protein structure
is the sequence of a chain of amino acids

Amino acids

Beta sheet

Alpha helix

Secondary protein structure
occurs when the sequence of amino acids is linked by weak hydrogen bonds into one or two structural forms

Beta sheet

Tertiary protein structure
occurs when certain attractions are present between alpha helixes and beta sheets producing a unique 3D structure

Alpha helix

Quaternary protein structure
is a protein consisting of more than one amino acid chain, each participating in the final 3D shape

Figure 4.3 The Four Levels of Protein Structure The proper folding of proteins is necessary for full functional capability. Purification methods must guarantee that proper folding is retained.

Primary structures

The 20 commonly occurring amino acids are the building blocks that make up proteins. Ten to 10,000 amino acids can be linked together in a head-to-tail fashion. The sequence in which amino acids are linked is known as the *primary structure*. Altering a single amino acid in the sequence can mean that the protein loses all function. Genetic diseases are often the result of these protein mutations.

Secondary structures

Secondary protein structures occur when chains of amino acids fold or twist at specific points, forming new shapes due to the formation of hydrogen bonds between the amino acids. The most common shapes, alpha helices and beta sheets, are described in detail in the section on protein folding. Both the helix and sheet structures exist because they are the most stable structures the protein can assume. In other words, these forms take the least energy to maintain.

Tertiary structures

Tertiary protein structures are three-dimensional polypeptides (large molecules made up of many similar, smaller molecules) that are formed when secondary structures combine and are bound together. The bonds that hold together tertiary structures occur between amino acids capable of forming secondary bonds (like cysteine, which can form disulfide bonds with another adjacent cysteine crosslinking the protein into a unique shape). An example of a protein with tertiary structure is the enzyme ribulose biphosphate carboxylase (RubisCo). This enzyme can absorb energy directly from the sun, and it plays an essential role in photosynthesis. Without it, life would not exist.

Quaternary structures

Quaternary protein structures are unique, globular three-dimensional complexes built of several polypeptides. Hemoglobin, which carries oxygen in the blood, is an example of a protein with quaternary structure.

Protein Folding

Everything that is important about a protein—its structure, its function—depends on folding. Folding describes how different strands of amino acids take their shapes; for example, sickle cell anemia results from a misfolding due to a single amino acid replacement at a strategic location in the primary structure. If the protein is folded incorrectly, not only will the desired function of the protein be lost but the resulting misfolded protein can be detrimental. For example, the plaque that forms in Alzheimer's disease accumulates because the misfolded protein cannot be broken down by the enzymes present in brain cells.

The first breakthrough in understanding the fundamental forms of proteins came in 1951. Researchers described two regular structures—dubbed α (alpha) and β (beta)—that are the most common results of the protein folding process. Both arrangements depend on the hydrogen bonds that tie together the molecules built of chains of amino acids. Researchers have learned that the tangled "plaques" that are part of the damage done to the brain by Alzheimer's disease may be the result of errors in the protein folding process. Cystic fibrosis, mad cow disease (bovine spongiform encephalitis, or BSE), many forms of cancer, and heart attacks have all been linked to clumps of incorrectly folded proteins. Because protein folding that occurs naturally presents problems, it is easy to understand why one of the biggest challenges biotechnology faces is understanding and controlling the folding in the manufacturing process.

Quick chemistry reminder

A hydrogen atom is a simple proton orbited by an electron. Nitrogen and oxygen are both "electron-hungry" structures. When these two elements come into contact with hydrogen atoms, much of the hydrogen atom's electron cloud is pulled away, leaving the hydrogen positively charged. If the positively charged hydrogen atom comes into contact with a negatively charged atom, there is an attraction between the two. This attraction is known as the hydrogen bond.

In the alpha-helix arrangement, the amino acids form a right-handed spiral. The hydrogen bonds stabilize the structure, linking an amino acid's nitrogen atom to the oxygen atom of another amino acid. Because the links occur at regular intervals, the spiraling chain is formed.

In the beta-sheet structure, the hydrogen bonds also link the nitrogen and oxygen atoms; however, because the atoms belong to amino acid chains that run side by side, a flat sheet is essentially formed. The sheets can either be "parallel" (if the chains all run in the same direction) or "antiparallel" (in which case the chains alternate in direction). One of the fundamental elements of protein structure, the beta-turn, occurs when a single chain loops back on itself to form an anti-parallel beta sheet. Even though they are not entirely random, other arrangements are known as random coils, even though they might better be described as unperiodic coils.

No matter what structure the protein takes, it is important to remember that structures are fragile. Those hydrogen bonds can be broken easily, damaging a valuable protein.

Glycolsylation

After a protein is synthesized on the ribosome, more than 100 **posttranslational modifications** (the most common of these is glycolsylation) occur. In **glycolsylation,** carbohydrate (sugar molecule) units are added to specific locations on proteins (see Figure 4.4). This change can have a significant effect on the protein's activity: It can increase solubility and orient proteins into membranes and may extend the active life of the molecule in the organism. Because glycolsylation can affect the bioactivity of proteins, it can affect the amount of protein produced in an organism.

Protein Engineering

At times during protein engineering, it is valuable to introduce specific, predefined alterations in the amino acid sequence. This can be done with directed molecular evolution technology. A major biotech company, for example, induces mutations randomly into genes and then selects the organisms (bacteria) with the protein product (enzyme) that has the highest activity. In this way, they have been able to produce organisms (and industrial enzymes) that tolerate more than 1.0 M cyanide concentration. This is significant because it would never have occurred by "natural" molecular evolution. No natural environment has experienced cyanide at this level. The resulting selected organisms can be used to remediate cyanide contamination resulting from mining and other industrial waste accumulation. This would never happen in a "natural selection" event because the environment would not change that much to select for

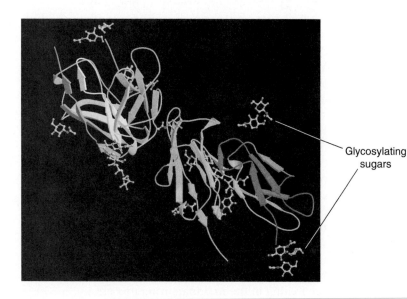

Glycosylating sugars

Figure 4.4 Three-Dimensional Protein Structure with Glycolsylated Side Chains Glycolsylation occurs within eukaryotic cells and probably extends the life of the protein.

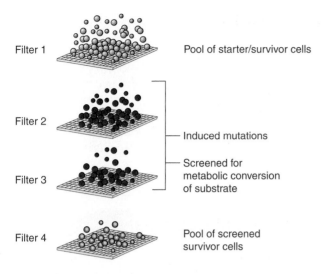

Filter 1 — Pool of starter/survivor cells

Filter 2

Induced mutations

Screened for
metabolic conversion
of substrate

Filter 3

Filter 4 — Pool of screened
survivor cells

**Figure 4.5 Directed Molecular Evolution
Technology** Genes with valuable proteins can be subjected to
mutational events. This process generates a diverse group of
novel gene sequences. The gene product is screened for unique
properties. After a measurable predetermined improvement is
achieved, the process can be repeated until the maximum
function is obtained. Unlike evolution by natural selection, this
process focuses on the properties of the protein, not the
organism, and can achieve changes that may never occur in
nature. Maxygen, Inc. (www.maxygen.com) has been a leader
in commercializing this technology.

these types of bacterial survivors. Directed molecular
evolution requires introducing specific changes in the
nucleotide sequences of a particular gene, as seen in
Figure 4.5.

Such newly modified genes can then be introduced into a host cell, where the required amino acid
sequence is produced by the host system. This
technique allows researchers to create proteins with
specific enhancements. Unlike naturally occurring
mutation, directed molecular evolution focuses
only on mutations that occur in a specific gene and
selects the best proteins from that gene, irrespective of
the potential benefits it may have for the original
organism. For example, when *E. coli* manufactures
human insulin, there is no benefit to the bacteria.
(For more information about directed molecular
evolution, visit Maxygen Corporation's website at
www.maxygen.com.)

In addition to naturally occurring and mutation-
produced proteins, biotechnology is also creating
entirely new protein molecules. These molecules,
designed and built in the laboratory, indicate that
it might be possible to invent proteins that are tailored for specific applications. Faulty protein folding
does occur usually as diseases caused by infectious
protein particles, called **prions** (Figure 4.6). These
infectious proteins attract normal cell proteins and
induce changes in their structure, usually leading to
the accumulation of useless proteins that damage
cells. Prion diseases can occur in sheep and goats
(scrapie) and cows (bovine spongiform encephalitis,
or "mad cow" disease). Human forms of these brain-
destroying diseases include kuru and **transformable spongiform encephalitis (TSE).** All these
diseases involve changes in the conformation of

**Figure 4.6 Prions are Misfolded
Proteins** Misfolding of protein that occurs in prion
disorders can be duplicated in the lab to produce large
quantities of prion proteins, which can be studied to
create diagnostic kits.

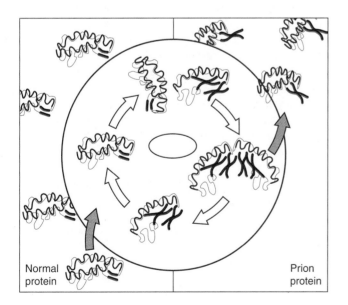

Normal
protein

Prion
protein

prion precursor protein, a protein normally found in mammalian neurons as a membrane glycoprotein. The inability to detect the disease until the infected animal is either sick or dead has seriously complicated control of these diseases. In response, biotechnology research has sought to form synthetic infectious particles that can be studied (see Figure 4.6), which will aid in developing detection and control methods.

4.4 Protein Production

By now, two things should be evident: (1) Proteins are valuable, and (2) proteins are complex and fragile products. With these points in mind, we now examine the real work of biotechnology in the production of proteins.

Producing a protein is a long and painstaking process, and at every stage there are many methods of

---

TOOLS OF THE TRADE

Piecing Together the Human Proteome

Proteomes, the collection of proteins associated with a specific life function, have become more important since the discovery of the human genome. As mentioned in previous chapters, the cost of bringing a drug to market is about $500 million and usually takes between five to eight years to accomplish. Because most of the drugs produced by the biotechnology industry are proteins (such as growth factors, antibodies, and synthetic hormones) that replace missing or nonfunctional proteins in humans, companies are particularly interested in developing protein microarrays. Similar to DNA microarrays, these miniature devices detect proteins associated with disease or those present in abnormal concentrations. In their infancy, these biochips were also used to evaluate a biotechnology company's proposal to develop replacement proteins, before they spent large amounts of time and money.

Protein microarrays have many advantages. Because the number of genes discovered by the Human Genome Project is about a third of what was expected, it is commonly believed that human genes code for more than one protein (see Chapter 3 for an explanation of how this occurs). Most current drugs function at the protein level, interacting with receptors, triggering events, and targeting other proteins in cells of the body. We all have experienced the benefit of proteins that stimulate our immune system to recognize disease organisms, without getting the disease: vaccination. Protein structural differences make them harder to detect than different DNA molecules.

Besides the fact that there are more different proteins than genes, proteins have a broad range of concentrations (from picograms to micrograms) in the body. As discussed earlier in this chapter, proteins have drastically different chemical properties, owing to the types and amounts of the various kinds of amino acids. Because the shape of each protein usually determines its

function (or nonfunction), proteins are also more vulnerable to structural change than is DNA. If we are to develop microarrays that can detect different proteins in different amounts, all these factors must be taken into consideration.

Microarrays can be constructed of glass slides coated with a material that binds proteins. Usually these slides have an attached antibody that is specific to the protein to be detected, as well as a signaling mechanism that indicates capture has occurred. Cambridge Antibody Technology (United Kingdom) has an antibody library of more than 100 billion antibodies collected from the blood of healthy individuals. Packard BioScience (Connecticut) has developed slides coated with acrylamide that allows the attached protein to maintain its three-dimensional shape while embedded in the material coating the slides. Application of the fluids to be detected is based on ink-jet technology and has reached 20,000 spots per standard microscope slide. Finally, Ciphergen Biosystems (California) has adapted mass spectroscopy to perform rapid on-chip separation, detection, and analysis of proteins directly from biological samples. So, what is left to do?

As we will see in Chapter 11, there are countless proteins from numerous diseases to discover. The ability to diagnose a disease and determine the most effective treatment will depend on the ability to purify proteins and develop antibodies that are related to disease. Most of the biochemistry of disease processes is yet to be discovered, and most disease-related conditions are based on the actions of proteins. The opportunity to change human conditions that were previously unchangeable depends on discovering and purifying the important proteins of life's proteome. Take the time to access the websites of some of the companies that develop these devices to keep up with this important technology (e.g., www.ciphergen.com or www.packardbioscience.com or www.biacore.com).

production from which to choose. Experts would probably describe the job of producing proteins as tedious at best and exasperating at worst. We refer to the two major phases used in producing proteins as **upstream processing** and **downstream processing.** Upstream processing includes the actual expression of the protein in the cell. During downstream processing, the protein is first separated from other parts of the cell and isolated from other proteins. Then its purity and functional abilities are verified. Finally, a stable means of preserving the protein is developed. Note that choices made during upstream processing can simplify—or complicate—downstream processing.

Protein Expression: The First Phase in Protein Processing

We begin a detailed discussion of protein processing by looking at the first decision to be made in upstream processing: selecting the cell to be used as a protein source. Microorganisms, fungi, plant cells, and animal cells all have unique qualities that make them good choices in certain circumstances.

Microorganisms

Microorganisms are an attractive protein source for several reasons. First, the fermentation processes of microorganisms are well understood. Also, microorganisms can be cultured in large quantities in a short time. In industrial applications, this ability to generate the product on a large scale is often essential. Microorganisms are attractive also because they are relatively easy to alter genetically.

A number of methods of recombinant DNA technology can be used to increase the level of production of a microbial protein. One method is introducing additional copies of the relevant gene to the host cell. Another introduces the relevant gene into the organism when control of expression has been placed under a more powerful transcriptional promoter (see Chapter 3).

The bacterial species most commonly used to produce genetically engineered proteins is *E. coli*. Because early research into bacterial genetics focused on *E. coli* as a model system, we now understand the genetic characteristics of *E. coli* well.

In some instances, the foreign gene (called **cDNA** or **complementary DNA**) for the desired protein product is attached directly to a complete or partial *E. coli* gene. The result is that the genetically engineered *E. coli* then produces the desired protein, but it is in the form of a **fusion protein,** meaning that the target protein is fused to a bacterial protein, and so an additional step is required to break the target protein away from the rest of the fusion protein. The fused bacterial

Table 4.3 ADVANTAGES AND DISADVANTAGES OF RECOMBINANT PROTEIN PRODUCTION IN *E. COLI*

Advantages	Disadvantages
E. coli genetics are well understood	Proteins contained in intracellular inclusion bodies must be disrupted
Almost unlimited quantities of proteins generated	Proteins cannot be folded in ways needed for many proteins active in mammalian systems
Fermentation technology is well understood	Some proteins are inactive in humans

protein is usually an enzyme that will bind to its substrate that can be attached to a purification column (see description of affinity columns). The majority of proteins synthesized naturally by *E. coli* are intracellular (within the cell). In most cases, the resultant protein accumulates in the cell cytoplasm in the form of insoluble clumps called **inclusion bodies,** which must be purified from the other cell proteins before they can be used.

There are some limitations to the use of microorganisms in protein production. Bacteria like *E. coli* are prokaryotic (primitive one-celled organisms). More advanced eukaryotic cells are found in protozoans and multicellular organisms. Prokaryotes are unable to carry out certain processes, such as glycosylation. For this reason, some proteins can be produced only by eukaryotic cells, as seen in Table 4.3.

Although it is possible to conduct the entire process in a flask in the laboratory, genetically engineered microorganisms can also be grown in large-scale **bioreactors.**

Computers monitor the environment in the bioreactors, keeping oxygen levels and the temperature ideal for cell growth. Cell growth is monitored carefully because there is often a phase of growth when the concentration of the protein is highest. Activating a gene in a recombinant organism requires correct timing after the organism has completed synthesizing important natural proteins needed for its metabolism (as shown in Figure 4.7).

Fungi

Fungi are the source of a wide range of proteins used in products as diverse as animal feeds and beer. Naturally existing proteins found in some fungi are nutritious and being used as a food source—like yeast in bread and other foods. In addition to the naturally occurring proteins, many species of fungi are good sources of engineered proteins. Unlike bacteria, fungi

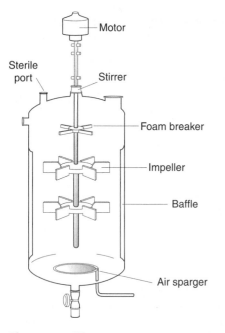

Motor

Sterile
port

Stirrer

Foam breaker

Impeller

Baffle

Air sparger

Figure 4.7 Bioreactor   Large volume cell culture is commonly performed in a self-contained, closed (sterile) system until the products are harvested. Through sterile ports, workers can adjust pH, gas concentrations, and other variables based on input from internal sensors.

are eukaryotic, and because of that they are capable of some posttranslational modification and can fold human proteins correctly, as illustrated in Table 4.4.

Plants

Plant cells can also be the site of protein expression. In fact, plants are an abundant source of naturally occurring, biologically active molecules, and 85% of all current drugs can be found in plants. One example of a plant-derived protein produced on an industrial scale is the **proteolytic** (protein degrading) enzyme **papain.** Papain, or vegetable pepsin, is a protease used as a meat-tenderizing agent. It digests the collagen present in connective tissue and blood vessels that makes meat tough. The advantage provided by plants' abilities to produce enzymes will be undoubtedly used

Table 4.4 SOME RECOMBINANT PROTEINS FROM FUNGI	
Protein	**Fungi**
Human interferon	*Aspergillus niger, A. nidulans*
Human lactoferrin	*A. oryzae, A. niger*
Bovine chymosin	*A. niger, A. nidulans*
Aspartic proteinase	*A. oryzae*
Triglyceride lipase	*A. oryzae*

in the near future for drug production. Plants can be modified genetically to make them produce specific desired proteins that do not occur naturally. This process encourages rapid growth and reproductive rates of plants, which can be a distinct advantage. For example, tobacco, the first plant to be genetically engineered, can produce a million seeds from a single plant. Once the hard work of introducing the required genetic material is completed, a million new "plant protein factories" can fill the fields, as described in Chapter 6.

There are also disadvantages to using plants as protein producers. Not all proteins can be expressed in plants, and, because plants have tough cell walls, the process of extracting proteins from the plant cells can be time consuming and difficult. Finally, although plant cells can often properly glycosylate proteins, the process is slightly different from what occurs in animal cells. This may rule out using plants as biofactories for expression of some proteins (we discuss transgenic plants in greater detail in Chapter 6).

Mammalian cell systems

Sometimes it is possible to culture animal cells, growing them in a medium until it is time to harvest the proteins. This process is challenging because the nutritional requirements of animal cells, even in a petri dish, are more complex than those of microbial cells. Animal cells also grow relatively slowly, and a larger number of cells are required to seed bioreactors. In addition, because animal cells take longer to grow, the opportunity for animal cell cultures to become contaminated is greater than in other culture systems. Despite all these issues, in many cases, animal cells are still the best, if not the only, choice for certain protein products, largely due to the patented processes that have been approved for production of these proteins.

Whole-animal production systems

Cells in culture are not the only option when using animal cells; sometimes living animals are the protein producers. Consider, for example, the technique used to harvest monoclonal antibodies. Monoclonal antibodies react against only one target, making them valuable in diagnostic and therapeutic applications (see Chapter 7 for a complete explanation of how they are produced). Antibodies are proteins produced in reaction to an **antigen** (invading virus or bacterium). Antibodies can combine with and neutralize the antigen, protecting the organism. The production of antibodies is part of the immune response that helps living things resist infectious diseases. When monoclonal antibodies are the goal, mice are injected with an antigen. The mouse secretes the desired antibody, or the antibody-producing tissue from the mouse can be fused with cancer cells, and when fluid from the tumor is collected, the monoclonal antibodies can be purified from it. Another method of whole-animal protein production uses the milk or eggs from transgenic animals (animals that contain genes from other organisms). These animal products contain the proteins from the recombinant gene that was introduced and can be purified from the milk or egg proteins. Transgenic animals look and act like their normal barnyard relatives. A transgenic sheep like Tracy in Chapter 7 eats hay and gives milk. The big difference is that her milk includes valuable proteins not ordinarily produced by the animal. (Chapter 7 explains monoclonal antibody production in transgenic animals.)

Insect systems

Insect systems are another avenue of protein production from animal cells. Until recently, the expression techniques depended on insect cell cultures, but new techniques that use insect larvae are being developed. In either case, **baculoviruses** (viruses that infect insects) are used as vehicles to insert DNA that cause the desired proteins to be produced by the insect cells. However, there are instances in which the posttranslational modification of proteins is slightly different in insects than it is in mammals, so this may prevent the use of insect expression systems.

4.5 Protein Purification Methods

Once the protein is produced, downstream processing begins. First, the protein must be harvested. If the protein is intracellular, the entire cell is harvested; however, if it is extracellular, the protein is excreted into the culture medium that is collected.

Harvesting is just the beginning of downstream processing. Then the protein must be purified, which involves separating the target protein from the complex mixture of biological molecules.

Note that "purity" is a relative term. Generally, the FDA requires that a sample be composed of 99.99% of the target protein. Separating the proteins from all the other cellular contents is not easy, and isolating the target protein from the other proteins in the sample can be even more difficult. To understand the process of purifying proteins, we look at some steps commonly followed in this process.

Preparing the Extract for Purification

The media, or culture filtrate, harvested from a large fermenter can fill a swimming pool, and the target protein usually represents less than 1% of the pool.

Even if the protein is being expressed on a much smaller scale, finding the essential protein can seem like finding a small needle in a large haystack.

If the protein is intracellular, the first task is **cell lysis,** or disrupting the cell wall to release the protein. There are many methods for doing this, including freezing and thawing the cells (which disrupts cell membranes releasing cell contents), using detergents to dissolve the cell walls, and implementing such mechanical options as ultrasonics or grinding with tiny glass beads. Given the fragility of proteins, freeing them from the cell without degrading them entirely is challenging. The disruption process releases the protein of interest as well as the entire intracellular content of the cell. The cell is mechanically broken down or ruptured.

After the cells have been ruptured, detergents and salts may be added to the mixture. The detergents reduce the hydrophobic orientation of the proteins, which makes separating them later on the basis of size and molecular weight much easier. Salts may be added to reduce the interactions between the molecules and to keep the proteins in solution.

Stabilizing the Proteins in Solution

After the mixture is prepared, the proteins must be stabilized. Recall that it is important to maintain the bioactivity of the protein and that proteins are relatively fragile molecules. As a consequence, precautions must be taken to protect the protein during the purification process.

Temperature is a threat to proteins. If you have ever cooked an egg and seen the whites immediately solidify, you have witnessed the powerful effect of heat on proteins. Another edible example is the mozzarella on a pizza; those long strings of cheese are actually evidence that the protein molecules have been broken down. Even heat as moderate as room temperature limits the active lifespan of proteins. Because the goal is an active protein, and because the purification process takes time, purification is often done at temperatures barely above freezing. Outright freezing and freeze-thaw cycles can destroy proteins, however.

Natural proteases that can digest the target proteins in the sample are another threat. Protease inhibitors and antimicrobials can be added to prevent the protein molecules from being dismantled, but then these additives need to be removed later in the purification process. When the goal is producing pharmaceutical protein, the addition of protease is prohibited by the FDA because it is not an active component and can lead to degradation and potentially harmful reactions.

Still another potential threat is mechanical destruction of the protein by foaming or shearing of the proteins into useless fragments. Once again, additives can help prevent foaming and sheering from destroying the protein, but the additives must be removed later in the purification process.

As you can see, it takes a balancing act to extract and purify the proteins successfully. Although it is essential that the protein be purified, it is equally important that the protein maintain its biological activity. Unfortunately, some powerful purification methods are also powerful enough to damage the target protein.

Separating the Components in the Extract

Similarities between proteins permit us to separate proteins from nonprotein material such as lipids (fats), carbohydrates, and nucleic acids, which are also released when a cell is disrupted. Differences between individual proteins are then used to separate the target proteins from other proteins.

Protein precipitation

Proteins often have hydrophilic amino acids on their surfaces that attract and interact with water molecules. That characteristic is used as the basis for separating proteins from other substances in the extract. Salts, most commonly ammonium sulfate, can be added to the protein mixture to **precipitate** the proteins (to cause them to settle out of solution). Ammonium sulfate precipitation is frequently the first step in protein purification. It separates the proteins away from nonproteins and also results in a protein precipitate that is quite stable.

Problems associated with ammonium sulfate precipitation make it a poor choice in some industrial situations. Ammonium sulfate is highly reactive when it contacts stainless steel, for example, and many industrial purification facilities are made of stainless steel. Other solvents frequently used to promote protein precipitation include ethanol, isopropanol, acetone, and diethyl ether. These solvents cause proteins to precipitate by removing water from between the protein molecules, as does ammonium sulfate.

Filtration (size-based) separation methods

There are a variety of ways to separate molecules on the basis of size and density. **Centrifugation** is one such method. Centrifuging separates samples by spinning them at high speed. With this process, the proteins can usually be isolated in a single layer. Small-volume centrifuges are capable of processing only a few liters each run. Large reactors can produce hundreds or thousands of liters of extract that require processing. (See Figure 4.8.) Industrial-scale centrifugation is normally achieved using continuous-flow

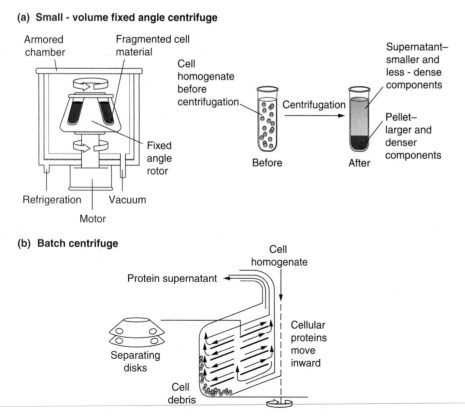

(a) Small - volume fixed angle centrifuge

Armored chamber

Fragmented cell material

Cell homogenate before centrifugation

Centrifugation

Before

After

Supernatant– smaller and less - dense components

Pellet– larger and denser components

Fixed angle rotor

Refrigeration   Vacuum

Motor

(b) Batch centrifuge

Cell homogenate

Protein supernatant

Separating disks

Cell debris

Cellular proteins move inward

Figure 4.8 Fixed Angle and Batch Centrifugation   Fixed angle centrifuges (a) can develop extremely high gravitational forces (g forces) but are limited to smaller quantities and must be run separately for each batch. Batch centrifuges (b) were developed to allow continuous flow of materials and separation of cell debris from cell proteins.

centrifuges that allow continuous processing of the extract from a bioreactor. This accelerates the separation process compared with small-volume centrifugation.

Filters of various sizes and types can also be used to separate proteins. Sometimes a series of filtration techniques are used in sequence, first separating out all of the cellular matter and later separating out the larger protein molecules from smaller particles. In **membrane filtration,** thin membranes of nylon or other engineered substances with extremely small pore structures are used to filter out all of the cellular debris from a solution. **Microfiltration** removes the precipitates and bacteria. **Ultrafiltration** uses filters that can catch molecules such as proteins and nucleic acids. Some ultrafiltration processes can actually separate large proteins from smaller ones. (Refer to www.amicon.com to view some of these devices.) One of the main shortcomings of membrane filtration systems is their tendency to clog easily. On the plus side, these filtration systems take less time than centrifugation.

Diafiltration and **dialysis** are filtration methods that rely on the chemical concept of equilibrium, the migration of dissolved substances from areas of higher concentration to areas of lower concentration. As shown in Figure 4.9, dialysis depends on the ability of some molecules to pass through semipermeable membranes while others are halted or slowed because of their size. This purification step is often required to remove the salts, solvents, and other additives used earlier in the process. The salts are then replaced with buffering agents that help stabilize the proteins during the remainder of the purification process. Diafiltration adds a filtering component to dialysis.

Chromatography

The initial steps in any purification process liberate the protein from the cell, remove undesired contami-

nants and particulates, and concentrate the proteins. **Chromatography** methods allow us to sort out the proteins by size or by how they tend to cling to or dissolve in various other substances. In chromatography, long glass tubes are filled with resin and a buffered salt solution. The protein extract is added and flows through the column. Depending on the resin used, the protein either sticks to the resin beads or passes through the column while the beads act as a filtration system.

Size-exclusion chromatography (SEC) uses gel beads as a filtering system. The larger protein molecules quickly work their way around the gel beads while other, smaller molecules pass through more slowly because they can slip through tiny holes in the gel beads, as shown in Figure 4.10. The gels are available in a variety of pore sizes, and the necessary gel for proper filtration use depends on the molecular weight of contaminants to be filtered out of the extract. This method can make only preliminary separations, however, and can pose problems in industrial settings because it requires very long columns.

Ion exchange (IonX) **chromatography,** an extremely useful protein purification and concentration method, relies on electrostatic charge (static cling) to bind the proteins to the gel beads in the column. While the proteins cling to the resin, other contaminants pass through and out of the column, as shown in Figure 4.11. The proteins can then be eluted (released from the gel) by changing the electrostatic charge; this is done by rinsing the column with salt solutions of increasing concentrations. The bound protein is then released from its attachment and collected.

Affinity chromatography relies on the ability of most proteins to bind specifically and reversibly to uniquely shaped compounds called **ligands.** Ligands are small molecules that bind to a particular large molecule. You might think of their fitting together

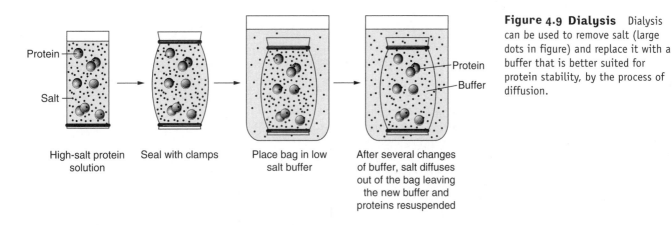

Figure 4.9 Dialysis Dialysis can be used to remove salt (large dots in figure) and replace it with a buffer that is better suited for protein stability, by the process of diffusion.

High-salt protein solution | Seal with clamps | Place bag in low salt buffer | After several changes of buffer, salt diffuses out of the bag leaving the new buffer and proteins resuspended

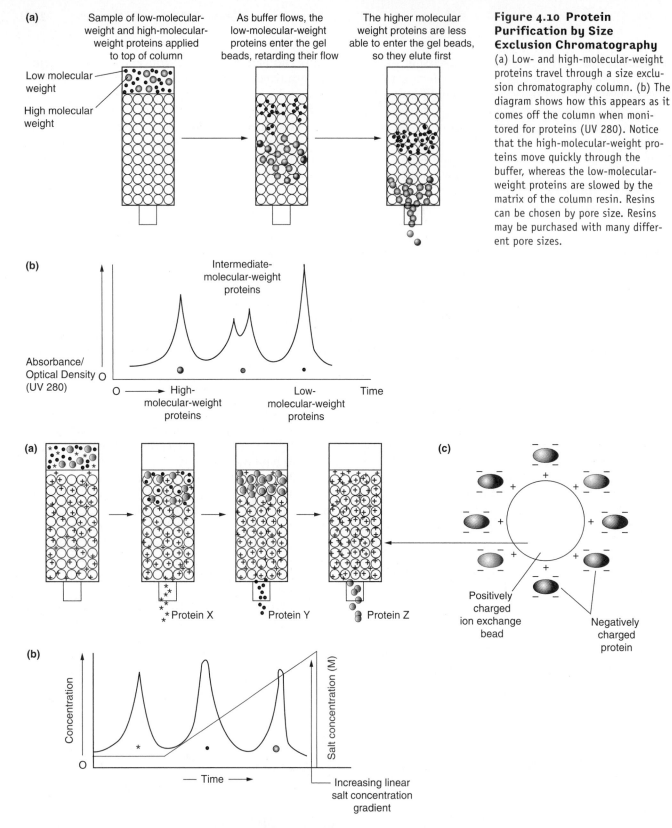

(a)

Sample of low-molecular-weight and high-molecular-weight proteins applied to top of column

Low molecular weight

High molecular weight

As buffer flows, the low-molecular-weight proteins enter the gel beads, retarding their flow

The higher molecular weight proteins are less able to enter the gel beads, so they elute first

Figure 4.10 Protein Purification by Size Exclusion Chromatography
(a) Low- and high-molecular-weight proteins travel through a size exclusion chromatography column. (b) The diagram shows how this appears as it comes off the column when monitored for proteins (UV 280). Notice that the high-molecular-weight proteins move quickly through the buffer, whereas the low-molecular-weight proteins are slowed by the matrix of the column resin. Resins can be chosen by pore size. Resins may be purchased with many different pore sizes.

(b)

Intermediate-molecular-weight proteins

Absorbance/Optical Density (UV 280)

High-molecular-weight proteins

Low-molecular-weight proteins

Time

(a)

Protein X Protein Y Protein Z

(c)

Positively charged ion exchange bead

Negatively charged protein

(b)

Concentration

Salt concentration (M)

Time

Increasing linear salt concentration gradient

Figure 4.11 Ion Exchange Chromatography (a) Charged amino acids bind to ionic resin beads. Increasing the ionic strength of the buffer displaces proteins (based on their binding strength) after they have bound to the column resin. (b) Protein X is released at the lowest concentration of the displacing salt gradient; Protein Y has a higher binding strength and is released second; Protein Z has the highest binding strength and requires a high salt concentration to displace it from the ion exchange beads of the column. (c) Anion exchange resin is positive; cation exchange resins are negatively charged.

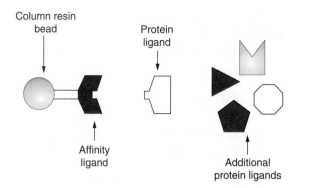

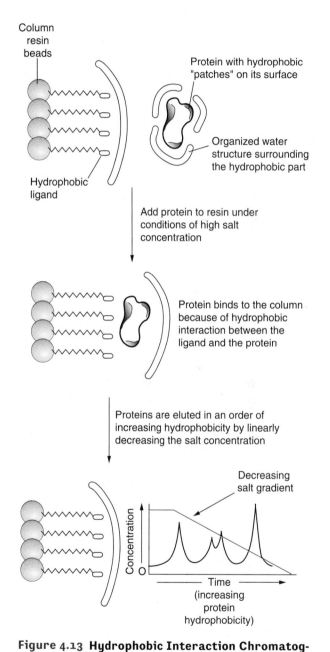

Figure 4.12 Affinity Chromatography Affinity ligands are designed to bind specifically to unique three-dimensional chemical components of the protein being purified. The non-bound protein ligands wash through the column. Increasing the ionic strength of the buffer can displace the bound protein (after preferential binding) and regenerate the affinity column. The displaced (pure) protein can be collected and concentrated.

as a key fitting into a lock (see Figure 4.12). After the proteins have bound to the resin beads, a buffer solution is used to wash out the unbound molecules. Finally, special buffer solutions are used to cause desorption (to break the ligand bonds) of the retained proteins. Fusion proteins, as mentioned earlier, can be used in affinity chromatography because the substrate (ligand) of the bacterial protein can be part of the affinity column, attracting the fused protein (bacterial enzyme protein) to the column. Affinity chromatography is most often used in later stages of the purification process.

As we have seen, amino acids are either attracted to or repelled by water molecules. In **hydrophobic interaction chromatography (HIC),** the proteins are sorted on the basis of their repulsion of water. The column beads in this method are coated with hydrophobic molecules, and the hydrophobic amino acids in a protein are attracted to the similar chemicals in the beads as shown in Figure 4.13.

Iso-electric focusing is often used in quality control during the purification process to identify two similar proteins that are difficult to separate by any other means. Each protein has a specific number of charged amino acids on its surface in specific places. Because of this unique combination of charged groups, each protein has a unique electric signature known as its **iso-electric point (IEP),** where the charges on the protein match the pH of the solution. This IEP can be used to separate otherwise similar proteins from one another. Iso-electric focusing is the first dimension of two-dimensional electrophoresis.

Two-dimensional electrophoresis, which separates proteins based on their electrical charge and size, is essentially the combination of two methods. In this technique, researchers introduce a solution of cell proteins onto a specially prepared strip of polymer. When the strip is exposed to an electrical current, each protein in the mixture settles into a layer according to its charge. Next the strip is aligned with a gel and again exposed to electrical current. As the proteins migrate through the gel, they separate according to their molecular weight.

Figure 4.13 Hydrophobic Interaction Chromatography Increasing the nonpolar concentration of the buffer can cause the hydrophobic portions of proteins to combine with a hydrophobic ion exchange column resin. Further reducing the ionic concentration displaces the protein from the column resin and replaces its attachment with a nonpolar solvent. Fractions can be collected and protein concentration determined based on detection by spectrophotometric analysis at 280 nm.

Analytic methods

High-performance liquid chromatography (HPLC)

adds a twist to the previously described chromatography methods, which all depend on gravity flow or very low pressure pumps to move the extract through the columns. Such low-flow methods can take several hours to process a single sample. In contrast, HPLC systems use greater pressure to force the extract through the column in a shorter time. HPLC systems have limitations. Less protein is separated, so the technique is more useful in analytical situations than in mass production.

Mass spectrometry (mass spec) is a highly sensitive method used to identify trace elements. In fact, it is used frequently on the outflow of HPLC systems. All mass spectrometers do three things: suspend the sample molecule into a charged gas phase, separate the molecules based on their mass-to-charge ratios, and detect the separated ions. A sample as small as a picogram (one billionth of a gram) can be analyzed by this process, which is illustrated in Figure 4.14. A definitive fingerprint indicates the identity and size of

most protein fragments. An important application of this process is in protein sequencing. A larger protein can be digested into fragments (peptides) and analyzed.

4.6 Verification

At each step of the purification process, it is important to verify that the protein of interest, the target protein, has not been lost and that the concentration efforts have been successful. **SDS-PAGE** (polyacrylamide gel electrophoresis) is often used for these checks. In this method, sodium dodecyl sulfate (SDS), a detergent, is added to a sample of the protein mixture, and the mixture is heated. The sulfate charges are distributed evenly along the denatured (heated) protein, making separation dependent on the size (number of charges) of the protein. After this treatment, the protein sample is loaded onto a special gel matrix (PAGE) where it forms a single band at a specific location, depending on its molecular size and mass, as can be seen in Figure 4.15.

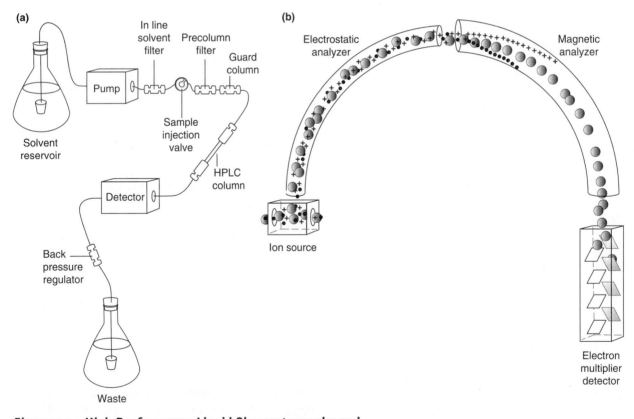

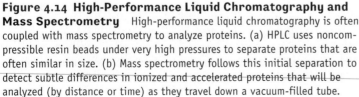

Figure 4.14 High-Performance Liquid Chromatography and Mass Spectrometry High-performance liquid chromatography is often coupled with mass spectrometry to analyze proteins. (a) HPLC uses noncompressible resin beads under very high pressures to separate proteins that are often similar in size. (b) Mass spectrometry follows this initial separation to detect subtle differences in ionized and accelerated proteins that will be analyzed (by distance or time) as they travel down a vacuum-filled tube.

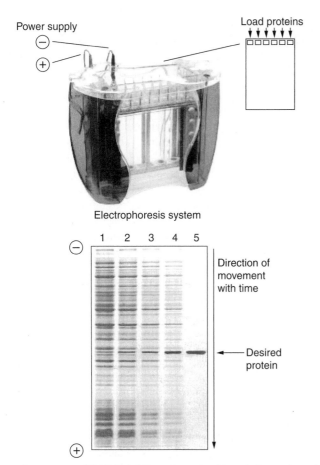

Power supply

Load proteins

Electrophoresis system

Direction of movement with time

Desired protein

Figure 4.15 SDS-PAGE Polyacrylamide gel electrophoresis of proteins that have uniform charges (produced by heating in sodium dodecyl sulfate) can be separated by their size, based on their migration in a vertical electrical field. This procedure commonly follows each major step in purification to verify that the band of desired protein is becoming more concentrated (and has not been lost in the separation). Notice that the 65-kilodalton protein has the highest concentration in the final step (lane 2) of this procedure (lane 1) when proteins of known size are used for comparison. The smallest proteins will reach the + pole first.

By adding a dye that combines with proteins, such as **Coomassie stain,** a clearly colored band results, and a known size marker can be compared with the stained sample. When the sample and the known marker are equivalent, we have evidence that the protein of interest is indeed being concentrated. As this test is run at each step in the purification process, the colored band should become increasingly intense, giving evidence that the protein has not been lost during the purification process.

A specific detection method for proteins separated by SDS-PAGE is Western blotting. Similar to Southern blotting (see Chapter 3), some protein is transferred from the gel to a nitrocellulose membrane and detected with a specific antibody that recognizes that protein by its unique structure. An attached enzyme to the antibody can be used to convert a fluorescent substrate, and a permanent record of the detection can be obtained. For this procedure, an electric current is applied to the gel so the separated proteins transfer through the gel and onto the membrane in the same pattern as they separate on the SDS-PAGE. All sites on the membrane that do not contain blotted protein from the gel can then be nonspecifically "blocked" so antibody (serum) will not nonspecifically bind to them. To detect the antigen blotted on the membrane, a primary antibody (serum) is added at an appropriate dilution and incubated with the membrane. If there are any antibodies present that are directed against one or more of the blotted antigens, those antibodies will bind to the protein(s) while other antibodies will be washed away at the end of the incubation. To detect the antibodies that have bound, anti-immunoglobulin antibodies coupled to a reporter group, such as the enzyme alkaline phosphatase, are added. Finally, after excess second antibody is washed free of the blot, a substrate is added that precipitates upon reaction with the conjugate, resulting in a visible band where the primary antibody bound to the protein. (See Figure 4.15.) If you would like to see an interpretation of a Western blot, access http://www.biology.arizona.edu/immunology/activities/western_blot/west2.html for an excellent tutorial from the Arizona State Immunology Department.

Antibody detection of a specific protein can also be carried out using enzyme-linked immunosorbent assay (ELISA). The most commonly used ELISA requires two antibodies: one to capture the unique protein and one attached to an enzyme to produce a color reaction (see Figure 4.12). The first antibody to the unique protein is plated on to a multiwell ELISA plate, the protein is added, and after a series of wash and blocking steps, the second antibody (enzyme attached) is added. The addition of the substrate permits a color reaction to occur if the antibody has bound. ELISA plates are designed to capture many important proteins currently under research. To see an animation of an ELISA plate test and tutorial from the Arizona State Immunology Department, access http://www.biology.arizona.edu/immunology/activities/elisa/technique.html.

4.7 Preserving Proteins

After the target protein has been isolated and collected, it must be saved in a manner that will preserve its activity until it can be used. One means of preserving the protein is **lyophilization,** or freeze drying. In this process, the protein, which is usually a liquid product, is first frozen. Then a vacuum is used to hasten the evaporation of water from the fluid. In lyophilization, ice crystals sublimate to water vapor directly, without melting into liquid water first. The containers are sealed after the water is removed, leaving the dried proteins behind. Lyophilization has been demonstrated to maintain protein structure, making it a commonly chosen method for preservation of biotechnologically derived proteins. Many freeze-dried proteins may be stored at room temperature for prolonged periods of time.

4.8 Scale-up of Protein Purification

Protein purification protocols are usually designed in the laboratory, on a small scale. These purification techniques, and this level of productivity, is feasible if the product is only needed in small amounts. Demands for monoclonal antibodies, for example, are in the range of grams per year. A single laboratory-scale bioreactor can usually produce adequate quantities. However, other proteins are required in much larger quantities, which means that the production methods must be scaled up.

Scaling-up is not always easy to accomplish. Laboratory methods that may work on a small scale may not always be adaptable to large-scale production. Furthermore, changes in the purification process can invalidate earlier laboratory-scale clinical studies. For example, when the FDA approves a bioengineered protein, it also approves the process for producing it. To change the process, it may be necessary to seek FDA approval once again. For this reason, bioprocess engineers are involved in the early stages of purification and work to ensure that scaling up the process will be possible later on.

4.9 Postpurification Analysis Methods

During research, it is often helpful to look closely at the purified protein. The goal may be to better understand the molecular structure of a specific protein or to change the structure to change the protein's function. The following methods are used to meet these ends.

Protein Sequencing

Recall from our discussion of protein structures that each protein has a specific sequence of amino acids, known as the primary sequence. To understand the protein completely, it is important to determine this primary sequence. Automated protein sequencers make this task possible. Automated protein sequenc-

Q Can you select the best purification scheme? An example of a purified protein with enormous potential for human benefit is described in Chapter 11. The protein is hemoglobin from the blood of cows. The purified hemoglobin combines with oxygen and carries it to the tissues of the body without being enclosed within cells. This "artificial blood" acts as a substitute for whole blood and does not appear to have any of the rejection-stimulating properties of transfused whole cow blood. As you would expect, many problems are associated with extracting hemoglobin: How can a large quantity of blood be reduced down to the hemoglobin protein without damaging it? How can the hemoglobin protein be separated from the many other proteins found in cow blood? How can the fact that hemoglobin contains iron, whereas other blood proteins do not, be used to develop an extraction method that is unique to this protein? What can be done to verify that the protein did not get lost in each of the purification steps?

A Sequence the techniques that have been explained in such a way that this intracellular protein can be purified from other proteins that are of similar molecular weight inside the blood cell. You need to verify your choices, explaining why you selected each technique in that sequence. There is more than one possible sequence, but if you access the website of the Biopure Corporation (www.biopure.com), the biotechnology company that has accomplished this purification, you should be able to learn more about its properties to improve individual steps in your chosen sequence.

ing has changed recently from the earlier Edmond degradation method to the current mass spectrometry method (see Figure 4.14). In contrast to digesting each amino acid from the end one at a time (a timely process), in the mass spectrometry method, peptide masses are identified by their unique signatures (retention times). By converting peptides into ions and subjecting them to mass spectrometry, it is possible to identify many proteins in a matter of minutes.

X-ray Crystallography

X-ray crystallography is used to determine the complex tertiary and quaternary structures of proteins. The method requires pure crystals of proteins that have been carefully dehydrated from solutions. Bombarded with X-rays, a pure protein creates specific shadow patterns based on its configuration (see

Chapter 2). Computer analysis of these patterns allows the generation of "ribbon" diagrams, which not only describe the structure of the proteins but also make it possible to plan potential modifications of the protein molecule to improve function. Protein crystallography is usually a requirement for approval by the FDA because it verifies that the process produces the product being approved.

4.10 Proteomics

Many diseases are the result of flaws in protein expression, but not all of those diseases can be understood simply in terms of genetic mutation. Because proteins undergo posttranslational modification, the puzzle can be far more complicated. A new scientific discipline, **proteomics,** is dedicated to understanding the complex relationship of disease and protein expression. In proteomics, **proteomes** (the PROTEin complement to a genOME) are compared under healthy and diseased states. The variations of protein expression are then correlated to onset or progression of a specific disease. The goal of proteomic research is the discovery of protein markers that can be used in new diagnostic methods and in the development of targeted drugs for treating disease. For example, this research has led to promising breakthroughs in lung cancer, mad cow disease, and the treatment of nerve damage.

Proteomics applies several of the techniques described earlier: Two-dimensional electrophoresis is used to separate proteins; mass spectrometry can be used to verify the protein's identity; and then the protein is characterized using amino acid sequencing. Some of this labor-intensive work may be assisted by automation in the future. **Protein microarrays** are a set of proteins immobilized on a surface—usually a glass slide—that has been coated with a reagent that will indicate binding by color change (see Figure 4.16).

Antibodies Detect Protein Binding

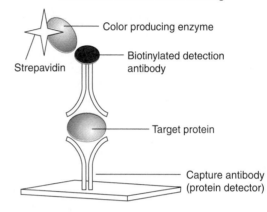

Figure 4.16 Protein Microarrays Protein microarrays that will uniquely bind to single proteins are being perfected. When the protein binds, it will release a fluorescent signal. The ability to detect the presence of unique proteins (proteomics) will permit proper decisions in diagnosis and/or therapy to be made.

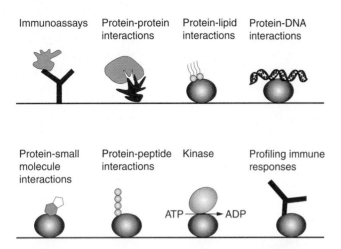

Immunoassays Protein-protein interactions Protein-lipid interactions Protein-DNA interactions

Protein-small molecule interactions Protein-peptide interactions Kinase Profiling immune responses

ATP ⟶ ADP

Figure 4.17 Protein Microarrays Can Detect More Than Just Proteins The capacity of protein microarrays has increased. Functional protein microarrays have recently been applied to the discovery of protein interactions (i.e., protein-protein, protein-lipid, protein-DNA, protein-drug, and protein-protein interactions), expanding knowledge of the significance of protein interactions to normal cell function. Binding antibodies or other proteins to the matrix allows detection when the binding occurs.

The capacity of these arrays has increased, but not as much as DNA arrays, due to the difficulty in producing these arrays. A single DNA array can monitor the expression of an entire genome. Functional protein microarrays have recently been applied to the discovery of protein interactions (i.e., protein-protein, protein-lipid, protein-DNA, protein-drug, and protein-protein interactions; see Figure 4.17).

QUESTIONS & ACTIVITIES

Answers can be found in Appendix 1.

1. What can the public database on proteins tell us?

2. How does directed molecular evolution technology differ from mutations that occur naturally?

3. How has proteomic research benefited from the results of the Protein Structure Initiative?

4. Why are proteins being primarily searched through cDNA products?

5. If organisms produce proteins on their own, why should companies be allowed to patent proteins?

6. If you are purifying a small protein from an SEC column, which fractions do you want to collect?

7. Do proteins strongly bound to an IonX column elute first or last from the column?

8. Is affinity chromatography more or less selective at separating proteins than IonX chromatography?

9. List the protein separation methods primarily used in analytic (rather than production) methods.

10. If you carry out an SDSPAGE analysis after each step in a protein purification sequence and found that the last step resulted in the lack of a band at the location you expected, what would this mean?

References and Further Reading

Burden, D., and Whitney, D. (1995). *Biotechnology: Proteins to PCR*. Boston: Birkhauser.

Huh, Won Ki, Falvo J.V., Gerke, L.C., et al. (2003). Global Analysis of Protein Localization in Budding Yeast. *Nature*, 425(6959): 686–691. (First comprehensive view of protein activity in higher organisms)

Microarray Technology Empowers Proteomics. (2001, December). *Genomics and Proteomics*, pp. 44–46.

Walsh, G., and Headon, D. (1994). *Protein Biotechnology*. New York: Wiley.

Zhang, Z., Gildersleeve, J., Yang, Y-Y., et al. (2004). A New Strategy for the Synthesis of Glycoproteins. *Science*, 303(5656): 371–373. (Engineered bacteria make glycosylated myoglobin.)

Visit www.pearsonhighered.com/biotechnology to download learning objectives, chapter summary, "Keeping Current" web links, glossary, flashcards, and jpegs of figures from this chapter.

Chapter 5

Microbial Biotechnology

After completing this chapter you should be able to:

- Provide examples of how prokaryotic cell structure differs from eukaryotic cell structure.

- Describe practical features of bacteria that make them useful as model organisms and tools for applications in biotechnology.

- Provide examples of how yeast can be used in biotechnology.

- Define fermentation and discuss the difference between alcohol and lactic acid fermentation. Provide examples of common foods and beverages produced by each fermentation process.

- List examples of medically important proteins that are produced in bacteria using recombinant DNA technology.

- Describe how microorganisms play an important role in the development and production of many vaccines, and provide examples of different types of vaccines.

- Discuss why studying microbial genomes can be valuable to scientists.

- Define environmental genomics.

- Define bioterrorism, identify microorganisms that may be used as bioweapons, and discuss strategies that may be used to detect and combat bioweapons.

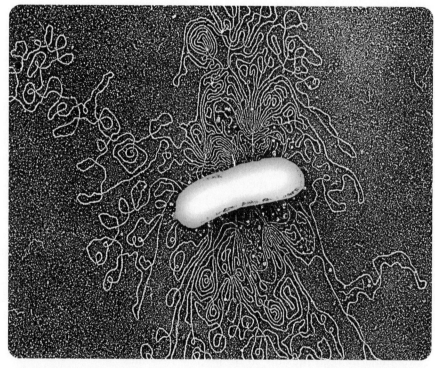

Microbes, such as the *E. coli* cell shown in the center of this photo, have a long history of applications in biotechnology. The cell wall and plasma membrane of this cell have been burst open to reveal the bacterial chromosome.

Just when Italian art restorers were at a loss to save one of the world's largest collections of 14th- and 15th-century medieval frescoes (paintings done on plaster) decorating a cemetery in Pisa near the Leaning Tower, soil bacteria called *Pseudomonas stutzeri* came to the rescue. Restorers were trying to clean the frescoes of paint-clouding grime contained within glue that had been used to repair them after damage in World War II. Traditional chemical reagents only further damaged the paintings. Knowing that *P. stutzeri* could degrade nitrogens, plastic resins, and pollutants such as tetrachloroethylene, scientists cultured a strain of these microbes in the lab and applied them to the frescoes. Within 6 to 12 hours, the bacteria had cleaned approximately 80% of the glue residue and the remaining residue was removed with a light washing. The next proposed project for these **microorganisms** is removing black crust from limestone and marble monuments in Greece.

Microorganisms, also called **microbes,** are tiny organisms that are too small to be seen individually by the naked eye and must be viewed with the help of a microscope. Although the most abundant microorganisms are bacteria, microbes also include viruses, fungi such as yeast and mold, algae, and single-celled organisms called protozoa. Almost all of these microbes have served interesting roles in biotechnology.

Bacteria have existed on the earth for over 3.5 billion years and they greatly outnumber humans. It has been estimated that microbes comprise over 50% of the earth's living matter. Yet less than 1% of all bacteria have been identified, cultured, and studied in the laboratory. We are literally surrounded by bacteria. They live on our skin, in our mouths, and in our intestines; they are in the air and on virtually every surface we touch. Bacteria have also adapted to live in some of the harshest environments on the planet: polar ice caps, deserts, boiling hot springs, and under extraordinarily high pressure in deep-sea vents miles under the ocean's surface.

The rich abundance of bacteria and other microbes provides a wealth of potential biotechnology applications. Well before the development of gene-cloning techniques, humans used microbes in biotechnology. In this chapter, we primarily discuss the important roles bacteria have played in both old and new practices of biotechnology. In addition we look at the many applications of yeast. Frequently the use of microorganisms as biotechnology "tools" depends on their cell structure. Therefore we begin by reviewing prokaryotic cell structure and then consider a range of applications of microbial biotechnology. We conclude by discussing the dangers of microbes as agents of bioterrorism.

5.1 The Structure of Microbes

Before you can consider the many applications of microbial biotechnology, you must be able to distinguish microorganisms from plant and animal cells. Recall from Chapter 2 that cells can be broadly classified based on the presence (eukaryotes) or absence (prokaryotes) of the nucleus that contains a cell's DNA. Therefore, cells classified as *eukaryotic* include plant and animal cells, fungi such as yeast, algae, and single-celled organisms called protozoans, which include amoebas like those you may have studied in high school biology. Unlike eukaryotic cells, *prokaryotic cells* lack most membrane-bound organelles, such as a nucleus. Prokaryotes include the **Domains** (domains are taxonomic categories above the kingdom level for example: Archaea, Bacteria, and Eukarya) **Bacteria** and **Archaea,** organisms that share properties of both eukaryotes and prokaryotes. The cellular structure of microorganisms is important in determining both where they will thrive and how they can be used in biotechnology. For example, many Archaea live in extreme environments such as very salty conditions or hot environments, and as a result they have very unique metabolic properties. Moreover, structural features of bacteria in particular make them excellent microorganisms to use for biotechnology research.

Bacterial cells are much smaller (1–5 µm; 1 µm = 0.001 millimeters) than eukaryotic cells (10–100 µm) and have a much simpler structure. Refer to Figure 2.1 for a diagram of prokaryotic cell structure. Bacterial cells also exhibit the following structural features:

- DNA is not contained within a nucleus and typically consists of a single circular chromosome that lacks histone proteins.
- Bacteria may contain plasmid DNA.
- Bacteria lack membrane-bound organelles.
- The cell wall that surrounds the cell (plasma) membrane is structurally different from the plant cell wall. Composed of a complex polysaccharide and protein substance called **peptidoglycan,** the cell wall forms a rigid outer barrier that protects the cells and determines their shape. In Archaea, this structure does not contain peptidoglycans.
- Some bacteria contain an outer layer of carbohydrates that form a structure called the *capsule.*

Most bacteria are classified by the **Gram stain,** a technique in which dyes are used to stain the cell wall of bacteria. Gram-positive bacteria stain purple and have simple cell wall structures rich in peptidoglycans, whereas gram-negative bacteria stain pink and have complex cell wall structures with less peptidoglycan. Bacteria do not form multicellular tissues like animal and plant cells, although some bacteria can associate with each other to form chains or filaments of many connected cells.

Bacteria vary in their size and shape. The most common shapes include spherical cells called **cocci** (singular, coccus), rod-shaped cells called **bacilli** (singular, bacillus), and corkscrew-shaped spiral bacteria (Figure 5.1). As you study microbes, you will learn that the name of bacteria frequently provides you with a tip about the shape of those cells. For instance, *Staphylococci* are spherical bacteria that live on the surface of our skin. *Bacillus anthracis* is a rod-shaped bacterium that causes the illness anthrax.

The single circular chromosome that comprises the genome of most bacteria is relatively small. Bacterial chromosomes average in the range of 2 million to 4 million base pairs in size compared with 200 million base pairs for a typical human chromosome. As you learned in Chapter 3, some bacteria also contain plasmids in addition to their chromosomal DNA. Plasmid DNA often contains genes for antibiotic resistance and genes encoding proteins that form connecting tubes called *pili* (see Figure 2.1), which allow bacteria to exchange DNA between cells. Plasmid DNA is an essential tool for molecular biologists because it can be used to carry and replicate pieces of DNA in recombinant DNA experiments.

Bacteria grow and divide rapidly. Under ideal growth conditions, many bacterial cells divide every 20 minutes or so, whereas eukaryotic cells often grow for 24 hours or much longer before they divide. Therefore, under favorable growth conditions in the laboratory, a small population of bacteria can divide rapidly to produce millions of identical cells. Because bacteria are so small, millions of cells can easily be grown on small dishes of agar or in liquid culture media. When grown on culture plates, each bacterial cell typically divides to form circular-shaped colonies that contain thousands or millions of cells (Figure 5.2a). For many applications in biotechnology, bacteria are often grown in fermenters that can hold several hundred or thousand liters of liquid culture media (Figure 5.2b).

It is also relatively easy to make mutant strains of bacteria that can be used for molecular and genetic studies. Mutants can be created by exposing bacteria to X-rays, ultraviolet light, and a variety of chemicals that mutate DNA (mutagens). Literally thousands of mutant strains are available. For these reasons and many more, bacteria are not only the favorite organisms of many microbiologists but also ideal model organisms for studies in molecular biology, genetics, biochemistry, and biotechnology.

Yeast Are Important Microbes Too

Although the primary focus of this chapter is the applications of bacteria in biotechnology, **yeast** have served many important roles in biotechnology. In fact, archeologists have uncovered recipes on ancient Babylonian tablets from 4300 B.C. for brewing beer using yeast, which is one of the oldest documented applications of biotechnology. Much more recently, yeast-producing recombinant human antibodies have been generated. Yeast are single-celled eukaryotic microbes that belong to a kingdom of organisms called **fungi**. There are well over 1.5 million species of fungi, yet only around 10% of these have been identified and classified so there is significant potential for identifying more valuable products from fungi. For instance, fungi are important sources of antibiotics and drugs that lower blood cholesterol. In addition to having many structures that are similar to other eukaryotic cells such as plant and animal cells, yeast also contain a number of membrane-enclosed organelles in the cytoplasm, a cytoskeleton, and chromosome structures similar to human chromosomes. Yeast cells also have larger genomes than most bacteria. Moreover, mechanisms of gene expression in yeast resemble those in human cells. These features make yeast a very valuable model organism for studying chromosome structure, gene regulation, cell division, and cell cycle control.

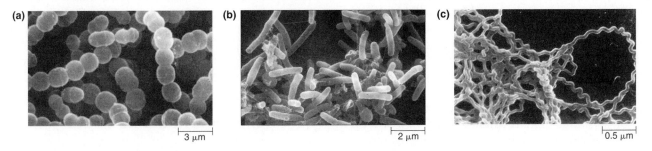

(a) 3 μm **(b)** 2 μm **(c)** 0.5 μm

Figure 5.1 Shapes of Bacteria The most common shapes of bacteria are (a) spherical, (b) rod shaped, and (c) corkscrew shaped.

(a)

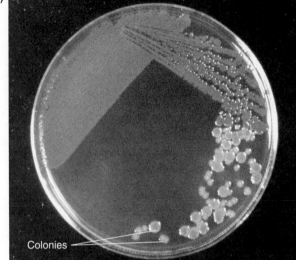

Colonies

(b)

Figure 5.2 Bacteria in Culture   (a) Most bacteria can be grown under a variety of conditions in liquid culture media or on solid media such as the petri dish shown here. On solid media, many bacteria grow in circular clusters called colonies, which typically begin with a single cell that divides rapidly to produce a spot visible to the naked eye. A single colony may contain millions of individual bacterial cells. (b) Bacteria can also be grown in large-scale quantities. The fermenters shown here contain several hundred liters of bacteria growing in liquid culture. These fermenters function as bioreactors that serve many purposes, such as growing bacteria for isolating recombinant proteins and culturing yeast to produce wine.

The different types of yeast vary greatly in size, but a majority are larger than bacteria and spherical, elliptical, or cylindrical in shape. Many can grow in the presence of oxygen **(aerobic conditions)** or in the absence of oxygen **(anaerobic conditions)** and under a variety of nutritional growth conditions. A wide number of different types of yeast mutants are also available. *Saccharomyces cerevisiae*, a commonly studied strain of yeast, was the first eukaryotic organism to have its complete genome sequenced. It has 16 linear chromosomes that contain over 12 million base pairs of DNA and approximately 6,300 genes. Several human disease genes have also been discovered in yeast and by studying them, scientists can learn a great deal about what these genes may do in humans and the diseases they may cause.

Recently a strain of yeast called *Pichia pastoris* has proven to be a particularly useful organism. *Pichia* grows to a higher density (biomass) in liquid culture than most laboratory strains of yeast, has a number of strong promoters that can be used for high production of proteins, and can be used in **batch processes** to produce large numbers of cells. In the next section we consider how microorganisms can be used by scientists as valuable tools for biotechnology research.

5.2 Microorganisms as Tools

Microorganisms in either their natural state or genetically modified forms have served as useful tools in a variety of fascinating ways.

Microbial Enzymes

Microbial enzymes have been used in applications from food production to molecular biology research. Because microbes are an excellent and convenient source of enzymes, some of the first commercially available enzymes isolated for use in molecular biology were DNA polymerases and restriction enzymes from bacteria. Initially isolated primarily from *E. coli,* DNA polymerases became available for a range of recombinant DNA techniques such as labeling DNA sequences to make probes and using the polymerase chain reaction (PCR) to amplify DNA.

In Chapter 3, you were introduced to *Taq* DNA polymerase as a **thermostable** enzyme. *Taq* is a heat-stable enzyme essential for PCR that was isolated from the hot-spring Archaean *Thermus aquaticus.* Because of their ability to grow and thrive under extreme heat, these microbes are called **thermophiles** (from the Greek words meaning "heat-loving"). Many similar thermostable enzymes have been identified in other thermophiles and are widely used in PCR and other reactions. Several companies have permission from the U.S. government to prospect geysers in Yellowstone National Park to identify other potentially valuable microorganisms that might contain novel and valuable enzymes. Such so-called **bioprospecting** projects are occurring all around the world. Recently a strain of yet-to-be named bacteria was isolated from a hydrothermal vent on the floor of the northeast Pacific Ocean. This strain has been shown to survive at 121°C, which is believed to be the record so far for the upper temperature limit at which life can exist.

The enzyme **cellulase,** produced by *E. coli,* degrades cellulose, a polysaccharide that forms the plant cell wall. Cellulase is widely used to make animal food more easily digestible. Have you ever owned a pair of stone-washed denim jeans? These soft and faded jeans are not produced by a literal washing with stones. Instead, the denim is treated with a mixture of cellulases from fungi such as *Trichoderma reesei* and *Aspergillus niger.* These cellulases mildly digest cellulose fibers in the cotton used to make the pants, resulting in a softer fabric. The protease **subtilisin,** derived from *Bacillus subtilis,* is a valuable component of many laundry detergents, where it functions to degrade and remove protein stains from clothing. Several bacterial enzymes are also used to manufacture foods, such as carbohydrate-digesting enzymes called amylases that are used to degrade starches for making corn syrup.

Bacterial Transformation

Recall from Chapter 3 that *transformation*—the ability of bacteria to take in DNA from their surrounding environment—is an essential step in the recombinant DNA cloning process. In DNA cloning, recombinant plasmids can be introduced into bacterial cells through transformation so that the bacteria can replicate the recombinant plasmids. Most bacteria do not take up DNA easily unless they are treated to make them more receptive. Cells that have been treated for transformation are called **competent cells.** One technique for preparing competent cells involves treating cells with an ice-cold solution of calcium chloride. Positively charged atoms (cations) in the calcium chloride disrupt the bacterial cell wall and membrane to create small holes through which DNA can enter. These cells can then be frozen at ultralow temperatures (−80 to −60°C) to maintain their competent state and then be used in the laboratory as needed.

Once competent cells are prepared, they can be transformed with DNA relatively easily, as shown in Figure 5.3a. Typically, the target DNA to be introduced into bacteria is inserted in a plasmid containing one or more antibiotic resistance genes. This plasmid vector is then mixed in a tube with the competent cells and the mixture placed on ice for a few minutes. The exact mechanism of transformation is not fully understood, but we do know that the cells must be kept cold during which time DNA sticks to the outer surface of the bacteria, and the cold conditions probably also serve to create gaps in the lipid structure of the cell membrane that allow for the entry of DNA. Cations in the calcium chloride are thought to play a significant role in neutralizing the negative charges of phosphates in the cell membrane and the DNA that would otherwise cause them to repel each other. The cells are then heated briefly (for about a minute) at temperatures between 37 and 42°C. During this brief heat "shock," DNA enters the bacterial cells. The heat shock produces a thermal gradient analogous to opening a door on a cold winter day. The heat sweeps into the cell bringing the DNA along with it.

After allowing these cells to grow in a liquid medium, they can be plated onto an agar medium containing antibiotics. Only those cells that were transformed with plasmid DNA containing the antibiotic-resistant genes will grow to produce colonies. Recall from Chapter 3 that this technique is called *antibiotic selection*. Plasmid DNA is replicated (cloned) by transformed bacteria, and genes carried on the plasmid are transcribed and translated into protein. Thus, the transformed bacterial cells now express recombinant proteins.

This process is called transformation because one can change the appearance or properties of a bacterial cell by introducing foreign genes. These are transformed cells because they have been genetically altered with new properties encoded by the DNA,

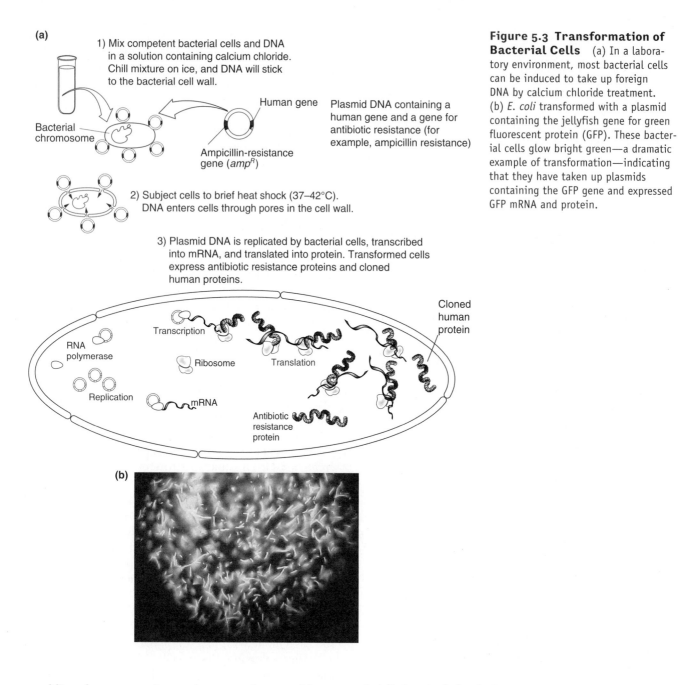

(a)

1) Mix competent bacterial cells and DNA in a solution containing calcium chloride. Chill mixture on ice, and DNA will stick to the bacterial cell wall.

Bacterial chromosome

Human gene

Plasmid DNA containing a human gene and a gene for antibiotic resistance (for example, ampicillin resistance)

Ampicillin-resistance gene (amp^R)

2) Subject cells to brief heat shock (37–42°C). DNA enters cells through pores in the cell wall.

3) Plasmid DNA is replicated by bacterial cells, transcribed into mRNA, and translated into protein. Transformed cells express antibiotic resistance proteins and cloned human proteins.

Cloned human protein

Transcription

RNA polymerase

Ribosome

Translation

Replication

mRNA

Antibiotic resistance protein

(b)

Figure 5.3 Transformation of Bacterial Cells   (a) In a laboratory environment, most bacterial cells can be induced to take up foreign DNA by calcium chloride treatment. (b) *E. coli* transformed with a plasmid containing the jellyfish gene for green fluorescent protein (GFP). These bacterial cells glow bright green—a dramatic example of transformation—indicating that they have taken up plasmids containing the GFP gene and expressed GFP mRNA and protein.

enabling them to produce substances they would not normally produce. For example, *E. coli* can be transformed with a gene called green fluorescent protein (GFP), which comes from jellyfish (Figure 5.3b). We look more closely at useful applications of GFP in Chapter 10. As we discuss in Section 5.3, bacterial cells have been transformed to express a large number of valuable genes including many human genes.

Electroporation

Another common technique for transforming cells is called **electroporation** (Figure 5.4). In this approach, an instrument called an electroporator produces a brief electrical shock that introduces DNA into bacterial cells without killing most of the cells. Electroporation offers several advantages over calcium chloride treatment for transformation. Electroporation is rapid, requires fewer cells, and can also be used to introduce DNA into many other cell types including yeast, fungi, animal, and plant cells. In addition, electroporation is a much more efficient process than calcium chloride transformation. A greater majority of cells will receive foreign DNA through electroporation than by calcium chloride treatment. Because of this, much less DNA can be used to transform cells (picogram amounts of DNA are sufficient). Regardless of how bacterial cells are transformed, once DNA of

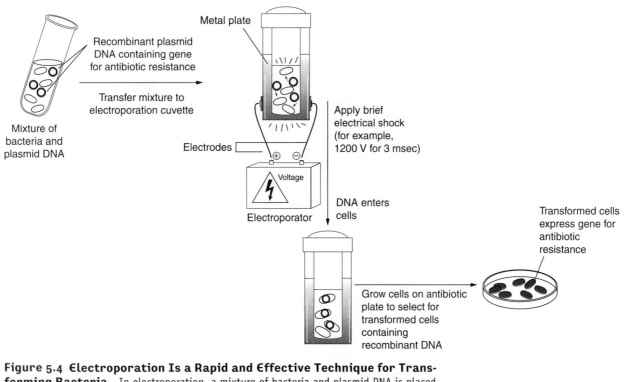

Figure 5.4 Electroporation Is a Rapid and Effective Technique for Transforming Bacteria   In electroporation, a mixture of bacteria and plasmid DNA is placed into a cuvette. By applying a brief electrical shock to the cuvette, DNA quickly enters cells. Cells containing recombinant DNA can then be selected for by growth on agar containing an antibiotic or another selection component.

interest is introduced into bacteria, a variety of useful techniques can be carried out.

Cloning and Expression Techniques

One reason for transforming bacterial cells is to replicate recombinant DNA of interest. In addition to replicating foreign DNA, transformed bacteria are also valuable because they can frequently be used to mass-produce proteins for a variety of purposes. One way to express and isolate a recombinant protein of interest is to make a fusion protein.

Creating bacterial fusion proteins to synthesize and isolate recombinant proteins

There are many techniques for using bacteria as tools for the synthesis and isolation of recombinant proteins. One popular approach involves making a **fusion protein.** There are a variety of ways to produce a fusion protein, but the basic concept of this technique is to use recombinant DNA methods to insert the gene for a protein of interest into a plasmid containing a gene for a well-known protein that serves as a "tag" (Figure 5.5). The tag protein then allows for the isolation and purification of the recombinant protein as a fusion protein. Plasmid vectors for making fusion

proteins are often called **expression vectors** because they enable bacterial cells to produce or express large amounts of protein. Commonly used expression vectors include those that synthesize proteins such as maltose-binding protein (shown in Figure 5.5), glutathione S-transferase, luciferase, green fluorescent protein, and β-galactosidase.

Expression vectors incorporate prokaryotic promoter sequences so that once the recombinant expression vector containing the gene of interest is introduced into bacteria by transformation, bacteria then synthesize mRNA and protein from this plasmid. The mRNA strands transcribed are hybrid molecules that contain sequence coding for the gene of interest and the tag protein. As a result, a fusion protein is synthesized from this mRNA consisting of the protein of interest joined to (fused) to the tag protein, in this case maltose-binding protein (Figure 5.5, step 2).

To isolate the fusion protein and separate it from other proteins normally made by the bacteria, cells are broken open (lysed) and homogenized to create a bacterial milkshake of sorts known as an *extract*. The extract is then passed through a tube called a *column*. One common approach is to fill a column with plastic beads coated with molecules that will bind to the tag protein portion of the fusion protein. Recall from

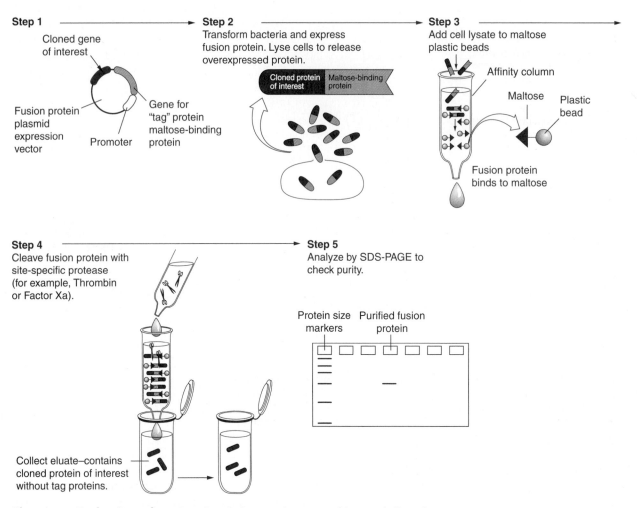

Step 1
Cloned gene
of interest

Fusion protein
plasmid
expression
vector

Promoter

Gene for
"tag" protein
maltose-binding
protein

Step 2
Transform bacteria and express
fusion protein. Lyse cells to release
overexpressed protein.

Cloned protein
of interest   Maltose-binding
protein

Step 3
Add cell lysate to maltose
plastic beads

Affinity column

Maltose   Plastic
bead

Fusion protein
binds to maltose

Step 4
Cleave fusion protein with
site-specific protease
(for example, Thrombin
or Factor Xa).

Collect eluate—contains
cloned protein of interest
without tag proteins.

Step 5
Analyze by SDS-PAGE to
check purity.

Protein size   Purified fusion
markers   protein

Figure 5.5 Fusion Proteins   To make a fusion protein, a gene of interest is ligated into an expression vector. Bacterial cells are transformed with recombinant DNA. Transformed cells express the fusion protein, which is isolated by binding it to an affinity column.

Chapter 4 that this technique is called affinity chromatography because the beads in the column have an attraction, or "affinity," for binding to the tag protein. For example, in Figure 5.5, plastic beads are attached to the sugar maltose, which will be bound by maltose-binding protein. Later, we'll see how antibody-coated beads can be used to isolate a fusion protein. Next, an enzyme treatment that uses protein-cutting enzymes called *proteases* cuts off and releases the protein of interest from the tag protein. Some techniques for making fusion proteins incorporate short peptide tags of just a few amino acids. For example, poly-His tags are a short string of the amino acid histidine. One benefit of this approach is that unlike maltose-binding protein and other tags that are large proteins, the small tags typically don't affect the structure and function of the protein they are fused to so they usually don't need to be removed. Fusion protein techniques are used to provide purified proteins to study protein structure and function and used to isolate insulin and

other medically important recombinant proteins (see Figure 5.9 later in the chapter).

E. coli and the gram-negative rod-shaped bacterium *Bacillus subtilis* are commonly used microbes for producing fusion proteins. In particular, *B. subtilis* is a favored microbe for many applications when producing fusion proteins for human proteins because it will secrete them into growth media where they can easily be harvested and purified. And unlike some bacteria, *B. subtilis* often processes proteins in such a way as to maintain their 3D folding and function.

Microbial proteins as reporters

According to recent estimates, close to three fourths of all marine organisms can release light through a process known as **bioluminescence.** For marine fish, bioluminescence in lines of cells along the side of a fish and in its fins can be used to attract mates in dark ocean environments. Bioluminescence in many marine species is created by bacteria such as *Vibrio fisheri* that use the marine organism as a host (Figure 5.6). Bacteria such as *Vibrio* have been used as biosensors to detect cancer-causing chemicals called carcinogens, environmental

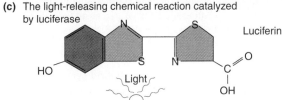

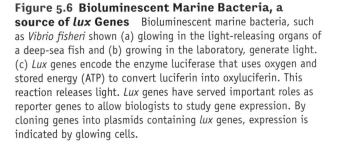

(c) The light-releasing chemical reaction catalyzed by luciferase

Luciferin

$Luciferin + ATP + O_2 \longrightarrow Oxyluciferin + AMP + CO_2 + PPi \text{ (inorganic phosphate)}$

Luciferase

Figure 5.6 Bioluminescent Marine Bacteria, a source of *lux* Genes Bioluminescent marine bacteria, such as *Vibrio fisheri* shown (a) glowing in the light-releasing organs of a deep-sea fish and (b) growing in the laboratory, generate light. (c) *Lux* genes encode the enzyme luciferase that uses oxygen and stored energy (ATP) to convert luciferin into oxyluciferin. This reaction releases light. *Lux* genes have served important roles as reporter genes to allow biologists to study gene expression. By cloning genes into plasmids containing *lux* genes, expression is indicated by glowing cells.

pollutants, and chemical and bacterial contaminants in foods. *Vibrio fisheri* and another marine bioluminescent strain called *Vibrio harveyi* create light through the action of genes called *lux* genes. Several *lux* genes encode protein subunits that form an enzyme called **luciferase** (derived from the Latin *lux ferre,* meaning "light bearer").

Luciferase is the same enzyme that allows fireflies to produce light. The *lux* genes have been cloned and used to study gene expression in a number of unique ways. For instance, by cloning *lux* genes into a plasmid, the *lux* plasmid can be used to produce fusion proteins. Also, *lux* genes can serve as valuable **reporter genes.** If inserted into animal or plant cells, the luciferase encoded by the *lux* plasmid cause these cells to fluoresce (Figure 5.6). In this manner, the *lux* plasmid is acting as a "reporter" to provide a visual indicator of gene expression.

Lux genes have recently been used to develop a fluorescent bioassay to test for tuberculosis (TB). TB is caused by the bacterium *Mycobacterium tuberculosis,* which grows slowly and can exist in a human for several years before the individual may develop TB (TB symptoms are discussed in Section 5.4). For the TB bioassay, scientists introduced *lux* genes into a virus that infects *M. tuberculosis.* Saliva from a patient who may be infected with *M. tuberculosis* is mixed together with the *lux*-containing virus. If *M. tuberculosis* is in the saliva sample, the virus infects these bacterial cells, which can be detected by their glowing. Similarly engineered strains of *E. coli* have been used to detect arsenic contamination in water. Reporter genes have many valuable roles in research and medicine. We consider them in more detail in Chapter 10. In the next section we explore a range of everyday applications that involve microbes.

5.3 Using Microbes for a Variety of Everyday Applications

Microbes are necessary for normal body functions and for many natural processes in the environment. Harnessing the great potential of microbes for making foods, developing and producing new drugs, and using microbes to detect and clean up wastes in our environment are not new applications. In this section we consider many modern applications of such microbial biotechnology.

Food Products

Microbes are used to make many foods including breads, yogurt, cheeses, and sauerkraut as well as beverages such as beer, wine, champagne, and liquors. As a child you probably learned the classic nursery rhyme

TOOLS OF THE TRADE

The Yeast Two-Hybrid System

The **yeast two-hybrid system** is currently one of the hottest techniques in cell and molecular biology. This innovative technique for studying proteins that interact with each other provides a way to understand protein function. For instance, suppose you identified several proteins that you believe may interact and work together as part of a metabolic pathway for the synthesis of an important hormone in the body. The yeast two-hybrid system is an excellent technique to use to determine if these proteins bind to and interact with each other.

As shown in Figure 5.7, the gene for one protein of interest is cloned and expressed as a fusion protein attached to the DNA binding domain (DBD) of another gene. These DNA-binding sites are commonly found on proteins, such as transcription factors that interact with DNA. This protein is often called the "bait" protein because it is used to find other proteins called "prey" that may bind to it. The gene for the second protein of interest is fused to another gene that contains transcriptional activator domain (AD) sequences. AD proteins, also required for the attachment of DNA-binding proteins to DNA, stimulate the binding domain of a bait protein to interact with DNA-binding sites such as a promoter sequence for a reporter gene.

This modified DNA is introduced into yeast cells. *Saccharomyces cerevisiae* is a commonly used strain. Neither the DBD fusion protein nor the AD fusion protein alone can stimulate transcription from this reporter gene. In Figure 5.7, the *lacZ* gene, which encodes the enzyme β-galactosidase, is shown as a common reporter gene whose activity can easily be measured using simple colorimetric tests. Only a combination of both proteins, a molecule forming a "hybrid" or complex of both proteins, will stimulate expression of the reporter gene. Because scientists can create mutations of both proteins of interest and see how these mutations affect the ability of the two proteins to interact, we can learn a lot about the structure and function of those proteins

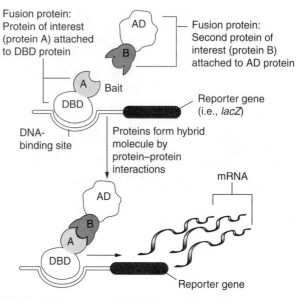

Hybrid protein can now stimulate transcription of reporter gene. Activity of protein from reporter gene mRNA can be measured.

Figure 5.7 The Yeast Two-Hybrid System for Studying Protein Interactions If a researcher wanted to know if two proteins (A and B) interact with one another, one protein could be expressed as a fusion protein attached to the DNA binding domain (DBD) of another protein (such as a transcription factor) and the second protein attached to an activator domain (AD) protein. Neither fusion protein alone is capable of binding to and stimulating transcription (mRNA production) of the reporter gene. The resulting hybrid protein created by protein–protein interactions of the two fusion proteins will stimulate transcription of the reporter gene, which can easily be measured as an indicator of protein interactions between the A and B proteins.

with this technique. The yeast two-hybrid system is a very powerful example of how microbial biotechnology can be used for a research application.

Little Miss Muffet. The tale of Miss Muffet sitting on her tuffet eating "curds and whey" might seem like an improbable way to discuss biotechnology, but the treat in Miss Muffet's bowl was the result of biotechnology. Curds and whey are formed from coagulated milk, and milk coagulation is an important step during cheese production. To make cheese, milk from cows, goats, or sheep is treated to help it coagulate (*curd*). The watery liquid that remains after curd forms is called *whey.*

One way to make cheese from coagulated milk is to treat the milk with an acidic solution, but the best-tasting cheeses are typically made using the enzyme **rennin.** In the early days of cheese production, rennin was traditionally obtained by extracting it from the stomachs of calves and other milk-producing species such as goats, sheep, horses, and even zebras and camels. Rennin coagulates milk to produce curd by digesting a family of proteins called *casein,* which is a major component of milk. Digested casein forms an insoluble mixture of proteins that clumps (coagulates) in a process similar to what happens when milk spoils.

In the 1980s, using recombinant DNA techniques, scientists cloned rennin and expressed it in bacterial cells and fungi such as *Aspergillus niger*. Recombinant rennin (now called **chymosin**) from microbes is widely used by cheese makers as an inexpensive substitute for extracting rennin from calves. In 1990, rennin was the first recombinant DNA food ingredient approved by the Food and Drug Administration (FDA). For some types of cheese, certain strains of bacteria called lactic acid bacteria *(Lactococcus lactis, L. acidophilus)* are used for coagulation. These bacteria degrade casein and use an enzyme called *lactase* to break down sugars in the milk that are eventually used by bacteria for **fermentation,** an aspect of bacterial metabolism that is essential for many common foods and drinks we enjoy.

Fermenting microbes

Fermentation is an important microbial process that produces many food products and beverages including a variety of breads, beers, wines, champagnes, yogurts, and cheeses. Fermenting microbes have very important roles in biotechnology. One of the earliest applications of microorganisms—the brewing of beer and wine—involves fermentation by yeast (Figure 5.8a). To appreciate how making beer, wine, and bread requires microbes, you need to know more about the process of fermentation.

Animal and plant cells and many microbes obtain energy from carbohydrates such as glucose by using electrons from these sugars to create a molecule called **adenosine triphosphate (ATP).** ATP production occurs as a series of reactions. The first major reaction, called glycolysis, converts glucose into two molecules called *pyruvate*. During this conversion, electrons are transferred from glucose to electron carrier molecules called *NAD+*, which capture electrons to produce NADH (Figure 5.8b). This molecule transports electrons to subsequent reactions in the process that results in ATP production. For certain bacteria and yeast, oxygen is an important part of these electron transport reactions. Microbes that use oxygen for ATP production are called **aerobes** because they undergo oxygen-dependent (aerobic) metabolism.

Many microbes live in areas where oxygen is rare or absent, such as the intestines of animals, deep water, or soil. Because these organisms must survive without oxygen, they have evolved the ability to derive energy from sugars in the absence of oxygen (anaerobic conditions). This is fermentation, and microbes that use fermentation are called **anaerobes.**

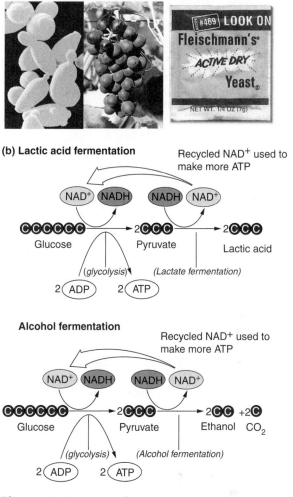

(a)

(b) Lactic acid fermentation

Alcohol fermentation

Figure 5.8 Fermentation Certain strains of yeast and bacteria are capable of producing energy from sugars (glucose) through fermentation. (a) Yeast such as *S. cerevisiae* (left) shown here causes bread dough to rise; other strains of *Saccharomyces* grow on grapevines and are important for making wine (center photo). (b) Anaerobic bacteria that undergo lactic acid fermentation make lactic acid (lactate) as a waste product; alcohol-fermenting bacteria create ethanol and carbon dioxide (CO_2) as waste products. Both processes are important for manufacturing many foods and beverages.

Fermentation is similar to glycolysis in that NAD^+ is used to capture electrons to make NADH and pyruvate; however, neither NADH nor pyruvate has anywhere to go. In aerobic metabolism, oxygen is required to use electrons from NADH and pyruvate to make ATP, but under anaerobic conditions there is little or no oxygen so the NADH and pyruvate cannot be used to make ATP. All organisms must recycle NADH into NAD^+. Fermentation enables many anaerobes to do this in the absence of oxygen, and some anaerobes are capable of either fermentation or aerobic respiration depending on the presence or absence of oxygen. In the absence of oxygen, anaerobes have evolved to acquire fermentation reactions as a way to solve the problem of recycling NADH into NAD⁺. Fermenting microbes use pyruvate as an electron acceptor molecule to take electrons from NADH to regenerate NAD⁺ (Figure 5.8b). Two of the most common types of fermentation are **lactic acid fermentation** and **alcohol (ethanol) fermentation.**

In lactic acid fermentation, electrons from NADH are used to convert pyruvate into lactic acid, also called lactate, and during alcohol fermentation, electrons from NADH convert pyruvate into ethanol. NAD⁺ is regenerated when electrons are removed from NADH and transferred to pyruvate to make lactate or ethanol in the final step of fermentation. During alcohol fermentation, carbon dioxide gas is also produced as a waste product. There are many strains of fermenting bacteria and yeast. Other types of fermentation create sauerkraut from cabbage and produce such useful products as acetic acid in vinegar, citric acid in fruit juices, and acetone and methanol, two chemicals often used in laboratories for cleaning glassware. Furthermore, these microbial products have the advantage of being produced more efficiently and cheaply than by other means. Lactic acid fermentation also occurs in human muscle cells during strenuous exercise. The burning you feel in your legs when you run because you are late to class is created by fermentation in muscle cells creating lactic acid.

So how are fermenting microbes used to make foods and beverages? To make beer and wine, many processes rely on wild strains of yeast that live on grape vines and on such domestic strains of yeast as *Saccharomyces cerevisiae,* which are very good at alcohol fermentation. Large barrels, or fermenters, containing crushed grapes and yeast are mixed together under carefully controlled conditions. Fermenting yeast converts sugars from the grapes into alcohol. Fermentation rates are monitored and carefully controlled by changing both the amount of oxygen in the fermenter and the temperature. By manipulating fermentation rates, wine makers can control the alcohol content of the brewing wine until the desired alcohol content and flavor are achieved. Bottles of champagne and other sparkling wines are capped while the yeast is still in the liquid and actively fermenting; carbon dioxide gas is trapped in the bottle and is released only when the bottle is opened, producing the characteristic champagne bottle "pop." Production of some wines relies also on lactic acid bacteria such as *Oenococcus oeni,* which will convert bitter-tasting malic acid that is formed during fermentation into softer-tasting lactic acid and thus provide a milder flavor and aroma to the wine.

Lactic acid-fermenting bacteria are used to produce cheeses, sour cream, and yogurts, and the sharp or sour flavors of these products is primarily due to lactic acid. Yogurt was first created in Bulgaria and has been around for centuries. Yogurt production typically involves a blend of bacteria that often includes strains of anaerobic lactic acid-fermenting microbes such as *Streptococcus thermophilus,* a strain called *Lactobacillus* (*Lactobacillus delbrueckii* and *Lactobacillus bulgaricus*), and another strain called *Lactococcus (Lactococcus lactis).* Active cultures of these lactic acid-fermenting microbes are added to mixtures of milk and sugar in a fermenter that is maintained at carefully controlled temperatures. Microbes in the mixture use sugars to produce lactic acid. Fruit and other flavorings may then be added to the yogurt before it is cooled to refrigeration temperature (4–5°C) to prevent changes in its composition. When you enjoy a spoonful of yogurt, you are also eating large numbers of fermenting microbes that are still in the yogurt. Lactic acid and other products of fermentation contribute to the sweet and sour taste of yogurt and assist in the coagulation of the yogurt.

Another lactic acid bacterium, *Lactobacillus sakei,* is found naturally on fresh meat and fish. *L. sakei* serves as a natural biopreservative in meat products such as sausage where it wards off growth of other undesirable microbes that will spoil the food or cause illness. In addition to lactic acid, this microbe produces molecules called *bacteriocins* that act as naturally occurring antimicrobial agents to kill other microbes.

Q Why is yeast important for making bread? Why does bread dough rise?

A Yeast is commonly used in a variety of doughs from pizza to Italian bread and croissants. Most of these yeasts are alcohol (ethanol) fermenters such as *Saccharomyces cerevisiae,* which is sold in grocery stores as small packets of baker's yeast. Yeast is added to a mixture of flour, water, and sugar to make dough. *S. cerevisiae* uses the sugar in the dough for ethanol fermentation and releases carbon dioxide as the sugar is degraded. Bubbles of carbon dioxide are trapped in the dough causing it to expand and rise. Baking the dough causes gas to be released from the dough, leaving behind holes in the bread; the ethanol evaporates as the bread is heated. Certain breads such as sourdough are undercooked, leaving a little ethanol in the bread, which gives it a bitter flavor. The next time you make dough, leave some in a plastic bag at room temperature for a day or two. What do you smell when you open the bag? The bitter fragrance is a result of the ethanol produced in the dough.

Therapeutic Proteins

The development of recombinant DNA technology quickly led to using bacteria to produce such important medical products as therapeutic proteins. Insulin was the first recombinant molecule expressed in bacteria for use in humans. Here we use insulin production as an example of how microbes can be used to make therapeutic proteins.

Producing recombinant insulin in bacteria

Insulin is a hormone produced by cells in the pancreas called *beta cells*. When insulin is secreted into the bloodstream by the pancreas, it plays an essential role in carbohydrate metabolism. One of its primary functions is to stimulate the uptake of glucose into body cells such as muscle cells, where the glucose can be broken down to produce ATP as an energy source. **Type I,** or **insulin-dependent diabetes mellitus** occurs when beta cells do not produce insulin. The lack of insulin results in an elevated blood glucose concentration that can cause a number of health problems such as high blood pressure, poor circulation, cataracts, and nerve damage. People with type I diabetes require regular injections of insulin to control their blood sugar levels.

Prior to recombinant DNA technology, insulin used to treat diabetes was purified from the pancreases of pigs and cows before being injected into diabetics. The purification process was cumbersome, expensive, and often produced impure batches of insulin. Also, purified insulin was ineffective in some individuals, and many others developed allergies to insulin from cows. In 1978, insulin was cloned into an expression vector plasmid, expressed in bacterial cells, and isolated by scientists at Genentech, a biotechnology company in San Francisco, California. In 1982, this recombinant human insulin, called Humulin, was the first biotechnology product to be approved for human applications by the FDA.

Bacteria do not normally make insulin, and producing human insulin in recombinant bacteria was a significant advance in biotechnology. It remains an outstanding example of microbial biotechnology in action. Human insulin consists of two polypeptides called the A (21 amino acids) and B (30 amino acids) chains or subunits; they bind to each other by disulfide bonds to create the active hormone (Figure 5.9). In the pancreas, beta cells synthesize both insulin chains as one polypeptide that is secreted and then enzymatically cut (cleaved) and folded to join the two subunits. When the human genes for insulin were cloned and expressed in bacteria, genes for each of the subunits were cloned into separate expression vector plasmids containing the *lacZ* gene encoding the enzyme

β-galactosidase (β-gal) and then used to transform bacteria (Figure 5.9).

Because the insulin genes are connected to the *lacZ* gene, when bacteria synthesize proteins from these plasmids, they produce a protein that contains β-gal attached to the human insulin protein to create a β-gal–insulin fusion protein. As we saw in Section 5.2, making a fusion protein enables scientists to isolate and purify a protein of interest such as insulin. Bacterial extracts are passed over an affinity column to isolate the β-gal–insulin fusion proteins (Figure 5.9). The fusion protein is chemically treated to cleave off the β-gal, releasing the insulin protein; then, purified A and B chains of insulin are mixed together under conditions that allow the two subunits to bind and form the active hormone. After further purification to conform to FDA guidelines, the recombinant hormone is ready for patient use as an injectable drug.

Shortly after insulin became available, growth hormone—used to treat children who suffer from a form of dwarfism—was cloned in bacteria and became available for human use. A short time later, a wide variety of other medically important proteins that were once difficult to obtain became readily available as a result of recombinant DNA technology and expressing proteins in bacteria. As shown in Table 5.1, many other therapeutic proteins with valuable applications for treating medical illness in humans have been expressed in and isolated from bacteria. A major category of medical products from recombinant bacteria are vaccines. We cover vaccines in Section 5.4.

Using Microbes Against Other Microbes

Antibiotics are substances produced by microbes that inhibit the growth of other microbes. Antibiotics are a type of **antimicrobial drug,** a general category defined as any drug (whether produced by microbes or not) that inhibits microorganisms. Penicillin was the first antibiotic to be used widely in humans, and its discovery is an excellent example of how some microbes protect themselves from others by making antimicrobial substances. Alexander Fleming was the microbiologist who, in 1928, discovered that colonies of the mold *Penicillium notatum* inhibited growth of the bacterium *Staphylococcus aureus*. When cultured together on a petri dish, *S. aureus* would not grow in a small zone of agar surrounding mold colonies. A dozen years later, scientists used *P. notatum* to isolate the drug they called penicillin, which was subsequently mass-produced and used to treat bacterial infections in humans.

A majority of antibiotics are isolated from bacteria, and most of these substances inhibit the growth of other bacteria. In the over 60 years since penicillin was

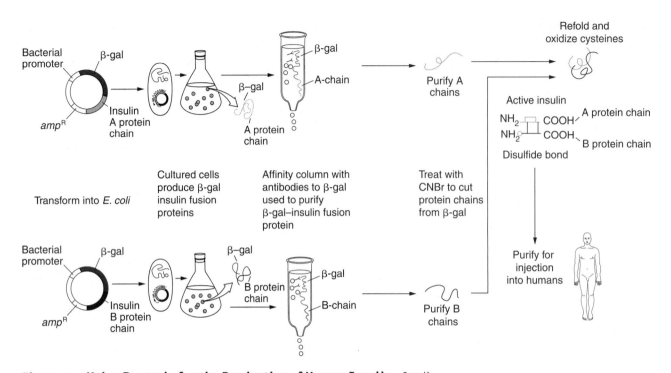

Figure 5.9 Using Bacteria for the Production of Human Insulin   Insulin was the first protein expressed in recombinant bacteria that was approved for use in humans. Insulin consists of two protein chains (A and B) produced from separate genes. To make recombinant insulin, scientists cloned the insulin genes into plasmids containing the *lacZ* gene, which encodes the enzyme β-galactosidase. Recombinant plasmids were used to transform bacteria, enabling them to produce β-gal–insulin fusion proteins. Affinity chromatography was used to isolate fusion proteins, which were then chemically treated to separate the cloned insulin from β-gal proteins. Purified forms of the A and B protein chains of insulin could then combine to form active insulin, which is given to people with diabetes to control blood sugar levels.

discovered, thousands of other antibiotic-producing microbes have been discovered, and hundreds of different antibiotics have been isolated. Table 5.2 shows examples of common antibiotics and their source microbes.

How do antibiotics and other antimicrobial drugs affect bacterial cells? Most of these substances act in a few key ways. Typically they either prevent bacteria from replicating or kill microbes directly, which of course also prevents affected cells from replicating. Antibiotics can damage the cell wall or prevent its synthesis (which is how penicillin acts), block protein synthesis, inhibit DNA replication, or inhibit the synthesis or activity of an important enzyme required for bacterial cell metabolism (Figure 5.10 on page 134).

You sneeze, your body aches, your nose is running, your throat is sore, and you can't sleep, but you still have an exam to take tomorrow afternoon. How will you do it? By going to your doctor and asking for antibiotics of course! But are antibiotics what you really need? Because antibiotics are only effective against bacteria, they do not work against viruses such

as those that cause flu. Also, bacterial resistance to antibiotics has become a major problem. Improper and overuse of antibiotics in humans and farm animals has led to dramatic increases in antibiotic-resistant bacteria, including some strains that do not respond at all to many antibiotics that were effective in the past.

Antibiotic-resistant strains of *S. aureus, Pseudomonas aeruginosa, Streptococcus pneumoniae, M. tuberculosis,* and many other deadly strains of human pathogens have already been detected in hospitals. Because most antibiotics attack a bacterial cell in a limited number of ways (Figure 5.10), resistance to one antibiotic often leads to resistance to many other drugs. Consequently, new antimicrobial drugs that are harmful to bacteria in different ways need to be developed for medical use as well as for treating food animals such as cows, pigs, and chickens. Marine microbiologists are discovering new strains of ocean microbes with novel antibiotics and anticancer compounds. From treetops to polar ice caps, deserts, and the ocean's depths, scientists are evaluating many microorganisms as potential sources of new antimicrobial substances.

Table 5.1 THERAPEUTIC PROTEINS FROM RECOMBINANT BACTERIA

Protein	Function	Medical Application(s)
DNase	DNA-digesting enzyme	Treatment of cystic fibrosis patients
Erythropoietin	Stimulates production of red blood cells	Used to treat patients with anemia (low number of red blood cells)
Factor VIII	Blood clotting factor	Used to treat certain types of hemophilia (bleeding diseases due to deficiencies in blood clotting factors)
Granulocyte colony-stimulating factor	Stimulates growth of white blood cells	Used to increase production of certain types of white blood cells; stimulate blood cell production following bone marrow transplants
Growth hormone (human, bovine, porcine)	Hormone stimulates bone and muscle tissue growth	In humans used to treat individuals with dwarfism. Improves weight gain in pigs and cows; stimulates milk production in cows.
Insulin	Hormone required for glucose uptake by body cells	Used to control blood sugar levels in patients with diabetes
Interferons and interleukins	Growth factors that stimulate blood cell growth and production	Used to treat blood cell cancers such as leukemia; improve platelet counts; some used to treat different cancers
Superoxide dismutase	An antioxidant that binds and destroys harmful free radicals	Minimizes tissue damage during and after a heart attack
Tissue plasminogen activator (tPA)	Dissolves blood clots	Used to treat heart attack patients and stroke victims
Vaccines (e.g., Hepatitis B vaccine)	Stimulate immune system to prevent bacterial and viral infections	Used to immunize humans and animals against a variety of pathogens; also used in some cancer tumor treatments

Table 5.2 COMMON ANTIBIOTICS

Antibiotic	Source Microbe	Common Uses of Antibiotic
Bacitracin	*Bacillus subtilis* (bacterium)	First aid ointment and skin creams
Erythromycin	*Streptomyces erythraeus* (bacterium)	Broad uses to treat bacterial infections especially in children
Neomycin	*Streptomyces fradiae* (bacterium)	Skin ointments and other topical creams
Penicillin	*Penicillium notatum* (fungus)	Injected or oral antibiotic used in humans and farm animals (cattle and poultry)
Streptomycin	*Streptomyces griseus* (bacterium)	Oral antibiotic used to treat many bacterial infections in children
Tetracycline	*Streptomyces aureofaciens* (bacterium)	Used to treat infections of the urinary tract in humans; commonly used in animal feed to reduce infections and stimulate weight gain

Another way to create new antimicrobial drugs is to study bacterial pathogens and identify toxins and properties that disease-causing bacteria use to create illness. By understanding factors involved in causing illness, scientists can develop new strategies to block bacterial replication. For instance, for certain bacteria,

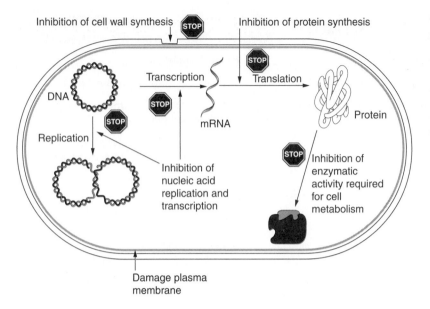

Inhibition of cell wall synthesis

Inhibition of protein synthesis

Transcription

Translation

DNA

mRNA

Protein

Replication

Inhibition of
nucleic acid
replication and
transcription

Inhibition of
enzymatic
activity required
for cell
metabolism

Damage plasma
membrane

Figure 5.10 Antibiotics and Other Antimicrobial Drugs Work Against Microbes in a Variety of Ways Antibiotics can kill microbes directly or prevent cell replication by inhibiting important events in the bacterial cell (such as DNA replication, protein synthesis, and synthesis of the bacterial cell wall) and by blocking enzymes required for cell metabolism.

their ability to cause disease requires that they attach (adhere) to human tissues. Once attached, bacterial cells can multiply and then produce sufficient toxins to cause illness. Scientists are working not only to identify proteins that bacteria use to attach to human tissues and cause disease but also to design what are called "inducer" microbes—harmless bacterial cells that have been genetically engineered to act as living factories to produce anti-adhesion molecules, enzymes, and other antimicrobial agents. The inducer microbes would release specific antimicrobial compounds into patients to destroy the disease-causing microorganism literally around the clock. Regardless of the antimicrobial strategy, microbes will continue to be important players in the fight against disease-causing and harmful microbes.

A Florida company called Oragenics recently received FDA approval to begin clinical trials of a recombinant form of *Streptococcus mutans* to evaluate its safety and efficacy for reducing tooth decay. Unlike naturally occurring strains of *S. mutans* found in the oral cavity, the recombinant strain cannot metabolize sugars to produce lactic acid. Because lactic acid dissolves enamel and dentin in teeth, it leads to cavities and tooth decay. Scientists will attempt to use recombinant *S. mutans* to colonize the oral cavity and replace natural strains of *S. mutans* and then determine if tooth decay is reduced. Recently, scientists at UCLA created a sugar-free lollipop containing an ingredient in licorice that kills *S. mutans* and these bacteria-killing lollipops are now available for purchase.

Field Applications of Recombinant Microorganisms

Genetically engineered bacteria have been used to produce many different recombinant products, and some of their most controversial uses have involved the release of recombinant bacteria for field applications.

Recombinant microbes in the field

Many natural strains of bacteria play an important role in the degradation of waste products and in bioremediation of polluted environments. New strategies in bioremediation also involve using genetically engineered microbes containing genes that help these organisms degrade wastes rapidly and efficiently when they are released in the environment.

The first field application of genetically engineered bacteria was developed at the University of California by plant pathologist Steven Lindow and colleagues. They identified a common strain of bacteria called *Pseudomonas syringae*, which makes bacterial proteins that stimulate ice crystal formation. Lindow's group created "ice-minus" bacteria by removing the ice protein-producing genes from *P. syringae*. They proposed that releasing ice-minus bacteria onto plants would cause the ice-minus bacteria to crowd out normal, ice-forming *P. syringae* and provide frost-sensitive crop plants with protection from the cold, thus extending the growing season and increasing crop yields.

Surrounded by a great deal of controversy, Lindow received approval in 1987 to test *P. syringae* on a crop of potatoes. Around the same time, other scientists received permission to test ice-minus bacteria on strawberries in a small town in California. This was the first time that genetically altered microbes were ever intentionally released into the environment in the United States. In both experiments, a majority of plants were damaged by activists concerned about the release of genetically altered microbes. Ice-minus *P. syringae* have shown some promise for frost protection, but they have not been as effective at crowding out the growth of normal ice-forming *P. syringae* as Lindow and others had hoped. Experiments with these strains are still being conducted. Recently the genome for a tomato-infecting strain of *P. syringae* has been sequenced, and scientists plan to use information about its more than 5,500 genes to help combat the spread of this microbe and the damage it causes to tomato crops.

Another example of field applications of genetically altered strains of bacteria involves *Pseudomonas fluorescens*. This strain is being engineered and experimented with to protect plants against root-eating insects that damage such agriculturally important plants as cotton and corn. Scientists have introduced a toxin-producing gene from the insect bacterial pathogen *Bacillus thuringiensis* (Bt) into *P. fluorescens*. Bt produces a number of toxins that work as effective insecticides when ingested by insects. One of the Bt toxin genes encodes the enzyme galactosyltransferase, which attaches carbohydrates to lipids and proteins in the insect gut, killing the insect. Bt-toxin—producing strains of *P. fluorescens* and other bacteria have been sprayed onto the leaves of plants to provide these crops with insect resistance. When insects eat these leaves, they ingest some of the genetically altered *P. fluorescens* and die. In Chapter 6, we discuss the role of Bt toxin in creating insect-resistant transgenic plants.

These examples briefly illustrate potential roles of recombinant microbes in field applications. As you will learn in Section 5.5, the use of recombinant microbes in the lab and in the field is likely to increase as scientists begin to unravel secrets of microbial genomes. In the next section we discuss how microbes are used to create vaccines, a very important and widely used application of biotechnology.

5.4 Vaccines

The use of antibiotics and vaccines has proven to be very effective for treating a number of infectious disease conditions in humans caused by microorganisms (Figure 5.11). These measures have generally worked well for treating disease-causing microbes in humans and animals; however, **pathogens** (disease-causing microbes) with resistance to many widely used antibiotics and vaccines have emerged and challenge the effectiveness of vaccines and antibiotics. Infectious diseases created by microbes affect everyone, and worldwide over 60% of the causes of death among children before age 4 are due to infectious diseases. Without question our ability to prevent, detect, and treat infectious diseases is an important aspect of microbial

YOU DECIDE

Microbes on the Loose

We have seen how recombinant microbes can be employed, from using genetically altered bacteria and making recombinant proteins for the treatment of human disease to releasing ice-minus bacteria. One of the many controversies surrounding microbial biotechnology is the prospect that recombinant microbes can enter the environment, through accidental introduction or intentional release, as in the ice-minus studies. If recombinant microbes are loose in the environment, how can we know what will ultimately happen to these organisms?

Because gene transfer between bacteria is a natural process that occurs in the wild, scientists are concerned about horizontal gene transfer, the spread of genes to related microbes. As a result of genetic recombination and the creation of new genes, new strains of microbes with different characteristics based on the genes they inherit may be produced. What would happen if recombinant microbes could transfer genetically altered genes into other microorganisms for which they were not originally intended? For instance, what would happen if ice-minus bacterial genes were transferred to strains of bacteria that are accustomed to living under cold conditions? Or how might the transfer of Bt genes from genetically altered microbes to common soil strains of *Pseudomonas* affect the natural growth of many soil insects with important natural roles?

Once recombinant microbes escape or are released into the environment, we cannot simply call them back into the lab if we do not like what they are doing in the field. Can we prevent the escape of genetically altered microbes in field experiments? It is a difficult if not impossible task when wind, rain, and other weather elements are involved. Should genetically engineered microbes be released even in "controlled" experiments that might result in beneficial applications of biotechnology? You decide.

biotechnology, and vaccines play a key role in this process. What is the difference between an antibiotic and a vaccine, and how do vaccines work?

The world's first vaccine was developed in 1796 when Edward Jenner demonstrated that a live cowpox virus could be used to vaccinate humans against smallpox. Smallpox and cowpox are closely related viruses. Smallpox epidemics ravaged areas of Europe, and an estimated 80% or more of Native Americans on the

Crude Death Rate for Infectious Diseases

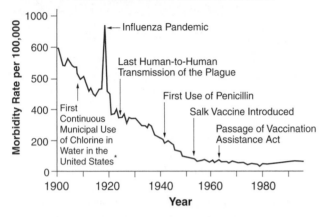

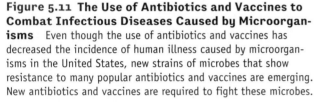

Figure 5.11 The Use of Antibiotics and Vaccines to Combat Infectious Diseases Caused by Microorganisms Even though the use of antibiotics and vaccines has decreased the incidence of human illness caused by microorganisms in the United States, new strains of microbes that show resistance to many popular antibiotics and vaccines are emerging. New antibiotics and vaccines are required to fight these microbes.

*The American Society for Microbiology Report: Congressional Briefing. Infectious Disease Threats, 2001.

East Coast of the United States died from smallpox infections carried by European settlers in North America. Cowpox produces blisters and lesions on the udders of cows and produces similar skin ulcers in humans. Based on a milkmaid's claim that cowpox infections protected her from smallpox, Jenner prepared his vaccine. He took fluid from cowpox blisters on the milkmaid and used needles containing this fluid to scratch the skin of healthy volunteers. His first "patient" was an 8-year-old boy. A majority of Jenner's volunteers did not develop cowpox or smallpox even when subsequently exposed to persons infected with smallpox. Exposure to cowpox fluid had stimulated the immune system of Jenner's volunteers to develop protection against smallpox.

These experiments demonstrated the potential of **vaccination** (named from the Latin word *vacca,* which means "cow")—using infectious agents to provide immune protection against illness. Although the United States stopped routine vaccinations for smallpox in 1972, by 1980, subsequent widespread applications of the vaccine had eradicated this disease. In the United States many vaccines are routinely given to newborns, children, and adults. Although you may not remember your first vaccination (which usually occurs sometime from 2 to 15 months of age), you were probably vaccinated with the **DPT vaccine,** which provides several years of protection against three bacterial toxins called *d*iphtheria toxin, *p*ertussis toxin, and *t*etanus toxin.

Another childhood vaccine is the **MMR** (*m*easles, *m*umps, *r*ubella) vaccine.

You were probably also vaccinated with **OPV** (*o*ral *p*olio *v*accine) for the poliovirus, a strain that infects neurons in the spinal cord causing devastating paralysis called poliomyelitis (polio). Like the smallpox vaccination, the OPV has dramatically decreased the incidence of polio. It has virtually been eliminated in North America, South America, and most of Europe; however, it still exists in some areas of the world. Polio, once a much more common disease, ravaged millions of children worldwide prior to 1954 when Jonas Salk developed the first vaccine for polio. Salk's original vaccine required injection; Albert Sabin developed the current version, which can be taken by mouth, in 1961. To understand how vaccines work, you need to be familiar with the basic aspects of the human immune system.

A Primer on Antibodies

The immune system in humans and other animals is extremely complex. Numerous cells throughout the body work together in intricate ways to recognize foreign materials that have entered our body and mount an attack to neutralize or destroy those materials. Foreign substances that stimulate an immune response are called **antigens.** They may be whole bacteria, fungi, and viruses or individual molecules such as proteins or lipids found on pollen. For instance, people with food allergies have immune responses to proteins, carbohydrates, and lipids in certain foods.

The immune system typically responds to antigens in part by producing antibodies. This response is called **antibody-mediated immunity.** When exposed to antigens, **B lymphocytes** (simply called **B cells**), which are a type of white blood cell or **leukocyte,** recognize and bind to antigen. **T lymphocytes (T cells)** play essential roles in helping B cells recognize and respond to antigen. After antigen exposure, B cells develop to form **plasma cells,** which produce and secrete antibodies. Most antibodies are released into the bloodstream, but there are also antibodies in saliva, tears, and the fluids lining the digestive system, among others. One purpose of antibody production is to provide lasting protection against antigens. During the process of B cell development, some B cells become "memory" cells, which have the ability to recognize foreign materials years later and in response grow and produce more plasma cells and antibodies that provide the body with long-term protection against antigens (Figure 5.12).

Antibodies are very specific for the antigen for which they were made, but how do these proteins protect the body against foreign materials? Many

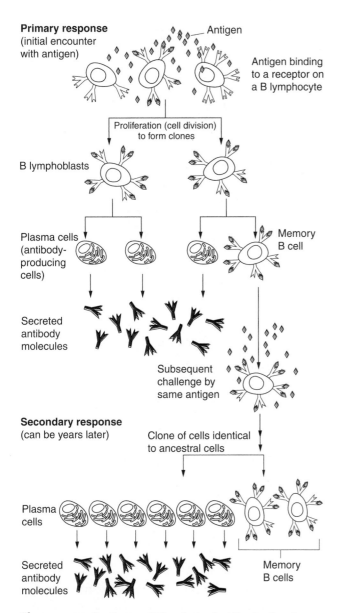

Primary response
(initial encounter
with antigen)

Antigen

Antigen binding
to a receptor on
a B lymphocyte

Proliferation (cell division)
to form clones

B lymphoblasts

Plasma cells
(antibody-
producing
cells)

Memory
B cell

Secreted
antibody
molecules

Subsequent
challenge by
same antigen

Secondary response
(can be years later)

Clone of cells identical
to ancestral cells

Plasma
cells

Secreted
antibody
molecules

Memory
B cells

Figure 5.12 Antigens Stimulate Antibody Production by the Immune System In response to the initial antigen exposure, which may be whole cells or individual molecules, B cells divide repeatedly to form many other B cells (clones). B cells differentiate into plasma cells that produce antibodies specific to the antigen. During this process, memory B cells are formed. If a person is exposed to the same antigen again in the future, even many years later, memory B cells can recognize and produce a stronger and more rapid response to the antigen.

antibodies bind to and coat the antigen for which they were made (Figure 5.13). After antigens are covered with antibodies, a type of leukocyte called a **macrophage** can often recognize them. Macrophages are cells that are very effective at phagocytosis (which literally means "cell eating," derived from the Greek terms *phago* = eating, *cyto* = cell). In phagocytosis,

macrophages engulf antigen covered with antibody; then, organelles in the macrophage called *lysosomes* unleash digestive enzymes that degrade the antigen (Figure 5.13). When the antigen is a foreign cell such as a bacterium, some antibodies are involved in mechanisms that rupture the cell through a process called *cell lysis*.

We are constantly being exposed to antigens that our immune system develops antibodies against. But sometimes our natural production of antibodies is not sufficient to protect us from pathogens such as smallpox, viruses that cause hepatitis, and HIV, the cause of AIDS. Biotechnology can help our immune systems by boosting our immunity through the use of vaccines.

How Are Vaccines Made?

Vaccines are parts of a pathogen or whole organisms that can be given to humans or animals by mouth or by injection to stimulate the immune system against infection by those pathogens. When people or animals are vaccinated, their immune system recognizes the vaccine as an antigen and responds by making antibodies and B memory cells. By stimulating the immune system, the vaccine has pressured the immune system into stockpiling antibodies and immune memory cells that can go to work on exposure to the real pathogen in the future should this occur. Vaccination is used in pets, farm animals, zoo animals, and even many wild animals.

Types of vaccines

So how are vaccines made? Three major strategies are generally used to create immune responses using vaccines. **Subunit vaccines** are made by injecting portions of viral or bacterial structures, usually proteins or lipids from the microbe, to which the immune system responds. A fairly effective vaccine against hepatitis B virus was one of the first examples of a subunit vaccine, and vaccines for tetanus and for anthrax are also subunit vaccines. **Attenuated vaccines** involve using live bacteria or viruses that have been weakened through aging or by altering their growth conditions to prevent their replication after they are introduced into the recipient of the vaccine. The Sabin vaccine for polio is an attenuated vaccine. So are the MMR, tuberculosis, cholera, and chickenpox (varicella) vaccines as well as many others. Lastly, **inactivated (killed) vaccines** are prepared by killing the pathogen and using the dead or inactive microorganism for the vaccine. A mixture of inactivated polio virus is used in the Salk vaccine against polio. The rabies vaccines that are administered by injection to dogs, cats, and humans and the influenza (flu) vaccines, which have become

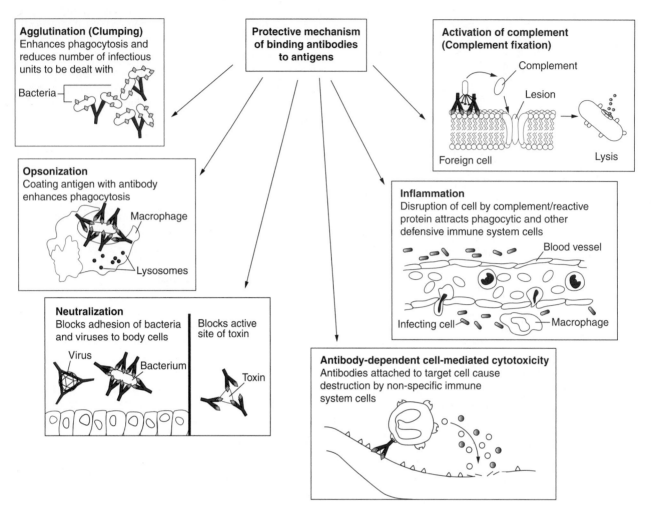

Figure 5.13 Mechanisms of Antibody Action Antibodies can inactivate and destroy antigens in a number of ways.

common in recent years, are also examples of inactivated vaccines. Inactivated flu vaccines can also be delivered as a nasal spray.

DNA-based vaccines have been attempted but so far they have not proven to generate sufficient immune stimulation. However, in 2005, the U.S. Department of Agriculture approved the world's first licensed DNA vaccine, a vaccine against West Nile virus (WNV). Developed by Fort Dodge Laboratories of Fort Dodge, Iowa, this vaccine is designed to protect horses from WNV, a mosquito-borne virus. Equine infections of WNV are on the rise, and about a third of horses infected with WNV will die or will need to be euthanized.

Immunity from vaccinations can fade with time, particularly for inactivated vaccines which often do not produce a strong immune response. As a result, many vaccines require immunization "booster" shots every few years to restimulate the immune system so it continues to provide protective levels of antibodies

and immune memory cells. For instance, the DPT vaccine is effective for about 10 years, as is the tetanus vaccine, and the flu vaccine that you may have received requires annual injections because new strains of influenza virus are developing each year.

Attenuated and inactivated vaccines were some of the first vaccines ever developed. Some subunit vaccines against bacteria were made prior to recombinant DNA technology by growing bacterial pathogens in liquid culture. Many bacteria release proteins into the surrounding media, and these proteins can be purified and mixed with compounds that will help stimulate an immune response when injected into humans. But as scientists have learned more about the molecular structure of many pathogens, attempts at making recombinant subunit vaccines have become more popular. For instance, hepatitis B is a bloodborne virus transmitted by exposure to body fluids, sexual intercourse, and contaminated blood transfusions. Hepatitis B causes deadly liver diseases. When vaccines for hepatitis B were first prepared, scientists isolated the virus from the blood of infected patients and then used biochemical techniques to purify viral proteins. These proteins were then injected into humans as a vaccine. The hepatitis B vaccine is recommended for international travelers, particularly people visiting Africa and Asia, and health care workers and others who may come in contact with hepatitis-infected persons or their body fluids. Currently a majority of subunit vaccines, including the vaccine for hepatitis B, are made using recombinant DNA approaches in which the vaccine is produced in microbes.

The recombinant subunit vaccine for hepatitis B, scientists cloned genes for proteins on the outer surface of the virus into plasmids. Yeasts transformed with these plasmids are used to express large amounts of viral protein as fusion proteins, which are then purified and used to vaccinate people against hepatitis B infections. This approach is a common strategy for producing subunit vaccines, although sometimes fusion proteins are expressed in bacteria or expressed in cultured mammalian cells. As you will learn in Section 5.5, many scientists are working on a microbial genome project to further characterize bacterial and viral genes that may help in the development of new vaccines.

In, 2005, the pharmaceutical company Merck received FDA approval of Gardasil, a recombinant subunit vaccine against cervical cancer and the first cancer vaccine to be approved by the FDA. Gardasil targets four specific strains of **human papillomavirus (HPV),** which cause about 70% of cervical cancers (HPV strains 16 and 18) and a large percentage of genital warts (caused by HPV strains 6 and 11). Cervical cancer affects 1 in 130 women, nearly half a million women worldwide, and approximately 70% of sexually active women will become infected with HPV during their lifetime. In the United States, more than 10,000 women each year contract cervical cancer and around 4,000 women die of the disease. Given as a series of 3 booster shots, Gardasil is designed as a prophylactic vaccine, for women aged 9 through 26, which means that it is taken to provide immune protection prior to exposure to HPV. Merck recommends that Gardasil be given to female teens prior to their becoming sexually active. Several states have pending legislation to require that schoolchildren be vaccinated with Gardasil.

Ideally, the immune system can be most effective during the early stages of exposure to an infectious agent when immune cells can attack the pathogens as soon as they enter the body. Disease-causing viruses use a number of complicated ways to infect cells, replicate, and cause disease. For instance, HIV typically infects human immune cells by binding to a cell and injecting its RNA genome (Figure 5.14). The enzyme reverse transcriptase copies the HIV genome into DNA. HIV and other viruses that transcribe their RNA genomes into DNA are called **retroviruses.** After the viral genome has been copied, it is transcribed to make RNA and translated to produce viral proteins that assemble to create more viral particles that are released from infected cells. We present this brief overview of viral replication because each stage essentially represents a potential target for antiviral drugs including some vaccines.

Bacterial and Viral Targets for Vaccines

Pathogens are changing all the time giving rise to both drug- and vaccine-resistant strains and new strains of disease-causing bacteria and viruses. More infectious microbes exist than there are vaccines. As a result, there are many research priorities for improving existing vaccines and producing new ones. Some of the diseases targeted for new or improved vaccines include hepatitis (A, B, C, and D), sexually transmitted diseases (herpes, gonorrhea, and chlamydia), HIV, influenza, malaria, and tuberculosis. Next, we consider a few of the many targets for new vaccines.

The flu is caused by a large number of viruses that belong to the **influenza** family of viruses. Even though most people experience flu symptoms that last for a few days and can be readily treated by over-the-counter medications, influenza kills approximately 500,000 to 1 million people worldwide each year. Because flu viruses mutate so rapidly, no one-size-fits-all vaccine protects against all strains. Currently, new flu vaccines are generated each year based on the three main flu virus strains that are expected to be prevalent during the upcoming flu season. Viruses for this vaccine are grown in eggs.

The World Health Organization (WHO)—an international group that monitors infectious diseases and epidemics—has established centers in over 80 countries so it can collect and screen samples of influenza for analysis and then develop vaccine treatment strategies. Infectious disease scientists are considering the development of a "global lab" to monitor strains of influenza viruses around the world, replicate these viruses, and then use recombinant DNA techniques to produce subunit vaccines in response to the new viral strains that are detected. This strategy is a surveillance and rapid-response approach to keeping up with new pathogens and producing vaccines as needed. In the future, it will likely be implemented for many different disease-causing organisms.

Because it mutates rapidly, influenza A represents one of the greatest potential threats to human health through a *pandemic*, a global outbreak. Pandemic

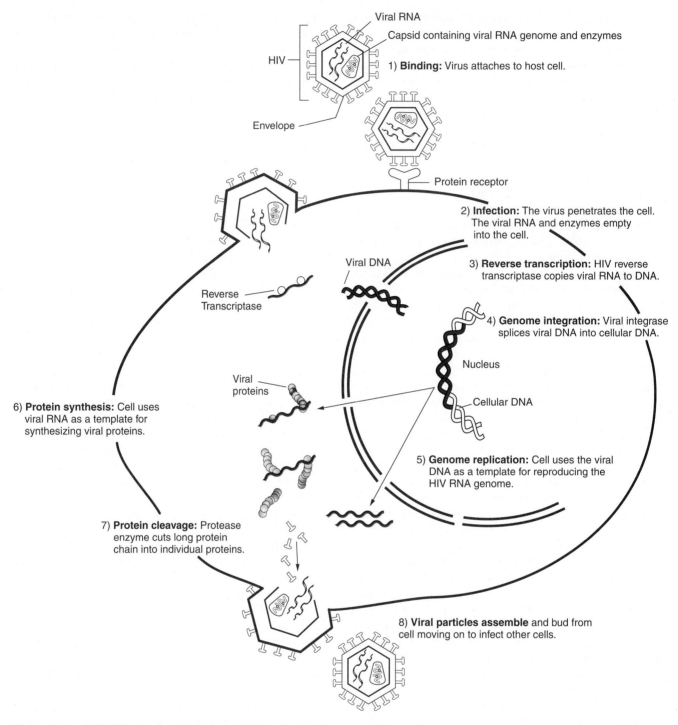

Figure 5.14 HIV Life Cycle Each stage of HIV replication is a potential target for antiviral drugs and some vaccines.

strains of influenza A have arisen in the past. In 1918, influenza virus killed at least 20 million people. Other pandemics occurred in 1957 and 1968, and epidemiologists predict that a future pandemic could be much worse than previous episodes. Of particular recent concern is the emerging strain of **avian flu,** which has received a lot of media attention because of its presence in chickens. Variations in influenza A are due to two glycoproteins called hemagglutinin (HA) and neuraminidase (NA) that project from the surface of the virus. Avian flu is an influenza A subtype called H5N1 because of the variation of these two proteins contained in this strain (abbreviated as H and N when used in strain names).

In 2003, strain H5N1 caused a pandemic in chickens in Asia that resulted in the killing of over 200 million birds in an effort to halt the spread. Like other viruses, scientists are concerned that this strain may mutate and make the jump into humans. Although a few isolated cases of human infection have occurred, widespread infection has not happened; however, bird to pig transmission of the virus has occurred and mutation of the virus in pigs could produce a strain that would infect humans. Because of the devastation such a virus could cause, the WHO has declared development of a vaccine to protect against H5N1 a high priority.

Tuberculosis (TB), caused by the bacterium *Mycobacterium tuberculosis,* is responsible for between 2 million and 3 million deaths each year. Inhaled particles of *M. tuberculosis* can infect lung tissue, creating lumpy lesions called *tubercles.* The spread of TB has been effectively controlled in many areas of the world by the use of antibiotics and vaccines. However, there has been a resurgence of TB because *M. tuberculosis* has proved to be very adept at evolving new strains that are resistant to treatment. Concern over such new strains is so great that in 1993, the WHO declared TB a global health emergency, and a number of research initiatives were launched. The Bill and Melinda Gates Foundation, together with other organizations, has provided over $30 million for these efforts. In 1998, the genome for *M. tuberculosis* was sequenced. As a result, new proteins have been discovered leading to the development of new TB vaccines, many of which are currently in clinical trials.

Malaria is caused by the protozoan parasite *Plasmodium falciparum* and transmitted by insects. Worldwide *Plasmodium* strains are developing resistance to the most commonly used antimalarial drugs. Although these drugs have been effective in parts of the world, the death rate from malaria is still unacceptable. Each year approximately half a billion cases of *Plasmodium* infections occur in children and cause nearly 3 million deaths. Researchers at Sanaria Inc., are developing an attenuated vaccine that will soon be ready for clinical trials. In addition, whole genome microarrays have been made for *Plasmodium* to help scientists identify new gene targets for inhibiting this parasite.

On another front, more than 33 million people are infected by HIV worldwide. The need for a vaccine to treat and stop the spread of HIV is critical if we are to curb this devastating disease and eventually eliminate the AIDS epidemic. Several vaccines for the prevention of HIV infections or the treatment of HIV-infected individuals have been tried in humans; most of these vaccines target viral surface proteins but to date none of these have lived up to their promise. Currently, one of the biggest obstacles facing HIV vaccine scientists is the high mutation rate of HIV. Consequently, the multi-subunit vaccines called *cocktails*— those that contain mixtures of many viral proteins, together with antiviral drugs that block viral replication—may be a more effective strategy to combat HIV than using either vaccines or antiviral drugs alone. Similar strategies are being developed for treating other viruses such as hepatitis B and C.

We continue our discussion on vaccine development and applications in Chapter 11. In the next section, we consider the many reasons for sequencing microbial genomes and the tools used to carry out this work.

5.5 Microbial Genomes

In 1995, the Institute for Genomic Research, which has played a major role in the Human Genome Project, reported the first completed sequence of a microbial genome when they published the sequence for *Haemophilus influenzae.* Since then, over 100 microbial genomes have been published, and work is being carried out on the genomes for several hundred other microbes. In 1994, as an extension of the Human Genome Project, the U.S. Department of Energy initiated the **Microbial Genome Program (MGP).** A goal of the MGP is to sequence the entire genomes of microorganisms that have potential applications in environmental biology, research, industry, and health such as bacteria that cause tuberculosis, gonorrhea, and cholera, as well as genomes of protozoan pathogens such as the organism *(Plasmodium)* that causes malaria. In early 2008, the National Institutes of Health announced plans for the **Human Microbiome Project,** a $115 million, 5-year project to sequence approximately 600 genomes of microorganisms that live on and inside humans.

Why Sequence Microbial Genomes?

Streptococcus pneumoniae, the bacterium that causes ear and lung infections including pneumonia, kills approximately 3 million children worldwide each year. Infections of *S. pneumoniae,* which can also cause bacterial meningitis, have been effectively treated since 1946 using vaccines that generate proteins and sugar molecules that coat the bacterium. But many of these vaccines are ineffective in young children, who are particularly susceptible to infection and serious health consequences. In 2001, the *S. pneumoniae* genome was completely sequenced, and many genes encoding previously undiscovered proteins on the surface of the bacterium were identified. Researchers are optimistic that this new understanding of the *S. pneumoniae* genome will lead to new treatments for pneumonia, including gene therapy approaches to rid children of infections that may persist for years.

This is just one dramatic example of the potential power of genomics at work. Revealing the secrets of bacterial genomes holds the promise for helping develop a basic understanding of the molecular biology of microbes. Sequencing microbial genomes will enable scientists to identify many secrets of bacteria, from genes involved in bacterial cell metabolism and cell division to genes that cause human and animal illnesses. In addition, researchers will find bacterial genes that may enable scientists to develop new strains of microbes that can be used in bioremediation and to reduce atmospheric carbon dioxide and other greenhouse gases, to find disease-causing organisms in food and water, to detect biological weapons, to synthesize plastics, to make better food products, and to produce genetically altered bacteria as biosensors for detecting harmful substances among many other examples.

Scientists have been studying the genetics of lactic acid bacteria for about 35 years to help them understand how these bacteria contribute to the flavor and texture of cheeses, milk, and other products we discussed earlier. Currently, genome projects have been completed for several dozen dairy-related lactic acid bacteria. These projects have already helped food scientists better utilize different strains to make specific cheeses with enhanced flavor characteristics and to refine culture conditions to maximize growth abilities of different microbes.

Of particular interest for many microbiologists are the genes involved in infection and disease-causing abilities for many bacterial pathogens. Understanding bacterial genomes will allow microbiologists to develop a greater understanding of how microbes contribute to normal health and how they result in disease. Determining the DNA sequence of a microbe will provide access to predicted protein sequences and protein structures. Our ability to sequence microbial genomes is also expected to lead to new and rapid diagnostic methods and ways to treat infectious conditions. For instance, if scientists sequence genes encoding cell surface proteins that coat a particular bacterial pathogen, they may be able to use these proteins to generate new diagnostic tools, vaccines, and antimicrobial agents.

Genome studies of gut and fecal microbes have revealed that hundreds of types of bacteria are involved in digesting food and in maintaining the health of our bowels. In addition, 1,200 different bacteriophages (recall from Chapter 3 that phages are viruses that infect bacteria) are found in the gut and we know virtually nothing about more than half of them. As a non-medical example of genomics in action, researchers at

Michigan State University completed the genome for *Fusarium graminearum,* an enemy of bread and beer manufacturers. This fungus infects wheat and barley crops rendering them useless, and outbreaks of this pathogen have resulted in billions of dollars of lost crops. Because traditional approaches for controlling *Fusarium* are not working, scientists hope that understanding its genome will lead to effective methods for preventing its growth, and work has already begun on inhibiting genes involved in the release of new fungus particles called spore pods.

Bacteria perform a wealth of biochemical activities, which is reflected in their genomes. Of the microbial genomes sequenced to date, approximately 45% of the genes identified produce proteins of unknown function, and approximately 25% of genes discovered produce proteins that are unique to the bacterial genome sequenced. Therefore, the potential for identifying new genes and proteins with unique properties that may have important applications in biotechnology is very high.

Microbial Genome Sequencing Strategies

Although microbes have substantially smaller genomes than humans, many of the same techniques used for the Human Genome Project have been applied to clone and sequence bacterial genomes. Figure 5.15 illustrates the shotgun cloning and sequencing strategies that have dominated the methods used to sequence bacterial genomes. Recall from Chapter 3 that shotgun approaches usually involve creating a genomic or a cDNA library of fragments of different sizes, randomly cloning these fragments into plasmids or other vectors (although sequencing can now be carried out without cloning fragments into vectors), and then sequencing these fragments. Eventually DNA sequences from these fragments can be compared, and the overlapping sequences can be used to piece together DNA fragments of entire chromosomes. DNA sequences are then subjected to a process called **annotation,** which involves searching databases to identify genes in microbial genomes that may have already been identified and assigning names and possible functions to as many of the predicted genes as possible. Annotation can also involve identifying the regulatory elements of a gene such as promoter and enhancer sequences.

Selected Genomes Sequenced to Date

Of the millions of different bacteria that have been identified, which ones are of greatest interest to microbial genome researchers? As shown in Table 5.3, the bacterial genomes that have received the most attention are those from microbes responsible for serious illnesses and diseases in humans. For example, recently the genome for *Pseudomonas aeruginosa* was completed. It is a major human pathogen causing urinary tract infections, a number of skin infections, and persistent lung infections that are a significant cause of death in cystic fibrosis patients. *P. aeruginosa* is a

1: Library Construction

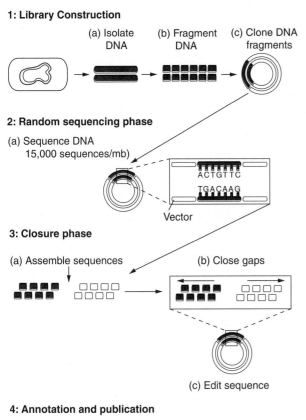

2: Random sequencing phase

3: Closure phase

4: Annotation and publication

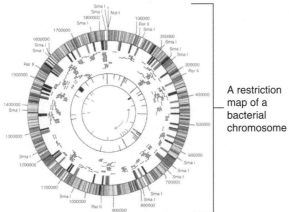

A restriction map of a bacterial chromosome

Figure 5.15 Random Sequencing of Microbial Genomes Most microbial genomes have been sequenced by preparing complementary DNA or genomic DNA libraries, randomly sequencing DNA clones from a library, and then assembling overlapping sequences to create a map of a bacterial chromosome.

particularly problematic bacterium because it is resistant to many antibiotics and disinfectants commonly used to treat other microbes. Learning more about the genes involved in the metabolism, replication, and the breakdown of compounds (such as antibiotics) in *P. aeruginosa* will be greatly helped by an understanding of its genome.

Another one of the first microbes targeted for genome studies was *Vibrio cholerae*, which is typically found in polluted waters in areas of the world with poor sanitary practices. This bacterium causes the disease **cholera,** which is characterized by severe diarrhea and vomiting, leading to massive fluid loss that can cause shock and even death. Many strains of antibiotic-resistant *V. cholerae* are causing recurring problems in Asia, India, Latin America, and even areas of the Gulf Coast in the United States. Understanding the genome of *V. cholerae* will help scientists identify toxin genes, genes for antibiotic resistance, and other genes that will augment our current methods for combating this microbe. Genome biologists are also focusing on microorganisms that may be used as biological weapons in a terrorist attack. In Section 5.7, we discuss why and how microbes can be used as bioweapons and what can be learned by studying their genomes.

The genomes for a number of bacteria that are important for food production have also been completed. For example, scientists recently sequenced the genome for *Lactococcus lactis,* a strain that is important for making cheese. An understanding of the genomes for *L. lactis* and other strains will provide for improvements in food biotechnology.

Sorcerer II: Traversing the Globe to Sequence Microbial Genomes

Environmental genomics, also called **metagenomics,** involves sequencing genomes for entire communities of microbes in environmental samples of water, air, and soils from oceans throughout the world, glaciers, mines—virtually every corner of the globe. Human genome pioneer J. Craig Venter left Celera in 2003 to form the J. Craig Venter Institute, and he has played a central role in establishing the field of environmental genomics. One of the institute's major initiatives involves a global expedition to sample marine and terrestrial microorganisms from around the world and to sequence their genomes. Called the Sorcerer II Expedition, Venter and his researchers are traveling the globe by yacht in a sailing voyage that has been described as a modern-day version of Charles Darwin's famous treks on the HMS *Beagle.* The Discovery Channel has even chronicled this journey.

Table 5.3 SELECTED MICROBIAL GENOMES

Bacterium	Human Disease Condition	Approximate Genome Size (megabases, mB)	Approximate Number of Genes
Bacillus anthracis	Anthrax	5.23	5,000
Borrelia burgdorferi	Lyme disease	1.44	853
Chlamydia trachomatis	Eye infections, genitourinary tract infections (e.g., pelvic inflammatory disease)	1.04	896
Escherichia coli 0157:H7	Severe foodborne illness (diarrhea)	4.10	5,283
Haemophilus influenzae	Serious infections in children (eye, throat, and ear infections, meningitis)	1.83	1,746
Helicobacter pylori	Stomach (gastric) ulcers	1.66	1,590
Listeria monocytogenes	Listeriosis (serious foodborne illness)	2.94	2,853
Mycobacterium tuberculosis	Tuberculosis	4.41	3,974
Neisseria meningitidis (MC58)	Meningitis and blood infections	2.27	2,158
Pseudomonas aeruginosa	Pneumonia, chronic lung infections	6.30	5,570
Rickettsia prowazekii	Typhus	1.11	834
Rickettsia conorii	Mediterranean spotted fever	1.30	1,374
Streptococcus pneumoniae	Acute (short-term) respiratory infection	2.16	2,236
Yersinia pestis	Plague	4.65	4,012
Vibrio cholerae	Cholera (diarrheal disease)	4.00	3,885

Sources: Sawyer, T. K. (2001). Genes to Drugs. *Biotechniques* 30(1): 164–168. TIGR Microbial Database (www.tigr.org/tdb/mdb/mdbcomplete) and Gold: Genomes OnLine Database (wit.integratedgenomics.com/GOLD).

A pilot study the institute conducted on the Sargasso Sea off Bermuda yielded around 1,800 new species of microorganisms and over 1.2 million novel DNA sequences. Samples of water from different layers in the water column are passed through high-density filters of various sizes to filter out microbes. DNA is then isolated from the microbes and used for shotgun cloning and then sequenced with computer-automated sequencers on board running nearly around the clock. This expedition has great potential for identifying new microbes and genes with novel functions, including commercially valuable genes. For example, the Sargasso Sea project identified hundreds of photoreceptor genes. Some microorganisms rely on photoreceptors for capturing light energy to power photosynthesis. Scientists are interested in learning more about photoreceptors to help develop ways in which photosynthesis may be used to produce hydrogen as a fuel source. Medical researchers are also very interested in photoreceptors because in humans and many other species, photoreceptors in the eye are responsible for vision. By early 2007, the Sorcerer II expedition sequenced approximately 6 billion base pairs (bp) of DNA from more than 400 uncharacterized microbial species. These sequences contained 7.7 million previously uncharacterized sequences, encoding more than 6 million different potential proteins.

Viral Genomics

Studying viral genomes is another hot area of research (Table 5.4). This is true in part because many deadly

Table 5.4 EXAMPLES OF MEDICALLY IMPORTANT VIRAL GENOMES THAT HAVE BEEN SEQUENCED

Virus	Human Disease or Illness	Year Sequenced
Ebola virus	Ebola hemorrhagic fever	1993
Hepatitis A virus	Hepatitis A	1987
Hepatitis B virus	Hepatitis B	1984
Hepatitis C virus	Hepatitis C	1990
Herpes simplex virus, type I	Cold sores	1988
Human immunodeficiency virus (HIV-1)	Acquired immunodeficiency syndrome (AIDS)	1985
Human papillomavirus	Cervical cancer	1985
Human poliovirus	Poliomyelitis	1981
Human rhinovirus	Common cold	1984
Severe acute respiratory coronavirus (SARS-CoV)	Severe acute respiratory syndrome (SARS)	2003
Variola virus	Smallpox	1992

viruses mutate quickly in response to vaccine and antiviral treatments. Antiviral drugs are designed to work in several ways. Some antiviral drugs block viruses from binding to the surface of cells and infecting cells; others block viral replication after the virus has infected body cells. Work on genomes for many of these will help scientists learn how viruses cause disease and develop new and effective antiviral drugs. For example, if genomics reveal that a certain protein is required for viral replication, scientists can design drugs that will specifically interfere with the function of that protein to block viral replication. In the next section we briefly describe how biotechnology can be used to detect microbes for a number of different reasons.

Assembling Genomes to Produce Human-Made Viruses

Researchers at Stony Brook University in New York made headlines when they assembled approximately 7,500 bp of synthetically produced DNA sequences that were used to synthesize proteins and lipids that assembled into a recreated polio virus: the first synthetically made virus. Shortly after this work was reported, another group synthesized a bacteriophage using a similar approach, and experiments are underway to do the same with bacterial genomes. These experiments in which genomes were assembled to produce a functional virus triggered a range of responses. Not surprisingly one fear raised was that bioterrorists could use the same strategies to recreate

viruses such as smallpox or Ebola (read about these bioweapons in the the next section on bioterrorism). Some scientists expressed excitement about a new age of "synthetic" biology and how this field will help us understand how genes work to build an organism. What do you think?

5.6 Microbial Diagnostics

We have repeatedly seen that microorganisms cause a number of diseases in humans, pets, and agriculturally important crops. Recent advances in molecular biology have enabled microbiologists to use a variety of molecular techniques to detect and track microbes, an approach called **microbial diagnostics**.

Bacterial Detection Strategies

Recent studies suggest that microbes, both bacterial and viral, may be involved in cardiovascular disease and chronic respiratory illnesses such as asthma. An important step toward developing treatment strategies involves tracking these microbes to learn which organisms are causing illness and to identify pathogens in clinical settings when a person has an illness.

Before the advent of molecular biology techniques, microbiologists relied on biochemical tests and bacteria cultured on different growth media to identify strains of disease-causing bacteria. For example, when doctors take a throat culture, they use a swab of bacteria from your throat to check for the presence of *Streptococcus pyogenes,* a bacterium that causes strep throat. Even though these and other similar techniques still have an important place in microbial diagnosis, techniques in molecular biology allow for the rapid detection of bacteria and viruses with great sensitivity.

Molecular techniques such as restriction fragment length polymorphism (RFLP) analysis, PCR, and DNA sequencing, which were discussed in Chapter 3, can be used for bacterial identification (Figure 5.16). If the genome of the pathogen is large and produces too many restriction enzyme fragments, which prevent

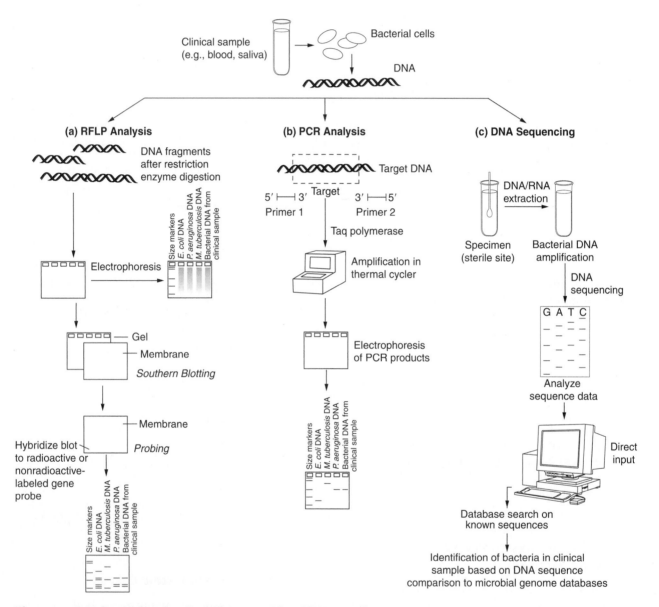

Figure 5.16 Using Molecular Techniques to Identify Bacteria   Many molecular techniques are available for identifying bacteria. (a) RFLP is one such technique. For some pathogens, isolated DNA (which may come from a clinical sample such as blood or saliva) can be subjected to restriction enzyme digestion and separation by agarose gel electrophoresis. Banding patterns of DNA fragments can be compared to reference strains of known bacteria to allow for a positive identification. In this example, bacterial DNA isolated from the clinical sample matches *P. aeruginosa.* (b) PCR can also be used for bacterial identification. PCR has the advantage of being much more sensitive than RFLP analysis; therefore, only small amounts of clinical samples and small amounts of DNA are required. The sensitivity of PCR also makes it possible to identify small amounts of DNA from just a few cells, allowing for early treatment of an infection. (c) DNA-sequencing strategies are also commonly used for microbial identification.

visualizing individual DNA bands on an agarose gel, DNA may be subjected to Southern blot analysis (Figure 5.16).

Many databases of RFLPs, PCR patterns, and bacterial DNA sequences are available for comparison of clinical samples. For example, if a doctor suspects a bacterial or viral infection, samples including blood, saliva, feces, and cerebrospinal fluid from the patient can be used to isolate bacterial and viral pathogens. DNA from the suspected pathogen is then isolated and subjected to molecular techniques such as PCR (Figure 5.16). PCR is an important tool for diagnostic testing in clinical microbiology laboratories and widely used to diagnose infection caused by microbes such as the hepatitis viruses (A, B, and C), *Chlamydia trachomatis* and *Neisseria gonorrhoeae* (both of which cause sexually transmitted diseases), HIV-1, and many other bacteria and viruses.

Tracking Disease-Causing Microorganisms

Scientists use molecular biology techniques to track patterns of disease-causing microbes and the illnesses and outbreaks of illnesses they may cause. Microbes play a very important role in the production of milk and related products. Recall that some bacteria carry out reactions that are absolutely essential for making yogurts and cheeses. But milk and dairy products are also susceptible to contamination with pathogenic microorganisms. Information about the microbes in milk can be used to determine the quality of milk and milk spoilage.

Bacterial contamination of food is a significant problem worldwide. For example, you may have heard of the bacteria *Salmonella,* which can contaminate meats, poultry, and eggs. *Salmonella* can infect the human intestinal tract causing serious diarrhea and vomiting, symptoms commonly called food poisoning. Researchers are experimenting with DNA-based techniques and approaches using antibodies to detect foodborne microbes such as *Salmonella.* For instance, a food or fluid sample can be passed over dye-containing antibodies attached to glass beads, and then chemiluminescence detection strategies can be used to determine if antibodies have bound to microbes in the food sample tested. Preliminary data suggest these techniques will provide a sensitive approach for detecting microbes in meats, fruits, vegetables, juices, and many other food products.

After successfully responding to a 1993 outbreak of meat contaminated with *E. coli*, the Centers for Disease Control and Prevention (CDC) decided to set up a network of DNA-detecting laboratories to expand its coverage and boost its response time. The CDC and the U.S. Department of Agriculture developed a network of cooperating labs called **PulseNet,** which enable biologists to rapidly identify microbes involved in a public health condition using DNA fingerprinting approaches. Results can be compared with a database to identify outbreaks of microbes in contaminated foods and to decide how to respond so that a minimal number of people are affected.

Approximately 76 million cases of foodborne disease due to microbes occur in the United States each year, causing well over 300,000 hospitalizations and approximately 5,000 deaths. In the United States alone, *E. coli* strain O157:H7 causes close to 20,000 cases of food poisoning each year. This infectious strain is lethal. PulseNet now monitors *E. coli* O157, *Salmonella, Shigella,* and *Listeria,* and it soon will be able to monitor non-foodborne diseases such as tuberculosis. Tuberculosis has been particularly difficult to track, especially with resistant strains beginning to show up from China, Russia, and India. Estimates suggest that a third of the world's population carries TB bacteria, but only 5 to 10% will show symptoms of infection. Improvements in detection may be the short-term solution until more is known about how the disease remains silent.

Microarrays for Tracking Contagious Diseases

Microarrays have created a range of new approaches for detecting and identifying pathogens and for examining host responses to infectious diseases. For example, microarray pioneer Affymetrics Inc. has developed the SARS-CoV GeneChip, which contains approximately 30,000 probes representing the entire viral genome for the coronavirus that causes **severe acute respiratory syndrome (SARS).** The SARS virus is a highly contagious respiratory virus that has infected approximately 9,000 people and killed nearly 900 since it was first detected in November 2002. Developing a vaccine for SARS-CoV is another priority, but until that happens scientists are using sensitive microarrays to detect the virus. Similar chips are being used to detect flu strain H5N1 that we discussed in the previous section.

Microarray approaches are also being used to study gene expression changes that occur when an organism is infected with a pathogen (Figure 5.17), providing a "signature" for infection by a particular organism. With these chips, patterns of genes that are stimulated or inhibited by a pathogen can be analyzed as a hallmark signature unique to that particular pathogen. Notice how the three pathogens used in Figure 5.17 stimulate different sets of genes in mice.

In the next section, we provide a brief glimpse of biological agents that can pose a threat as weapons and

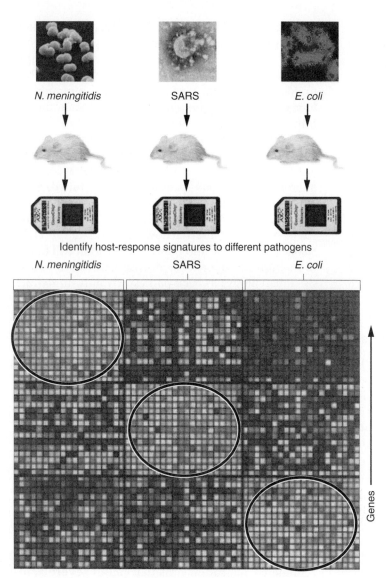

Identify host-response signatures to different pathogens

Figure 5.17 Host-Response Gene Expression Signatures for Pathogen Identification   In this animal model, mice were infected with *Neisseria meningitidis,* the bacteria that causes meningitis, SARS, or *E. coli,* and microarray analysis was carried out to examine changes in gene expression following infection by each pathogen. In this example, actively expressed genes are shown as light spots. Notice how each pathogen activates a specific subset of genes (circled) creating a gene expression "signature" that researchers can use to identify the infecting pathogen.

discuss how biotechnology can be used to detect and combat bioterrorism.

5.7 Combating Bioterrorism

The tragic events of September 11, 2001, were the most catastrophic attacks of terrorism on American soil. In the weeks that followed these horrific tragedies, America and the world were also served notice of a bioterrorism threat when letters contaminated with dried powder spores of the bacterium *Bacillus anthracis* were mailed to two senators, other legislators, and members of the press. Toxic proteins from *B. anthracis* cause significant damage to cells of

the skin, respiratory tract, and gastrointestinal tract depending on how a person is exposed to the microbe. As a result of this anthrax exposure, 5 people died and another 22 people became ill. These events raised our awareness that bioterrorism activities could cause devastating harm should biological agents be released in large quantities. **Bioterrorism** is broadly defined as the use of biological materials as weapons to harm humans or the animals and plants we depend on for food. Biotechnology by its very definition is designed to improve the quality of life for humans and other organisms. Unfortunately, bioterrorism represents a most extreme abuse of living organisms.

Bioterrorism has been a legitimate concern for centuries. In the 14th century, bodies of bubonic plague victims were used to spread the bacterium *Yersinia pestis* that caused bubonic plague during wars in Russia and other countries. This subsequently played a part in causing the Black Death plague of Europe. Early European settlers of the New World spread measles, smallpox, and influenza to Native Americans. Although the introduction of these diseases may not have been intentional, it led to the deaths of hundreds of thousands of Native Americans because they had virtually no immunity to these pathogens. Between 1990 and 1995, aerosols of toxins from the bacterium *Clostridium botulinum* were released at crowded sites in downtown Tokyo, Japan, although no infections resulted from these attempts. In the last few decades, many other lesser known and, fortunately, unsuccessful incidents have occurred around the world.

Microbes as Bioweapons

New strains of infectious and potentially deadly pathogens are evolving every day all around the world. The threat posed by these disease-causing microorganisms, which could be used as **bioweapons** of mass destruction, may conjure images of science fiction novels; however, the potential for a bioterrorism attack is real and of significant concern.

Even though thousands of different organisms that infect humans are potential choices as bioweapons, most experts believe that only a dozen or so organisms could feasibly be cultured, refined, and used in bioterrorism (Table 5.5). These agents include bacteria such as *Bacillus anthracis,* the gram-positive bacilli that causes anthrax, and deadly viruses such as smallpox and Ebola (Figure 5.18). The possibility that unknown organisms could be used as bioweapons is disquieting because they would probably be very difficult to detect and neutralize. However, a little-known microbe would probably be difficult to deliver as a bioweapon.

As a bioweapon, smallpox, which is a disease caused by the variola virus, is of concern for several reasons. Virtually all humans are susceptible to smallpox infection because widespread vaccination stopped over 20 years ago. At this time, when the last case of confirmed smallpox was reported, the WHO declared the disease to be eradicated throughout the world, in large part as a result of vaccines. Recent outbreaks of a monkey smallpox strain, however, may revive smallpox

---

Table 5.5 POTENTIAL BIOLOGICAL WEAPONS

Agent	Disease Threat and Common Symptoms
Brucella (bacteria)	Different strains of *Brucella* infect livestock such as cattle and goats. Can cause brucellosis in animals and humans. Prolonged fever and lethargy are common symptoms. Can be mild or life threatening.
Bacillus anthracis (bacterium)	Anthrax. Skin form (cutaneous) produces skin surface lesions that are generally treatable. Inhalation anthrax initially produces flulike symptoms leading to pulmonary pneumonia that is usually fatal.
Clostridium botulinum (bacterium)	Botulism. Caused by ingestion of food contaminated with *C. botulinum* or its toxins. Varying degrees of paralysis of the muscular system created by botulinum toxins are typical. Respiratory paralysis and cardiac arrest often cause death.
Ebola virus or Marburg virus	Both are highly virulent viruses that cause hemorrhagic fever. Symptoms include severe fever, muscle/joint pain, and bleeding disorders.
Francisella tularensis (bacterium)	Tularemia. Lung inflammation can cause respiratory failure, shock, and death.
Influenza viruses (a large highly contagious group)	Influenza (flu). Severity and outcome depend largely on the strain of the virus.
Rickettsiae (several bacteria strains)	Different strains cause diseases such as Rocky Mountain spotted fever and typhus.
Variola virus	Smallpox. Chills, high fever, backache, headache, and skin lesions.
Yersinia pestis (bacterium)	"Black" plague. High fever, headache, painful swelling of lymph nodes, shock, circulatory collapse, organ failure, and death within days after infection in a majority of cases.

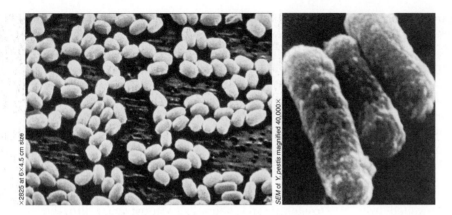

Figure 5.18 Deadly Microbes as Bioweapons Potential bioweapons might include the organisms that cause anthrax (*Bacillus anthracis*, left) and plague (*Yersinia pestis*, right).

as a concern. Following the anthrax events of 2001, based on concern about the use of smallpox as a bioweapon, the U.S. government started to work with biotechnology companies to mass-produce and stockpile supplies of smallpox vaccines.

YOU DECIDE

Should Pathogen Genome Sequences Remain in Public Databases?

Science as a process relies on scientists sharing data and information through presentations at conferences, publications, and Web resources such as databases. Now that genomes have been completed for many human pathogens such as those that cause cholera, anthrax, meningitis, and smallpox, there has been considerable debate about whether genome sequences for potential bioweapon microbes should be publicly available in DNA databases. Media reports on this subject raise concern that terrorists could use such sequence data to develop recombinant proteins for pathogen toxins as a way to produce bioweapons. Others have suggested that terrorists could use genome data to make more effective "superweapon" pathogens resistant to existing vaccines and drugs. However, many genome scientists have spoken in favor of keeping pathogen sequences in public databases, citing the importance of open access to information and claiming there is so much that we don't understand about the genomes of these pathogens that no human health threat is posed by making their sequences available. What do you think? Should pathogen genome sequences remain in public databases? You decide.

Targets of Bioterrorism

Antibioterrorism experts expect that bioterrorists will target cities or events where large numbers of humans gather at the same time. Experts, however, have had a poor track record of predicting where and how such acts will occur. At best, we can speculate that there may be many different potential ways to deliver a bioweapon to injure or kill humans. Bioterrorists are most likely to use a limited number of approaches to achieve quick and effective results. Widespread application of most agents might occur via some type of aerosol release in which small particles of the bioweapon are released into the air and inhaled. The aerosol could be created by grinding the bioweapon into a fine powder and producing a "silent bomb" that would release a cloud of the bioweapon into the air. This aerosol cloud would be gaslike, colorless, odorless, and tasteless. Such a silent attack could go undetected for several days. If exposed to a biological agent, in the days following an attack, a few people may develop early symptoms, which physicians might misdiagnose, or their symptoms may closely resemble common illnesses. Meanwhile, if the biological agent can be spread from human to human, larger numbers of people in other states and even other countries will become infected as infected individuals travel from place to place and spread their illness. Anti-bioterrorism experts also suspect that biological agents might be delivered by crop duster planes or disseminated into water supplies.

In addition to the direct threat to humans, experts are concerned with preventing bioweapon attacks that could cause severe damage to crops, food animals, and other food supplies (see Table 5.6). Not only could such an attack affect human health, but this approach could also cripple the agricultural economy of a country if food animals such as cows were infected. If there were general concerns about the safety of beef and milk, many consumers would likely shy away from buying these products for fear of contamination.

Table 5.6 POTENTIAL BIOLOGICAL PATHOGENS FOR A BIOWEAPONS ATTACK ON FOOD SOURCES

Disease	Target/Vector	Agent
Animal diseases		
Foot-and-mouth disease	Livestock	Foot-and-mouth virus
African swine fever virus	Pigs	African swine fever
Plant diseases		
Stem rust for cereals (fungus)	Oat, barley, wheat	*Puccinia* spp.
Southern corn leaf blight (fungus)	Corn	*Bipolaris maydis*
Rice blast (fungus)	Rice	*Pyricularia grisea*
Potato blight (fungus)	Potato	*Phytophthora infestans*
Citrus canker (bacterium)	Citrus	*Xanthomonas axonopodis* pv. *citri*
Zoonoses		
Brucellosis (bacterium)	Livestock	*Brucella melitensis*
Japanese encephalitis (flavivirus)	Mosquitoes	Japanese encephalitis virus
Cutaneous anthrax (bacterium)	Livestock	*Bacillus anthracis*

Source: Gilmore, R. (2004): U.S. Food Safety Under Siege? *Nature Biotechnology,* 22: 1503–1504.

Using Biotechnology Against Bioweapons

As was evident during the anthrax incidents of 2001, the United States is generally unprepared for an attack with biological weapons. Numerous agencies including the American Society for Microbiology, the U.S. Department of Agriculture, the CDC, the Department of Health and Human Services, and Congress have worked to develop legislation for minimizing the dangers of bioterrorism and responding to possible strikes. For instance, the U.S. Postal Service has implemented technologies for sanitizing mail by using X-rays or ultraviolet light.

In 2004, the U.S. federal government appropriated approximately $6 billion to be spent over 10 years to combat biological and chemical terrorism through an initiative called Project BioShield. A main goal of BioShield is to develop and purchase substantial quantities of vaccines and drugs to treat or protect Americans from bioweapons. Even with increased budgets, new laws, and international treaties, no measures can guarantee that bioterrorism will never occur or that we can detect and protect people against illness from such an attack if it were to occur. Worldwide efforts to prevent bioterrorism are essential, but how can biotechnology help to detect bioweapons and respond to an attack if it does occur?

Field tests are an essential step in the detection of an attack. Some field tests involve antibody-based tests such as ELISAs to determine if pathogens, or specific molecules from a pathogen, are present in an air or water sample. Such units were put in place around the Pentagon during the anthrax scare of 2001, the Gulf War, and the wars in Afghanistan and Iraq. These units draw in air and use antibodies to detect pathogens in the air. This technique is flawed because many of these instruments are not very sensitive and cannot detect small quantities of a pathogen. In fact, these sensors have been known to detect harmless microbes that live naturally in the environment. Newer, more sensitive biosensors must be developed.

Variations of protein-based assays include rapid handheld immunoassays developed by navy scientists and protein microarrays used by the military to detect airborne pathogens such as anthrax spores and the smallpox virus (Figure 5.19). Designed for use in lab settings or even quick diagnostic tests in field settings (some field-based microarrays use diamond-coated surfaces that make them particularly durable), these arrays detect whole pathogen or pathogen components such as proteins or spores.

Another detection approach involves extracting nucleic acids from environmental samples such as air, soil, or water as well as body fluids and tissue samples. Then PCR is performed with broad-based sets of primers specific for different pathogens. PCR can also be very helpful in tracking the source of a bioterrorism sample of bacteria by comparing amplified DNA from one pathogen with other known samples to check for

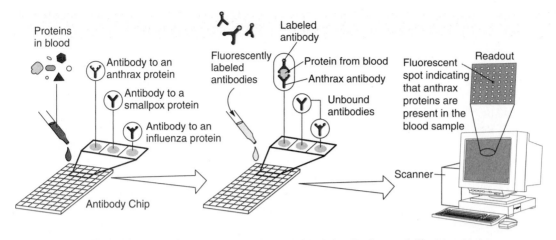

1) Apply blood from a patient to a chip, or array, consisting of antibodies assigned to specific squares on a grid. Each square includes multiple copies of an antibody able to bind to a specific protein from one organism and so represents a distinct disease-causing agent.

2) Apply fluorescently labeled antibodies able to attach to a second site on the proteins recognizable by the antibodies on the chip. If a protein from the blood has bound to the chip, one of these fluorescent antibodies will bind to that protein, enclosing it in an antibody "sandwich."

3) Feed the chip into a scanner to determine which organism is present in the patient's body. In this case, the culprit is shown to be *Bacillus anthracis*, the bacterium that causes anthrax.

Figure 5.19 Protein Microarrays for Detecting Bioweapon Pathogens

similarities and differences in the DNA sequence of different strains. The genome for one *B. anthracis* strain has already been sequenced, and plans are underway to sequence the genome of over 12 different strains of *B. anthracis* from around the world. This information may be used to help investigators track strains, catch future bioterrorists, and develop drugs and vaccines.

Emergency response teams and cleanup crews will also be called to action to evaluate the extent of contamination and determine appropriate cleanup procedures based on the bioweapon released (Figure 5.20). Should an attack occur, treatment drugs such as antibiotics will be needed. Countries must build a stockpile of drugs and vaccines that can be widely distributed to large numbers of people if necessary. The basic problem with vaccines is that they must be administered *before* exposure to bioweapons in order to be effective—giving them to infected individuals after an attack is useless. Since 2001, the CDC has increased the U.S. stockpile of smallpox vaccine. Mandatory vaccination of certain military personnel and a voluntary campaign to vaccinate health care workers and those first responders most likely to come in contact with the virus was also instituted. However, even drugs and vaccines may be ineffective if a bioterrorist attack involves organisms that have been engineered against most conventional treatments or if an unknown organism is used as the bioweapon. For instance, you may recall that the antibiotic ciprofloxacin, commonly referred to as "cipro," was in high demand during the anthrax threats of 2001.

Figure 5.20 Battling Bioterrorism   Hazardous material workers from the U.S. Coast Guard decontaminate a co-worker after working inside the Hart Senate Office Building to cleanse the building of anthrax spores during the anthrax attacks in 2001. Battling bioterrorism will require the coordinated efforts of many individuals, from scientists, physicians, and politicians to emergency response teams and cleanup personnel.

An unfortunate reality is that somewhere in this world, someone may be working to plan an attack with biological weapons. Will we be prepared? The vulnerability and grief many Americans felt on September 11 and the months that followed only served to increase fear and concerns about attacks using

CAREER PROFILE

Microbial Biotechnology Offers Many Career Options

Biotechnology is a growing industry comprised of a large number of startup or small companies, several large companies (including pharmaceutical companies now entering this market), several contract research companies, and various contract production companies. These companies are developing and manufacturing biological products such as therapeutic proteins and vaccines, and microbial applications play very important roles in this process. Many groups are needed to develop and produce FDA-approved biologics, including research, analytical, process development, manufacturing, and support groups such as quality control and quality assurance.

Microbial biotechnology research and development provides opportunities for all education levels. Even though advanced degrees are more common for upper-level positions, people with bachelor's degrees hold a significant number of positions. Owing to the wide variety of products, companies look for people with backgrounds in all areas of biology (molecular biology, protein biochemistry, microbiology, bioengineering, and chemical engineering). These people will design the product and determine how to test the product for identification, efficacy, and potency. The biologists and engineers in these groups will develop methods of production on a small scale that can then be used for large-scale production such as batch culturing recombinant microbes in fermenters.

Strict regulation in the biotech industry necessitates that manufacturing be a very precise process. Thus, well-trained manufacturing personnel, who can have a variety of educational backgrounds from high school to bachelor's degrees, are needed. Companies typically train personnel for these positions themselves. Quality control personnel test products and conduct tests on process samples to ensure that the products are produced properly. They collect and test a variety of samples ranging from water and culture media to chemical raw materials and final products. Educational background again can vary from a high school education to a bachelor's degree. Often companies provide training for personnel, but for the more advanced analytical procedures, individuals with an understanding of and experience with bioanalytical methods (e.g., high performance liquid chromatography [HPLC], gel analysis, blotting techniques, and PCR) will have an advantage.

Quality assurance personnel are charged with ensuring that products are made according to strict guidelines. They review the manufacturing process record (batch record) before and after manufacturing; they not only must remain current with regulations and be the liaison with the FDA and other regulatory agencies but must also be familiar enough with the production process to serve in this position. A background and understanding of microbiology is very helpful, but an interest in legal matters and regulations and good writing skills are also essential.

Source: Contributed by Daniel B. Rudolph (Process Engineer, Lonza Group LTD, Baltimore, MD).

---

biological agents. But the lives of those lost in these tragic and sad moments in American history will not have been in vain if we develop a greater awareness of bioterrorism and implement strategies to prevent and defuse bioweapon attacks.

QUESTIONS & ACTIVITIES

Answers can be found in Appendix 1.

1. Explain how prokaryotic cell structure differs from eukaryotic cell structure by describing at least three structural differences. Provide specific examples of how prokaryotic cells have served important roles in biotechnology.

2. Describe how yeasts differ from bacteria, and describe the role(s) of yeast in at least two important biotechnology applications.

3. Discuss the three major ways in which vaccines can be made to fight a virus or bacterium, and describe how each vaccine works.

4. Why are anaerobic microbes important for making many foods?

5. If you were in charge of protecting citizens from a bioterrorism attack, what biotechnology strategies would you consider for monitoring infectious agents? Do you think vaccination against a bioweapon such as the *Variola vaccinia* virus that causes smallpox should be mandatory for all citizens even if the vaccine causes life-threatening side effects in some people?

6. How can information gained from studying microbial genomes be used in microbial biotechnology? Provide three examples.

7. Search the FDA website (http://www.fda.gov/) and the CDC website (http://www.cdc.gov/) to determine (1) which influenza strains are approved for vaccine development for the year, (2) which companies have been approved to manufacture the vaccine, and (3) how many doses of the vaccine will be prepared.

8. We have become accustomed to being vaccinated against pathogens such as those that cause measles and polio. However, some states are working on proposals to require administering Gardasil, the vaccine for the sexually transmitted human papillomavirus, to all teens. What do you think about this proposal? Should Gardasil be required? What does this proposal presume about teen sexual activity?

9. What is a fusion protein and how can it be used to isolate a recombinant protein of interest?

10. Name three therapeutic recombinant proteins produced in bacteria and explain what they are used for.

References and Further Readings

Broadbent, J. R., and Steele, J. L. (2005). Cheese Flavor and the Genomics of Lactic Acid Bacteria. *ASM News,* 71: 121–128.

Gilmore, R. (2004). U.S. Food Safety Under Siege? *Nature Biotechnology,* 22: 1503–1504.

Goodwin, S., and Phillis, R. W. (2003). *Biological Terrorism.* M. A. Palladino, Ed. San Francisco: Benjamin Cummings.

Hill, S. A. (2006). *Emerging Infectious Diseases.* M. A. Palladino, Ed. San Francisco: Benjamin Cummings.

Tortora, G. J., Funke, B. R., and Case, C. L. (2006). *Microbiology: An Introduction, 9e.* San Francisco: Benjamin Cummings.

Ulmer, J. B., Valley, U., and Rappuoli, R. (2006). Vaccine Manufacturing: Challenges and Solutions. *Nature Biotechnology,* 24: 1377–1383.

Venter, J. C., Remington, K., Heidelberg, J. F., et al. (2004). Environmental Genome Shotgun Sequencing of the Sargasso Sea. *Science,* 304: 66–74.

Wein, L. M., and Liu, Y. (2005). Analyzing a Bioterror Attack on the Food Supply: The Case of Botulinum Toxin in Milk. *Proceedings of the National Academy of Science,* 102: 9984–9989.

Wessner, D. R. (2006). *HIV and AIDS.* M. A. Palladino, Ed. San Francisco: Benjamin Cummings.

Young, J. A., and Collier, R. J. (2002). Attacking Anthrax. *Scientific American,* 286: 48–59.

Visit www.pearsonhighered.com/biotechnology to download learning objectives, chapter summary, "Keeping Current" web links, glossary, flashcards, and jpegs of figures from this chapter.

Chapter 6

Plant Biotechnology

After completing this chapter you should be able to:

- Describe the impact of biotechnology on the agricultural industry.

- Discuss the limitations of conventional crossbreeding techniques as a means of developing new plant products.

- Explain why plants are especially suitable for genetic engineering.

- List and describe several methods used in plant transgenesis, including protoplast fusion, the leaf fragment technique, and gene guns.

- Describe the use of *Agrobacterium* and the Ti plasmid as a gene vector.

- Define antisense technology and give an example of its use in plant biotechnology.

- List some crops improved by genetic engineering.

- Outline the environmental impacts, both pro and con, of biotechnologically enhanced crops.

- Analyze the health concerns raised by opponents of biotechnology.

- Outline several ways that biotechnology might reduce hunger and malnutrition around the world.

Bioengineered plants being evaluated for quality control at a plant biotechnology company.

The red juicy tomatoes on sale at the grocery store are a true feat of engineering. Countless generations of selective breeding transformed a puny acidic berry into the delicious fruit we know today. In the last few decades, conventional hybridization (by cross-pollination) has produced tomatoes that are easy to grow, quick to ripen, and resistant to disease. Additionally, pioneering efforts in biotechnological research have created tomatoes that can stay on store shelves longer without losing flavor. The future holds the possibility of even more amazing transformations for the tomato: It could someday supplement or possibly replace inoculations as a means of vaccination against human disease.

In this chapter, we consider the role of biotechnology in the agriculture industry. First, we survey the agriculture industry to better understand the motives driving biotech research and development. We then look more closely at the specific methods used to exchange genes in plants. We also learn how bioengineering can protect crops from disease, reduce the need for pesticides, and improve the nutritional value of foods. Next, we examine the future of plant-based biotechnology products, from pharmaceuticals to petroleum alternatives. Finally, we consider the environmental and health concerns surrounding plant biotechnology.

Figure 6.1 Arial View of Terraced Agricultural Operation   Agriculture is the biggest industry in the world producing $1.3 trillion of products per year.

6.1 Agriculture: The Next Revolution

Over the past 40 years, the world population has nearly doubled while the amount of land available for agriculture has increased by a scant 10%. Yet we still live in a world of comparative abundance. In fact, world food production per person has increased 25% over the past 40 years. How has it been possible to feed so many people with only a marginal increase in available land? Most of that improved productivity has depended on crossbreeding methods developed hundreds of years ago to provide animals and plants with specific traits. Recently, however, the development of new, more productive crops has been accelerated by the direct transfer of genes as shown in Figure 6.1.

Plant transgenesis (transferring genes to plants directly) allows innovations that are impossible to achieve with conventional hybridization methods. A few developments that have significant commercial potential are plants that produce their own pesticides, plants that are resistant to herbicides, and even bioproducts like plant vaccines. Because producing transgenic proteins is relatively easy and the quality of the proteins is reasonably good, future research and development in these areas is especially bright. For example, through classical breeding, the average strength of cotton fibers

has been steadily increasing by about 1.5% per year; however, biotechnology has dramatically accelerated this pace. By inserting a single gene, the strength of one major upland cotton variety increased by 60%.

In 2005, the 10-year anniversary of commercialized biotech crops, the one-billionth biotech acre was planted. Farmers in 17 countries are growing more than 200 million acres of crops improved through biotechnology. American farmers planted 111 million acres of genetically modified corn, soybeans, and cotton in 2004, a 17% increase from the year before. About 86% of the soybeans, 78% of the cotton, and 46% of the corn in the United States are genetically engineered to resist pests and/or herbicides. Since 1996 there has been a 4000% increase in biotech crop acreage worldwide, with industrialized and developing countries rapidly utilizing the technology. New biotech crops such as alfalfa, wheat, and potatoes will come to the market within 10 years, as well as high oleic acid soybeans and vitamin A-enriched rice. Beyond foods and feed crops, plant-produced vaccines and bioplastics, as well as enhanced phytoremediation plants (see Chapter 9) will reach the market in the next 10 years.

Although food crops are only one aspect of biotechnology's impact, they have been the focus of considerable controversy worldwide. On the one hand, hunger continues to plague much of the world. This reality is a compelling argument for the rapid development of more productive and nutritious crops. On the other, some sectors are concerned that experimentation could be harmful to the environment and human health.

The debate is far from over. To develop an informed opinion, decision makers must understand the science behind these new products, analyze the products themselves, and be knowledgeable about the regulations that exist to monitor biotechnological research. In any case, it is unlikely that the revolution in agricultural biotechnology will stop. Protests or not, biotechnology plant products will play a key role in our society. The next section describes the methods used to create new agricultural products.

6.2 Methods Used in Plant Transgenesis

Conventional Selective Breeding and Hybridization

Genetic manipulation of plants is not new. Ever since the birth of agriculture, farmers have selected plants with desired traits. Even though careful crossbreeding has continued to improve plants through the millennia, giving us large corn cobs, juicy apples, and a host of other modernized crops, the methods of classical plant breeding are slow and uncertain. Creating a plant with desired characteristics requires a sexual cross between two lines and repeated backcrossing between hybrid offspring and one of the parents. Isolating a desired trait in this fashion can take years. For instance, Luther Burbank's development of the white blackberry involved 65,000 unsuccessful crosses. In fact, plants from different species generally do not hybridize, so a genetic trait cannot be isolated and refined unless it already exists in a plant strain.

Transferring genes by crossing plants is not the only way to create plants with desirable features. **Polyploid** (multiple chromosome sets greater than normal, usually more than 2N) plants have been used for many years as a means to increase desirable traits (especially size) of many crops including watermelons, sweet potatoes, bananas, strawberries, wheat, and others. The utilization of the drug colchicine (which does not allow the cell to divide after it has double chromosomes) followed by hybridization is a means of introducing commercially important features of related species into potentially new cultivated crops (refer to Figure 10.9). This process results in hybrid plants where whole chromosome sets from related plants can be transferred rather than single genes. The additional chromosomes usually result in production of enlarged fruit that results in the commercially available fruits and vegetables mentioned earlier, which are much larger than native wild varieties. These fruits convey new traits that may be commercially beneficial.

Biotechnology promises to circumvent these historical limitations. Scientists today can transfer specific genes for desirable traits into plants. The process is quick and certain because plants offer several unique advantages to genetic engineers:

1. The long history of plant breeding provides plant geneticists with a wealth of strains that can be exploited at the molecular level.
2. Plants produce large numbers of progeny, so rare mutations and recombinations can be found more easily.
3. Plants have better regenerative capabilities than animals.
4. Species boundaries and sexual compatibility are no longer an issue.

Cloning: Growing Plants from Single Cells

Plant cells are different from animal cells in many ways, but one characteristic of plant cells is especially important to biotechnology: Many types of plants can regenerate from a single cell. The resulting plant is a genetic replica—or clone—of the parent cell. (Animals can be cloned too, but the process is more complicated. Chapter 7 discusses animal cloning in detail.) This natural ability of plant cells has made them ideal for genetic research. After new genetic material is introduced into a plant cell, the cell rapidly produces a mature plant, and the researcher can see the results of the genetic modification in a relatively short time. Next we consider some of the methods used to insert genetic information into plant cells.

Protoplast Fusion

When a plant is injured, a mass of cells called a **callus** may grow over the site of the wound. Callus cells have the capability to redifferentiate into shoots and roots, and a whole flowering plant can be produced at the site of the injury. You may have taken advantage of this capability if you have ever "cloned" a favorite houseplant by rooting a cutting.

The natural potential for these cells to be "reprogrammed" makes them ideal candidates for genetic manipulation. Like any plant cells, however, callus cells are surrounded by a thick wall of cellulose, a barrier that hampers any uptake of new DNA. Fortunately, the cell wall can be dissolved with the enzyme cellulase, leaving a denuded cell called a **protoplast.** The protoplast can be fused with another protoplast from a different species, creating a cell that can grow into a hybrid plant. This method, called **protoplast fusion,** as shown in Figure 6.2, has been used to create broccoflower, a fusion of broccoli and cauliflower, as well as other novel plants.

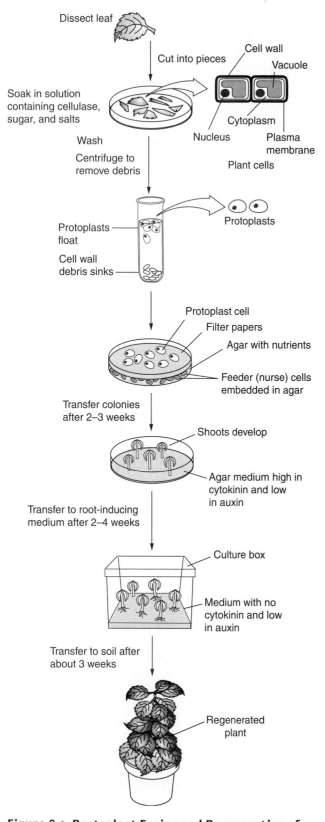

Figure 6.2 Protoplast Fusion and Regeneration of a Hybrid Plant After dissecting a plant leaf, it is possible to create protoplasts by digesting the cell wall with the enzyme cellulase. To create a hybrid plant, fuse protoplasts from different plants by culturing the fused cells in sterile media that stimulates shoots (cytokinin) and roots (auxin).

Leaf Fragment Technique

Genetic transfer occurs naturally in plants in response to some pathogenic organisms. For instance, a wound can be infected by a soil bacterium called *Agrobacterium tumefaciens* (*Agrobacter*). This bacterium contains a large circular double-stranded DNA molecule called a *plasmid,* which triggers an uncontrolled growth of cells (tumor) in the plant. For this reason, it is known as a **tumor-inducing (TI) plasmid.** The resulting tumor is known as **crown gall.** If you have ever seen a swelling on a tree or rose bush, you may have seen *Agrobacter*'s effects (see Figure 6.3).

The bacterial plasmid gives biotechnologists an ideal vehicle for transferring DNA. To put that vehicle to use, researchers often employ the **leaf fragment technique.** In this method, small discs are cut from a leaf. When the fragments begin to regenerate, they are cultured briefly in a medium containing genetically modified *Agrobacter,* as shown in Figure 6.4. During this exposure, the DNA from the TI plasmid integrates with the DNA of the host cell, and the genetic payload is delivered. The leaf discs are then treated with plant hormones to stimulate shoot and root development before the new plants are planted in soil.

The major limitation to this process is that *Agrobacter* cannot infect **monocotyledonous** plants (plants that grow from a single seed embryo) such as corn and wheat. **Dicotyledonous** plants (plants that grow from two seed halves), such as tomatoes, potatoes, apples, and soybeans, are all good candidates for the process, however.

Gene Guns

There is another option for inserting genes into *Agrobacter*-resistant crops. Instead of relying on a microbial vehicle, researchers can use a **gene gun** to literally blast tiny metal beads coated with DNA into an embryonic plant cell, as shown in Figure 6.5. The process is rather hit and miss—and more than a little messy—but some of the plant cells will adopt the new DNA.

Gene guns are typically used to shoot DNA into the nucleus of the plant cell, but they can also be aimed at the **chloroplast,** the part of the cell that contains chlorophyll. Plants have between 10 and 100 chloroplasts per cell, and each chloroplast contains its own bundle of DNA. Whether they target the nucleus or the chloroplast, researchers must be able to identify the cells that have incorporated the new DNA. In one common approach, they combine the gene of interest with a gene that makes the cell resistant to certain antibiotics. This gene is called a **marker gene** or a reporter gene. After firing the gene gun, the researchers collect the cells and try to grow them in a

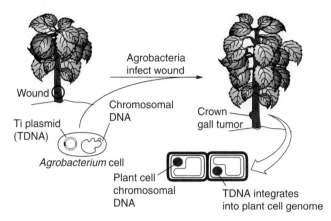

Figure 6.3 Process of Crown Gall Formation in Plants The Ti plasmid of *Agrobacter* normally causes "crown galls" or tumors in susceptible plants. Through genetic engineering, it is possible to silence the tumor-inducing genes and insert desirable genes into the plasmid, making it a vector for gene transfer to plants.

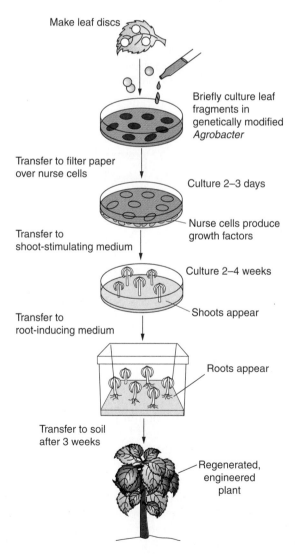

Figure 6.4 Transfer of Genetically Modified Ti Plasmid to Susceptible Plants Through Tissue Culture

medium that contains the antibiotic. Only the genetically transformed cells will survive. The antibiotic-resistant gene can then be removed before the cells grow into mature plants, if the researcher so desires. (For a more detailed explanation of gene markers, see Chapter 5.)

Chloroplast Engineering

The chloroplast can be an inviting target for bioengineers. Unlike the DNA in a cell's nucleus, the DNA in a chloroplast can accept several new genes at once. Also, a high percentage of genes inserted into the chloroplast will remain active when the plant matures. Another advantage is that the DNA in the chloroplast is completely separate from the DNA released in a plant's pollen. When chloroplasts are genetically modified, there is no chance that transformed genes will be carried on the wind to distant crops. This process is shown in Figure 6.6 on page 161.

Antisense Technology

Recall the tomato. It is red and juicy and tasty—and extremely perishable. When picked ripe, most tomatoes will turn to mush within days. But the Flavr Savr tomato, introduced in 1994 after years of experimentation, can stay ripe for weeks. Genetic engineering improved an already good thing.

Ripe tomatoes normally produce the enzyme **polygalacturonase** (or PG), a chemical substance that digests pectin in the wall of the plant. This digestion induces the normal decay that is part of the natural plant cycle. Researchers at Calgene (now a division of Monsanto, Inc.) identified the gene that encodes PG, removed the gene from plant cells, and produced a complementary copy of the gene. Using *Agrobacter* as a vector organism, they transferred the new gene into tomato cells. In the cell, the gene encoded an mRNA molecule **(antisense molecule)** that unites antisense RNA with and inactivates (complementary sequence) the normal mRNA molecule (the sense molecule) for PG production. With the normal mRNA inactivated, no PG is produced, no pectin is digested, and natural "rotting" is considerably slower. This process is shown in Figure 6.7 on page 161. The Flavr Savr tomato was the first genetically modified food approved by the Food and Drug Administration. Although the Flavr Savr tomato was not an economic success and is no longer available, it is common to find other varieties of genetically modified food on the market today.

We can expect to see more antisense developments in the near future. For example, DNA technologists are working on a potato that resists bruising. They have removed the gene responsible for producing an

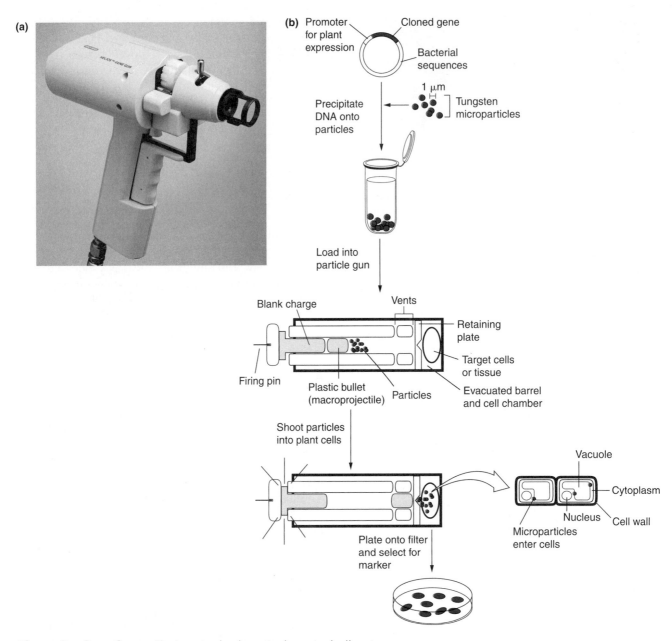

Figure 6.5 Gene Guns The tungsten beads are 1 micrometer in diameter. The DNA is coated on the surfaces of the beads and fired from guns with velocities of about 430 meters per second. The targets can include suspensions of embryonic cells, intact leaves, and soft seed kernels.

enzyme promoting color changes in peeled potatoes. It may sound like a subtle improvement, but market analyses have shown that consumers prefer buying potatoes that are not bruised by handling. Simply put, it is a small change that could translate to greater profits for potato farmers (and happier potato purchasers). In related research, genes from a chicken have been spliced into potatoes to increase the protein content. This improvement in the nutritional value of a common food could help many people worldwide get the protein they need in their diets.

A recent use of this technology has led to the silencing of genes in the opium poppy to prevent the production of morphine (and heroin). Investigators in the Netherlands used interfering RNA (see Chapter 3) to silence a terminal enzyme in the plant that synthesizes morphine. The metabolic shutdown led to an increase in other metabolites that have pharmaceutical activities in humans, acting as hair growth stimulants and possibly as anticancer agents. Whether this research will be utilized in poor countries that depend on opium sales to provide a legal product for sale

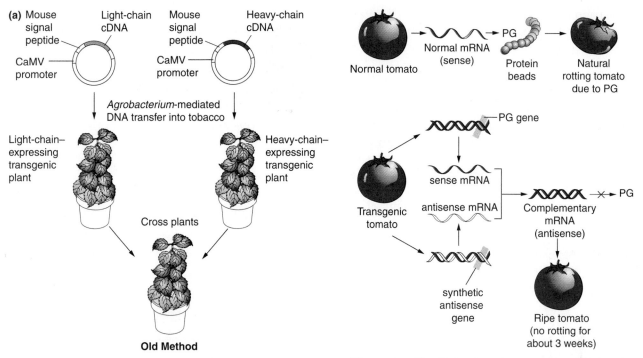

Figure 6.7 The Flavr Savr Tomato, One of the First Commercial Transgenic Plant Products Tomatoes that slowly rot are produced by first isolating the gene that encodes polygalacturonase. This gene encodes normal (sense) mRNA that is translated into PG. After inducing it to produce a cDNA (complimentary DNA) counterpart (see Chapter 2), it is possible to insert the PG into a vector for transfer to tomatoes. Transgenic plants with the new insert will produce PG mRNA and antisense mRNA, canceling the production of PG (see antisense technology). This slows the rotting and produces a tomato that will last about three weeks after ripening.

remains to be seen. (See Chapter 13 for the ethical considerations that surround these and other food crop innovations.) The explanation of how some of these innovations are accomplished is found in the next section.

6.3 Practical Applications in the Field

Vaccines for Plants

Crops are vulnerable to a wide range of plant viruses. Infections can lead to reduced growth rate, poor crop yield, and low crop quality. Fortunately, farmers can protect their crops by stimulating a plant's natural defenses against disease. One option is to inject the plants with a vaccine. Just like a shot for chickenpox or polio, these vaccines contain dead or weakened strains of the plant virus. The vaccine turns on the

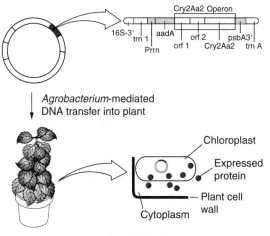

Figure 6.6 Old Versus New Method for Cloning Multiple Genes into Plants In the past, when more than one gene was to be expressed, two plants, each with its own inserted gene, had to be produced. (a) Standard crossing by pollen transfer was required to produce the hybrid plant. (b) Now it is possible to insert more than one gene by inserting them into the chloroplast DNA. Brandy DeCosa and fellow research scientists have inserted three insect-resistant genes. Orf1, orf2, and Cry2Aa2 were inserted into the 16S ribosome gene (an actively expressed gene in chloroplasts) of a chloroplast. Trn1 and trnA are known flanking regions of the 16S site, and aadA is a selection site for antibiotic resistance, needed for selection of successfully transformed cells. Prrn is a primer binding site for the PCR replication process that produces the DNA in quantity, and psbA3 is a site that stabilizes the foreign genes in the chloroplast DNA. Using chloroplasts for gene expression removes the danger of pollen transfer to nontarget plants (and possibly beneficial insects) because chloroplasts are not present in the pollen of plants.

plant's version of an immune system, making it invulnerable to the real virus.

Vaccinating an entire field is no easy task, but it is also no longer necessary. Now, instead of injecting the vaccine, the vaccine can be encoded in a plant's DNA. For example, researchers have recently inserted a gene from the tobacco mosaic virus (TMV) into tobacco plants. The gene produces a protein found on the surface of the virus and, like a vaccine, turns on the plant's immune system. In this case, tobacco plants with this gene are immune to TMV. Figure 6.8 shows this process.

Genetic vaccines have already proven themselves in a wide variety of crops. The development of disease-resistant plants has revitalized the once-ravaged papaya industry in Hawaii. Disease and pest-resistant strains of potatoes offer many advantages to both growers and consumers.

Genetic Pesticides: A Safer Alternative?

For the last 35 years, many farmers have relied on a natural bacterial pesticide to prevent insect damage. **Bacillus thuringiensis (Bt),** which was registered as a plant pesticide in 1961, produces a crystallized protein that kills harmful insects and their larvae. The crystalline protein (from the Cry gene) deteriorates the cementing substance that fuses the lining cells of the digestive tract of certain insects. The insects die in a short period of time due to "autodigestion." The Cry gene that causes this event is the subject of an expanding market of "insect-resistant" genetically engineered plants. By spreading spores of the bacterium across their fields, farmers can protect their crops without harmful chemicals.

With the introduction of biotechnology, instead of spreading the bacterium across their fields, farmers can spread Bt genes. Plants that contain the gene for the Bt toxin have a built-in defense against certain insects. This biotechnologically enhanced pesticide has been successfully introduced into a wide range of plants, including tobacco, tomato, corn, and cotton. In fact, most cotton seeds planted today contain the gene for Bt toxin, which effectively kills cotton-infesting insects by damaging their digestive systems when they eat the leaves. A photo of Bt with its insecticidal protein is shown in Figure 6.9a.

The widespread use of the Bt gene is one of the most successful stories in biotechnology. It is also one of the biggest sources of controversy. Cornell researchers conducted a laboratory experiment in 1999 that suggested the pollen produced by bioengineered corn could be deadly to monarch butterflies. The results were expected. Researchers had known for years that,

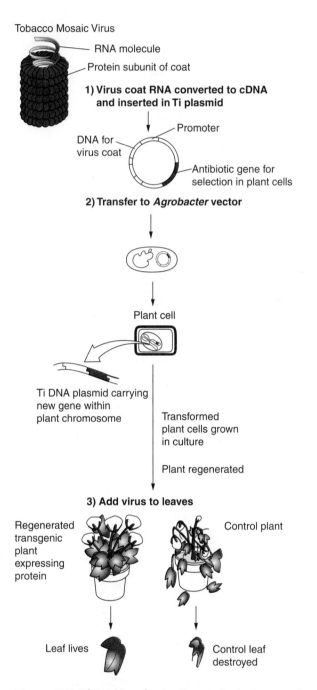

Figure 6.8 Plant Vaccines Transgenic plants expressing the TMV coat protein (CP) were produced by *Agrobacterium*-mediated gene transfer. When the generated plants expressing the viral CP were infected with TMV, they exhibited increased resistance to infection. Whereas control plants developed symptoms in 3 to 4 days, transgenic CP-expressing plants resisted infection for 30 days.

in large doses, the toxin naturally produced by *B. thuringiensis* could be harmful to butterflies. Still, the report set off a firestorm of controversy. It was the first tangible evidence that genetically altered food could harm the environment, and the monarch butterfly

(a)

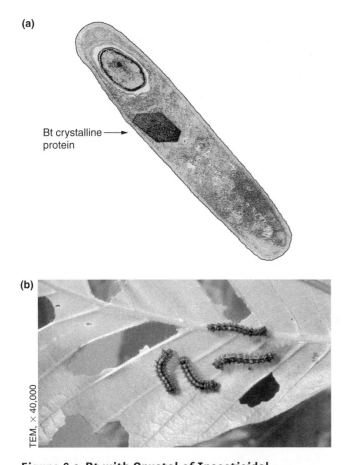

Bt crystalline protein

(b)

TEM, × 40,000

Figure 6.9 Bt with Crystal of Insecticidal Protein Figure (a) shows the crystalline protein as a crystal within the *Bacillus thuringiensis* bacterium. Genetically engineered plants expressing the Cry gene are "insect resistant" due to the production of a small amount of this bacterial protein. Insect larva (b) that would normally consume plant tissue will die if they consume the crystalline protein (Cry gene product).

quickly became the unofficial mascot of opponents of genetic engineering.

When researchers took their experiments out of the laboratory and into the field, many of their concerns quickly faded. Several studies found that few butterflies in the real world would be exposed to enough pollen to cause any harm. In fact, butterflies are unlikely to ingest toxic amounts of pollen even if they feed on milkweed plants less than 1 meter from the typical genetically modified cornfield. Still, scientists speculate that a small percentage of butterflies will inevitably get dusted with a lethal amount of pollen. Some of the monarchs that survive the exposure may be unfit for their long migrations. On the whole, however, concerns that genetically altered corn could devastate monarchs seem to be misplaced. After two years of study, the Agricultural Research Service (a division of the U.S. Department of Agriculture) announced in 2002 that Bt toxin posed little risk to monarch butterflies in real-world situations.

Safe Storage

The field is not the only place where crops are vulnerable to insects. In the United States, millions of dollars worth of crops are lost every year owing to insect infestations during storage. Such damage is especially devastating in developing countries with scarce food supplies. In these regions, a few infestations could lead to widespread starvation.

Once again, biotechnology may provide a solution. Studies show that transgenic corn that expresses **avidin**—a protein found in egg whites—is highly resistant to pests during storage. The protein blocks the availability of biotin, a vitamin that insects require to grow. Applying this technology on a wide scale could save many lives in developing countries.

Herbicide Resistance

Traditional weed killers have a fundamental flaw: They tend to kill desirable plants along with the weeds. Today, biotechnology allows farmers to use herbicides without threatening their livelihood. Crops can be genetically engineered to be resistant to common herbicides such as **glyphosphate.** This herbicide works by blocking an enzyme required for photosynthesis. Through bioengineering, scientists have created transgenic crops that produce an alternate enzyme that is not affected by glyphosphate. This approach has been especially successful in soybeans. Most soybeans grown today contain herbicide-resistant genes. The process is shown in Figure 6.10.

Farmers who plant herbicide-resistant crops are generally able to control weeds with chemicals that are milder and more environmentally friendly than typical herbicides. This development is significant because, before the advent of resistant crops, U.S. cotton farmers spent $300 million per year on harsh chemicals to spray on their fields. (This total does not include the human factor of laboring in the sun to weed between the cotton plants, which is now unnecessary.)

Stronger Fibers

As mentioned earlier, classical breeding has only been able to increase the average strength of cotton fibers by about 1.5% per year, whereas biotechnology through gene insertion increased the strength of one major upland cotton variety by 60%. Stronger fibers mean softer, more durable clothes for consumers and greater profits for farmers.

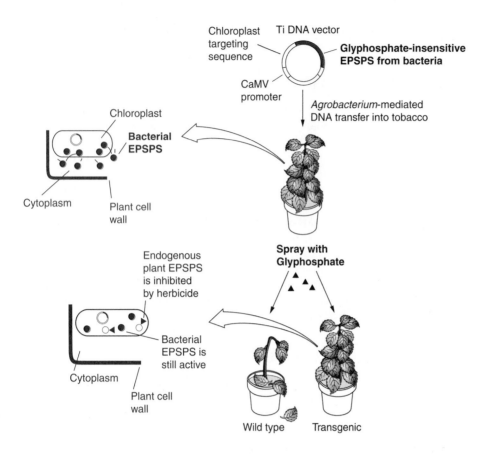

Chloroplast targeting sequence

Ti DNA vector

Glyphosphate-insensitive EPSPS from bacteria

CaMV promoter

Agrobacterium-mediated DNA transfer into tobacco

Chloroplast

Bacterial EPSPS

Cytoplasm

Plant cell wall

Endogenous plant EPSPS is inhibited by herbicide

Spray with Glyphosphate

Bacterial EPSPS is still active

Cytoplasm

Plant cell wall

Wild type   Transgenic

Figure 6.10 Engineering Herbicide-Resistant Plants   Glyphosphate is a compound that blocks a key enzyme in photosynthesis. Plants sprayed with this compound (e.g., Roundup) will die because a key photosynthesis enzyme is blocked. Genetically engineered plants that express an alternate gene (EPSPS) can bypass this blockage with an enzyme that uses an alternative chemical pathway. Transferring the TDNA vector with the EPSPS gene to plants and activating the CaMV promoter (with the help of a chloroplast targeting sequence) can result in incorporation of the EPSPS gene in the plants, making them immune to the glyphosphate. This process has been used on many commercial plant varieties, making them herbicide resistant and allowing the grower to spray susceptible weeds and resistant plants (and substituting glyphosphate for more harmful herbicides).

Enhanced Nutrition

Of all the potential benefits of biotechnology, nothing is more important than the opportunity to save millions of people from the crippling effects of malnutrition. One potential weapon against malnutrition is **Golden Rice,** rice that has been genetically modified to produce large amounts of beta carotene, a provitamin that the body converts to vitamin A. According to recent estimates, 500,000 children in the world will eventually become blind because of vitamin A deficiency. Currently, health workers carry doses of vitamin A from village to village in an effort to prevent blindness. Simply adding the nutrient to the food supply would be much more efficient and, in theory, much more effective (see Figure 6.11).

Biotechnology may not, however, be the magic bullet that ends malnutrition. Although promising, genetically modified foods have their limitations. For instance, the provitamin in Golden Rice must dissolve in fat before it can be used by the body. Children who do not get enough fat in their diets may not be able to reap the full benefits of the enriched rice.

Researchers know they cannot solve every problem. They also know that every little bit helps. Spurred on by the success of Golden Rice, researchers are developing rice that provides extra iron and protein, two other nutrients desperately needed in many impoverished areas.

The Future: From Pharmaceuticals to Fuel

Recall from Chapter 4 that plants can be ideal protein factories. A single field of a transgenic crop can produce a large amount of commercially valuable proteins. At this time, transgenic corn has the highest protein yield per invested dollar of any bioreactor organism. The possibilities are practically endless.

Figure 6.11 Golden Rice Is Only One of the Genetically Modified Crops Benefiting Developing Countries   Developing countries will drive increases in world food demand.

In the not-so-distant future, farmers will grow medicine along with their crops. It is already possible to harvest human growth hormone from transgenic tobacco plants. Plants can also manufacture vaccines for humans. Edible vaccines can be produced by introducing a gene for a subunit of the virus or bacterium. The plant expresses this protein subunit and it is eaten with the plant. When the subunit antigen enters the bloodstream, the immune system produces antibodies against it, providing immunity. The need for inexpensive vaccines that do not require refrigeration was first requested by the World Health Organization in the early 1990s and has resulted in studies of vaccines in bananas, potatoes, tomatoes, lettuce, rice, wheat, soybeans, and corn. Researchers at Cornell University have recently created tomatoes and bananas that produce a human vaccine against the viral infection hepatitis B. Researchers are actively studying the tomato as another source of pharmaceuticals. Through engineering of the chloroplast (abundant in green tomatoes), scientists hope to create an edible source of vaccines and antibodies, as shown in Table 6.1. Other potential biotechnological products include plant-based petroleum for fuel, alternatives to rubber, nicotine-free tobacco, caffeine-free coffee, biodegradable "plastics," stress-tolerant plants for agricultural and forest production, and industrial fibers. Early trials are currently underway for a number of pharmaceuticals

Table 6.1

Beneficial Product Traits	Crops
Bt crops are protected against insect damage and reduce pesticide use. Plants produce a protein—toxic only to certain insects—found in *Bacillus thuringiensis*.	Corn, cotton, potatoes **Future:** sunflower, soybeans, canola, wheat, tomatoes
Herbicide-tolerant crops allow farmers to apply a specific herbicide to control weeds without harm to the crop. They give farmers greater flexibility in pest management and promote conservation tillage.	Soybeans, cotton, corn, canola, rice **Future:** wheat, sugar beets
Disease-resistant crops are armed against destructive viral plant disease with the plant equivalent of a vaccine.	Sweet potatoes, cassava, rice, corn, squash, papaya **Future:** tomatoes, bananas
High-performance cooking oils maintain texture at high temperatures, reduce the need for processing, and create healthier food products. The oils are either high oleic or low linoenic. In the future, they will also be high stearate.	Sunflower, peanuts, soybeans
Healthier cooking oils have reduced saturated fat.	Soybeans
Delayed-ripening fruits and vegetables have superior flavor, color, and texture; are firmer for shipping; and stay fresh longer.	Tomatoes **Future:** raspberries, strawberries, cherries, tomatoes, bananas, pineapples
Increased-solids tomatoes have superior taste and texture for processed tomato pastes and sauces.	Tomatoes
rBST is a recombinant form of a natural hormone, bovine somatotropin, which causes cows to produce milk. rBST increases milk production by as much as 10–15%. It is used to treat over 30% of U.S. cows.	rBST (milk production)
Food enzymes, including a purer, more soluble form of chymosin used to curdle milk in cheese production, are used to make 60% percent of hard cheeses. Replaces chymosin of rennet from slaughtered calves' stomachs.	Chymosin (in cheese)—the *first* biotechnology product in food
Nutritionally enhanced foods will offer increased levels of nutrients, vitamins, and other healthful phytochemicals. Benefits range from helping developing nations meet basic dietary requirements to boosting disease-fighting and health-promoting foods.	**Future:** protein-enhanced sweet potatoes and rice; high-vitamin-A canola oil; increased antioxidant fruits and vegetables

Source: BIO member survey.

including alpha interferon, animal vaccines, Newcastle disease vaccine for chickens, lactoferrin (gastric disorders), thyroid-stimulating hormone, tooth decay prevention, Gaucher disease enzyme, and human insulin.

The 2007 Biofuels initiative increased federal funding 60% over the 2006 budget with the stated intention of replacing 30% of U.S. current fuel with bioethanol by 2030 (see Figure 6.12). Biofuels can be produced almost anywhere in the world from homegrown raw materials. Unfortunately, shifting a higher percentage of the U.S. corn crop (an available source of glucose from corn grain) to biofuel production has already escalated the price of soft drinks, corn syrup, and animal feeds. Ethanol production from corn has other disadvantages. Runoff of large amounts of nitrogen fertilizer, as well as pesticides and herbicides, bleeds into aquifers and rivers.

Lignocellulose, the most abundant plant product in the world, can be obtained from wheat straw, corn husks, prairie grass, discarded rice hulls, and trees. The technology to produce biofuels from these sources efficiently will come from biotechnology. Breaking apart the chains of sugar molecules that make up cellulose (as well as lignocellulose and hemicellulose) is already being accomplished by the enzymes from fungi, the gut of termites, and it can also come from the labs of genetic engineers. The bioethanol refineries that have sprung up all over the Midwest (resulting largely from subsidies and incentives) can be used to convert sugars from any cellulosic source. Because it takes 7 gallons of gasoline to produce 10 gallons of corn kernal ethanol, biotechnology is needed to convert the readily available cellulosic sources now.

Two major factors limit the possibilities of plant biotechnology: our knowledge of genes and the imagination of researchers. As more and more genomes are being sequenced, the first barrier is quickly disappearing.

Scientists are discovering genes at a dizzying pace. More importantly, they are learning how the expression of multiple genes helps shape an organism, a field known as **functional genomics.** Such research will open the door to even more amazing applications of biotechnology in the future.

Metabolic Engineering

The second wave of genetically modified plants will apply functional genomics in the form of **metabolic engineering,** the manipulation of plant biochemistry to produce nonprotein products or to alter cellular properties. These products may be alkaloids such as quinine (a drug still in common use), lipids such as long chain polyunsaturated fatty acids to reduce cholesterol in foods, polyterpenes such as new kinds of rubber, aromatic components such as S-linalool (the enticing aroma in fresh tomatoes), pigment production like blue delphinidin in flowers, and biodegradable plastics such as polyhydroxyanakanoates. This challenge is extremely ambitious because it involves transfer of more than one gene and more finite regulation. Work on this new technology began after publication of the *Arabidopsis* (mustard) genome sequence, when it was found that more than 5% of the genes were transcription factors (see Chapter 3 for a refresher). Switching on multiple genes (called gene stacks) will require much better knowledge of how multiple gene sets are regulated naturally in plants. Research institutions and those in industry are extremely interested in gene stacks because of the potential to better control products, develop new products, and regulate existing characteristics of commercial plants. If you are interested in exciting research this is a good area to consider. If you have strong feelings about manipulating plant genes, you may want to read the next section carefully.

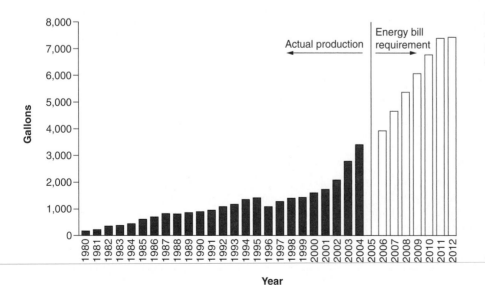

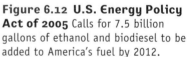

Figure 6.12 U.S. Energy Policy Act of 2005 Calls for 7.5 billion gallons of ethanol and biodiesel to be added to America's fuel by 2012.

Q Are some genetically modified crops healthier than nongenetically modified crops?

A Yes, those that have been enriched in vitamins and certain metabolic precursors, like Golden Rice (see table).

GENETICALLY MODIFIED CROPS WITH HEALTH-PROMOTING PROPERTIES

Crop	Added GM Substance	Benefit	Transgene Source
Rice	Provitamin A	Vitamin A supplement	Daffodil, *Erwina*
Rice	Iron	Iron supplement	*Phaselous, Aspergillus*
Canola	Vitamin E	Antioxidant	*Arabidopsis*
Tomato	Flavinoids	Antioxidant	Petunia
Sugar beet	Fructans	Low calorie	*Helianthus tuberosis*

6.4 Health and Environmental Concerns

Ever since the inception of transgenic plants, people have worried about potentially harmful effects to humans and the environment. In an age when "natural" is often equated with "safe," these decidedly unnatural plants carry an air of danger. Activists staged protests against companies producing genetically modified plants (GMOs, or genetically modified organisms). See Figure 6.13, which was taken in front of a biotechnology conference in 2000.

Such fears have the power to shake up an industry. In 2000, potato-processing plants in the Northwest stopped buying genetically modified potatoes. There was never any sign that these potatoes—engineered to be pest resistant—were inferior or dangerous. They looked and tasted just like nongenetically modified potatoes, and farmers did not need to use gallons of chemicals to get them to grow. They could survive aphids and potato bugs but not the tide of public opinion (see Figure 6.14).

- 2,500 lbs of waste
- 150,000 gallons of fuel to transport product
- 5,000,000 lbs of formulated product
- 180,000 containers
- 4,000,000 lbs of raw material
- Energy from 1,500 barrels of oil
- 3,800,000 lbs of inerts

Figure 6.13 A Parade in Boston Protesting Genetically Modified Food at the 2000 BIO Conference

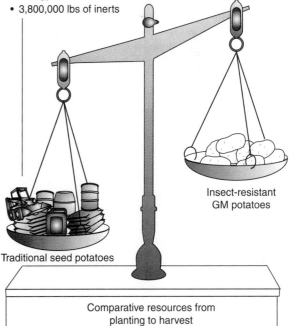

Insect-resistant GM potatoes

Traditional seed potatoes

Comparative resources from planting to harvest

Figure 6.14 Comparative Resources Needed to Control the Colorado Potato and Leaf Roll Virus in U.S. Potatoes Considerably more raw materials and energy are expended in the production and application of this insecticide compared with having insecticidal genes in the plant itself.

TOOLS OF THE TRADE

Excision of Reporter Genes

Researchers know that a gene has been transferred to a plant cell because antibiotic-resistant genes (for example, kanamycin) are usually used as "reporters" for commercial plant engineering. They allow only those transformed plant cells that can live on the antibiotic media to be selected (see Chapter 3). The presence of antibiotic genes (and small amounts of antibiotics) in plants has caused some public concern as described in Section 6.4. However, we know that it is possible to remove specific genes after transformation and selection because four scientists at Rockefeller University developed the process. It involves the use of a promotor that can be activated (see Chapter 3) to stimulate excision through naturally existing mechanisms in the plant embryo or plant organ tissue *after* the antibiotic selection for the transformed cells has occurred.

What are some opponents of plant biotechnology saying, and what are some of the other points of view? What does science say about their claims?

Concerns About Human Health

Every plant contains DNA. Whenever you munch a carrot or a bite into a slice of bread, you're eating more than a few genes. Opponents of genetic engineering have nothing against genes per se. Instead, they fear the effects of *foreign* genes, bits of DNA that would not naturally be found in the plant. A 1996 report in the *New England Journal of Medicine* seemed to confirm at least some of those fears. The study found that soybeans containing a gene from the Brazil nut could trigger an allergic reaction in people who were sensitive to Brazil nuts. Because of this discovery, this type of transgenic soybean never made it to market.

We can look at this incident in two different ways. Opponents say that this case of the soybean clearly demonstrates the pitfalls of biotechnology. They envision many scenarios where novel proteins trigger dangerous reactions in unsuspecting customers. Supporters see it as a success story: The system detected the unusual threat before it ever reached the public.

At this time, most experts agree that genetically modified foods are unlikely to cause widespread allergic reactions. According to a recent report from the American Medical Association, very few proteins have the potential to trigger allergic reactions, and most of them are already well known to scientists. The odds of an unknown allergen "sneaking" into a genetically modified food on the grocery shelf are very small. In fact, biotechnology may someday help prevent allergy-related deaths. Researchers are now working to produce peanuts that lack the proteins that can trigger violent allergic reactions.

Allergies are not the only concern. Some scientists have speculated that the antibiotic-resistant genes used as markers in some transgenic plants could spread to disease-causing bacteria in humans. In theory, these bacteria would then become harder to treat. Fortunately, bacteria do not regularly scavenge genes from our food. According to a recent report in the journal *Science,* there is only a "minuscule" chance that an antibiotic-resistant gene could ever pass from a plant to bacteria. Furthermore, many bacteria have already evolved antibiotic-resistant genes.

If you scan the anti-biotechnology literature, you'll see many more accusations. Headlines such as "Frankenfoods may cause cancer" are common. To this date, however, science has not supported any of these concerns. The National Academy of Sciences recently reported that the transgenic food crops on the market today are perfectly safe for human consumption.

Concerns About the Environment

Recall from the section on genetic pesticides that recent studies have put to rest fears that bioengineered corn could kill large numbers of monarch butterflies. However, worries about the environment have not disappeared. For one thing, genetic enhancement of crops could lead to new breeds of so-called super-weeds. Just as genes for antibiotic resistance could theoretically spread from plants to bacteria, genes for pest or herbicide resistance could potentially spread to weeds. Because many crops—including squash, canola, and sunflowers—are close relatives to weeds, crossbreeding occasionally occurs, allowing the genes from one plant to mix with the genes of the other. At this time, however, few experts predict any sort of explosion of genetically enhanced weeds. Further studies are needed to gauge the full extent of this threat and to develop ways to minimize the risk.

The potential ecological hazards of biotechnologically enhanced crops must be weighed against the clearly established benefits. First and foremost, biotechnology can dramatically reduce the use of chemical pesticides, as seen in Figure 6.14. According

The StarLink Episode

In 2000, traces of genetically modified StarLink corn turned up in taco shells that were sold in grocery stores. Intended for animal consumption, the corn contained the gene for herbicide resistance, which breaks down in the soil or the stomach of the cattle. On the surface, this news was not shocking. Many processed foods on the market shelves contain genetically altered corn or soybean products. This particular type of corn, however, had never been approved for human consumption because of lingering concerns over potential allergic reactions. The Environmental Protection Agency (EPA)—the agency that regulates the use of all pesticides—had only approved StarLink for animal feed and industrial use. The discovery of StarLink corn in the food supply triggered a massive recall of potentially "tainted" products. Soon after the recall, Aventis, the company that produced StarLink, struck a deal with the EPA and agreed to stop planting the corn.

Had this "unapproved pesticide" really done any harm? The Centers for Disease Control and Prevention took immediate steps to find out. A handful of people had complained of allergic-type reactions after eating the genetically altered corn, and the CDC closely investigated each case. A total of 28 people were found to have had symptoms consistent with an allergic reaction. However, blood tests showed that none of these people were sensitive to the Bt protein.

How should this episode have been handled? Was the public adequately protected from harm? Did the government overreact? What is the best balance between regulations and commercial interests? You decide.

to the National Center for Food and Agriculture Policy, farmers who planted biotechnologically enhanced cotton in 1998 were able to cut back their use of pesticides by more than a million pounds.

On the whole, biotechnology does not seem to be taking us to the brink of ecological disaster. Indeed, the National Academy of Sciences recently reported that biotechnologically enhanced crops pose no greater environmental threat than traditional crops.

Regulations

Biotechnology is not a lawless frontier. As we have already seen, several different agencies regulate the production and marketing of genetically modified foods. The FDA regulates foods on the market, the U.S. Department of Agriculture oversees the growing practices, and the Environmental Protection Agency controls the use of Bt proteins and other so-called pesticides. The approach of these agencies has changed over the years, especially in the case of the FDA, but they are actively involved in approving plant crops. (Chapter 12 discusses the regulation of plant biotechnology in much greater detail.)

In 1992—the early days of the biotechnological revolution—the FDA announced that genetically altered food products would be regulated by the same tough standards that are applied to regular foods—nothing more, nothing less. Even though they were not bound by law, food companies voluntarily consulted with the FDA before marketing any product. In 2001, the agency suggested a stricter, more formal approach. Under the proposed rules, companies must notify the FDA at least 120 days before a genetically altered food reaches the market. The manufacturer must also provide evidence that the new product is no

Today's Agriculturalist

Agriculturists in the 21st century will understand genetic engineering (and the associated regulations), botany, statistics, chemistry, and global imaging systems. In other words, the labor-intensive agriculture industry that we know today will require the tools of science. American farmers and ranchers have already embraced technology, allowing them to become the most efficient producers in the world. Education is the key to the decision making that will ultimately decide the quantity and quality of food and fiber available to the world. A national study sponsored by the USDA and Purdue University projected about 58,000 job openings per year in

the field of agricultural science, considerably more than the available number of college graduates in agriculture.

Agricultural biotechnology is one of the newest applications of biotechnology, with industries interested in employees with skills in biotechnology and agriculture. The ability to continue to produce the food needed to feed a human population that is doubling every 45 years from about 2% of the total land surface of the world cannot occur without advancements in technology, and biotechnology has delivered the most change to agriculture in the last decade. Agricultural biotechnology has a great future for those willing to develop the needed skills.

more dangerous than the food it replaces. The determining factor for plant-based foods or products will probably be the attitude of consumers. The Center for Food Safety and the Grocery Manufacturers Association and Food Products Association have informed the USDA of "its strong opposition to the use of food crops to produce plant manufactured pharmaceuticals in the absence of controls and procedures that assure essentially 100% of the food supply." There are many nonfood plants and contained growth plant systems that will satisfy this concern, and no manufacturer has violated these rules to date.

QUESTIONS & ACTIVITIES

Answers can be found in Appendix 1.

1. How do you detect whether the crops that you eat, say corn or soybeans, have been enhanced through biotechnology?

2. What are the environmental benefits from agricultural biotechnology?

3. What are some of the consumer benefits from agricultural biotechnology?

4. What are the grower benefits from agricultural biotechnology?

5. Are biotechnologically enhanced crops labeled?

6. Will insect-protected crops promote the development of insect resistance?

7. How do you know there will not be any long-term consequences to humans or animals from consumption of food or feed that contain products from biotechnologically enhanced plants?

8. Under what circumstances are gene guns primarily used as the means to transfer genes into plants?

9. Golden rice has been responsible for saving many lives and preventing disease in children born in developing countries. How much has this technology cost these countries?

10. Explain how gene stacks can be used to produce insect-resistant plants through biotechnology.

References and Further Reading

Biotechnology Industry Organization, 1225 Eye Street NW, Suite 400, Washington, DC 20005, 202.962.9200; info@bio.org.

Comello, V., and Beachy, R. (1999). A Leader in Revitalizing Plant Science. *R and D Magazine*, 41: 18–22.

DeCosa, B., Moar, W. S., Eung-Bum, L., et al. (2001). Overexpression of the Bt cry2Aa2 Operon in Chloroplasts Leads to Formation of Insecticidal Crystals. *Nature Biotechnology*, 19: 69–74.

Dove, A. (2002). Antisense and Sensibility. *Nature Biotechnology*, 20: 121–124.

Fox, J. (2006). Turning Plants into Protein Based Factories. *Nature Biotechnology*, 24: 1191–1193.

Grusak, M. A. (2005). Golden Rice Gets Boost from Maize. *Nature Biotechnology*, 23: 429–430.

Herrara, S. (2005). Syngenta's Gaff Embarrasses Industry and the White House. *Nature Biotechnology*, 23: 514.

International Service for the Acquisition of Agri-Biotech Applications, Clive James, Chairman, Ithaca, NY.

Jensen, D. (2002). Coping with an Up and Down Bio-Job Market. *GEN*, 22: 18–20.

Kelly, E. (2002). Novel Technologies for Bioindustrial Enzymes. *GEN*, 22: 18–58.

Langridge, W. H. R. (2000, September). Edible Vaccines. *Scientific American*, pp. 26–31.

Martineau, B. (2001). *First Fruit: The Creation of the Flavr Savr™ Tomato and the Birth of Genetically Engineered Foods*. New York: McGraw-Hill.

Memelink, J. (2004). Putting the Opium in Poppy to Sleep. *Nature Biotechnology*, 22: 1526.

Ostlie, K. (2001). Crafting Crop Resistance to Corn Rootworms. *Nature Biotechnology*, 19: 624–625.

Ragauskas, A. J., Williams, C. K., Davison, B. H., et al. (2006). The Path Forward for Biofuels and Biomaterials. *Science*, 311: 484–489.

Romeis, J., Meissle, M., and Bigler, F. (2006). Transgenic Crops Expressing *Bacillus thuringiensis* Toxins and Biological Control. *Nature Biotechnology*, 24: 63–71.

Schubert, C. (2006). Can Biofuels Finally Take Center Stage? *Nature Biotechnology*, 24: 777–784.

Werner, L. (2007). Genetically Modified Foods. In M. A. Palladino, ed., *Biological Terrorism*. San Francisco: Benjamin Cummings.

Zuo, J., Niu, Q. W., Moller, S. G., et al. (2002). Chemical-Regulated Site-Specific DNA Excision in Transgenic Plants. *Nature Biotechnology* 19: 157–161.

Employment Opportunities for College Graduates in the Food and Agricultural Sciences, 2000–2005, available at www.usda.gov/reports.

Visit www.pearsonhighered.com/biotechnology to download learning objectives, chapter summary, "Keeping Current" web links, glossary, flashcards, and jpegs of figures from this chapter.

Chapter 7

Animal Biotechnology

After completing this chapter you should be able to:

- List some of the medical advances made using animal research models.

- Explain what makes a good animal model for genetic studies.

- Describe two alternatives to the use of animal models and discuss their limitations.

- Discuss some of the ethical concerns surrounding animal research.

- Outline the process used to clone Dolly the sheep.

- Discuss some of the limitations to all current cloning processes.

- List some products that can be produced using transgenic animals as bioreactors.

- Explain how knockout animals can be used to provide information about genetic disorders and other diseases.

- Describe the process used to create monoclonal antibodies.

Pigs may become a source of human organs through biotechnology.

7.1 Introduction to Animal Biotechnology

Dolly the sheep looked like a sheep and acted like a sheep. But Dolly was not a normal sheep: She spent considerable time in front of television cameras because she was a clone, an identical twin animal engineered from a cell from another sheep.

Some people feel uneasy seeing the words *animal* and *engineering* in the same sentence. For them, the words conjure up images of legless cows, featherless chickens, or other freakish creatures. But biotechnologists are not in the business of creating bizarre new animals. Genetically engineered animals are used to develop new medical treatments, improve our food supply, and enhance our understanding of all animals, including humans. They can also be used for other purposes that some may consider inhumane or unethical, but market forces have largely prevented widespread introduction of such controversial practices.

Animals offer biotechnological opportunities, but they also present many tough scientific and ethical challenges. Researchers who successfully meet those challenges have the potential to make important contributions to science and to society. For example, animal health products and services generated over $5 billion in 2005, a benefit to nonhuman animals.

This chapter reviews many aspects of animal biotechnology. We begin by looking at the use of animals in research and then visit some of the issues at the leading edge of biotechnology: cloning, transgenics, and the use of animals as bioreactors.

7.2 Animals in Research

White mice are an icon of scientific research; along with lab coats and microscopes, they represent the business of discovery. Although animals have been central to research methods for decades, their use is controversial. Next, we explore the role of animals in research (and the resulting debated issues).

Animal Models

The differences between mice and humans are obvious, but there are many genetic and physiological similarities between the two. For this reason, research conducted on mice—or other animals, for that matter—can tell us much about ourselves. Research using animals has been key to most medical breakthroughs in the past century. Without question, such research has prevented much human suffering. Here are a few examples:

- Without the polio vaccine, which was developed using animals, thousands of children and adults would die or suffer debilitating side effects from the disease each year.

- Without cataract surgery techniques, which were perfected on animals, more than a million people would lose sight in at least one eye this year.

- Without dialysis, which was tested on animals, tens of thousands of patients suffering from end-stage kidney disease would die.

- Virtually every major medical advance in the last century has depended on research with animals.

- Biotechnology has developed 111 USDA-approved veterinary biologics and vaccines that treat heartworm, arthritis, parasites, allergies, and heart disease as well as vaccines for rabies and feline HIV that veterinarians use daily.

As shown in Figure 7.1, biomanufacturers are quite interested in transgenic animals.

According to regulations enforced by the Food and Drug Administration (FDA), new drugs, medical procedures, and even cosmetic products must pass safety tests before they can be marketed. Safety testing involves rigorous scientific methodology known as **phase testing.** The FDA requires manufacturers to conduct a statistically significant number of trials on cell cultures, in live animals, and on human research participants before such products can be made available to the general public. If initial tests on cell cultures indicate dangerous levels of toxicity, the product is never tested on live animals. The choice of animal to

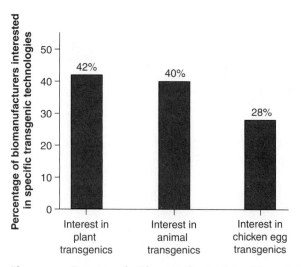

Figure 7.1 Transgenic Biomanufacturing The majority of biomanufacturers are focusing on either chicken or animal transgenic technologies (transferring genes into animal eggs) as a means to meet the manufacturing demand for proteins from biotechnology.

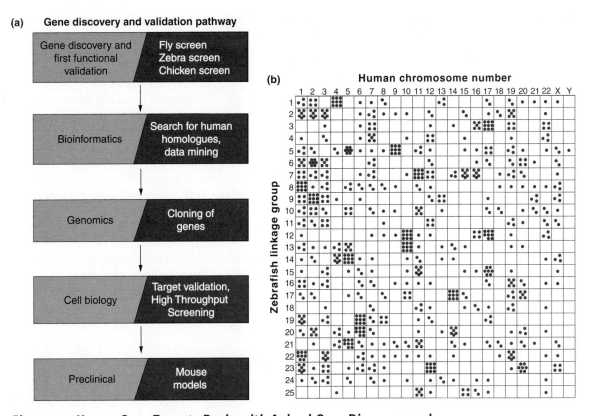

Figure 7.2 Human Gene Targets Begin with Animal Gene Discovery and Testing   Gene discovery (a) usually begins with analysis of genetic patterns in insects (fly screen), zebrafish (zebra screen), or chickens (chicken screen) to determine whether a gene homologous to humans exists in an animal that can be tested (often by silencing, or knockout). (b) The Oxford University research grid illustrates results from one of these homology tests. Dots represent similar genes in humans and zebrafish plotted by chromosome position in the two species. Boxes containing more than one dot represent conserved sequences. Early testing in animals with homologous genes can save time and drug testing in humans.

test is usually determined by their genetic similarity to humans (see Figure 7.2a). Two or more species are used in the tests because different effects may be revealed in different animals, as seen in Figure 7.2b. Toxicity and unexpected nontarget effects can often be detected by using more than one species with genes homologous to humans. The rate of absorption, specific chemical metabolism, and time required to excrete (eliminate from the body) the substance can all be examined in animal models. If significant problems are detected in the animal tests, human research participants are never recruited to participate in trials at that level. Unintended results can occur, although they are rare, but participants are informed of this potential as a requirement of informed consent.

You might wonder why testing on cells *in vitro* (cells cultured outside the body) would not be sufficient. The answer is that new drugs and medical procedures have effects beyond single cells in tissues and organs and must be tested in animals to determine the impact of the treatment on an entire organism. For example, a drug may be a powerful agent able to kill cancer cells, but it may also be destructive to unrelated organs.

One drug with unexpected side effects is Propecia (finasteride), which is used to encourage hair growth. This drug carries a clear warning that pregnant women should not handle broken or crushed pills. That warning and a special protective coating on the pills themselves are the direct result of information collected during testing on animal subjects. Those tests revealed serious birth defects (malformations of the reproductive organs) in the male offspring born to animal mothers given large oral doses of the drug. Even though merely handling the pills is unlikely to result in similar defects in human infants, the warnings were put in place before the drug was marketed. In this case, experiments using animal subjects helped prevent unexpected disasters. Thanks to animal testing, pregnant women were never given the drug.

The animals used most often in research are pure-bred mice and rats (of which the genetics are known and controlled), but other species are also used. Researchers have developed model animal systems using both vertebrates (the zebrafish) and invertebrates (the fruit fly) and worms (the nematode). An extremely valuable research animal is the tiny zebrafish (*Brachydanio rerio*), a popular and hardy aquarium fish shown in Figure 7.3. Zebrafish are small (adult length is 3 cm, about the length of a paper clip), making it possible to house large numbers of animals in small spaces. Spawning is virtually continuous by this active, warm-water fish, with only about three months between generations. The average number of progeny per female per week can exceed 200. Zebrafish eggs complete embryogenesis (early organ development) in about 120 hours; the gut, liver, and kidney are developed in the first 48 to 72 hours. Because the fish grow so rapidly, scientists can test for toxicity or adverse effects in about five days. If a drug is toxic to a zebrafish, it is probably toxic to humans.

Zebrafish are ideal for studies on development and for genetic research. The rapid growth of an easily visible embryo inside the zebrafish's egg, as seen in Figure 7.3a, makes studies efficient. The eggs lend themselves readily to gene transfer, there is no need to place eggs inside a donor mother for gestation, and the embryos are transparent, making it possible to study cell division under the microscope in its first hours.

But there are times when fish or mice are not the best test animals. For example, the lung and cardiovascular systems of dogs are similar to those of humans, making them a better choice for the study of heart disease and lung disorders. HIV and AIDS research is conducted on monkeys and chimpanzees because they are the only known animals that share humans' vulnerability to the virus. Although cats, dogs, and primates like monkeys and chimps are used in specific instances when their particular biology is important to research, they make up less than 1% of the total number of research animals. In addition, the number of those animals used in experiments has been declining for the past 20 years because of the increased use and availability of alternative methods of testing, which are discussed next.

Alternatives to Animal Models

Whenever possible, researchers use cell cultures and computer-generated models in initial testing. Both of these tools are significantly less expensive than research animals and can save time as well.

As mentioned earlier, cell culture studies are often used as a preliminary screen to check on the toxicity of substances (see Figure 7.4). In addition, fundamental questions about biology are often best answered by collecting evidence at the cellular level. Clues about which compounds might make the best drugs are often the result of *in vitro* research (see Figure 7.5). But, as already noted, cell cultures cannot provide

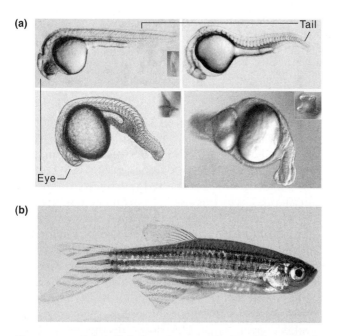

Figure 7.3 Zebrafish Are Commonly Used in Animal Research Zebrafish are fast-growing alternatives to other animals for gene studies and toxicity analyses. Transplanting genes into their embryos (a) allows study of the developmental changes (like blood vessel growth) before they become adults (b). The transparent embryos, fast growth rate, and ease of study make zebrafish common animal models for testing.

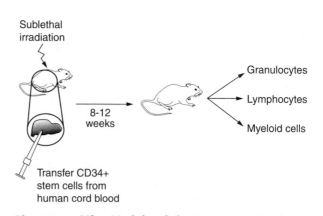

Figure 7.4 Mice Models of the Human Immune System After destroying the immune system of a developing mouse it is possible to replace it by injecting human stem cells. In this way it is possible to study therapies designed to affect this human blood cell forming system in mice.

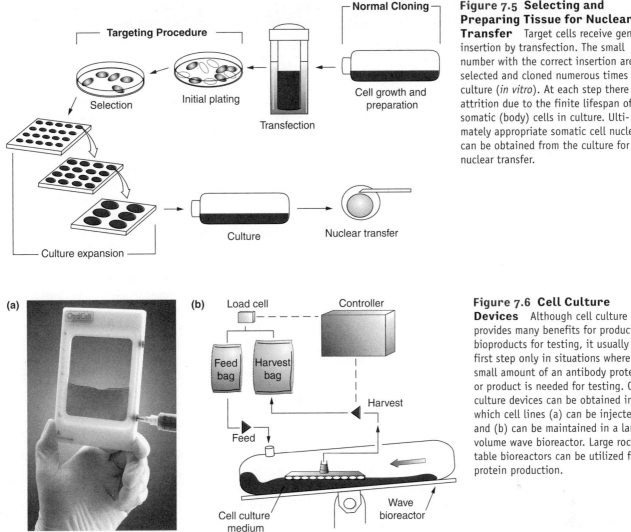

Figure 7.5 Selecting and Preparing Tissue for Nuclear Transfer Target cells receive gene insertion by transfection. The small number with the correct insertion are selected and cloned numerous times in culture (*in vitro*). At each step there is attrition due to the finite lifespan of somatic (body) cells in culture. Ultimately appropriate somatic cell nuclei can be obtained from the culture for nuclear transfer.

Figure 7.6 Cell Culture Devices Although cell culture provides many benefits for producing bioproducts for testing, it usually is a first step only in situations where a small amount of an antibody protein or product is needed for testing. Cell culture devices can be obtained in which cell lines (a) can be injected and (b) can be maintained in a large volume wave bioreactor. Large rocker table bioreactors can be utilized for protein production.

information about the potential impacts on an entire living organism (see Figure 7.6).

Computers are also used to simulate specific molecular and chemical structures and their interactions. Computer models are becoming increasingly sophisticated, but computers are limited by their programming. They can provide clues, but they cannot provide final answers. Unless we already know the impact on a living system, we cannot create a computer program that can simulate that system. Computers are excellent for processing data and helping us see the patterns that emerge, but they cannot react unless we have programmed them to do so, although they can reduce animal usage in research. For example, a computer model to test the efficiency of Propecia would have resulted in a clear message that it was a good drug; the computer model could not be expected to address the problem of unexpected birth defects because programmers would not have known in advance of the likely side effects.

Regulation of Animal Research

Animal research is heavily regulated. The federal Animal Welfare Act sets specific standards concerning the housing, feeding, cleanliness, and medical care of research animals. Before a study using animals can begin, researchers are required to prove the need to employ animals. They also must select the most appropriate species and devise a plan for using as few animals as possible to conduct the research. An oversight committee with the host institution reviews the research while it is in progress to make certain that institutional and federal standards are being upheld. Government agencies regularly monitor the conditions in the laboratories (see Good Laboratory Practices, later in this chapter). To receive funding for research from the National Institutes of Health, the FDA, or the Centers for Disease Control and Prevention, researchers must follow the standards of care set out in *The Guide for the Care and Use of Laboratory Animals,* formulated by the

TOOLS OF THE TRADE

Human Organs for Transplantation, from Animals

Until recently, the ability to create mammals with sophisticated genetic manipulations was available only in mice. Only a limited number of human stem cell lines were available to develop cloned organs that did not initiate rejection after they were transplanted (and ultimately lead to the life-threatening condition they were supposed to have solved). Homologous recombination is a relatively rare event, requiring researchers to carefully screen many DNA integration events. Dolly the sheep introduced an alternative approach to genetic recombination in embryonic stem cell lines by targeting genes *in vitro* and manipulating the somatic cells before they stopped dividing. Targeted gene disruption is an important technique in animal biotechnology, but the process is not easy.

After collecting somatic cells, researchers must genetically modify them *in vitro*, culture them long enough to isolate the rare homologous recombination events, and then reconstruct an embryo from these cells that can be used for nuclear transfer implanting, as seen in Figure 7.5. If the reconstructed embryo lives long enough to reproduce, the genetic modification can be propagated.

Investigators at PPL Therapeutics have achieved a breakthrough with this technology by removing from pigs a gene that commonly causes hyperacute rejection of pig-to-human organ transplantation. A common carbohydrate structure exists on the cell surfaces of almost all mammals other than humans, apes, and Old World monkeys. The presence of this structure is the consequence of a gene, α (1,3) galactosidase. Humans do not have this gene and recognize its function on organ surfaces as an antigen. As a consequence, humans produce massive quantities of antibodies to reject the foreign carbohydrate. If scientists are able to target this gene (and potentially other rejection-related genes) *in vitro*, they can produce much-needed donor human organs.

National Academy of Sciences. Given the cost of research, the availability of grant monies from those agencies can be essential, and thus most institutions are eager to follow the guidelines.

In summary, animal models play an essential role in scientific inquiry. Still, researchers strive to achieve the "Three Rs" of animal research:

- *Reduce* the number of higher species (cats, dogs, primates) used.
- *Replace* animals with alternative models whenever possible.
- *Refine* tests and experiments to ensure the most humane conditions possible.

Chapter 12 focuses on the regulation of biotechnology in more detail.

Veterinary Medicine as Clinical Trials

Not all animal research is conducted in laboratories. Veterinarians, who have devoted their lives to the care of animals, also participate in research. Research at veterinary clinics has led to new cancer treatments for pets. Because animal and human cancers are strikingly similar, information gleaned from one species might be used to treat the other. For example, the BRCA1 gene found in 65% of human breast tumors is similar to the BRCA1 gene in dogs. Clinical trials of cancer therapies on pets can pave the way for human trials.

Randomized trials in both dogs and humans have shown that hyperthermia (heating) of a tumor site combined with radiation can be more effective in local tumor control than radiation therapy alone. Researchers at Duke University are using hyperthermia along with gene therapy to destroy solid tumors. They inject directly into the tumors a gene that produces interleukin-12 (IL-12). They then apply heat only to the tumor. The heat turns on the heat-shock control gene within the tumor, promoting the production of IL-12, which helps destroy the cancerous tissue. If successful, this therapy, now being tested on cats, could be phased into human trials faster than it might have been had the early trials been conducted on cultured cells (and the cats with these tumors could benefit from the treatment too).

In a related example, researchers at the University of Wisconsin are using dogs to develop genetically modified vaccines to treat skin cancer. They culture an animal's own tumor and genetically modify the tumor cells in the laboratory so that they will express certain **cytokines** (stimulating factors) to enhance the animal's own immune system. Researchers found that in dogs with stage 1 melanoma, about 75% were cured of their disease.

Bioengineering Mosquitoes to Prevent Malaria

Even after decades of attempts to control malaria, more than a million people die from the disease each year, and hundreds of millions more become infected. As soon as a new tool emerges, a new form of resistance ensues: DDT-resistant mosquitoes have rendered DDT powerless, and the number of drug-resistant parasites rises each year. Two teams of scientists may have solved the problem of resistance through biotechnology. One team from Case Western Reserve University has transferred a gene into mosquitoes that prevents the parasite from traversing the midgut, blocking the continuation of its life cycle. Another team from the University of California at Irvine has developed an antibody that prevents the parasite from entering the mosquito's salivary gland, which the parasite must penetrate to develop and spread. Researchers plan to use both of these methods because the parasite is particularly good at developing resistance against single changes. The release of these mosquitoes into nature (after approval) is at least 10 years away, but it signals the solution to a problem that other methods have been unable to tackle successfully.

7.3 Clones

Creating Dolly: A Breakthrough in Cloning

Sheep, like other higher animals, contain genes from two parents. The mixture of genes comes down to chance. To prove this, just look at the children in any large family: Some may have curly hair and others have straight hair; some may be able to "roll" their tongues and others cannot. One child may have a genetic disorder that the others do not have. This uncertainty is eliminated with cloning. When cloning is used to reproduce an animal, there is only one genetic contributor. The offspring has exactly the same genetic programming as the parent. If the donor sheep has long soft wool, the lamb clone will too.

Embryo twinning (splitting embryos in half) was the first step toward cloning. The first successful experiment produced two perfectly healthy calves that were essentially artificially created twins. This procedure is commonly practiced in the cattle industry today where there is an advantage in producing many cattle with the same strong blood lines. Embryo twinning efforts also resulted in the birth of Tetra, a healthy rhesus monkey produced from a split embryo. The embryo-splitting procedure is relatively easy, but it has limited applications. Even though the end result is identical twins, the exact nature of those twins is the result of

Q What is a clone?

A A clone is a cell or collection of cells that is genetically identical to another cell or collection of cells. Animals that reproduce asexually are clones of a parent. Identical twins are clones of each other. Clones are nothing new, but scientists have created new types of clones that could revolutionize health care and food production. After reviewing hundreds of studies of the safety of cloned animals, the FDA in January 2008 approved the food produced from cloned cattle, pigs, and goats as safe. This means that these animal products can be sold in markets and restaurants without special labels. Over the next five years, the market for cloned animals in the U.S. is expected to reach nearly $50 million annually, according to industry analysts. Due to the high cost of the cloning process, the percentage of cloned animals in the marketplace is expected to be low, but use of clones for bull semen production is expected to double or triple. The use of cloned or genetically identical stud cattle for semen production is a cost-saving prospect that the industry will apply quickly, since ordinary cattle impregnated with prize bull sperm from clones can produce greater numbers of animals with desirable characteristics.

old-fashioned mixing of the genetic material from two parents. You may end up with animals that have the desired characteristics, but you have to wait until they are grown to find out. This is why media darling Dolly from the late 1990s—who was created from an adult cell, not an embryo—was such a breakthrough. Dolly was the exact duplicate of an adult with known characteristics.

To create a clone from an adult, the DNA from a donor cell must be inserted into an egg. In the first step, cells are collected from the donor animal and placed in a low-nutrient culture solution. The cells are alive, but starved, so they stop division and switch off their active genes. The donor cells are then ready to be introduced into a recipient egg, as seen in Figure 7.7.

The egg itself is prepared by **enucleation.** A pipette gently suctions out the DNA congregated in the nucleus of the egg. The researcher must then choose a method for delivering the desired genetic payload into the egg. One method is illustrated in Figure 7.7. Another method, known as the Honolulu technique, involves injecting the nucleus of the donor cell directly into the middle of the enucleated egg cell. No matter which technique is used, manipulating the microscopic cells can be extremely difficult.

After the DNA has been delivered to the egg, biology takes over. The new cell responds by behaving

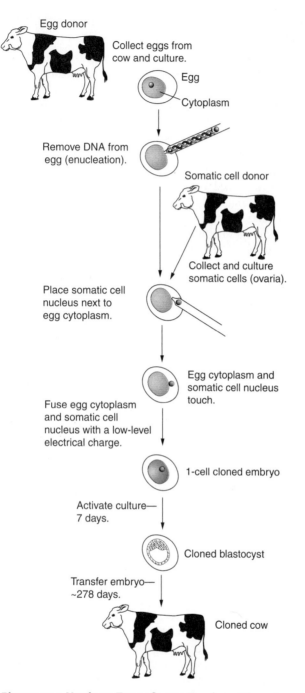

Egg donor

Collect eggs from cow and culture.

Egg

Cytoplasm

Remove DNA from egg (enucleation).

Somatic cell donor

Collect and culture somatic cells (ovaria).

Place somatic cell nucleus next to egg cytoplasm.

Egg cytoplasm and somatic cell nucleus touch.

Fuse egg cytoplasm and somatic cell nucleus with a low-level electrical charge.

1-cell cloned embryo

Activate culture— 7 days.

Cloned blastocyst

Transfer embryo— ~278 days.

Cloned cow

Figure 7.7 Nuclear Transfer to Produce Cloned Animal Cloning usually begins with removal of the egg nucleus from a fertilized egg and replacement with another nucleus (possibly from an adult animal like a cow). If the process is successful, then desirable characteristics can be obtained with a high degree of certainty. Success is not always easy (especially because it takes 278 days in cattle to find out if it worked!).

as if it was an embryonic cell, rather than an adult cell. If all goes well, cell division occurs, just as it would in an ordinary fertilized egg. For the first few days, the embryo grows in an incubator. Then, when

the new embryo is ready, it is transferred into a surrogate mother for gestation. The surrogate will give birth to a clone that is genetically identical to the genetic donor.

The list of species successfully cloned in this fashion now includes sheep, goats, pigs, cattle, and an endangered cow-like animal called a gaur. In the winter of 2002, a kitten named Cc: (for carbon copy) made headlines worldwide. The healthy, energetic kitten opened up a new prospect for many people. Cloning no longer seemed like a strange scientific experiment being played out in laboratories and research farms; people began to think about how a clone could be part of their lives. It is not surprising that many people are now banking cells from treasured cats, dogs, cattle, and horses in the hope of one day having a clone of their own.

The Limits to Cloning

There are limits to the power of cloning technology. First, the donor cell must come from a living organism. Free-ranging dinosaurs à la the film *Jurassic Park* are likely to remain science fiction. Some people are intrigued by the possibility that cells collected from frozen mammoths discovered in Siberia might be cloned and fostered by a modern-day elephant. Most experts, however, think the chance of finding a viable mammoth cell, managing to clone it, and succeeding in fostering it in an elephant or other distant relative is laughably remote.

It is also important to realize that, even though clones are exact genetic duplicates, they are not exactly identical to the original in other ways. Animals are shaped by their experiences and environments as well as their genes. Even identical human twins are not the same individuals and have different personalities. For this reason, those people who are banking cells from beloved pets might well be disappointed. For example, a dog's personality was shaped in part by its relationship to its litter mates and by the events of its life. Although we may be able to reproduce a genetically identical animal, there is no way to duplicate those early experiences that molded the dog's behaviors. Even racehorses, which are carefully bred to have essential genetic characteristics, are shaped by factors other than genetics. The fun of running and the will to win may not be genetically encoded. For that reason, it is unlikely there will be a rush to clone a winning horse.

More significantly, the present success rate for cloning is actually quite low, although it is improving. Dolly was the result of 277 efforts. Of those attempts, only 29 implanted embryos lived longer than six days, and only Dolly was born. Cc:, the kitten, represents the

YOU DECIDE

Genetics Savings and Clone

Shortly after the California company Genetics Savings & Clone (GSC) made Cc:, they produced two more clone cats, Baba Ganoush and Tabouli (Figure 7.8). In 2004, the company then made headlines when it revealed it had sold a cloned kitten to a woman in Texas, and for a mere $50,000 GSC would clone anyone's cat. Within a few years GSC reduced the price to $32,000, and by late 2006 it announced it would close at the end of the year and provide refunds for anyone who had put down a deposit for a cloned cat. After six years in business, GSC cloned five kittens but sold only two. Many questioned whether a company could be profitable selling cloned pets given the costs of cloning. Others questioned the ethics of cloning pets for human enjoyment or to create a genetically identical clone of a deceased pet. What do you think? Should pets be cloned? If a pet in your family died, would you want an identical clone if it was affordable? You decide.

only success out of 87 implanted clone embryos. Cloning failures do not make the evening news, but they are a significant barrier to the feared "animal assembly line" cloning, which is not the aim of researchers. Scientists are still trying to understand some of the unexpected problems that arise during the process. Many clones are born with defects including enlarged tongues, kidney problems, intestinal blockages, diabetes, and crippling disabilities resulting from shortened tendons. In addition, the mortality rate of the surrogate mothers is much higher than in normal births. No clear pattern to these perplexing problems is apparent.

An even more challenging problem is the possibility that clones may be old before their time. This problem came to light when analyses indicated that Dolly's cells seemed to be *older* than she was. This observation bears a little explanation. Every cell in your body contains the chromosomes that determine your genetic makeup. At the end of each chromosome is a string of highly repetitive DNA called a *telomere* (see Chapter 2 for a review of telomeres). As your cells replicate, the telomere becomes shorter with each division. In this sense, there really is a biological clock inside living cells. A cell's chromosomal age can be read by inspecting the length of the telomeres. As the telomere unravels, signs of aging appear in the organism. Tests showed that Dolly's telomeres were not like those of other lambs born the year she was; instead, they were

about the same length as her donor's, or three years older than Dolly. Dolly was diagnosed with arthritis, a possible sign of premature aging. She eventually developed a virus-induced lung tumor that led to her being euthanized in 2003. But this particular respiratory virus commonly infects sheep, and scientists at the Roslin Institute found no evidence that indicated the tumor originated because Dolly was a clone. Other clones may be in for a similar fate. Cloned mice, for example, have been found to have much shorter lifespans than typical mice. There is some speculation that lifespan may vary from species to species, but all the evidence seems to indicate that a shorter lifespan may be part of the destiny of a clone.

This cloud over cloning has a potential silver lining. Researchers investigating the problem of shortened telomeres in clones learned more about telomerase, an enzyme that regenerates the telomere. This enzyme could be a key to understanding and preventing many age-related diseases in the future, as well as possibly providing an answer to shortened clone life.

The Future of Cloning

When Dolly's novel birth was announced in the media, there was a flurry of speculation and concern about the eventual role of cloning, such as the widespread replacement of natural birth processes. Even though there have been steady improvements in the techniques used to clone animals, it is still a young science and subject to much experimentation. Some researchers have even managed to make clones from clones. DNA from frozen cells has been used successfully in the cloning process, opening up new possibilities that will depend on cryogenically preserved cells.

Among the early speculation was the notion that clones could be used to provide replacement body parts for their "parents"—the donor of the original cells. Even though a genetic match would probably reduce the chance of clone tissue being rejected by the donor-turned-transplant recipient, this is not a practical solution to the problem of transplant organ shortages. Aside from the profound ethical questions about such a practice, it would take years for the clone to be mature enough to donate organs. Far more promising biotechnological solutions are on the horizon: xenotransplantation, the use of organs from other species, some of which may have been made more suitable through the use of transgenics, is expected to become a viable source for organs within the next five years. (Xenotransplantation is discussed in more detail in Chapter 10.) Other scientists have had early success using cell cultures to grow replacement skin, eyes, and rudimentary kidneys for animals ranging from tadpoles to cattle.

So why are clones necessary? There are many reasons why it is advantageous to have multiple animals with the exact genetic makeup. Clones can be used in medical research, where their identical genetics makes it easier to sort out the results of treatments without the confounding factor of different genetic predispositions. Clones may also provide a unique window on the cellular and molecular secrets of development, aging, and diseases.

In the case of endangered animals, there is a compelling reason to look to cloning to sustain breeding populations. Steps have already been taken to use cloning to preserve animals. Interspecies embryo transfer has been used successfully to provide surrogate mothers for African wildcats *(Felis silvestria libyca)*. Noah, the gaur calf, was a clone born to a domestic cow surrogate. Noah later died of an infection unrelated to the cloning process. Despite that disappointment, efforts are now underway to clone pandas, using more common black bears as hosts.

Cloning can also be used directly to improve agricultural production. It traditionally takes six to nine years to develop and commercialize new breeds of pigs. Cloning could reduce that time to only three years, and this economic value of cloning in the pork industry could be considerable. This is the business plan of ProLinia, Inc., a new agricultural biotechnological genetics supplier spun from expertise in the Animal Science Department of the University of Georgia. ProLinia estimates that it will take three to eight years to get cloning integrated into commercial hog producer systems. The first step will be establishing a "genetic nucleus farm" that uses quantitative genetics and molecular genetics to select a particular combination of traits for the breeding population. Next, multiplication farms increase the number of pigs, which then go to commercial farms.

Figure 7.8 Cloned Kittens At the annual Cat Fanciers' Association/IAMS New York cat show in 2004, 4-month-old cloned cats Tabouli and Baba Ghanoush and their genetic donor, Tahini, were introduced. The Bengal kittens were cloned by Genetic Savings & Clone of Sausalito, California.

to a cell in culture, and the effects on that one cell were observed. When cloning technology became available, transgenics took a new course. Rather than manipulate a single cell, genes could be added to a large number of cells. These cells are then screened to discover which exhibit the desired genes. After these cells are found, they each can be used to grow a complete animal from a single cell, using cloning technology (see Figure 7.8).

7.4 Transgenic Animals

Chapter 6 covered transgenics as it applies to plant biotechnology. The process of introducing new genetic material is not limited to plants. It can be used on animals as well. In Chapter 4, we saw some of the differences between protein produced from microbial (prokaryotic) sources and other sources. Because the protein structural differences are minimal, cell and animal culture offer a good solution to the need for proper structure. Some of the most exciting developments in biotechnology are the result of transgenic research in animals.

The first experiments in animal transgenics were conducted on the cellular level. A new gene was added

Transgenic Techniques

A variety of methods can be used to introduce new genetic material into animals, and new methods continue to be developed and refined. Here is a brief overview of some techniques:

- *Retrovirus-mediated transgenics* is accomplished by infecting mouse embryos with retroviruses before the embryos are implanted. The retrovirus acts as a vector for the new DNA. This method is restricted in its applications because the size of the **transgene** (or *trans*ferred *gene*tic material) is limited, and the virus's own genetic sequence can interfere with the process, making transgenics a rather hit-and-miss project (retroviruses are discussed in Chapter 3).

- The **pronuclear microinjection** method introduces the transgene DNA at the earliest possible stage of development of the zygote (fertilized egg). When the sperm and egg cells join, the DNA is injected directly into either the nucleus of the sperm or the egg. Because the new DNA is injected directly, no vector is required; therefore, there is no external genetic sequence to muddle the process (see Figure 7.9).

- In the **embryonic stem cell method,** embryonic stem cells are collected from the inner cell mass of blastocysts. They are then mixed with DNA, which has usually been created using recombinant DNA methods. Some of the ES cells will absorb the DNA and be transformed by the new genetic material. Those transformed ES cells will then be injected into the inner cell mass of a host blastocyst.

- A similar method, **sperm-mediated transfer,** uses "linker proteins" to attach DNA to sperm

cells. When the sperm cell fertilizes an egg, it carries the valuable genes along with it.

- Finally, gene guns, which were discussed in Chapter 6, can also be used on animal cells.

Improving Agricultural Products with Transgenics

Researchers have used gene transfer to improve the productivity of livestock. By introducing genes that are responsible for faster growth rates or leaner growth patterns, animals can be raised to market more quickly, and the meat they provide can be healthier for those who eat it. The process is known as selective improvement.

Traditional crossbreeding techniques have been used for years to create chickens that grow from chick to market size as quickly as possible. Despite continuing efforts, nobody has been able to achieve this in fewer than 42 days. Breeders discovered that birds that took less than 42 days to grow did not produce eggs. Although eggs are not the primary goal of those raising chickens for meat, it is very hard to raise any chickens at all without some eggs to hatch. Forty-two days does not seem like a long time, but every extra day represents additional costs: Extra food is needed for each day, and more waste is produced as well. Americans consume 10 billion chickens per year, and any reduction in the time required to produce those billions of chickens would embody significant savings for producers. These potential savings are motivating some current research in which genes from fast-growing meat-producing chicken breeds are being swapped into eggs produced by egg-laying breeds. These researchers have met with success on a limited scale. Now the challenge is to develop methods that will work on the scale of industrial farming. In the next few years, researchers hope to devise equipment that will accurately inject the desired traits into 30 or 40 billion chicken embryos a year.

Eggs themselves could even be made healthier for human consumption through transgenics. Eggs are an inexpensive source of high-quality protein, but many people avoid them because they are rich in cholesterol. Tweaking the genes responsible for cholesterol production could result in healthier, lower cholesterol eggs.

The dairy industry is also a target for genetic improvement. Researchers have used transgenics to increase milk production, make milk richer in proteins, and lower the fat content. Herman, a transgenic bull (produced by trangenics and nuclear transfer), carries the human gene for lactoferrin. This gene is responsible for the higher iron content (essential to

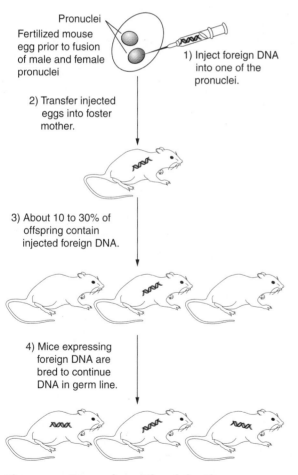

Pronuclei
Fertilized mouse egg prior to fusion of male and female pronuclei

1) Inject foreign DNA into one of the pronuclei.

2) Transfer injected eggs into foster mother.

3) About 10 to 30% of offspring contain injected foreign DNA.

4) Mice expressing foreign DNA are bred to continue DNA in germ line.

Figure 7.9 Pronuclear Microinjection Cloning Injecting DNA into one of the pronuclei (female pronucleus and male pronucleus combine at fertilization) so that it is incorporated can be used to produce clones that bear the new DNA.

the healthy development of infants) found in human milk. Herman's offspring also carry the gene for lactoferrin. Consequently, in much the same way that human polyclonal antibodies are produced in cattle (see Figure 7.10), Herman's daughters may be the first of many cows able to produce milk better suited for human children.

Another important improvement is the reduction of diseases in animals raised for food production. During the epidemic of foot-and-mouth disease in England in 2000, herds of dairy and beef cattle as well as sheep and goats were destroyed. The loss of entire

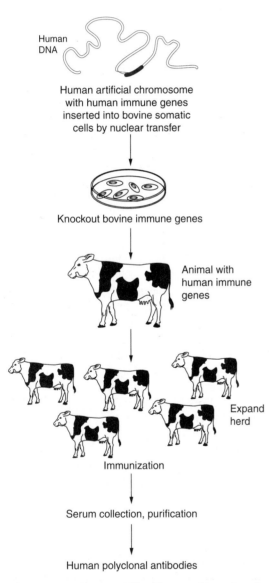

Human DNA

Human artificial chromosome with human immune genes inserted into bovine somatic cells by nuclear transfer

Knockout bovine immune genes

Animal with human immune genes

Expand herd

Immunization

Serum collection, purification

Human polyclonal antibodies

Figure 7.10 Human-Like Plasma from Cattle Human plasma for patients is in short supply and could be obtained from cattle cloned to carry human genes for blood factors and antibody genes. Human artificial chromosomes carrying these genes have been successfully transferred by procedures described in Chapter 4.

herds was catastrophic to farmers and devastating to that nation's agricultural industry. In the United States, concerns over the spread of foot-and-mouth disease resulted in the confiscation and destruction of sheep that had been imported from England, even though they did not necessarily test positive for the disease. Spurred by the threat of future disease outbreaks, researchers are studying disease-prevention genes and hope to develop animals that are resistant to foot-and-mouth disease. Similar processes may someday help prevent cholera in hogs and Newcastle disease in chickens.

Researchers at the U.S. Department of Agriculture in Maryland turned to genetic engineering to create dairy cows that can resist mastitis, an inflammation of milk glands caused by bacterial infections. Bacteria such as *Staphylococcus aureus* infect mammary glands, resulting in a highly contagious condition that spreads rapidly and results in losses of billions of dollars for the dairy industry. Scientists created transgenic cows containing a gene from *Staphylococcus simulans*, a relative of *S. aureus*, which produces a protein called lysostaphin that kills *S. aureus*. Preliminary data indicate that these animals do indeed show resistance to mastitis. But researchers must be cautious. They know that these animals will not necessarily be fully protected from infections by *S. aureus* or other microbes and that these cows represent a first step toward improving the resistance of cows to bacteria that cause mastitis. Whether the dairy industry will embrace these and other GM cows remains to be seen.

Researchers at the University of Guelph in Ontario, Canada, recently developed the **EnviroPig,** a transgenic pig expressing the enzyme phytase in its saliva. It is deemed environmentally friendly because it produces substantially less phosphorous in its urine and feces than nontransgenic pigs. Phosphorous is the major pollutant generated on pig farms. Phytase degrades phosphates in the pig's food, reducing the amounts of phosphorous waste products excreted by about 30%.

The safety of the food that reaches our tables can also be improved by transgenic technology. Human food poisoning may be reduced in the future by transferring antimicrobial genes to farm animals. This application will reduce the thousands of food poisoning deaths in the United States each year and also mean that fewer antibiotics will be used in agriculture. That is good news because overuse of antibiotics is leading to the development of resistant strains of *Salmonella, Listeria,* and *E. coli.*

Researchers are also turning to transgenics to create animals that will better serve hungry and poor populations around the world. In many areas, available arable land is limited, and little can be set aside for grazing animals. Researchers are directing their attention to the problem of transforming animals into more efficient grazers, so that more animals can be sustained on the same land. To this end, they are looking at the genetic structure of wild grazers, like African antelopes, that thrive in environments where domestic livestock would fail. It may also be possible to create cattle that are better able to do hard work such as pulling plows even while they live on less. As researchers work toward improving livestock, they are also concerned with identifying and conserving the genetic heritage of traditional livestock breeds.

Transgenic Animals as Bioreactors

Recall from Chapter 4 that proteins are an important biotechnological product and that whole animals can serve as "bioreactors" to produce those proteins. An animal bioreactor works as follows: First, the gene for a desired protein is introduced via transgenics to the target cell. Second, using cloning techniques, the cell is raised to become an adult animal. That adult may produce milk or eggs that are rich in the desired protein, thanks to the introduced gene.

An example of an application of this technology is Biosteel, an extraordinary new product that may soon be used to strengthen bulletproof vests and suture silk in operating rooms. Fundamentally, Biosteel is made from spider web protein, which is one of the strongest fibers on earth. In the past, spider silk has never been a viable industrial product because spiders produce too little silk to be useful. Thanks to transgenics, spiders are not the only source of spider silk proteins. The gene that governs the production of spider silk has been successfully transferred into goats, and those goats have reproduced, passing the new gene to their offspring. Now there is a whole herd of "silk-milk"

goats. There is no need to monitor these bioreactors. These goats, like the one shown in Figure 7.11, eat grass, hay, and grain and naturally produce the milk that now contains the proteins that make up spider silk. Purification of the silk protein from the goat milk occurs using methods described in Chapter 4, which the company then used to weave into thread for fabrication into the commercial applications mentioned earlier.

Transgenic animals are also being used to produce pharmaceutical proteins. A human gene called ATryn, which is responsible for an anticlotting agent in humans, has been transferred to goats. As a result, the goat's milk contains a protein that prevents blood from clotting. A herd of 100 goats would be capable of producing $200 million worth of this precious substance. The transgenic method is faster, more reliable, and cheaper than other more complex ways of synthesizing pharmaceutical proteins. The product was approved for use in 2006 by the European Medicines Evaluation Agency (EMEA), but the FDA has no similar evaluation procedure. The market for patients with hereditary antithrombin deficiency which is treated by ATryn is not large, but individuals who might suffer problems from clots during surgery or childbirth may be candidates for this drug product from goat's milk.

You may be wondering why transgenic researchers use goats. Goats reproduce more quickly and are cheaper to raise than cattle. Because they also produce abundant milk, they are an excellent choice for transgenic product development. Many diseases need short-term treatment of larger quantity recombinant proteins that can be made amply available from transgenic animal sources. Until gene therapy becomes broadly utilized, transgenic-derived therapeutic proteins are the most attractive alternative for replacement therapy in genetic disorders such as hemophilia. As you may

Figure 7.11 Goats with Transgenes That Produce Valuable Proteins in their Milk Transgenic animals (bioreactors), including goats, cattle, and chickens, offer the advantage of lower material and product costs for transgenic protein production.

know, hemophiliacs lack the functional clotting protein antithrombin III. This protein is being produced in transgenic goats and being tested for FDA approval. Using transgenic animal bioreactors for the production of complex therapeutic proteins provides lower production costs, higher production capacities, and safer pathogen-free products.

Milk-producing animals are not the only animal bioreactors, however. Eggs are another efficient method of producing biotechnological products in animals. Because poultry have an efficient reproductive system, it is easy to generate thousands of eggs from a small flock in a relatively short time. Each hen produces some 250 eggs per year. With each egg white containing about 3.5 grams of protein, replacement of even a small fraction of this protein with recombinant protein creates a highly productive "hen bioreactor." Some medical products such as lysozyme and attenuated virus vaccines are already derived from eggs. Later in this chapter, we discuss the use of chickens as antibody producers.

Knockouts: A Special Case of Transgenics

Adding genes is only one side of the transgenic story. We can also subtract the influence of a gene. Researchers are working with transgenic mice—so-called knockout mice—to better understand what happens when a gene is eliminated from the picture. **Knockouts** have been genetically engineered so that a specific gene is disrupted. Simply put, an active gene is replaced with DNA that has no functional information. When the gene is knocked out of place by the useless DNA, the trait controlled by the gene is eliminated from the animal.

Knockout mice begin as embryonic stem cells with specially modified DNA. The DNA is prepared using recombinant DNA techniques either to clip away a specific portion, leaving a deletion in the genetic code, or to replace one portion of the DNA with a snippet of DNA that has no function. If you think of DNA as an arrangement of letters forming a sentence that provide directions, the knockout process would delete, and possibly replace, a word. If we begin with "Open the door and place the mail on the table" and knock out a portion of the message, we might end up with "Open the mail on the table." The resulting action would be different from the action that the original instruction was intended to produce. A slightly different knockout and replacement could result in "Open the elephant and place the mail on the table." In this case, nothing would be accomplished because the sequence is nonsensical. Similarly, deleting or replacing a bit of DNA can result in a change in the traits expressed, including termination of the expression of vital proteins.

After the DNA is modified, it is added to embryonic stem cells where it recombines with the existing gene on a chromosome, essentially deleting or replacing part of the genetic instructions in the cell. This process is called **homologous recombination,** as seen in Figure 7.12. The modified ES cells are then introduced into a normal embryo, and the embryo is implanted in an incubator mother. Interfering RNA (see Chapter 2 for a refresher) is also capable of gene silencing and being utilized as an alternative to homologous recombination in many applications.

The genetically modified organism—generally a mouse pup—is not the end of the story. This offspring of the mouse pup is a **chimera.** Some of its cells are normal and some are knockouts. The activity of the normal cells often conceals the deficit caused by the missing or replaced gene. Two generations of crossbreeding are required to produce animals that are complete knockouts. The reward is worthwhile in terms of genetic research. When we know what gene has been knocked out and we see the result of that change, we have a clear idea of the function of that gene. Being able to see clearly what happens when a gene is damaged or absent will provide new insights

YOU DECIDE

Monkey Knockouts

ANDi is the first step on the path to developing a knockout monkey that could be a better animal model for studying diseases and their treatments. Genetically modified primates could potentially be used to study breast cancer, HIV, or any number of diseases.

Not everyone agrees that creating a knockout monkey ought to be a scientific goal. Some point out that mice, and even zebrafish, share the majority of human genes and have been used effectively in research. Others argue that a knockout monkey could present a threat to public health because a pathogen that could make the leap from the study animals to the human population might develop. They point to the possibility that HIV may have been a pathogen that made an unexpected and unassisted leap from wild monkeys to human beings. Others are uncomfortable with the use of primates in any sort of research. They argue that, precisely because these primates are so like us, they may share self-awareness that makes such research ethically repugnant.

What do you think? Should we press forward to develop a knockout monkey? Or are primates too closely related to humans to be used in this way? You decide.

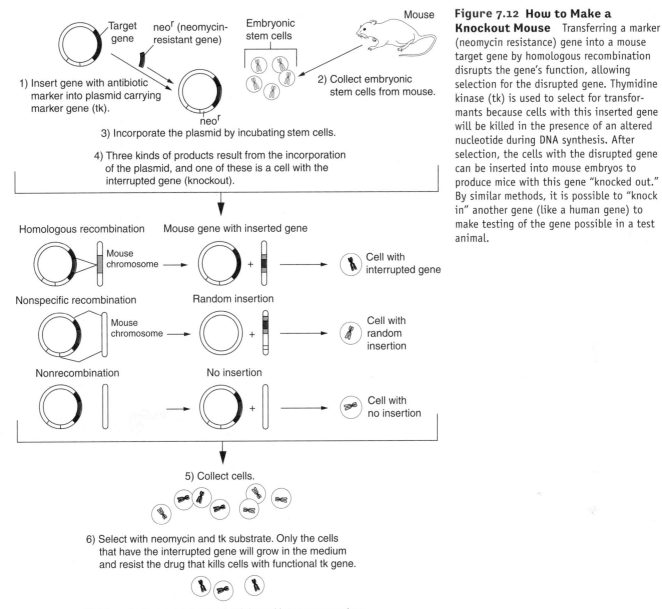

1) Insert gene with antibiotic marker into plasmid carrying marker gene (tk).

2) Collect embryonic stem cells from mouse.

3) Incorporate the plasmid by incubating stem cells.

4) Three kinds of products result from the incorporation of the plasmid, and one of these is a cell with the interrupted gene (knockout).

Homologous recombination

Mouse gene with inserted gene

Mouse chromosome → Cell with interrupted gene

Nonspecific recombination

Random insertion

Mouse chromosome → Cell with random insertion

Nonrecombination

No insertion

→ Cell with no insertion

5) Collect cells.

6) Select with neomycin and tk substrate. Only the cells that have the interrupted gene will grow in the medium and resist the drug that kills cells with functional tk gene.

7) The cells that survive are microinjected into mouse embryo.

Figure 7.12 How to Make a Knockout Mouse Transferring a marker (neomycin resistance) gene into a mouse target gene by homologous recombination disrupts the gene's function, allowing selection for the disrupted gene. Thymidine kinase (tk) is used to select for transformants because cells with this inserted gene will be killed in the presence of an altered nucleotide during DNA synthesis. After selection, the cells with the disrupted gene can be inserted into mouse embryos to produce mice with this gene "knocked out." By similar methods, it is possible to "knock in" another gene (like a human gene) to make testing of the gene possible in a test animal.

into a host of genetic disorders, including type I diabetes, cystic fibrosis, muscular dystrophy, and Down syndrome. A good example of this is the breast cancer mouse, which was patented in 1988 by Harvard University scientists and has been used extensively to test new breast cancer drugs and therapies (offered at minimal or no cost to researchers). The ability to test new therapies in these breast cancer mice, rather than human patients, has saved many cancer patients countless hours of unnecessary suffering.

A common approach is now to compare the effect of a potential drug product on knockout animals and **knock-in** animals. Animals can have a human gene inserted to replace their own counterpart by homologous recombination to become knock-ins. In a knockout mouse, the drug would have no or minimal effects because the gene product was not present, but it would have effects in the presence of the human gene in the knock-in mouse. If effects are seen in the knockout animal, they are off target and can represent potential side effects of the candidate drug (saving human suffering without having to test humans).

Thus far, mice and zebrafish are the most commonly used knockout animals. Research is also being done on knockout primates. The first steps toward developing a potentially invaluable animal model have already resulted in genetically modified monkeys. Researchers at the Oregon Regional Primate Research Center at the Oregon Health Sciences University in Portland have introduced the gene for green fluorescent protein into a rhesus monkey named ANDi (the name stands for "inserted DNA," spelled backward). The GFP gene, which had its origin in a jellyfish, does not cause ANDi to glow in the dark. In fact, he looks exactly like a regular little rhesus monkey. It takes a special microscope to see the "glow" in ANDi's cells. Not everyone agrees with the idea of using monkeys for research, even though they have considerable biological similarity to humans. The demonstration that a gene can be inserted into monkeys was a major accomplishment to prove the method can work in higher animals, even if it never results in a significant medical application.

7.5 Producing Human Antibodies in Animals

In Chapter 4, we reviewed the value of antibodies in fighting diseases. The power of antibodies makes them among the most valuable protein products. We first look at how the antibodies are produced, using animals as bioreactors.

Monoclonal Antibodies

Researchers have long dreamed of harnessing the specificity of antibodies for a variety of uses that target particular sites in the body. The use of antibody-targeting devices led to the concept in the 1980s of the "magic bullet," a treatment that could effectively seek and destroy tumor cells wherever they reside. One major limitation in the therapeutic use of antibodies is the production of a specific antibody in large quantities. Initially, researchers screened **myelomas,** which are antibody-secreting tumors, for the production of useful antibodies. But they lacked a means to program the myeloma to manufacture an antibody to their specifications. This situation changed dramatically with the development of monoclonal antibody technology. Figure 7.13 illustrates the procedure for producing monoclonal (made from a single clone of cells) antibodies (Mabs).

First a mouse or rat is inoculated with the antigen (Ag) to which an antibody is desired. After the animal produces an immune response to the antigens, its spleen is harvested. The spleen houses antibody-producing cells, or lymphocytes. These spleen cells are fused en masse to a specialized myeloma cell line that no longer produces an antibody of its own. The resulting fused cells, or hybridomas, retain the properties of both parents. They grow continuously and rapidly in culture like the myeloma (cancerous) cell, and they produce antibodies specified by the fused lymphocytes from the immunized animal. Hundreds of hybridomas can be produced from a single fusion event. The hybridomas are then systematically screened to identify the clone that produces large amounts of the desired antibody. After this antibody is identified, it will be produced in large quantities.

Monoclonal antibody products are now used to treat cancer, heart disease, and transplant rejection, but the process has not always been smooth. An early limitation to the success of therapeutic antibodies was immunogenicity. The classical method for producing Mabs using mouse hybridoma cells provoked an immune response in human patients. Mouse hybridomas produce enough "mouse" antigens that they still evoke an undesirable immune response in humans. Researchers called this the **human anti-mouse antibody (HAMA) response.** Patients quickly eliminated mouse Mabs so that the therapeutic effects were short lived. The solution is to make Mabs more human. Removing mouse antigens or creating chimeric cells is costly and time-consuming, and the resulting Mabs retain enough mouse proteins to provoke a HAMA response. Researchers are working on solving the problem of the HAMA response and are producing Mabs in different organisms, as shown in Table 7.1. There are now 17 therapeutic Mabs approved by the FDA (as of 2006). All but three of these have been engineered to reduce immunogenicity, and most have some human genetic sequence. Transgenic mice producing humanized monoclonal antibodies are under study in at least 33 current trials for new drugs.

The FDA approved the first antibody to be produced in animals that does not activate the human immune system to rejection in 2007. Panitumab, produced from transgenic mice with "knock-in" human immune genes was manufactured by Amgen in Thousand Oaks, California. The process involved inactivating the mouse antibody machinery (heavy and light chain proteins of antibodies) and introducing the human equivalent of these antibody-producing genes by homologous recombination into the inactivated (deleted) regions. The work was done in mouse embryonic stem cells, followed by introduction of these stem cells into mouse embryos to produce founder animals. The transgenic mice (xenomice) produce a fully human antibody against epidermal growth factor receptor (as a treatment for people with advanced colorectal cancer), which can be purified and does not produce the HAMA response.

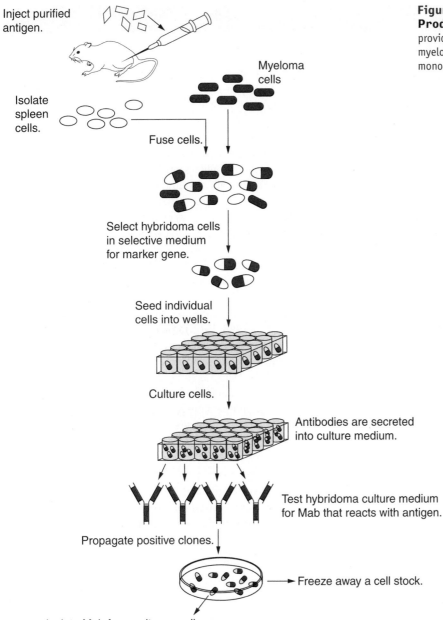

Inject purified antigen.

Isolate spleen cells.

Myeloma cells

Fuse cells.

Select hybridoma cells in selective medium for marker gene.

Seed individual cells into wells.

Culture cells.

Antibodies are secreted into culture medium.

Test hybridoma culture medium for Mab that reacts with antigen.

Propagate positive clones.

Freeze away a cell stock.

Isolate Mab from culture medium.

Figure 7.13 Monoclonal Antibody Production An antigen-stimulated mouse will provide spleen cells that can be fused with myeloma (cancer) cells to continuously produce monoclonal antibodies in a culture.

Table 7.1 COMPARISON OF DIFFERENT TRANSGENIC CLONES FOR BIOPRODUCT PRODUCTION

Source	% Raw Protein	(kg/yr)	Protein ($/g)	Scale-up Time (yr)	Facility Cost ($)	Previous Approvals	Future
Chicken egg	10	0.06/hen	0.02	—	1 million	Vaccines	Mabs
Pigs	—	—	—	3–8	?	Pork	Mabs, xenoorgans
Cattle	—	—	—	1–2	?	—	Blood proteins
Corn	8	0.5/acre	1.00	1–2	?	Corn syrup	Mabs

Animal technicians assist scientists who use laboratory and farm animals in biotechnology research or product testing. They perform medical procedures on animals and care for research animals before and after surgery; they may also perform some surgeries. Technicians not only assist in restraining the animal during examinations and inoculations but also may participate in making daily animal observations, breeding, and weaning. Keeping detailed records of animal health is an important part of this job. Writing standard operating procedures (SOPs) for the handling of animals may be a part of the job.

Animal technicians usually have an associate's degree in a veterinary science or a related program. Knowledge of biology and math is also preferred. Many biotechnological positions prefer or require that applicants be certified by the American Association for Laboratory Animal Science (AALAS). For information about AALAS, visit their website at www.aalas.org.

Salary ranges are skilled entry level, and technicians must be prepared to work nights, weekends, and holidays (because animals need daily care). Most companies offer excellent benefit packages, flexible schedules, and work release for continuing education. Extra hours are commonly available in this position.

Veterinarians play a major role in the health care of all types of pets, livestock, zoo animals, sporting animals, and laboratory animals. Some veterinarians use their skills to protect humans against diseases carried by animals and conduct clinical research on human and animal health problems. Others work in basic research, broadening the scope of fundamental theoretical knowledge, and in applied research, developing new ways to use knowledge. Prospective veterinarians must graduate from a four-year program at an accredited college of veterinary medicine with a doctor of veterinary medicine (D.V.M. or V.M.D.) degree and obtain a license to practice. All of these positions require a significant number of credit hours (ranging from 45 to 90 semester hours) at the undergraduate level. Competition for admission to veterinary school is keen; about one in three applicants was accepted in 1998. Veterinarians who seek board certification in a specialty must also complete a two- to three-year residency program that provides intensive training in specialties, such as internal medicine, oncology, radiology, surgery, dermatology, anesthesiology, neurology, cardiology, ophthalmology, and exotic small animal medicine. The median annual salary of veterinarians was $60,910 in 2000.

Eggs as Antibody Factories

As previously mentioned, eggs can be reservoirs of valuable proteins. Appropriately enough, proteins harvested from eggs have proven to be especially useful for enhancing the growth of chickens.

For more than 50 years, antibiotics have been used as growth promoters in chickens and other livestock. Even though these drugs do help animals pack on pounds, they come with a significant drawback. Studies have shown that the low-level use of antibiotics in livestock can produce drug-resistant bacteria that are dangerous and potentially lethal when transmitted to humans. Researchers are now using polyclonal antibody proteins from eggs to give chickens the growth-enhancing benefits of antibiotics without the risks.

QUESTIONS & ACTIVITIES

Answers can be found in Appendix 1.

1. Why are myeloma cells used in hybridomas?

2. Why would a human immune system reject a mouse-produced antibody?

3. How have some biotechnology companies produced antibodies that do not stimulate a HAMA-like response?

4. Discuss some of the ethical concerns about using animals in research, including their justifications.

5. How do knockout and knock-in animals provide a better prediction of how a drug will work in humans?

6. Explain the use of hyperthermia in cancer treatment in animals.

7. Why are zebrafish commonly being used to test new human drugs?

8. Explain how Mabs are used.

9. Explain the concept of "reduce, replace, and refine" in the use of animals in biological research.

10. Explain sperm-mediated transfer as a transgenic method.

References and Further Reading

Chan, A. W., Chong, K. Y., Martinovich, C., et al. (2001). Transgenic Monkeys Produced by Retroviral Gene Transfer into Mature Oocytes. *Science, 291*: 309–312.

Choi, C. (2006). Old MacDonald's Pharm. *Scientific American, 295*: 24.

Lewis, R. (2001). This Little (Cloned) Piggie Went to Market. *The Scientist, 14*: 10.

Lonberg, N. (2005). Human Antibodies from Transgenic Animals. *Nature Biotechnology, 23*: 1117–1125.

McCoy, H. (2001). Zebrafish and Genomics. *GEN, 21*: 1.

Morrow, K., Jr. (2000). Antibody Production Technologies. *GEN, 20*: 1.

Rainard, P. (2005). Tackling Mastitis in Dairy Cows. *Nature Biotechnology, 23*: 430–432.

Sedlak, B. (2001). Anticancer Tools and Techniques. *GEN, 21*: 1.

Shah, A. (2004). Mouse Model Makeovers Knock Genes Around. *Drug Discovery and Development,* 44–48.

Van Cott K. E., and Velander, W. H. Transgenic Animals as Drug Factories: A New Source of Recombinant Protein Therapeutics. Pharmaceutical Engineering Institute, Department of Chemical Engineering, Virginia Tech, Blacksburg, VA 24061, USA.

Waters, D. J., and Wildasin, K. (2006). Cancer Clues from Pet Dogs. *Scientific American, 295*: 94–101.

Visit www.pearsonhighered.com/biotechnology to download learning objectives, chapter summary, "Keeping Current" web links, glossary, flashcards, and jpegs of figures from this chapter.

Chapter 8

DNA Fingerprinting and Forensic Analysis

After completing this chapter you should be able to:

- Define DNA fingerprinting.

- Outline the process of collecting and preparing a DNA sample to be used as evidence.

- List some factors that can degrade DNA evidence and some of the precautions required to maintain the reliability of DNA evidence.

- Describe the steps in RFLP analysis.

- Describe the PCR method.

- Explain how DNA fingerprints are compared and evaluated.

- Compare two or more DNA fingerprints to determine if they are matches.

- Describe the use of DNA fingerprinting techniques in establishing familial relationships.

- List some of the uses of DNA profiling in biological research.

- Discuss some of the ethical issues surrounding DNA fingerprinting.

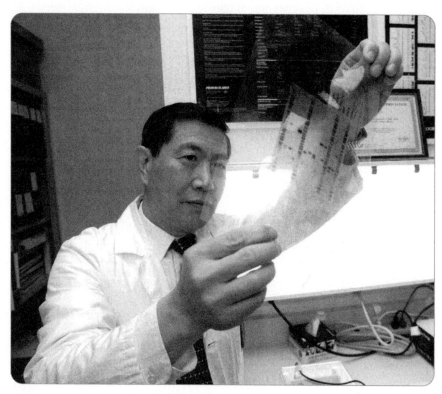

A biotechnician examines DNA fingerprints looking for similarities and differences between individuals.

8.1 Introduction to DNA Fingerprinting and Forensics

Forensic science is the intersection of law and science. It can be used to condemn the guilty or exonerate the innocent, and it can also be used to help recreate crimes. Many court cases hinge on scientific evidence. Throughout the years, scientists have developed new technologies to uncover facts in criminal investigations, and the law has been quick to embrace the technologies as they become available.

For example, in the late 1800s, efforts to fight crime were given a boost by new technology: photography. Thanks to the invention of the camera, it was possible to depict criminals in custody so accurately that these images could be used later as references. Photographs were certainly an improvement over verbal descriptions and hand-drawn "wanted" posters, but they still had severe limitations. Criminals found many ways to alter their appearance—cutting their hair, growing beards, wearing eyeglasses—that could make identification based on a photograph almost impossible.

A little more than 100 years ago, scientists discovered that the tiny arches and whorls in the skin of human fingertips could be used to establish identity. After a single bloody print on the bottom of a cash box helped solve a murder in England, the process of inking a suspect's fingers and collecting a set of prints became routine. The Federal Bureau of Investigation (FBI), the Central Intelligence Agency (CIA), and other law enforcement agencies amassed huge collections of these prints. At first, clerks were responsible for painstakingly examining the prints visually, looking for matches. With the development of the computer, however, the sorting process became less tedious and more reliable. But fingerprints are not always available. Fingerprints can be wiped away, and gloves can be worn to keep from leaving fingerprints behind.

In 1985, a revolutionary new technology emerged as an important forensics tool. Instead of counting on smudged fingerprints left at the scene of a crime, investigators could look at a new kind of "fingerprint," the unique signature found in each person's genetic makeup. In this chapter, we look at a description of DNA fingerprints and discuss what makes each person's DNA unique. Then we learn the processes used to collect DNA samples and to produce DNA fingerprints for analysis. We compare DNA profiles and learn what it takes for a DNA match to occur. Next we consider one of the best known cases involving DNA evidence, the Simpson and Goldman murders, which illustrates the value—and vulnerabilities—of DNA as evidence. We then consider how DNA profiles can be used to establish familial relationships, focusing on

the use of mitochondrial DNA to provide proof of kinship. Finally, because DNA profiles are not limited to humans, we look at the uses of DNA analysis to establish the origin of valuable plant crops, to help enforce laws to protect wildlife, and to perform biological research.

8.2 What Is a DNA Fingerprint?

We know that every human being carries a unique set of genes. The chemical structure of DNA is always the same, but the order of the base pairs in chromosomes differs in individuals. The novel assemblage of the 3 billion nucleotides formed into 23 pairs of chromosomes gives each of us our unique genetic identity.

We also know that every cell contains a copy of the DNA that defines the organism as a whole, even though individual cells have different functions (cardiac muscle cells keep our hearts beating, neurons transmit the signals that are our thoughts, T lymphocyte cells fight infections, etc.). These two aspects of DNA—the uniform nature of DNA in a single individual and the genetic variability between individuals—make DNA fingerprinting possible. Because every cell in a body shares the same DNA, cells collected by swabbing the inside of a person's cheek will be a perfect match for those found in white blood cells, skin cells, or other tissue.

How Is DNA Typing Performed?

Only a tenth of 1% of DNA (about 3 million bases) differs from each person. These variable regions can generate a DNA profile of an individual, using samples from blood, bone, hair, and other body tissues. In criminal cases, this generally involves obtaining samples from crime-scene evidence and a suspect, extracting the DNA, and analyzing it for the presence of a set of specific DNA regions.

There are two main types of forensic DNA testing: RFLP and PCR (polymerase chain reaction)-based testing. Generally, RFLP testing requires larger amounts of DNA and the DNA must not be degraded. Crime-scene evidence that is old or only present in small amounts is often unsuitable for RFLP testing. Warm, moist conditions may accelerate DNA degradation rendering it unsuitable for RFLP in a relatively short period of time.

PCR-based testing requires less DNA than RFLP testing and is still effective if the sample is partially degraded. However, PCR-based tests are also extremely sensitive to contaminating DNA at the crime scene and within the laboratory. Each of these methods are used, although in different situations. We discuss both.

Fortunately, it is not necessary to catalog every base pair in an individual's DNA to arrive at a unique fingerprint. Instead, DNA profiling depends on a small portion of the genome. Every strand of DNA is composed of both active genetic information, which codes proteins (these portions are known as *exons*), and inactive DNA, which do not code for proteins. These inert portions contain repeated sequences of between 1 and 100 base pairs (see Chapter 3 for a review). These sequences, called **variable number tandem repeats (VNTRs),** are of particular interest in determining genetic identity. Every person has some VNTRs that were inherited from his or her mother and father. No person has VNTRs that are identical to that of either parent (this could only occur as a result of cloning). Instead, the individual's VNTRs are a combination of repeats of those of the parents' DNA regions in tandem, as seen in Figure 8.1. The uniqueness of an individual's VNTRs provides the scientific marker of identity known as a DNA fingerprint.

DNA fingerprints are usually restricted to detecting the presence of **microsatellites,** which are one to six nucleotide repeats that are dispersed throughout chromosomes. Because these repeated regions can occur in many locations within the DNA, the probes used to identify them complement the DNA regions that surround the specific microsatellite being analyzed. The small size of these repeated segments have resulted in the term **short tandem repeat,** or **STR.** The FBI has chosen 13 unique STRs for testing for DNA profiles contained within their **Combined DNA Index System,** or **CODIS.**

8.3 Preparing a DNA Fingerprint

Specimen Collection

Crime scene investigators routinely search for sources of DNA: dirty laundry, a licked envelope, a cigarette butt, a used coffee cup in the sink, or anything else that might be a source for human cells. Tiny blood stains, a smear of dried semen, or a trace of saliva is often all it takes to crack a case.

Every living thing has DNA, so every crime scene is full of sources of possible contamination. For this reason, scrupulous attention to detail is required while collecting and preserving evidence. To protect the

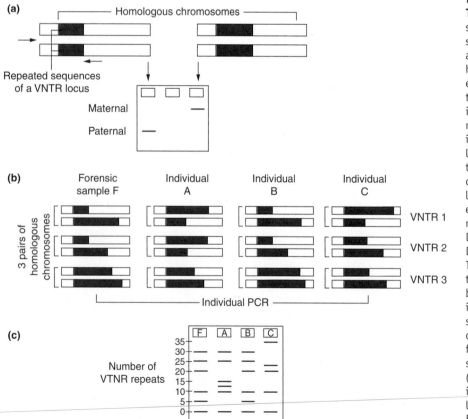

Figure 8.1 Variable Number Tandem Repeats (VNTRs) A VNTR sequence comes from short repeated sequences, such as GTGTGT . . . , which are found in various locations in the human genome. The number of repeats in each run is highly variable in the population, ranging from 4 to 40 in different individuals and are also referred to as microsatellites. (a) Each individual usually inherits a different variant of each VNTR locus from his or her parents; therefore, two unrelated individuals do not usually contain the same pair of sequences. The length of the DNA, and its position after electrophoresis, depends on the exact number of repeats at the locus. Only the DNA between the arrows (5′ to 3′ on each DNA strand) will be replicated by PCR. (b) The same three VNTR loci are analyzed from three individuals (A, B, and C), giving six bands for each person. The overall pattern is different for each. This band pattern can serve as a "DNA fingerprint." Note that individuals A and C can be eliminated from further inquiries and B remains a clear suspect. (c) The actual fingerprints (as electrophoresis gel bands) show each individual's DNA fingerprint (note the four long bands and two short bands in sample F are 30, 25, 20, 10, 5, and close to zero).

evidence, workers at a crime scene must take the following precautions:

- Wear disposable gloves and change them frequently.
- Use disposable instruments (like tweezers or swabs). If disposable instruments are not available, make certain that the instruments are thoroughly cleaned before and after handling each sample.
- Avoid talking, sneezing, and coughing to prevent contamination with micro-droplets of saliva.
- Avoid touching any item that might contain DNA (like a face, nose, or mouth) while handling evidence.
- Air-dry the evidence thoroughly before packaging. Mold can contaminate a sample.

Additional precautions for protecting specific types of evidence are listed at the FBI website.

The enemies of evidence are everywhere. Sunlight and high temperatures can degrade the DNA. Bacteria, busy doing their natural work as decomposers, can contaminate a sample before or during collection. In addition, evidence should not be stored in plastic bags because they can retain damaging moisture (specifically designed evidence bags prevent moisture damage).

DNA fingerprinting is a comparative process. DNA from the crime scene must be compared with known DNA samples from the suspect. The ideal specimen used to compare evidence is 1 mL or more of fresh whole blood treated with an anticlotting agent called ethylenediaminetetraacetic acid (EDTA). This quantity of whole blood contains an adequate amount of leukocytes with nuclear DNA for the process. Unlike erythrocytes that have no nuclear DNA, leukocytes each contain the normal complement of human chromosomes. If such a clean specimen cannot be attained from the known source, all is not necessarily lost—DNA has been retrieved and successfully analyzed from samples that were a decade old by amplifying the DNA used for comparison by the PCR.

Extracting DNA for Analysis

After the sample is collected from the known source, a lab technician is responsible for determining its genetic profile. First, the technician extracts the DNA from the sample. DNA can be purified either chemically (using detergents that wash away the unwanted cellular material) or mechanically (using pressure to force the DNA out of the cell). Once the DNA is extracted, the technician must follow several steps to transform the unique signature of that DNA into visible evidence.

Restriction Fragment Length Polymorphism (RFLP) Analysis

Because it would be time-consuming to analyze 3 billion base pairs, a method that focuses on VNTRs is used. To isolate the VNTRs, DNA is treated with an enzyme called a **restriction endonuclease** that targets and cuts the helix of DNA wherever a specific sequence appears in the strands (as described in Chapter 3 and shown in Figure 8.2). This process is called **restriction fragment length polymorphism (RFLP).** The restriction endonucleases used are found in bacteria like *E. coli,* where they occur as part of the bacteria's natural defense system against viral infections. There are hundreds of restriction endonucleases, and each recognizes a different sequence. By using several of these enzymes, either in sequence or in combination, the strand of DNA can be cut into fragments for comparison.

After the DNA is fragmented, technicians use electrophoresis to separate the pieces. Recall from Chapter 4 that in electrophoresis, negatively charged DNA fragments travel through a gel medium toward a positively charged electrode. The movement of the fragments is slowed by the pores in the gel. Smaller, lighter fragments move more quickly so those fragments will travel farther through the gel. The result is a gel with DNA sorted on the basis of the size of the genetic fragment. The gel is then treated chemically or heated to denature the DNA and deconstruct the double helix, leaving two single strands. Single-stranded DNA is capable of binding to single-stranded probes that can be used to detect unique DNA sequences.

The Southern Blot Technique

After the DNA is separated and sorted by electrophoresis, technicians transfer the fragments from the gel to a nitrocellulose or nylon membrane. The transfer process is known as the Southern blot technique, which was discussed at length in Chapter 3. The membrane "blots" up the DNA from the gel like a paper towel blots up a spill, but with a great deal more precision. The exact arrangement and position of the cut pieces of DNA is preserved on the transfer membrane (in other words, the "fingerprint"), as seen in Figure 8.3.

After the DNA is permanently fixed to the nitrocellulose or nylon membrane, the membrane is incubated with a radioactive (or fluorescent) short strand of complementary DNA called a *probe*. Probes are single-stranded fragments of DNA or RNA containing the complementary code for a specific sequence (usually 17 or more) of bases. The targeted area on the DNA sample is called a *locus*. A **single-locus probe** targets a

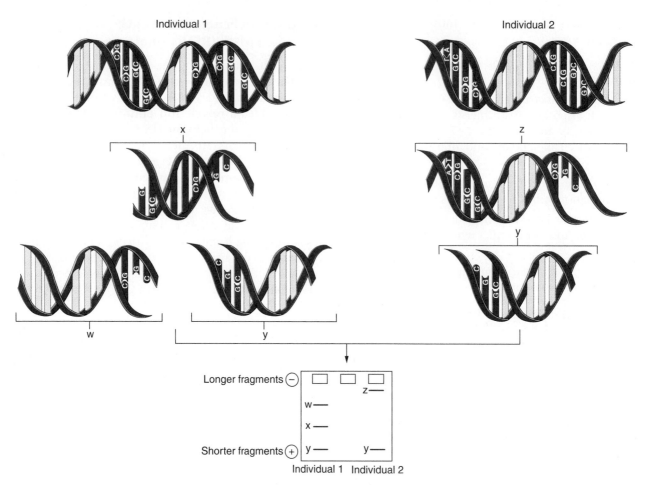

Figure 8.2 Bands Created by Cutting DNA with Restriction Enzymes at Specific Sites   Restriction enzymes recognize a short sequence of nucleotides where they will cut the DNA. Notice how the enzyme has recognized two cut sites in the DNA (producing three fragments) sequence in individual 1 but, owing to a mutation, has recognized only one sequence in individual 2 (producing only two fragments). The electrophoresis gel pattern (DNA fingerprint) below the DNA diagrams show that the two cuts produced three bands (w, x, y) at three locations, and one cut produced only two bands (y, z, with z being considerably larger than y). More bands produced using a larger number of enzymes can be sufficient to distinguish individuals with reliability (compared with their cultural population).

sequence that appears in only one position on the genome, as shown in Figure 8.4. Several single-locus probes can be used together. In contrast, a **multilocus probe** attaches to sequences in many places in the genome (the advantages of both are discussed later). The binding of the DNA fragment with its complementary probe is called *hybridization.* Hydrogen bonds make possible the merger of the two. Because hydrogen bonds are rather picky, the probe only binds to complementary fragments. After hybridization, any part of the probe that does not bond with the DNA in the sample is washed away.

Now comes the part of the process that provides us with visible evidence. X-ray (or photo) film is placed on the nitrocellulose. Because the target DNA is radioactive (or fluorescent) and emits particles, thanks to the probe, an image forms on the photographic film. This image, called an autoradiograph (or autorad for short), looks a little like a bar code. Each DNA sample produces an image as bands located in specific positions. It is then possible to compare two or more autoradiographs to see if the bands match. The patterns produced depend on the probes used. A single-locus probe produces a pattern of only one or two bands. This band or two of genetic information can result in a random match in about 1 in 10,000 cases (depending on the population used for comparison). A multilocus probe provides a more detailed genetic profile. The result can be an image that includes more than 40 bands for comparison. The odds that DNA samples

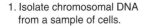

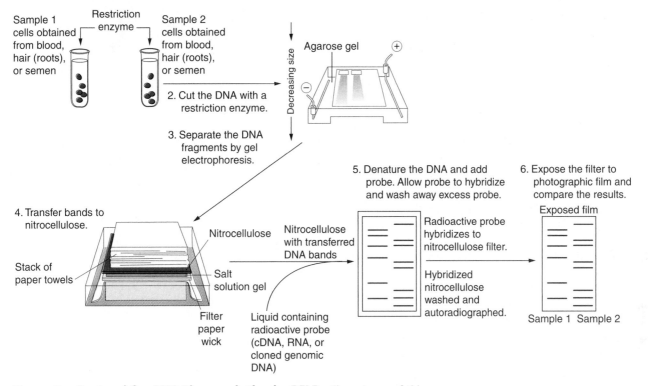

Figure 8.3 Protocol for DNA Fingerprinting by RFLP The outcome of this experiment indicates that the two samples came from different individuals (even one dissimilar band represents difference). The details of Southern blotting are covered in Chapter 3 (Figure 3.15).

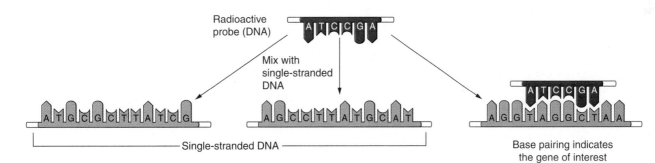

Figure 8.4 DNA Autoradiography The DNA probe is a short sequence of DNA usually 11 or more bases in length (calculate the number of possible combinations of 4 bases in a sequence of 11 . . . it is unlikely to occur randomly). If the probe binds to the DNA carrying a fluorescent (or radioactive) tag by complementary base pairing, the labeled DNA can be visualized, and the light emitted exposes photographic film creating an autoradiogram of a distinctive pattern.

from two people will match all 40 or more bands are astronomically low (assuming that the two people are not identical twins). RFLP is not used as much as it once was in human analysis because it requires rela-tively large amounts of DNA. Contaminating agents, such as dirt or mold, interfere with RFLP analysis of a given DNA sample, thus reducing the usefulness of the technique.

PCR and DNA Amplification

Several thousand cells are required to do an RFLP analysis, more than can often be collected at a crime scene. Imagine a case in which the only evidence is the saliva on an envelope flap or a blood stain invisible to the naked eye. In those cases, PCR can be used to "grow" or amplify the sample into an amount that can be analyzed.

PCR is much like a photocopier for DNA. The first step involves locating the portions of DNA that can be useful for comparison. Primers or short pieces of DNA are the tool used to find those portions on each DNA strand in the 5′ to 3′ direction (see arrows in Figure 8.4). They behave much like probes and seek out complementary sections of the DNA. Once the complementary segment of DNA is located, copying begins using a device called a *thermal cycler,* as illustrated in Figure 3.8 and described in Chapter 3. With the aid of the enzyme *Taq* polymerase and DNA nucleotides, cycles of heat and cold cause the DNA to separate and replicate repeatedly. In about three hours, the DNA segment can be increased by millions or billions. That rapid growth presents us with a possible problem. Unfortunately, PCR is very sensitive to contamination and a small error in field or laboratory procedure can result in the duplication of a useless DNA sample a million times.

Dot Blot (or Slot Blot) Analysis

Assuming that the sample is a good one, it is now time to move on to analysis. Because the DNA produced using the PCR process is identical to the original sample, electrophoresis is not needed to sort and separate it. Instead, the DNA is applied to specially prepared blot strips. Each dot on the strip contains a different DNA probe from human DNA (human leukocyte antigen, or HLA). The DNA probes present in the "dots" on the nitrocellulose strip are attached to an enzyme complex that can convert a colorless substrate into a colored one if binding occurs. In this case, the probes are not radioactive, but rather chemically reactive, and easier to perform than RFLPs. If human DNA binds to its complementary probe on the strip, the dot changes color, as shown in Figure 8.5. The PM plus DQA1 (PE Applied Biosystems) typing kit targets six genetic loci. All six are copied in the initial PCR. The products from this reaction are then placed onto two separate typing strips. One strip is for DQ alpha and the other types the remaining five loci. There are several steps in a DQ alpha PCR Dot Blot test: (1) DNA from 50 or more cells is extracted. (2) The DNA from the sample is copied by PCR resulting in amplification of the original target sequence. (3) The amplified DNA is now treated

(a) Dot Hybridization (also called reverse dot hybridization)

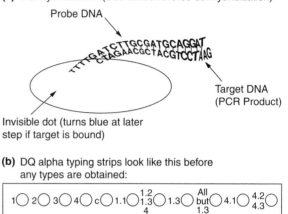

Probe DNA

Target DNA (PCR Product)

Invisible dot (turns blue at later step if target is bound)

(b) DQ alpha typing strips look like this before any types are obtained:

(c) Interpretation: Genotype is 1.1, 2 alleles

Figure 8.5 Dot Blot Process The PM plus DQA1 (PE Applied Biosystems) typing kit targets six genetic loci (DQA is a gene in the human antibody complex). All six DNA sites are copied by PCR using DNA primers for this DNA. The products from this reaction are then placed onto two separate typing strips. One strip is for DQ alpha, and the other types the remaining five genetic loci.

with a variety of probes that are bound to a blot. From the pattern of probes that the amplified DNA binds to, a potential DNA type can be seen.

STR Analysis

STR technology is used to evaluate specific regions (loci) within nuclear DNA. Variability in STR regions can be used to distinguish one DNA profile from another. The FBI uses a standard set of 13 specific STR regions for CODIS, a software program that operates local, state, and national databases of DNA profiles from convicted offenders, unsolved crime scene evidence, and missing persons. The odds that two individuals will have the same 13-loci DNA profile is more than one in a billion.

8.4 Putting DNA to Use

DNA fingerprinting is similar to old-fashioned fingerprint analysis in an important way. In both cases, the evidence collected from the crime scene is compared with evidence collected from a known source. During evaluation of the samples, analysts are looking for alignment of the bands or dots in the fingerprint. If the

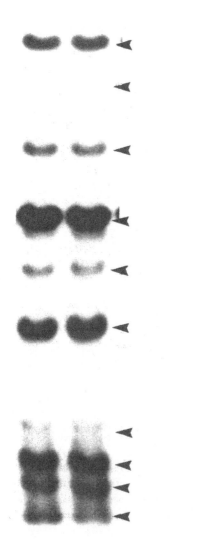

Figure 8.6 A DNA Fingerprint of Identical Twins When no problems occur in loading and running a DNA gel, the bands should line up exactly for identical DNA (or the DNA of identical twins). Frequently, known size fragment DNA, or marker DNA, is loaded every five lanes to check for band shifting. Band shifting can interfere with testing and occurs for a number of reasons. Loading marker DNA (cut to known lengths) allows comparison of the location of the same size DNA fragments across the gel, correcting for band shifting. This electrophoresis gel image shows no band shifting.

samples are from the same person (or from identical twins), the bands or dots should line up exactly (see Figure 8.6). All tests are based on exclusion. In other words, testing continues only until a difference is found. If no difference is found after a statistically acceptable amount of testing, the probability of a match is high.

As we will see, such clear-cut visual evidence can be invaluable to crime investigators. Next we consider a few examples of DNA profiling in action.

The Narborough Village Murders

The first reported use of genetic fingerprinting in a criminal case involved the sexual assault and murder of a schoolgirl in the United Kingdom in 1983. Sir Alex Jefferies and his colleagues, Dr. Peter Gill and Dr. Dave Werrett, developed techniques for extracting DNA and preparing profiles using old samples of human tissue. Gill also developed a method for separating sperm from vaginal cells, which allows fingerprints to be run on vaginal cells first and then sperm cells, in order to provide comparison samples. As shown in Figure 8.7, a detergent can break open vaginal cells but leave sperm cells. Without these developments, it would be difficult to use DNA evidence to solve rape cases because sperm cells could not be located to compare them against known samples.

In the Narborough murders, Jefferies and his colleagues compared DNA evidence collected in the case with a semen sample from a similar rape/murder that had taken place earlier. This analysis indicated that both crimes had been committed by the same man. At this point, the police had a prime suspect. However, when the DNA evidence was compared with a DNA sample from the suspect's blood, the DNA did not match. The prime suspect was excluded from consideration.

The investigation continued. The police conducted the world's first mass screening of DNA, collecting 5,500 samples from the district's male population. Using simple blood-typing tests, all but 10% of this large group were quickly excluded (using the RFLP analysis methods described earlier). After many grueling hours of analysis, the investigation had come to a dead end because none of the remaining profiles matched that of the rapist/killer.

Then came the lucky break. A man was overheard saying he had given a sample in the name of a friend. When the man who had evaded the mass testing was apprehended, his DNA was analyzed. The pattern of his DNA was an exact match for the DNA in the semen specimens from the crime. The suspect confessed to both crimes and was sentenced to life in prison.

This case highlights one of the important limitations to the use of DNA evidence: Unless there is a known sample to be used as a comparison, identity cannot be established. In the example shown in Figure 8.8, the blood samples from the victim and the suspect are known. By comparing these two known DNA fingerprints, investigators were able to place the suspect at the scene of the crime based solely on the DNA fingerprints of the blood found on the suspect's clothes.

The Forest Hills Rapist

DNA evidence was first used in the United States in 1987 and has since helped resolve thousands of

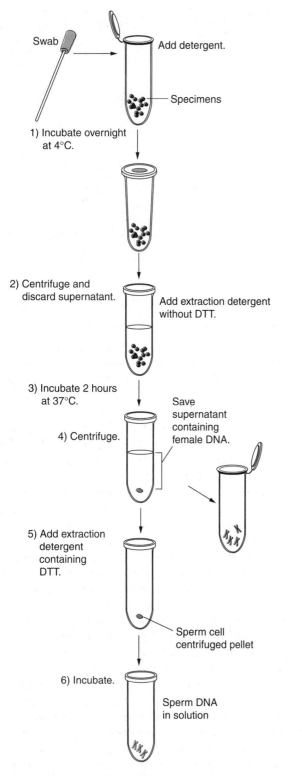

Figure 8.7 Isolation of Sperm DNA or Vaginal Cell DNA from Mixed Sources The sperm DNA can be fingerprinted to compare with the profile obtained from the female from a vaginal swab. The buffer containing Dithiothreitol (DTT) will break the sperm cell membrane and can be used to detect sperm DNA after vaginal DNA has been analyzed separately in mixed evidence (due to the ability of the sperm cell membrane to resist disruption until the addition of DTT).

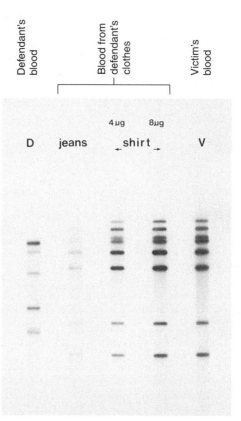

Figure 8.8 DNA Fingerprints from a Murder Case DNA bloodstains on the defendant's clothes (jeans, shirt) match the DNA fingerprints of the victim but not the defendant. This indicates that the victim's blood got on the defendant's clothes, placing the defendant at the scene of the crime.

cases. A particularly important use of DNA evidence is to refute other, erroneous, evidence that has sometimes led to false conviction. DNA evidence is especially valuable when used to expose faulty eyewitness testimony. Eyewitness testimony, which might seem to be the gold standard of evidence, is actually quite fallible.

For example, in 1988, Victor Lopez, the so-called Forest Hills rapist, was tried for the sexual assault of three women. All three women had described their assailant as a black man when they reported the crime to the police. Because Lopez was not black, concerns were raised that this was a case of mistaken identity. Was Lopez an innocent man falsely accused by the system? The blood of the accused was analyzed and compared with sperm left at the scene, as seen in Figure 8.9. The DNA was a match. Lopez was found guilty of the attacks, despite the contradictory eyewitness testimony. Now let's look at how much of what we have learned is admissible as evidence.

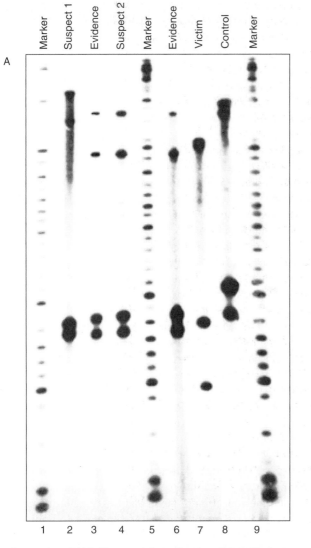

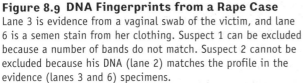

Figure 8.9 DNA Fingerprints from a Rape Case
Lane 3 is evidence from a vaginal swab of the victim, and lane 6 is a semen stain from her clothing. Suspect 1 can be excluded because a number of bands do not match. Suspect 2 cannot be excluded because his DNA (lane 2) matches the profile in the evidence (lanes 3 and 6) specimens.

Terrorism and Natural Disasters Force Development of New Technologies

World Trade Center

Shortly after the twin towers of the World Trade Center in New York City were destroyed by the terror attacks of September 11, 2001, forensic scientists came together and rapidly accelerated efforts to use DNA-based techniques to identify the remains of victims. Issues such as the tremendous amount of debris at the site, dangerous working conditions, and heat and microbial decomposition of remains coupled with the hundreds of thousands of tissue samples (primarily bone fragments) at the site from the nearly 3,000 indi-

viduals lost made it evident that new strategies would need to be employed to quickly prepare and organize DNA profiles and compare them with DNA profiles from relatives. How would scientists establish DNA identity for those who perished in over 1.5 million tons of rubble?

State agencies such as the New York Department of Health and Medical Examiner's Lab, and federal agencies, including the U.S. Department of Defense, National Institutes of Health, and the U.S. Department of Justice, immediately responded to help with this task. Within 24 hours after the disaster, the New York City Police Department had established collection points throughout the city where family members could file missing person's reports and provide cheek cell swabs for DNA isolation. Personal items (combs, toothbrushes) from the missing were also collected for DNA profiling.

Myriad Genetics, Inc., Gene Codes Forensics, DNA-VIEW, Celera Genomics, Bode Technology Group, and Orchid Genescreen were among the companies assisting in this effort. Several companies were involved in developing new software programs to help match samples submitted from family members to DNA profiles obtained from WTC victims. Because tissue samples primarily consisted of small bone fragments and teeth, which provided fragmented DNA due to DNA degradation by the intense heat at the site, forensic scientists primarily used STR, mtDNA, and SNP analysis of DNA fragments to develop profiles.

DNA analysis was conducted on over 15,000 tissue samples, although less than 1,700 of the estimated 2,819 people who died at the site were ultimately identified. This tragedy forced the development of new forensic strategies for analyzing and organizing remains recovered from the site and most importantly provided closure for a number of families that lost loved ones in the attack.

South Asian Tsunami

The South Asian tsunami of December 2004 was a tragedy that claimed over 225,000 lives and devastated areas of Indonesia, Sri Lanka, and Thailand. The Michigan-based company Gene Codes modified their software system called **Mass Fatality Identification System (M-FISys)** to help with the Thailand Tsunami Victim Identification effort. Because M-FISys was essentially built in response to the 9/11 tragedy, Gene Codes did not have to write entirely new software and they were able to customize M-FISys as necessary. In addition to analyzing mitochondrial DNA (see Section 8.6), M-FISys incorporated male-specific variations in the Y chromosome called Y-STRs to aid in the identification of individuals. Within three months, approximately 800 victims had been identified.

YOU DECIDE

Crime-Fighting Tool or Invasion of Privacy?

In the Narborough murders, many individuals were tested because they all fit a general description: They were young men who lived in the area of the murders. This strategy led to the conviction of the killer, but mass testing, especially on the basis of a general description, is extremely controversial. Opponents point out that more than 5,000 innocent men were called on to give a sample of their blood during the investigation, and the guilty individual very nearly escaped detection despite these efforts. Such mass testing is now prohibited both in the United States and the United Kingdom (without due cause). However, many people still believe that even the routine quest for DNA evidence during a criminal investigation continues to undermine the fundamental right to privacy.

Of particular concern are DNA-profiling databases. Computer-searchable collections of DNA data are now authorized by all 50 states. Many states have also registered the DNA profiles they collect on CODIS, the Combined DNA Index System run by the FBI. In addition to profiles of convicted offenders, CODIS contains a compilation of unidentified DNA profiles taken from crime scenes, and some advocates are calling for adding DNA profiles of all people arrested for any offense. Some states have even proposed that DNA profiles be taken from all individuals at the time of birth to create databases that might be used to identify criminals by matching DNA profiles from their relatives.

In early 2007, the federal government proposed that DNA profiles from more than 400,000 illegal immigrants and from detainee captives in the war on terrorism be added to CODIS.

By comparing the databases, law enforcement officers can identify possible suspects when no prior suspect exists. (Visit the CODIS FBI website, http://www.fbi.gov/hq/lab/codis, to see how the information is used in solving crimes.)

Opponents do not argue with the potential usefulness of the databases, but they are concerned that the technology could be abused. They point out that taking a blood sample is a more invasive process than taking a set of fingerprints. In some cases, the courts have agreed and determined that DNA collection is a violation of state and federal laws prohibiting unreasonable search and seizure. There is also the possibility that DNA information could someday be used to discriminate against job seekers or those applying for health insurance. Even though profiles are based on "junk" DNA and provide no information about genetic diseases or other traits, the original sample *does* contain an individual's complete DNA. On this basis, some demand that all samples be destroyed after investigation of a specific case is completed.

Proponents of DNA collection point out that the databases are regulated and secure, and samples are not obviously identified with the name of the source. Only trained professionals collect the blood samples, and the procedure presents no significant health risk to the donor.

Is routine collection of blood and the compilation of DNA databases a reasonable tool in the effort to fight crime or an unwarranted invasion of privacy? Should investigators be allowed to conduct the type of mass screening used in the Narborough case? You decide.

Recently, Gene Codes established the DNA Shoah Project (*shoah* is the Hebrew name for the Holocaust), an effort to use M-FISys to establish a genetic database of Nazi-era Holocaust survivors with an overall goal to try and reunite an estimated 10,000 postwar orphans around the world.

8.5 DNA and the Rules of Evidence

Before DNA fingerprints could ever be used in a court of law, DNA fingerprinting had to meet the overarching legal standards regarding the admissibility of evidence.

Courts use five different standards to determine whether scientific evidence should be allowed in a case. The test used depends on the jurisdiction. When a new technique or method is used to collect, process, or analyze evidence, it must meet one or several of these standards before the evidence is allowed.

- The relevancy test (Federal Rules of Evidence 401, 402, and 403) essentially allows anything that is deemed to be relevant by the courts.
- The *Frye* standard (1923) requires that the underlying theory and techniques used in gathering evidence have "been sufficiently used and tested within the scientific community and have gained

general acceptance." This is known as a general acceptance test.

- The *Coppolino* standard (1968) allows new or controversial science to be used if an adequate foundation can be laid, even if the profession as a whole is not familiar with the new method.

- The *Marx* standard (1975) is basically a common-sense test that requires that the court be able to understand and evaluate the scientific evidence presented. This is sometimes referred to as the no-jargon rule.

- The *Daubert* standard (1993) requires special pretrial hearings for scientific evidence. Under the Daubert standard, any scientific process used to gather or analyze evidence must have been described in a peer-reviewed journal.

The goal is to make certain that scientific methods and expertise used to provide evidence are trustworthy.

DNA Fingerprinting and the Simpson and Goldman Murders

DNA analysis was a relatively new forensic tool when the Los Angeles Police Department used it in the most infamous trial in recent history. In 1994, Nicole Brown Simpson and Ronald Goldman were murdered, and Nicole Simpson's ex-husband, O. J. Simpson, was a suspect. Forty-five samples were collected for DNA analysis, including known blood samples from the two victims and the suspect as well as blood drops found at the crime scene, in the suspect's home, and in his car. In addition, two bloody gloves, one found at the crime scene and the other alleged to have been found at O. J. Simpson's house, were analyzed for DNA. During pretrial proceedings, it was announced that the DNA collected at the crime scene matched that of O. J. Simpson.

Defense lawyers immediately attacked the procedures used in collecting, labeling, and testing the evidence. During the trial, the defense showed a videotape of the sample-collection methods and drew on expert testimony to establish doubt about the reliability of the evidence submitted. The defense stressed that contamination could have occurred when a technician touched the ground, when plastic bags were used to store wet swabs, and when sample collection tweezers were rinsed with water between touching samples. While on the stand, one prosecution witness admitted to mislabeling a sample. The possibility that the evidence might be tainted was obvious to both the court and the jury. As a result, the DNA evidence, which had been expected to make the case for the prosecution, was not effective. O. J. Simpson was found not guilty. When the chain of evidence is broken—when the rules of evidence are not followed—DNA samples lose their value in court.

Human Error and Sources of Contamination

One of the greatest threats to DNA evidence is human error. Earlier, we reviewed some of the precautions that crime scene investigators take to preserve and collect DNA evidence. A sneeze, improper storage, failure to label every single sample—these so-called small matters can result in the destruction of evidence. Even if the DNA evidence is not degraded by careless handling or bad conditions, it can be disregarded if the "chain of custody" is suspect. The chain of custody requires that the collection of evidence must be systematically recorded and access to the evidence must be controlled. Crime scene sample collection presents problems, but the collection of samples in more controlled environments, like the morgue, is also problematic. Studies of morgue tables and instruments have found that the DNA of at least three individuals is often present. That DNA could certainly confuse the results of any analysis undertaken on samples collected in that environment.

DNA evidence is also vulnerable to damage during the analysis itself (e.g., DNA from the technician's body or from other sources can be inadvertently added to the sample). Defined standards of laboratory practice and procedure can help guard against errors during forensic DNA analysis. When DNA evidence was a new idea, laboratories were not regulated and some serious errors were made. Consider the case of *New York v. Castro,* which involved the murder of a woman and her 2-year-old child. The prosecution presented autorads produced with the Southern blot method of the suspect's DNA and crime scene DNA and claimed that they were a match. However, as an expert witness for the defense pointed out, the bands most certainly did not match. In this case, technicians made errors during their analysis and evaluation of the autorads. Fortunately, because the standards for the admissibility of evidence are high and the defense had access to DNA experts as well, these errors did not lead to a fatal miscarriage of justice. The suspect could not be linked to the crime with this evidence.

One step to ensure the reliability of DNA is to make certain the laboratories that process the samples adhere to consistent high standards. The American Society of Crime Laboratories Directors (ASCLD), the National Forensic Science Technology Center (NSFTC), and the College of American Pathologists (CAP) all provide accreditation to forensic laboratories. Proficiency testing of technicians has become a basic requirement for employment. These tests include "blind" tests when the technician is unaware that the sample being processed is actually a test sample submitted by the certifying organization. In addition, the FBI has developed a standard operating procedure for the handling and DNA typing of evidence in criminal cases. With clear guidelines in place and thorough training of those responsible for collecting and preserving evidence, the reliability of DNA evidence should increase in court proceedings.

DNA and Juries

The use of DNA evidence also presents another challenge. To be useful, the science must make sense to the jury evaluating the case. Because DNA evidence is statistical in nature, the results can be confusing to a person without sufficient education, especially when large numbers are involved. When members of a jury panel hear there is "1 chance in 50 billion" that a random match might occur, it is possible they may focus on that one possibility and discount the overwhelming odds against it happening. Oddly enough, the same jurors might accept a conventional fingerprint as more reliable evidence, even though studies have concluded that such evidence is actually *less* statistically valid. Making DNA evidence clear and comprehensible to jurors is not an easy task. And if the DNA evidence is not clearly understood, the evidence may be disregarded.

Q Should DNA tests be extended to everyone convicted of a crime?

A DNA evidence may be damning to the guilty, but it provides salvation to the innocent. Hundreds of people have been released from prison after being exonerated by DNA testing. The number is bound to climb thanks to the Innocence Protection Act of 2001, a law that gives convicted individuals access to DNA testing, prohibits states from destroying biological evidence as long as a convicted offender is imprisoned, prohibits denial of DNA tests to death row inmates, and encourages compensation for convicted innocents. In October 2004, President George W. Bush signed the Justice for All Act, which grants any inmate convicted of a federal crime the right to petition federal court for DNA testing to support a person's claim of innocence. The Innocence Project at Cordoza School of Law in New York has been a strong crusader for the Innocence Protection Act. Using DNA evidence, the project has proved that 95 people were wrongly imprisoned. Ten of those individuals cleared were on death row. DNA evidence is much more reliable than most other forms of evidence, but DNA fingerprinting can be expensive. So the answer to our original question is probably yes, if more states establish their own DNA testing labs and the cost of performing these tests continues to fall. No one wants to see innocent people convicted of crimes they did not commit.

8.6 Familial Relationships and DNA Profiles

Crime solving is not the only forensic application of DNA fingerprints. Because DNA is shared by members of the same family, relationships can be conclusively determined by comparing samples between two individuals. An obvious use of this is paternity testing. Every year, roughly 400,000 paternity suits are filed in the United States, Canada, and the United Kingdom (this number has been increasing at about 10% since testing started in 1988). Given samples from the child and adults involved, verifying the child's parentage and resolving child support or custody disputes is relatively easy. Recently, new reproductive technologies, including *in vitro* fertilization, artificial insemination with donor sperm, and surrogacy have introduced a few quirks into custody and paternity issues. DNA testing has helped untangle cases in the courts that involved sperm mixups in fertility clinics, for example. Thanks to amniocentesis, as shown in Figure 8.10, it is possible to verify a child's parentage even before birth.

Mitochondrial DNA Analysis

Mitochondrial DNA analysis (mtDNA) can also be used to examine the DNA from samples that cannot be analyzed by RFLP or STR. Nuclear DNA must be extracted from samples for use in RFLP, PCR, and STR, but mtDNA analysis uses DNA extracted from a cellular organelle called a *mitochondrion*. Older biological samples that lack nucleated cellular material, such as hair, bones, and teeth, cannot be analyzed with STR and RFLP but they can be analyzed with mtDNA. All mothers have the same mitochondrial DNA as their daughters because the mitochondria of each new embryo comes from the mother's egg cell. The father's sperm contributes only nuclear DNA, and no mitochondria. Comparing the mtDNA profile of unidentified remains with the profile of a potential maternal relative can determine whether they share the same mtDNA profile and are related. mtDNA remains virtually the same from generation to generation, changing only about 1% every million years due to random mutation. Consequently, relationships can be

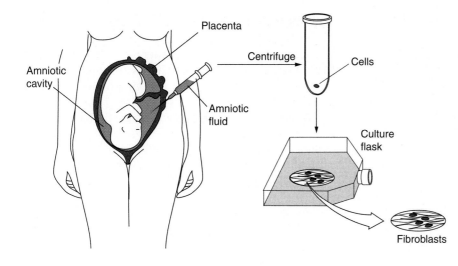

Figure 8.10 DNA for Paternity Testing For disputed paternity suits, it is possible to draw a few fetal cells from the fluid that surrounds the fetus (amniotic fluid) without harming the growing fetus. When cultured, these cells can be a source for DNA extraction and fingerprinting. When compared with the DNA of the suspected father, exclusion or inclusion of the individual can be determined.

traced through the unbroken material line, as shown in Figure 8.11.

Y-Chromosome Analysis

The Y chromosome is passed directly from father to son, making the analysis of genetic markers on the Y chromosome useful for tracing relationships among males or for analyzing biological evidence involving multiple male contributors. The Y chromosome can be analyzed with nuclear DNA (usually by PCR) methods.

mtDNA evidence was essential in reuniting families torn apart by the military government of Argentina during the war in the Falklands in the early 1980s. During this repressive regime, the military junta routinely arrested and questioned anyone suspected of subversive activity. Among those arrested were a number of young pregnant women. Roughly 15,000 of those arrested and questioned simply disappeared. After the collapse of the junta in 1983, the mothers and grandmothers of these "disappeared" young people began to gather together in the Placo de Mayo, the main square of the city of Buenos Aires. As they gathered, protesting and sharing stories, they began to be aware that a part of the horrible crime might well be ongoing. Babies born to their daughters while in prison had been taken and were

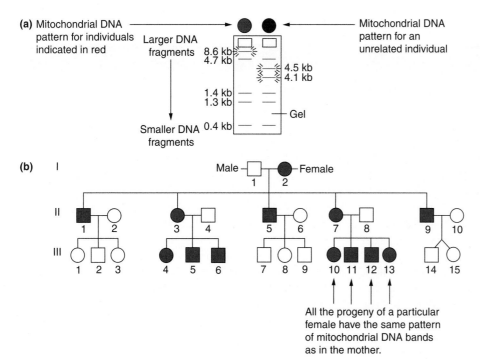

Figure 8.11 Maternal DNA Pattern of Inheritance Key genes for mitochondrial function (cell respiration) are located in a small DNA ring in human mitochondria. Because mitochondria are contributed by the egg (only) before fertilization, DNA can be traced through the maternal line with fingerprinting of the mitochondrial DNA. Some genes in mitochondria show variations due to mutations (see top of gel as 6.6, 4.7 vs. 4.5, 4.1 kilobases); others do not.

being raised by others—often by the families of those who had murdered their parents.

The mothers of Placo de Mayo sought help from the American Association for the Advancement of Science (AAAS), asking if genetic testing could be used to prove the identity of the stolen children. Through the perseverance of these mothers and grandmothers and the scientists who volunteered their help, the AAAS was able to prove that at least 51 children had been illegally adopted. The DNA evidence was used in the courts to restore the children to their living relatives.

Family ties established by DNA have also been used to identify the remains of historic figures. For many years, two communities—Kearney, Missouri, and Granbury, Texas—both claimed to be the final resting place of the notorious outlaw Jesse James. In 1995, a body was exhumed from Mount Olivet Cemetery in Kearney. Unfortunately, the acidity and moisture in the soil made it impossible to collect an adequate specimen of DNA from the bones. There was another source of DNA, however: teeth and hairs that had been collected from the body's original burial site on the James farm in Granbury.

So which body was Jesse James? Comparison of DNA retrieved from the teeth and hair were all identical, making it clear that the samples, at least, were all from the same person. The next step was collecting comparison DNA samples from living relatives of Jesse James. Analysis uncovered similarities between the samples from the teeth and hair and the samples from the living relatives. The evidence collected supported Kearney's claim that the Mount Olivet Cemetery is the final resting place of Jesse James.

8.7 Nonhuman DNA Analysis

Not every legal case is a matter of human identity. Many legal questions—as well as scientific quandaries—have been answered by the genetic profiles of plants and animals.

Ginseng, for example, is a valuable herbal product; the market for it is estimated to be $3 million in the United States alone, and the demand is high in other countries as well. Currently two major herbal products are referred to as ginseng. One is native to North America; the other is native to Asia. Oddly, Asian ginseng is the type most often sold in American health food stores. Most American ginseng—the rarer and more valuable of the two varieties—is exported to Asia. The two types of ginseng look almost identical, but they have very different reputations. Asian ginseng purportedly boosts energy, whereas American ginseng is prized for its supposed ability to calm nerves. Manufacturers are now using DNA sequencing to help make the important distinction between these two varieties. Confirming the origin of the product helps ensure quality control and protects the market for American ginseng products. This same type of analysis is performed for other plants as well. A DNA profile used in quality control of squash seeds is shown in Figure 8.12. This quality control process allows the seed supplier to guarantee a specific degree of genetic hybridity, increasing the seed value to growers, who depend on these traits to sell their high-value crops.

Another interesting "paternity" case of the plant world involved the ancestry of cabernet sauvignon

---

TOOLS OF THE TRADE

DNA Profiles Are Advancing Biological Research

In addition to their use in legal proceedings, DNA profiles from animals have been used in direct biological research. In 2002, it was reported that DNA analysis of the remaining bison herds, both public and private, indicate that the majority of bison have some domestic livestock as ancestors. There is no outward sign of the "hybrid" nature of these animals, but their heritage is clearly evident in their DNA. As a result of this research, steps can be taken to preserve the remaining "pure" herds.

Using DNA evidence, scientists have found that chickadees, barn swallows, indigo buntings, and other species are not monogamous. They examined the DNA of the parents and their offspring to determine if some of the babies in the nest actually had different fathers.

In addition, mitochondrial DNA (mtDNA) analysis has proven to be a rich source of new information about evolutionary biology. Mitochondria are estimated to mutate at the rate of 2 percent to 4 percent per million years, which allows scientists to trace gene frequency changes through time. Using frequency changes in mtDNA, scientists are able to differentiate between similar animal species and examine the role of base substitution during evolution. Only a nanogram of mtDNA is required for direct sequencing identification. In fact, the "Eve hypothesis," which used mtDNA evidence to trace a majority of people now living on earth to a single female ancestor from ancient Africa, is based on mtDNA analysis.

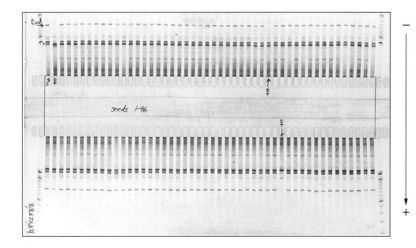

Figure 8.12 DNA Profile of Squash Hybrids Used as a Quality Control Analysis The anode is in the center of this gel with cathodes at the top and bottom. When the DNA from 96 seeds is loaded at the center of the gel, it will migrate to the cathodes during electrophoresis. The male parent plant donated the upper DNA band, and the female donated the lower of the two bands of the hybrid offspring plant being tested. The hybrid plants all have both bands (one exception gives 98.9% quality assurance of purity in 96 seeds tested). This level of hybridity (quality control) will increase the value of the seed to a grower depending on a high level of hybridity for the growing requirements.

grapes, which are prized for wine production. Hybrid grapes are considered inferior by some wine purists and legally excluded from bearing the prestigious distinction *appellation d'origine contrôlée* in France. When scientists examined the DNA of cabernet sauvignon plants, it was evident that two other varieties of grapes, cabernet franc and sauvignon blanc, were clearly the ancestors of cabernet sauvignon. This finding challenges the old notion that varieties derived from cross-breeding are inherently less valuable as wine grapes.

DNA evidence has also been used to generate the genetic profile of animals. For example, in Pennsylvania, DNA fingerprinting was used to prove that a hunter killed a bear illegally. The bear-hunting season in that state is designed to protect pregnant sows. A law making it illegal to kill bears in their dens ensures that a large portion of the breeding female population will be protected from loss during the season. In this case, a witness reported seeing a hunter discharge his rifle into a bear's den. The hunter had registered his kill at the appropriate check station (without disclosing the circumstances under which it was shot), where one of the harvested bear's premolars was removed as specified by state hunting regulations, to verify the sex and age of the kill. As a part of their investigation of this eyewitness report, authorities collected blood samples from the den and compared the DNA from the den with that from the check station. The tests revealed that the DNA samples were from the same animal, even though the hunter, in an effort to evade prosecution, had claimed to have killed the bear 5 miles from the den. Because of the DNA evidence, he was found guilty.

DNA profiling is regularly conducted by wildlife management authorities in such cases. Figure 8.13,

---

CAREER PROFILE

Laboratory Technician

Laboratory technicians are commonly employed in DNA fingerprinting labs to perform tests on evidence collected from crime scenes. These technicians must be able to follow directions with great accuracy. They must take great care to keep the work area clean to prevent contamination. In fact, all technicians working in law enforcement labs must pass a quantitative analysis course as a condition of employment. Technicians frequently may be required to work in a special "clean room" environment wearing sanitary gowns. Salaries range from $21,000 to $26,000 for an entry-level worker to $25,000 to $42,000 for more experienced technicians.

Most employers look for individuals with a bachelor's degree in biology, biochemistry, or molecular biology. Some employers hire applicants with a specialized associate's degree in biotechnology and laboratory experience. Hands-on laboratory experience is a critical skill in this position and may be gained through coursework, internships, and on-the-job training. However, technicians also need good math and communication skills. Writing is especially important because lab notebooks are considered legal documents. Strong computer skills are also necessary. Opportunities for advancement to research associate, especially for those with a bachelor's degree, are good.

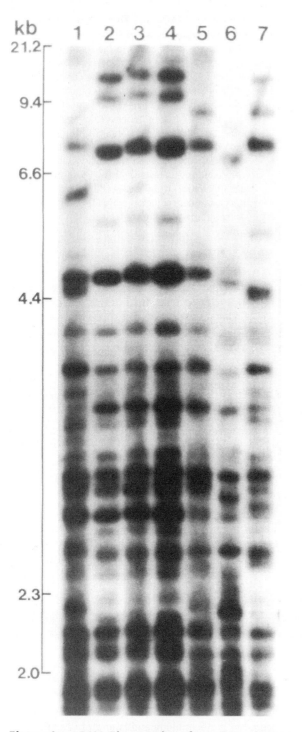

Figure 8.13 DNA Fingerprints from Four Moose Samples RFLP autoradiograms like these can be used to determine whether more than one pattern exists. In this analysis, there are four different fingerprints from a hunter with a tag for taking only one animal (notice similar bands in lanes 2, 3, 4, but not in others at the top of the gel). Different band patterns verify that they are not all the same (as claimed by the hunter with only one moose permit).

for example, shows DNA profiles for four moose that were killed by a hunter with a permit to kill only one moose. Armed with the tools of DNA fingerprinting, wildlife managers have improved their ability to prove that game has been taken beyond the limits set by regulations.

DNA Tagging to Fight Fraud

Several companies envision that DNA labels can provide a highly specific barrier to product counterfeiting and piracy by using DNA as an authentication label hidden in a wide variety of products. Virtually anything of value, from expensive fine art to sports memorabilia, could have DNA incorporated in it. Then the DNA labels could be detected using PCR or hybridization techniques to verify authenticity. For example, footballs in the 2003 Super Bowl were encoded with DNA to authenticate them as official products of the National Football League. During the 2000 Sydney Summer Olympics, DNA Technologies of Halifax, Nova Scotia, tagged 34 million labels with specific DNA strands and then the labels were attached to Olympic-licensed merchandise. Random inspections of merchandise vendors in Sydney revealed that nearly 15% of merchandise sold as officially licensed was in fact counterfeit, contributing to nearly $1 million in lost sales.

Earlier in this chapter you learned why DNA is a good forensic marker. These same reasons make DNA a good physical marker that is difficult to counterfeit. Typically, DNA for labels incorporate random single-stranded sequences from 20 to several thousand nucleotides long. These strands are then mixed into ink, cloth, threads, or other materials used in the product to be tagged. To determine whether an item is authentic, the portion thought to contain the DNA tag is removed and analyzed by PCR, or probes are added to the component that will fluoresce if they hybridize with the single-stranded tags in the product.

Whether DNA tagging ever becomes a widespread tool to combat counterfeiting remains to be seen, but for now it appears to be yet another promising biotechnology application of DNA.

QUESTIONS & ACTIVITIES

Answers can be found in Appendix 1.

1. Explain what is meant by a polymorphic DNA locus.

2. How do polymorphic DNA sequences (e.g., STRs) affect someone?

3. How many RFLP bands found in a child's DNA should occur in the DNA fingerprint of the father? The mother?

4. Why is PCR-based DNA fingerprinting used in forensics more commonly than RFLP?

5. How could contaminants from the technician possibly affect a RFLP?

6. Why is eyewitness testimony considered less reliable than DNA evidence (give an example)?

7. How was it possible to prove that cabernet sauvignon was a cross between cabernet franc and sauvignon blanc?

8. Why have dot/slot blots become more popular than RFLP in exclusion attempts?

9. What can be the result of overloading a lane with DNA in a gel?

10. Give at least one example of how mtDNA evidence has been used.

References and Further Reading

Barritt, J. A., Brenner, C. A., Matter, H. E., et al. (2001). Mitochondria in Human Offspring Derived from Ooplasmic Transplantation. *Forensic Science,* 46: 513–516.

Hagmann, M. A Paternity Case for Wine Lovers. *Science,* 285: 1470.

Harbison, S. A., Hamilton, J. F., and Walsch, S. J. (2001). New Zealand DNA Databank: Its Development and Significance as a Crime Solving Tool. *Science and Justice,* 41: 33–37.

Hollow, T. (2001). Reforming Criminal Law, Exposing Junk Forensic Science. *Scientist,* 15: 12.

Nowak, R. (1994). Forensic DNA Goes to Court with OJ. *Science,* 265: 1352–1354.

Jeffreys, A. J., Wilson, V., and Thein, S. L. (1985). Hypervariable 'Minisatellite' Regions in Human DNA. *Nature,* 314: 67–73.

Klovdahl, A. S., Grawiss, E., Yaganehdoost, A., et al. (2001). Networks and Tuberculosis: An Undetected Community Outbreak Involving Public Places. *Social Science and Medicine,* 52: 681–694.

Michele, T. M., Cronin, W. A., Graham, N. M., et al. (1997). Transmission of Mycobacterium Tuberculosis by a Fiberoptic Bronchoscope. Identification by DNA Fingerprinting. *Journal of the American Medical Association,* 278: 1093–1095.

Soares-Vieira, J. A., Billerbeck, A. E., Iwamura, E. S., et al. (2000). Parentage Testing on Blood Crusts from Firearms Projectiles by DNA Typing Settles an Insurance Fraud Case. *Journal of Forensic Sciences* 45: 1142–1143.

Stone, A. C., Starrs, J. E., and Stoneking, M. (2001). Mitochondrial DNA Analysis of the Presumptive Remains of Jesse James. *Forensic Science,* 46: 173–176.

Visit www.pearsonhighered.com/biotechnology to download learning objectives, chapter summary, "Keeping Current" web links, glossary, flashcards, and jpegs of figures from this chapter.

Chapter 9

Bioremediation

After completing this chapter you should be able to:

- Define bioremediation and describe why it is important.

- Describe advantages of bioremediation strategies over other types of cleanup approaches.

- Name common chemical pollutants that need to be cleaned up, and provide examples of ways in which chemicals enter different zones of the environment.

- Distinguish between aerobic and anaerobic biodegradation, and provide examples of microbes that can contribute to bioremediation.

- Explain why studying genomes of organisms involved in bioremediation is an active area of research.

- Define phytoremediation, and explain how it can be used to clean up the environment.

- Discuss how *in situ* and *ex situ* approaches can be used to bioremediate soil and groundwater.

- Discuss the roles of bioremediation at a wastewater treatment plant.

- Provide examples of how genetically modified organisms can be used in bioremediation.

Landfills around the world are overflowing with tons of garbage that we generate each day. Bioremediation techniques hold great promise for cleaning up chemicals in the environment, reducing waste in landfills, and even producing energy from society's garbage.

Our environment is being threatened with alarming frequency. The air we breathe, the water we drink, and the soil we rely on to grow plants for food are all being contaminated as a direct result of human activities. The average American generates approximately 4 pounds of solid trash per day—over 1,400 pounds of trash in a year per person. Yet household wastes are a relatively small part of the problem. Pollution from industrial manufacturing wastes as well as from chemical spills, household products, and pesticides has led to contamination of the environment. An increasing number of toxic chemicals are presenting serious threats to the health of environments throughout the world and the organisms that live there.

Just as biotechnology is considered to be a key to identifying and solving human health problems, it is also a powerful tool for studying and correcting the poor health of polluted environments. In this chapter we consider how biotechnology can help solve some of our pollution problems and create cleaner environments for humans and wildlife through bioremediation.

9.1 What Is Bioremediation?

The use of living organisms such as bacteria, fungi and plants to break down or degrade chemical compounds is called **biodegradation.** It takes advantage of natural chemical reactions and processes through which organisms break down compounds to obtain nutrients and derive energy. Bacteria, for example, metabolize sugars to make adenosine triphosphate (ATP) as an energy source for cells. In addition to degrading natural compounds to obtain energy, many microbes have developed unique metabolic reactions that can be used to degrade human-made chemicals. **Bioremediation** is the process of cleaning up environmental sites contaminated with chemical pollutants by using living organisms to degrade hazardous materials into less toxic substances.

Bioremediation is not a new application. Humans have relied on biological processes to reduce waste materials for thousands of years. In the simplest sense, the outhouse—which relied on natural microbes in soil to degrade human wastes—was an example of bioremediation. Similarly, sewage treatment plants have used microbes to degrade human wastes for decades. But, as you will learn in this chapter, modern applications in bioremediation involve a variety of new and innovative strategies to clean up a wide range of toxic chemicals in many different environmental settings.

Taking advantage of what many microbes already do is only one aspect of bioremediation. A key purpose of bioremediation is to improve natural mechanisms and increase rates of biodegradation to accelerate cleanup processes. In this chapter, we explore some of the ways that scientists can stimulate microbes to degrade a wide variety of wastes in many different situations. Another important aspect of bioremediation is the development of new approaches for the biodegradation of waste materials in the environment, which can involve using genetically modified microorganisms.

Why Is Bioremediation Important?

Our quality of life is directly related to the cleanliness and health of the environment. We know that environmental chemicals can influence our genetics and that some chemicals can act as mutagens leading to human disease conditions. Clearly, there is reason to be concerned about both short- and long-term chemical exposure and the consequences of environmental chemicals on humans and other organisms.

According to some estimates, over 200 million tons of hazardous materials are produced in the United States each year. Accidental chemical spills can and do occur, but these events typically are contained and cleaned up rapidly to minimize impact on the environment. More problematic, however, are illegal dumping practices and sites contaminated through neglect such as abandoned warehouses where stored chemicals may leak into the environment. In 1980, the U.S. Congress established the **Superfund Program** as an initiative of the **U.S. Environmental Protection Agency (EPA)** to counteract careless and even negligent practices of chemical dumping and storage, as well as concern over how these pollutants might affect human health and the environment. The primary purpose of the Superfund Program is to locate and clean up hazardous waste sites to protect U.S. citizens from contaminated areas.

One in every five Americans lives within 3 to 4 miles of a polluted site treated by the EPA. In the over 25 years since Superfund began, the EPA has cleaned up more than 700 sites in the United States. Nevertheless, well over 1,000 sites await cleanup, and Department of Energy estimates suggest as many as 220,000 sites need remediation with many new sites identified each year. Estimates for the cleanup costs of currently identified polluted areas in the United States are in excess of $1.5 trillion. The extent of contamination of other sites and the number of sites requiring cleanup will undoubtedly push that estimate much higher. Clearly, environmental pollution in the United States is an important problem that is receiving a lot of attention. In many other countries, environmental pollution is an even greater problem. To learn more about the Superfund Program, visit the Superfund website listed on the Companion Website. Through this site you can also check on contaminated environments close to where you live that are on the Superfund priority list for cleanup.

Through the National Institute of Environmental Health Sciences, a division of the National Institutes of Health, the United States started a program called the **Environmental Genome Project.** A primary purpose of this project is to study and understand the impacts of environmental chemicals on human disease. This includes the study of genes that are sensitive to environmental agents, learning more about detoxification genes, and identifying single-nucleotide polymorphisms that may be indicators of environmental impacts on human health. Ultimately, this project will generate genome data that will enable scientists to carry out epidemiological studies so they better understand how the environment contributes to disease risk and learn how specific diseases are influenced by environmental exposure to chemicals.

We know that pollution is a problem that can affect human health, and bioremediation is an important approach for cleaning up the environment. However, there are many ways to clean up pollutants, so why use bioremediation? We could physically remove contaminated material such as soil or chemically treat polluted areas, but these processes can be very expensive and, in the case of chemical treatments, they can create more pollutants that require cleanup. A major advantage of bioremediation is that most approaches convert harmful pollutants into relatively harmless materials such as carbon dioxide, chloride, water, and simple organic molecules. Because living organisms are used for the cleanup, bioremediation processes are generally cleaner than other types of cleanup strategies.

Another advantage of bioremediation is that many cleanup approaches typically can be conducted at the site *(in situ)* of pollution. Because the contaminated materials do not need to be transported to another site, a more complete cleanup is often possible without disturbing the environment. In addition to cleaning up environments, biotechnological approaches are essential for detecting pollutants, restoring ecosystems, learning about conditions that can result in human diseases, and converting waste products into valuable energy. In the next section, we consider some of the basic principles of bioremediation in terms of some common chemical pollutants and the environments they pollute. We also discuss some of the microbes and reactions that are important for bioremediation.

9.2 Bioremediation Basics

Naturally occurring marshes and wetlands have excelled at bioremediation for hundreds of years. In these environments, plant life and microbes can absorb and degrade a wide variety of chemicals and convert pollutants into harmless products. Before we discuss how living organisms can degrade pollutants, we first consider areas of the environment that require cleanup and take a look at some of the common chemicals that pollute the environment.

What Needs to Be Cleaned Up?

Unfortunate as it may be, the answer to this question is that almost everything needs to be cleaned up. Soil, air, water, and sediment (a combination of soil and decaying plant and animal life located at the bottom of a body of water) are all environments affected by pollution. Soil, water, and sediments are the most common treatment environments that require cleanup by bioremediation, although new bioremediation approaches are being developed to detect and clean up air pollution. Each area presents its own complexities for cleanup because the type of bioremediation approach used typically depends on site conditions. For instance, approaches for cleaning up soil can be very different from those used to clean up water. Similarly, surface water often needs to be treated differently than groundwater.

Pollutants can enter the environment in many different ways and affect diverse components of the environment. In some cases, pollutants enter the environment through a tanker spill, a truck accident, or a ruptured chemical tank at an industrial site. Of course, depending on the location of the accident, the amount of chemicals released, and the duration of the spill (hours versus weeks or years), different parts or zones of the environment may be affected. Figure 9.1 provides an example of a leaking chemical tank at an industrial plant. It may initially contaminate surface and subsurface soils; however, if large amounts of chemicals are released and the leaky tank goes undetected for a long period of time, chemicals may move deeper into the soil. Following heavy rains, these same chemicals may create runoff that can contaminate adjacent surface water supplies such as ponds, lakes, streams, and rivers. Chemicals may also leak through the ground creating what is called **leachate.** Leachate can cause contamination of subsurface water (called groundwater), including aquifers—deep pockets of underground water that are a common source of drinking water.

Chemicals may also enter the environment through the release of pollutants into the air, which can become trapped in clouds and contaminate surface water and then the groundwater when it rains. Pollutants from industrial manufacturing, landfills, illegal dumps, pesticides used for agriculture, and mining processes also contribute to environmental pollution.

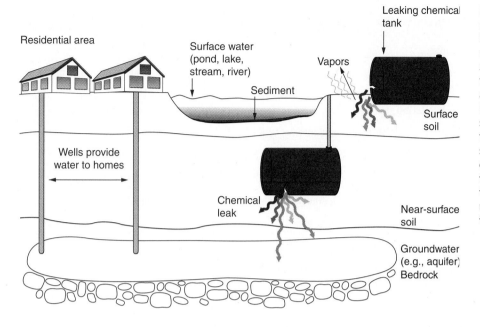

Residential area

Surface water (pond, lake, stream, river)

Sediment

Vapors

Leaking chemical tank

Surface soil

Wells provide water to homes

Chemical leak

Near-surface soil

Groundwater (e.g., aquifer)

Bedrock

Figure 9.1 Treatment Environments and Contamination Zones Chemical spills can create a number of treatment environments and zones of contamination that may be targets for bioremediation. A spill from a leaking chemical tank, which can be located above the ground or below the surface, can release materials that contaminate surface soil, subsurface soil, surface water, and groundwater. In this example, pollution of the aquifer threatens the health of individuals living in houses adjacent to the spill who rely on the aquifer as a source of drinking water.

Because the bioremediation approach used to clean up pollution depends on the treatment environment, cleaning up soil is very different from cleaning up water. How bioremediation is used also depends on the types of chemicals that need to be cleaned up.

Chemicals in the Environment

Techniques for using bioremediation to degrade human wastes at a sewage treatment plant are quite different (and somewhat simpler) than degrading the variety of chemicals that exist in the environment. Everyday household materials such as cleaning agents, detergents, perfumes, caffeine, insect repellents, pesticides, fertilizers, perfumes, and medicines appear in our wastewaters. Increasingly researchers are also finding that U.S. waterways contain prescription and over-the-counter drugs, including contraceptives, painkillers, antibiotics, cholesterol-lowering drugs, antidepressants, anticonvulsants, and anticancer drugs. Other chemicals that make their way into the environment are the products of industrial manufacturing processes or, as discussed earlier, the result of accidents.

Numerous chemicals from many different sources are common pollutants in the environment. Table 9.1 lists some of the most common categories of chemicals in our environment that require cleanup. Many of these chemicals are known to be potential mutagens and **carcinogens,** compounds that cause cancer. Although we do not discuss the health effects of chemical pollutants in any detail, most of these chemicals are known to cause illnesses ranging from skin rashes to birth defects and different types of cancer, as well as

to poison animal and plant life. Quite simply, the presence of pollutants in an environment leads to an overall decline of the environment along with the health of organisms living there.

In addition to the type of spill and the cleanup environment, the type of chemical pollutant also affects what cleanup organisms and approaches can be used for bioremediation. Throughout this chapter we consider strategies for cleaning up many of the pollutants listed in Table 9.1.

Fundamentals of Cleanup Reactions

Microbes can convert many chemicals into harmless compounds either through **aerobic metabolism**—those reactions that require oxygen (O_2)—or through **anaerobic metabolism**—reactions in which oxygen is not required. Both types of processes involve *oxidation* and *reduction reactions*. You must have a basic familiarity with oxidation and reduction reactions if you are to understand biodegradation.

Oxidation and reduction reactions

Oxidation involves the removal of one or more electrons from an atom or molecule, which can change the chemical structure and properties of a molecule. In the case of a chemical pollutant, oxidation can make the chemical harmless by changing its chemical properties. Oxidation reactions often occur together with **reduction reactions.** During reduction, an atom or molecule gains one or more electrons. Because oxidation and reduction reactions frequently occur together,

Table 9.1 TWENTY OF THE MOST COMMON CHEMICAL POLLUTANTS IN THE ENVIRONMENT

Chemical Pollutant	Source
Benzene	Petroleum products used to make plastics, nylon, resins, rubber, detergents, and many other materials
Chromium	Electroplating, leather tanning, corrosion protection
Creosote	Wood preservative to prevent rotting
Cyanide	Mining processes and manufacturing of plastics and metals
Dioxin	Pulp and paper bleaching, waste incineration, and chemical manufacturing processes
Methyl t-butyl ether (MTBE)	Fuel additive, automobile exhaust, boat engines, leaking gasoline tanks
Naphthalene	Product of crude oil and petroleum
Nitriles	Rubber compounds, plastics, and oils
Perchloroethylene/ tetrachloroethylene (PCE), trichloroethene (TCE), and trichloroethane (TCA)	Dry cleaning chemicals and degreasing agents
Pesticides (atrazine, carbamates, chlordane, DDT) and herbicides	Chemicals used to kill insects (pesticides) and weeds (herbicides)
Phenol and related compounds (chlorophenols)	Wood preservatives, paints, glues, textiles
Polychlorinated biphenyls (PCBs)	Electrical transistors, cooling and insulating systems
Polycyclic aromatic hydrocarbons (PAHs) and polychlorinated hydrocarbons	Incineration of wastes, automobile exhaust, oil refineries, and leaking oil from cars
Polyvinylchloride	Plastic manufacturing
Radioactive compounds	Research and medical institutions and nuclear power plants
Surfactants (detergents)	Manufacturing of paints, textiles, concrete, paper
Synthetic estrogens (ethinyl estradiol)	Female hormone (estrogen)-related compounds created by a variety of industrial manufacturing processes
Toluene	Petroleum component present in adhesive, inks, paints, cleaners, and glues
Trace metals (arsenic, cadmium, chromium, copper, lead, mercury, silver)	Car batteries and metal manufacturing processes
Trinitrotoluene (TNT)	Explosive used in building and construction industries

these electron transfer reactions are often called **redox reactions** (see Figure 9.2).

During redox reactions, molecules called **oxidizing agents**—also known as electron acceptors because they have a strong attraction for electrons—remove electrons during the transfer process. When oxidizing agents accept electrons, they become reduced. Oxygen (O_2), iron (Fe^{+3}), sulfate (SO_4^{-2}), and nitrate (NO_3^-) are often involved in redox reactions of bioremediation. Redox reactions are important for many cellular functions. For example, human body cells (and many other cell types) use oxidation and reduction reactions to degrade sugars and derive energy.

Aerobic and anaerobic biodegradation

In some environments, such as surface water and soil where oxygen is readily available, aerobic bacteria degrade pollutants by oxidizing chemical compounds. In aerobic biodegradation reactions, O_2 can oxidize a

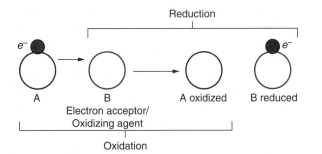

Figure 9.2 Oxidation and Reduction (Redox) Reactions Redox reactions are important for the bioremediation of many chemicals. In this oxidation reaction, an electron is transferred from molecule A to molecule B. Molecule B is acting as an electron acceptor or oxidizing agent. In redox reactions, molecule A is oxidized, and molecule B, which gains an electron, is reduced.

Aerobic biodegradation

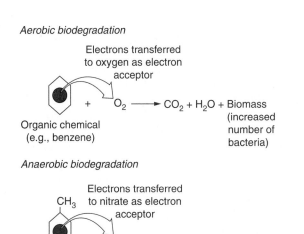

Figure 9.3 Aerobic and Anaerobic Biodegradation Aerobic bacteria (aerobes) use oxygen (O_2) as an electron acceptor molecule to oxidize organic chemical pollutants such as benzene. During this process, oxygen is reduced to produce water (H_2O), and carbon dioxide (CO_2) is derived from the oxidation of benzene. Energy from degrading the pollutant is used to stimulate bacterial cell growth (biomass). Similar reactions occur during anaerobic biodegradation, except that anaerobic bacteria (anaerobes) rely on iron (Fe^{+3}), sulfate (SO_4^{-2}), nitrate (NO_3^-), and other molecules as electron acceptors to oxidize pollutants.

variety of chemicals including **organic molecules** (those that contain carbon atoms) such as petroleum products (Figure 9.3). In the process, O_2 is reduced to produce water. Microbes can further degrade the oxidized organic compound to make simpler and relatively harmless molecules such as carbon dioxide (CO_2) and methane gas. Bacteria derive energy from

this process, which is used to make more cells, thus increasing **biomass.** Some aerobes also oxidize **inorganic compounds** (molecules that do not contain carbon) such as metals and ammonia.

In many heavily contaminated sites and deep subsurface environments such as aquifers, the concentration of oxygen may be very low. In subsurface soils, oxygen may diffuse poorly into the ground, and any oxygen there may be rapidly consumed by aerobes. Even though it is sometimes possible to inject oxygen into treatment sites to stimulate aerobic biodegradation in low oxygen environments, biodegradation may take place naturally via anaerobic metabolism. Anaerobic metabolism also requires oxidation and reduction; however, anaerobic bacteria (anaerobes) rely on molecules other than oxygen as electron acceptors (Figure 9.3). Iron (Fe^{+3}), sulfate (SO_4^{-2}), and nitrate (NO_3^-) are common electron acceptors for redox reactions in anaerobes (Figures 9.3 and 9.4). In addition, many microbes can carry out both aerobic and anaerobic metabolism. When the amount of oxygen in the environment decreases, they can switch to anaerobic metabolism to continue biodegradation. As you will learn in the next section, aerobes and anaerobes are both important for bioremediation.

The Players: Metabolizing Microbes

As you learned in Chapter 5, microorganisms are important for many applications in biotechnology. Scientists can use microbes—especially bacteria—as tools

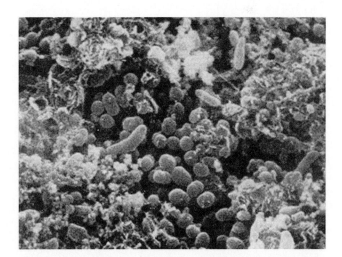

Figure 9.4 Anaerobic Bacteria Effectively Degrade Many Pollutants The dry cleaning agent called perchloroethylene (PCE) is a common contaminant of groundwater; however, anaerobic bacteria can use PCE as food. By growing bacteria on small particles of iron sulfide, which serve as electron acceptors providing the proper chemical environment for anaerobes, bacteria grow rapidly and thrive on PCE.

to clean up the environment. The ability of bacteria to degrade different chemicals effectively depends on many conditions. The type of chemical, temperature, zone of contamination (water versus soil, surface versus groundwater contamination, and so on), nutrients, and many other factors all influence the effectiveness and rates of biodegradation.

At many sites, bioremediation involves the combined actions of both aerobic and anaerobic bacteria to decontaminate the site fully. For instance, anaerobes usually dominate biodegradation reactions that are closest to the source of contamination, where oxygen tends to be very scarce but sulfates, nitrates, iron, and methane are present for use as electron acceptors by anaerobes. Farther from the source of contamination, where oxygen tends to be more abundant, aerobic bacteria are typically involved in biodegradation (Figure 9.5).

The search for useful microorganisms for bioremediation is often best carried out at polluted sites themselves. Anything living in a polluted site will have developed some resistance to the polluting chemicals and may be useful for bioremediation. **Indigenous microbes**—those found naturally at a polluted site—are often isolated, grown, and studied in a lab and then released back into cleanup environments in large numbers. Such microbes are typically the most common and effective "metabolizing" microbes for bioremediation. For instance, different strains of bacteria called *Pseudomonas*, which are very abundant in most soils, are known to degrade hundreds of different

chemicals. Certain strains of *E. coli* are also fairly effective at degrading many pollutants (recall that *E. coli* is an inhabitant of the human gut and an important microbe for many recombinant DNA techniques).

A large number of lesser-known bacteria have been used and are currently being studied for potential roles in bioremediation. For instance, in Section 9.6 we discuss possible applications of *Deinococcus radiodurans*, a microbe that shows an extraordinary ability to tolerate the hazardous effects of radiation. Recently, researchers at the University of Dublin discovered *Pseudomonas putida*, a strain of bacteria that can convert styrene, a toxic component of many plastics, into a biodegradable plastic. Scientists believe that many of the microbes that are most effective at bioremediation have not yet been identified. The search for new metabolizing microbes is an active area of bioremediation research.

Scientists are also experimenting with strains of algae and fungi that may be capable of biodegradation. Waste-degrading fungi such as *Phanerochaete chrysosporium* and *Phanerochaete sordida* can degrade toxic chemicals such as creosote, pentachlorophenol, and other pollutants that bacteria degrade poorly or not at all. Asbestos and heavy-metal degrading fungi include *Fusarium oxysporum* and *Mortierella hyaline*. Fungi are also very valuable in composting and degrading sewage and sludge at solid-waste and wastewater treatment plants, polychlorinated biphenyls (PCBs), and other compounds previously thought to be very resistant to biodegradation.

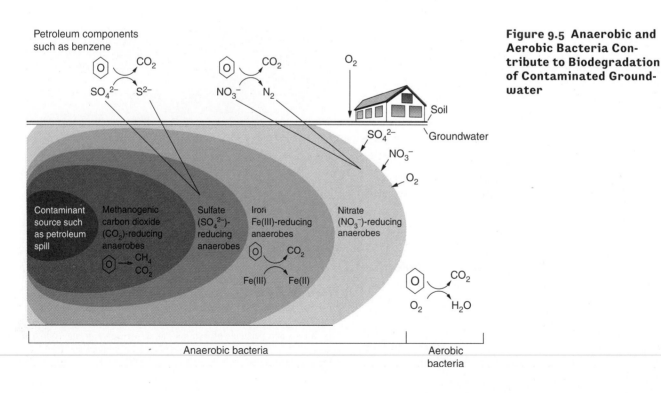

Figure 9.5 Anaerobic and Aerobic Bacteria Contribute to Biodegradation of Contaminated Groundwater

Bioremediation Genomics Programs

It should not surprise you to learn that many scientists are studying the genomes of organisms that are currently or may be used for bioremediation. Through genomics, scientists hope to identify novel genes and metabolic pathways that bioremediation organisms use to detoxify chemicals. Thus they will help scientists develop more effective cleanup strategies including improved strains of bioremediation organisms through genetic engineering (see Section 9.4.). It may also be possible to combine detoxifying genes from different microbes into different recombinant bacteria capable of degrading multiple contaminants at the same time, for example, PCBs and mercury.

As discussed in Chapter 5, the Department of Energy established the Microbial Genome Program (MGP), which has sequenced over 200 microbial genomes including many genomes for microbes involved in bioremediation. See Table 9.2 for a few examples of recently completed genomics projects involving bioremediation organisms. Be sure to visit the Genomes to Life link on the Companion Website, which you can use to access the MGP and other genomics studies involving bioremediation microbes.

Stimulating bioremediation

As discussed previously, bioremediation scientists typically take advantage of indigenous microbes to degrade pollutants. Depending on the pollutant, many indigenous bacteria are very effective at biodegradation. Scientists also use numerous strategies to assist microorganisms in their ability to degrade contaminants depending on the microorganisms involved, the environmental site being cleaned up, and the quantity and the type of chemical pollutants that need to be decontaminated.

Nutrient enrichment, also called **fertilization,** is a bioremediation approach in which fertilizers—similar to phosphorus and nitrogen that are applied to lawns of grass—are added to a contaminated environment to stimulate the growth of indigenous microorganisms that can degrade pollutants (Figure 9.6). Because living organisms need an abundance of key elements such as carbon, hydrogen, nitrogen, oxygen, and phosphorus for building macromolecules, adding fertilizers provides bioremediation microbes with essential elements to reproduce and thrive. In some instances manure, wood chips, and straw may be added to provide microbes with sources of carbon as a fertilizer. Fertilizers are usually delivered to the contaminated site by pumping them into groundwater or mixing them in the soil. The concept behind fertilization is simple. By adding more nutrients, microorganisms replicate, increase in number, and grow rapidly and thus increase the rate of biodegradation.

Bioaugmentation, or **seeding,** is another approach that involves adding bacteria to the contaminated environment to assist indigenous microbes with biodegradative processes. In some cases, seeding may involve applying genetically engineered microorganisms with unique biodegradation properties. Bioaugmentation is not always an effective solution in part because laboratory strains of microbes rarely grow and biodegrade as

Table 9.2 EXAMPLES OF BIOREMEDIATION GENOME PROJECTS UNDERWAY OR RECENTLY COMPLETED

Microorganism	Number of Genes (Year Genome Completed)	Bioremediation Applications
Accumulibacter phosphatis	2006 (not completed)	Major wastewater treatment plant microbe used for removing high phosphates loads from wastewaters and sludge
Alcanivorax borkumensis	2,755 (2006)	Hydrocarbon-degrading marine bacterium that is very effective at degrading many components of crude and refined oil
Dehalobacter restrictus	2006 (not completed)	Dechlorinate perchloroethylene (PCE)
Dehalococcoides ethenogenes	1,591 (2005)	Degrades hydrogen and chlorine. Only known organism to fully dechlorinate perchloroethylene (PCE) and trichloroethene (TCE). PCE and TCE cleanup in wastewaters; polychlorinated dioxin degradation.
Geobacter metallireducens	3,676 (2006)	Subsurface metal reduction, carbon cycling, generate electricity (see Figure 9.13)
Populus trichocarpa (poplar tree)	45,555 (2006)	First tree genome sequenced. Thought to have the largest number of genes for any organism sequenced to date. Potential use for reducing atmospheric carbon dioxide.

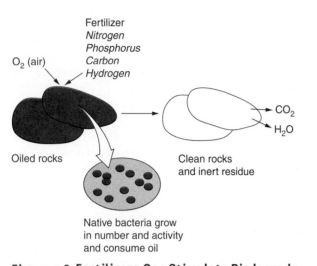

O$_2$ (air)

Fertilizer
Nitrogen
Phosphorus
Carbon
Hydrogen

Oiled rocks

CO$_2$
H$_2$O

Clean rocks
and inert residue

Native bacteria grow
in number and activity
and consume oil

Figure 9.6 Fertilizers Can Stimulate Biodegradation by Indigenous Bacteria Bioremediation of chemicals such as those present in oil can be accelerated by adding fertilizers (e.g., nitrogen and phosphorus). Fertilizers stimulate the growth and replication of indigenous bacteria, which degrade oil into inert (harmless) compounds such as carbon dioxide (CO$_2$) and water (H$_2$0).

well as indigenous bacteria, and scientists must be sure that seeded bacteria will not alter the ecology of the environment if they persist after the contamination is gone.

Phytoremediation

Although bacteria are involved in most bioremediation strategies, a growing number of approaches are utilizing plants to clean up chemicals in the soil, water, and air in an approach called **phytoremediation.** An estimated 350 species of plants naturally take up toxic materials. Poplar and juniper trees have been successfully used in phytoremediation, as have certain grasses and alfalfa. In phytoremediation, chemical pollutants are taken in through the roots of the plant as they absorb contaminated water from the ground (Figure 9.7). As an example, sunflower plants effectively removed radioactive cesium and strontium from ponds at the Chernobyl nuclear power plant in Ukraine. Water hyacinths have been used to remove arsenic from water supplies in Bangladesh, India. This is a significant technology considering arsenic concentrations exceed health standards in 60% of the groundwater in Bangladesh. After toxic chemicals enter the plant, the plant cells may use enzymes to degrade the chemicals. In other cases, the chemical concentrates in the plant cells so that the entire plant serves as a type of "plant sponge" for mopping up pollutants. The contaminated plants are treated as waste and may be burned or disposed of in other ways. Because high concentrations of pollutants often kill most plants, phytoremediation tends to work best where the amount of contamination is low—in shallow soils or groundwater.

Many scientists are also exploring ways in which plants can be used to clean up pollutants in the air, something many plants do naturally: for example, removing excess carbon dioxide (CO$_2$), the principal greenhouse gas released from burning fossil fuels, which contributes to global warming. Recently the

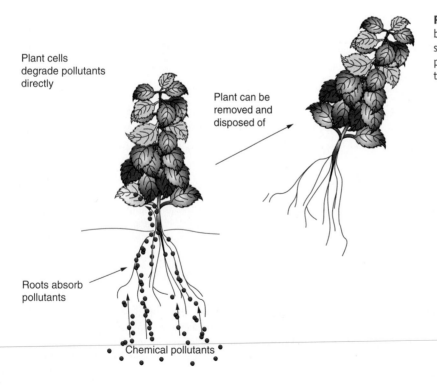

Plant cells
degrade pollutants
directly

Plant can be
removed and
disposed of

Roots absorb
pollutants

Chemical pollutants

Figure 9.7 Phytoremediation Plants can be a valuable addition to many bioremediation strategies. Some plants degrade environmental pollutants directly; others simply absorb pollutants and must be removed and disposed of.

first genome for a tree, the black cottonwood (a type of poplar), was sequenced. Poplars are commonly used for phytoremediation, and genetically engineered poplars have shown promise for capturing high levels of CO_2.

Phytoremediation can be an effective, low-cost, and low-maintenance approach for bioremediation. As an added benefit, phytoremediation can also be a less obvious and more eye-appealing strategy. For instance, planting trees and bushes can visually improve the appearance of a polluted landscape and clean the environment at the same time. Two main drawbacks of phytoremediation are that only surface layers (to around 50 cm deep) can be treated, and cleanup typically takes several years. In the next section we examine specific cleanup environments and different strategies used for bioremediation.

9.3 Cleanup Sites and Strategies

A wide variety of bioremediation treatment strategies exist. Which strategy is employed depends on many factors. Of primary consideration are the types of chemicals involved, the treatment environment, and the size of the area to be cleaned up. Consequently, some of the following questions must be considered before starting the cleanup process:

- Do the chemicals pose a fire or explosive hazard?
- Do the chemicals pose a threat to human health including the health of cleanup workers?
- Was the chemical released into the environment through a single incident, or was there long-term leakage from a storage container?
- Where did the contamination occur?
- Is the contaminated area at the surface of the soil? Below the ground? Does it affect water?
- How large is the contaminated area?

Answering these questions often requires the combined talents of molecular biologists, environmental engineers, chemists, and other scientists who work together to develop and implement plans to clean up environmental pollutants.

Soil Cleanup

Treatment strategies for both soil and water usually involve either removing chemical materials from the contaminated site to another location for treatment, an approach known as *ex situ* **bioremediation,** or cleaning up at the contaminated site without excavation or removal called *in situ* (a Latin term that means

"in place") **bioremediation.** *In situ* bioremediation is often the preferred method of bioremediation in part because it is usually less expensive than *ex situ* approaches. Also, because the soil or water does not have to be excavated or pumped out of the site, larger areas of contaminated soil can be treated at one time. *In situ* approaches rely on stimulating microorganisms in the contaminated soil or water. Those *in situ* approaches that require aerobic degradation methods often involve **bioventing,** pumping either air or hydrogen peroxide (H_2O_2) into the contaminated soil. Hydrogen peroxide is frequently used because it is easily degraded into water and oxygen to provide microbes with a source of oxygen. Fertilizers may also be added to the soil through bioventing to stimulate the growth and degrading activities of indigenous bacteria.

In situ bioremediation is not always the best solution, however. This approach is most effective in sandy soils that are less compact and allow microorganisms and fertilizing materials to spread rapidly. Solid clay and dense rocky soils are not typically good sites for *in situ* bioremediation, and contamination with chemicals that persist for long periods of time can take years to clean up this way.

For some soil cleanup sites, *ex situ* bioremediation can be faster and more effective than *in situ* approaches. As shown in Figure 9.8, *ex situ* bioremediation of soil can involve several different techniques depending on the type and amount of soil to be treated and the chemicals to be cleaned up. One common *ex situ* technique is called **slurry-phase bioremediation.** This approach involves moving contaminated soil to another site and then mixing the soil with water and fertilizers (oxygen is often also added) in large bioreactors in which the conditions of biodegradation by microorganisms in the soil can be carefully monitored and controlled. Slurry-phase bioremediation is a rapid process that works fairly well when small amounts of soil need to be cleaned and the composition of chemical pollutants is well known (Figure 9.8).

For many other soil cleanup strategies, **solid-phase bioremediation** techniques are required. Solid-phase processes are more time-consuming than slurry-phase approaches and typically require large amounts of space; however, they are often the best strategies for degrading certain chemicals. Three solid-phase techniques are widely used: composting, landfarming, and biopiles.

Composting can be used to degrade household wastes such as food scraps and grass clippings, and similar approaches are used to degrade chemical pollutants in contaminated soil. In a compost pile, hay, straw, or other materials are added to the soil to

Ex situ bioremediation

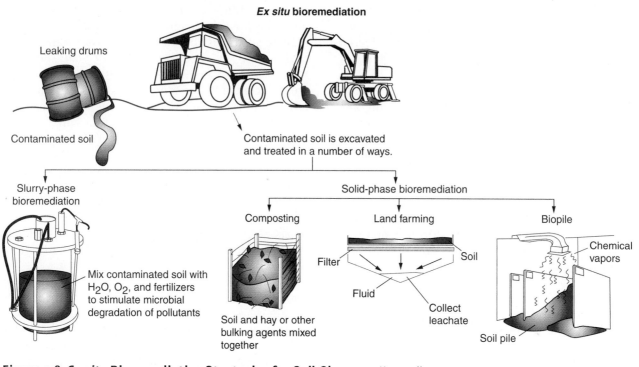

Figure 9.8 labels: Leaking drums; Contaminated soil; Contaminated soil is excavated and treated in a number of ways.; Slurry-phase bioremediation; Solid-phase bioremediation; Composting; Land farming; Biopile; Mix contaminated soil with H_2O, O_2, and fertilizers to stimulate microbial degradation of pollutants; Soil and hay or other bulking agents mixed together; Filter; Fluid; Soil; Collect leachate; Chemical vapors; Soil pile

Figure 9.8 *Ex situ* Bioremediation Strategies for Soil Cleanup Many soil cleanup approaches involve *ex situ* bioremediation in which contaminated soil is removed and then subjected to several different cleanup approaches.

provide bacteria with nutrients that help bacteria degrade chemicals. **Landfarming** strategies involve spreading contaminated soils on a pad so that water and leachates can leak out of the soil. A primary goal of this approach is to collect leachate so that polluted water cannot further contaminate the environment. Because the polluted soil is spread out in a thinner layer than it would be if it were below the ground, landfarming also allows chemicals to vaporize from the soil and aerates the soil so that microbes can better degrade pollutants.

Soil **biopiles** are used particularly when the chemicals in the soil are known to evaporate easily and when microbes in the soil pile are rapidly degrading the pollutants (Figure 9.9). In this approach, contaminated soil is piled up several meters high. Biopiles differ from compost piles in that relatively few bulking agents are added to the soil, and fans and piping systems are used to pump air into or over the pile. As chemicals in the pile evaporate, the vacuum airflow pulls the chemical vapors away from the pile and either releases them into the atmosphere or traps them in filters for disposal, depending on the type of chemical. Almost all *ex situ* strategies for cleaning up soil involve tilling and mixing soil to disperse nutrients, oxygenate the dirt, and increase the interaction of microbes with contaminated materials to increase biodegradation.

Figure 9.9 Soil Biopiles Polluted soil that has been removed from the cleanup site can be stored in piles and bioremediation processes monitored to ensure decontamination of the soil before determining whether the soil can be returned to the environment.

Bioremediation of Water

Contaminated water presents a number of challenges. In Section 9.5, we consider how surface water can be treated following large spills such as an oil spill. Wastewater and groundwater can be treated in many different ways, depending on the materials that need to be removed by bioremediation.

Wastewater treatment

Probably the most well-known application of bioremediation is the treatment of wastewater to remove human sewage (fecal material and paper wastes), soaps, detergents, and other household chemicals. Both septic systems and municipal wastewater treatment plants rely on bioremediation. In a typical septic system, human sewage and wastewaters from a single household move through the plumbing system out to a septic tank buried below the ground adjacent to the house. In the tank, solid materials such as feces and paper wastes settle to the bottom to be degraded by microbes while liquids flow out of the top of the tank and are dispersed underground across an area of soil and gravel called the septic bed. Within the bed indigenous microbes degrade waste components in the water.

One commercial application of bioremediation recommended to prevent clogged septic tanks is to add products such as Rid-X (Figure 9.10), which are flushed into the system periodically. These products contain freeze-dried bacteria that are rich in enzymes such as lipases, proteases, amylases, and cellulases, which in turn degrade fats, proteins, sugars, and cellulose in papers and vegetable matter, respectively. Adding microbes this way is an example of the bioaugmentation we discussed in the previous section, and this treatment helps the septic system degrade cooking fats and oils, human wastes, tissue paper products, and other materials that can clog the system.

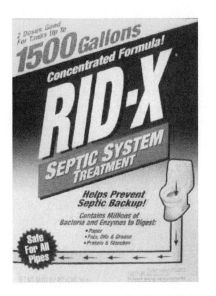

Figure 9.10 Septic Tank Additives Stimulate Bioremediation of Household Wastes

Wastewater (sewage) treatment plants are fairly complex and well-organized operations (Figure 9.11). Water from households that enters sewer lines is pumped into a treatment facility where feces and paper products are ground and filtered into smaller particles, which settle out into tanks to create a mud-like material called **sludge.** Water flowing out of these tanks is called **effluent.** Effluent is sent to aerating tanks where aerobic bacteria and other microbes oxidize organic materials in the effluent. In these tanks, water is sprayed over rocks or plastic covered with biofilms of waste-degrading microbes that actively degrade organic materials in the water.

Alternatively, effluent is passed into activated sludge systems—tanks that contain large numbers of waste-degrading microbes grown in carefully controlled environments. Usually these microbes are free floating within the water, but in some cases they may be grown on filters through which contaminated water flows. Eventually effluent is disinfected with a chlorine treatment before the water is released back into rivers or oceans.

Sludge is pumped into anaerobic digester tanks in which anaerobic bacteria further degrade the sludge. Methane-producing and carbon dioxide gas–producing bacteria are common in these tanks. Methane gas is often collected and used as fuel to run equipment at the sewage treatment plant. Tiny worms, which are usually present in the sludge, also help break down the sludge into small particles. Sludge is never fully broken down, but once the toxic materials have been removed, it is dried and can be used as landfill or fertilizer.

Scientists have recently discovered bacteria called *Candidatus 'Brocadia anammoxidans'* that possess a unique ability to degrade ammonium, a major waste product present in urine (Figure 9.12). Removing ammonium from wastewater before the water is released back into the environment is important because high amounts of ammonium can affect the environment by causing algal blooms and diminishing oxygen concentrations in waterways. Typically, wastewater plants rely on aerobic bacteria such as *Nitrosomonas europaea* to oxidize ammonium in a multistep set of reactions. However, *Candidatus 'Brocadia anammoxidans'* is capable of degrading ammonium in a single step under anaerobic conditions, a process called the anammox process. Wastewater treatment plants may soon be using this strain so that ammonium is removed from wastewater more efficiently.

Groundwater cleanup

With the exception of spills near coastal beaches, most large chemical spills such as oil spills occur in marine environments often far away from populated areas. However, freshwater pollution typically occurs closer

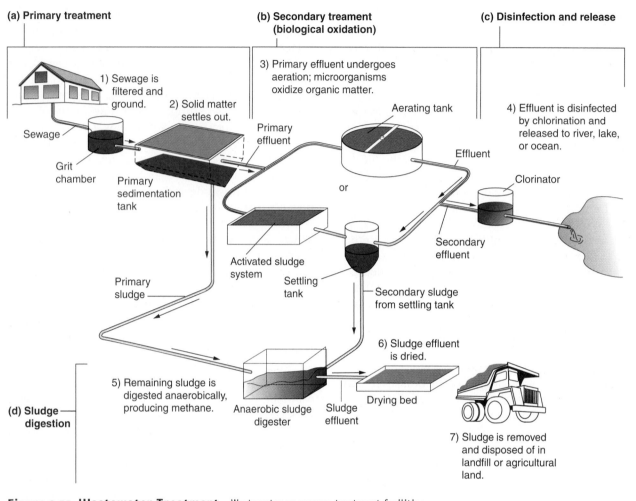

(a) Primary treatment

1) Sewage is filtered and ground.

2) Solid matter settles out.

Sewage

Grit chamber

Primary sedimentation tank

Primary sludge

(b) Secondary treament (biological oxidation)

3) Primary effluent undergoes aeration; microorganisms oxidize organic matter.

Aerating tank

Primary effluent

Effluent

or

Activated sludge system

Settling tank

Secondary effluent

(c) Disinfection and release

4) Effluent is disinfected by chlorination and released to river, lake, or ocean.

Clorinator

(d) Sludge digestion

5) Remaining sludge is digested anaerobically, producing methane.

Anaerobic sludge digester

Secondary sludge from settling tank

6) Sludge effluent is dried.

Drying bed

Sludge effluent

7) Sludge is removed and disposed of in landfill or agricultural land.

Figure 9.11 Wastewater Treatment Wastewater or sewage treatment facilities are well-planned operations that use aerobic and anaerobic bacteria to degrade organic materials such as human feces and household detergents in both the sludge and water (effluent).

to populated areas and poses a serious threat to human health by contaminating sources of drinking water, either groundwater or surface water such as reservoirs. Groundwater contamination is a common problem in many areas of the United States. Drinking water for approximately 50% of the U.S. population comes from groundwater sources, and, according to some estimates, a large percentage of groundwater supplies in the United States contain pollutants that may have an impact on human health. Polluted groundwater can sometimes be very difficult to clean up because contaminated water often gets trapped in soil and rocks and there is no easy way to "wash" aquifers.

Ex situ and *in situ* approaches are often used in combination. For instance, when groundwater is contaminated by oil or gasoline, these pollutants rise to the surface of the aquifer. Some of this oil or gas can be directly pumped out, but the portion mixed with

groundwater must be pumped to the surface and passed through a bioreactor (Figure 9.13). Inside the bioreactor, bacteria in biofilms growing over a screen or mesh degrade the pollutants. Fertilizers and oxygen are often added to the bioreactor. Clean water from the bioreactor containing fertilizer, bacteria, and oxygen is pumped back into the aquifer for *in situ* bioremediation (Figure 9.13).

Turning Wastes into Energy

Many landfills across the country are stressed to their limits, literally overflowing with trash from homes and businesses. The bulk of our household trash consists of food scraps, boxes, paper waste, cardboard packing containers from food, and similar items. A variety of chemical wastes such as detergents, cleaning fluids, paints, nail polish, and varnishes also make their way into the trash, despite the

Figure 9.12 Bioreactor Containing *Candidatus* **'Brocadia anammoxidans,' Anaerobic Bacteria That Can Degrade Ammonium** Novel metabolic properties enable these anaerobes to degrade ammonium from wastewater in a single step.

fact that most states are trying to reduce the amount of chemical wastes that can be disposed of as regular garbage.

Scientists are working on strategies to reduce waste, including bioreactors that contain anaerobic bacteria that can convert food waste and other trash into soil nutrients and methane gas. Methane gas can be used to produce electricity, and soil nutrients can be sold commercially as fertilizer for use by farms, nurseries, and other agricultural industries. They are also working on seeding strategies that may be used to reduce chemicals in landfills that would otherwise seep through the ground and contaminate local ground and surface waters. If successful, these applications of biotechnology may help reduce the amount of waste and greatly increase the usable space of many landfills.

Bioremediation scientists are also studying polluted sediments in sewage sludge and at the bottoms of oceans and lakes as an untapped source of energy. Sediment in lakes and oceans is rich in organic materials from the breakdown of decaying materials such as leaves and dead organisms. Within this "muck" are anaerobes that use organic molecules in the sediment to generate energy. The term *electicigens* is being used to describe electricity-generating microbes that have the ability to oxidize organic compounds to carbon dioxide and transfer electrons to electrodes. *Desulfuromonas acetoxidans* is an anaerobic marine bacterium that uses iron as an electron acceptor to oxidize organic molecules in sediment. Researchers are exploring ways in which electrons can be harvested from *D. acetoxidans* and other bacteria such as *Geobacter metallireducens* and *Rhodoferax ferrireducens* as a technique for capturing energy in *bacterial biobatteries* that can be used as fuel cells to provide a source of electricity (Figure 9.14). Researchers at the University of Massachusetts have demonstrated that electrigenic strains of microbes can generate electricity with high efficiency and although preliminary studies suggest that this technique has some promise, more research needs to be done (Figure 9.15).

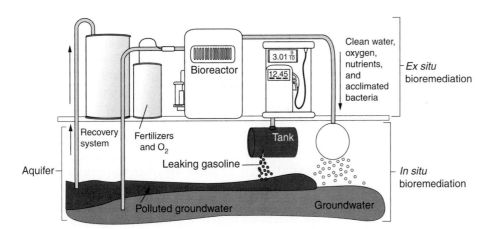

Figure 9.13 *Ex situ* **and** *in situ* **Bioremediation of Groundwater** Gasoline leaking into an underground water supply can be cleaned up using an aboveground (*ex situ*) system in combination with *in situ* bioremediation.

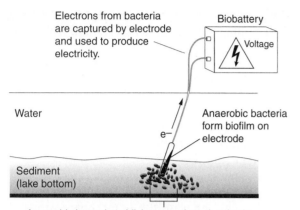

Figure 9.14 numbered labels:

Electrons from bacteria are captured by electrode and used to produce electricity.

Biobattery

Voltage

Water

e⁻

Anaerobic bacteria form biofilm on electrode

Sediment (lake bottom)

Anaerobic bacteria oxidizing organic molecules in sediment transfer electrons to electron acceptor molecules such as iron and sulfur.

Figure 9.14 Polluted Sediments May Be an Untapped Source of Energy   Scientists have found that anaerobic bacteria in sediment may be a source of energy. Because these bacteria use redox reactions to degrade molecules in sediment, electrons can be captured by electrodes, which can transfer electrons to generators to create electricity.

Figure 9.15 Microbe Powered Fuel Cells   Researchers at the University of Massachusetts have demonstrated that *Geobacter* fuel cells can effectively convert sugars into electricity.

Even though bioremediation strategies have effectively cleaned up many environmental pollutants, bioremediation is not the solution for all polluted sites. For instance, bioremediation is ineffective when the polluted environment contains high concentrations of very toxic substances such as heavy metals, radioactive compounds, and chlorine-rich organic molecules because these compounds typically kill microbes. Therefore, new strategies will need to be discovered and applied to tackle some of these cleanup challenges. Some of these new bioremediation strategies are likely to involve genetically engineered microorganisms, the topic of the next section.

9.4 Applying Genetically Engineered Strains to Clean Up the Environment

Bioremediation has traditionally relied on stimulating naturally occurring microorganisms; however, many indigenous microbes cannot degrade certain types of chemicals, especially very toxic compounds. For example, some organic chemicals produced during the manufacturing of plastics and resins are resistant to biodegradation and can persist in the environment for several hundred years. Many radioactive compounds also kill microbes, thus preventing biodegradation. To clean up some of these stubborn and particularly toxic pollutants, we may need to use bacteria and plants that have been genetically altered. The development of recombinant DNA technologies has enabled scientists to create genetically engineered organisms capable of improving bioremediation processes.

Petroleum-Eating Bacteria

The first effective genetically altered microbes for use in bioremediation were created in the 1970s by Ananda Chakrabarty and his colleagues at General Electric. This work was carried out before DNA cloning and recombinant DNA technologies were widely available, so how did Chakrabarty do this? He was able to isolate strains of *Pseudomonas* from soils contaminated with different types of chemicals including pesticides and crude oil. He then identified strains that showed the ability to degrade such organic compounds as naphthalene, octane, and xylene. Most of these strains could grow in the presence of these compounds because they contained plasmids that encoded genes for breaking down each component.

Chakrabarty mated these different strains and eventually produced a strain that contained several different plasmids. Together, the combined proteins produced by these plasmids could effectively degrade many of the chemical components of crude oil. For his work, Chakrabarty was awarded the first U.S. patent for a genetically altered living organism. However, this patent decision was very controversial and was held up in the courts for about 10 years. The primary issues being debated were whether life-forms could be patented and whether Chakrabarty's recombinant bacteria should be considered a product of nature or an

Microcosms Provide Major Benefits

Scientists are continually researching innovative ways to biodegrade different chemical compounds under a myriad of environmental conditions. Industry is continually creating new kinds of chemicals, and many of these will inevitably make their way into the environment. To stay one step ahead of new environmental pollutants, bioremediation researchers must continue to develop new cleanup strategies.

How can scientists study bioremediation of a new chemical that has never made it into the environment? They obviously cannot pollute large areas nor wait for a large-scale disaster like the Exxon *Valdez* spill before testing their theories about how to clean up this new chemical. One of the most practical approaches to learning about new bioremediation strategies is to make a **microcosm,** an artificially constructed test environment designed to mimic real-life environmental circumstances. Some microcosms consist of small bioreactors—about the size of a 5-gallon bucket—containing soil, water, pollutants, and microbes to be tested for their bioremediation abilities. A microcosm may be as small as a few grams of soil in a test tube, but they are more often scaled up to resemble larger environments. For example, large ponds, which may be indoors or outdoors, or soil plots that are prepared to prevent escape of pollutants outside of the test facility can be used as microcosms. By carefully designing microcosms, bioremediation researchers can attempt to simulate, in a small model scale, a site that needs to be cleaned up.

When indigenous or genetically engineered organisms with bioremediation potential have been identified, studies in bioreactors or on small isolated areas of land or water can be crucial for determining whether these organisms will effectively clean up pollution in larger settings. By carefully manipulating the environmental conditions in the microcosms, scientists can test the ability of organisms to degrade different pollutants under varying conditions including moisture, temperature, nutrients, oxygen, pH, and soil type.

Microcosm studies may even involve testing experimental microbes on polluted groundwater or contaminated soil that is placed into the microcosm so the rates of degradation can be monitored and the cleanup time can be evaluated. Scientists can also carry out experiments to study the bioremediation of mixtures of pollutants at the same time.

In an attempt to produce the best cleanup results, scientists will analyze data gathered and design new experiments. A cleanup approach that demonstrates success in a microcosm is not guaranteed to succeed in the field. Nevertheless, by testing bioremediation approaches in microcosms, much valuable time and money can be saved before deciding whether a cleanup approach is likely to have any chance at succeeding in cleaning up a polluted environment in the field.

---

invention. Eventually the Supreme Court ruled that the development of recombinant *Pseudomonas* was an invention worthy of a patent.

Chakrabarty's approach was not as effective as it might sound. Crude oil contains thousands of compounds, and his recombinant bacteria could degrade only a few of these. The majority of the chemicals in crude oil remain largely unaffected by recombinant organisms. Consequently, developing recombinant bacteria with different degradative properties is an intense area of research. In the future, a useful approach for cleaning up crude oil may be to release multiple bacterial strains, each with the ability to degrade different compounds in the oil.

Engineering *E. coli* to Clean Up Heavy Metals

Heavy metals including copper, lead, cadmium, chromium, and mercury can critically harm humans and wildlife. Mercury is an extremely toxic metal that can contaminate the environment. It is used in manufacturing plants, batteries, electrical switches, medical instruments, and many other products. Mercury, and a related compound called methylmercury, can accumulate in organisms through a process called **bioaccumulation.** In bioaccumulation, organisms higher up on the food chain contain higher concentrations of chemicals than organisms lower on the food chain. For instance, in a water supply, mercury may be ingested by small fish, which may then be eaten by birds, large fish, otters, raccoons, and other animals including humans. Large fish and birds need to eat a lot of small fish; therefore, they accumulate more mercury in their systems than small fish and birds that eat less. Similarly, if a person was to eat large fish as a primary source of food, that person would accumulate high amounts of mercury over time. Regular consumption of fish and shellfish contaminated with mercury and methylmercury poses serious health threats to humans including birth defects and brain damage.

For these reasons, in many areas of the United States, health officials suggest that pregnant women and young children eat only small amounts of certain types of fish, such as swordfish and fresh tuna, and restrict these meals to no more than one serving a week.

Because mercury is toxic at very low doses, most current strategies for removing mercury from contaminated water supplies do not remove enough mercury to meet acceptable standards. Scientists have developed genetically engineered strains of *E. coli* that may be useful for cleaning up mercury and other heavy metals. They have also identified naturally occurring metal-binding proteins in plants and other organisms. Two of the most well-characterized types of proteins—metallothioneins and phytochelatins—have a high capacity for binding to metals. For these proteins to function, however, metals must enter cells. Scientists have engineered *E. coli* to express transport proteins that allow for the rapid uptake of mercury into the bacterial cell cytoplasm where the mercury can bind to metal-binding proteins.

Some of these genetically altered bacteria can absorb mercury directly; others that bind mercury can be grown on biofilms to act as sponges for soaking up mercury from a water supply. The biofilms must be changed periodically to remove mercury-containing bacteria. Similarly, genetically engineered single-cell algae containing metallothionein genes and bacteria called cyanobacteria have shown promise for their ability to absorb cadmium, another very toxic heavy metal known to cause many serious health problems in humans.

Biosensors

Researchers have developed genetically engineered strains of the bacteria *Pseudomonas fluorescens* that can effectively degrade complex structures of carbon and hydrogen called polycyclic aromatic hydrocarbons (PAHs) and other toxic chemicals. Using recombinant DNA technology, scientists have been able to splice bacterial genes that encode enzymes that can metabolize these contaminants to reporter genes such as the *lux* genes from bioluminescent marine bacteria. Recall from Chapter 5 that *lux* genes are often used as reporter genes because they encode the light-releasing enzyme luciferase. As PAHs are degraded, bacteria release light that can be used to monitor biodegradation rates. Similar techniques are being used to develop **biosensors** from recombinant bacteria containing *lux* genes. These types of approaches are being widely used to develop many different types of biosensors capable of detecting a variety of environmental pollutants. In the future, genetically engineered microbes will probably play a greater role as biosensors

and as key players in many bioremediation applications. In the next section, we consider two of the most well-studied and highly publicized examples of bioremediation in action.

Genetically Modified Plants and Phytoremediation

In recent years scientists have also been working on genetically modifying plants to improve their phytoremediation capabilities. One promising area of this research has been developing plants that can remove chemicals from military explosives and weapons firing ranges that have contaminated soil and groundwater. Hexahydro-1,3,4-trinitro-1,3,5-triazine (commonly called royal demolition explosive, RDX) and 2,4,6-trinitrotoluene (TNT) are two of the most common chemical contaminants that result from the production, use, and disposal of explosives. The chemicals move readily through soils and cause contamination of groundwater. Both RDX and TNT are highly toxic to most organisms and pose significant health threats to wildlife and humans.

Notice that TNT was one of the top 20 chemicals listed in Table 9.1 and that the EPA lists both TNT and RDX as priority chemicals to remove from the environment. These contaminants are a major pollution problem worldwide. Incredibly, there are hundreds of tons of these compounds in sites around the world and tens of thousands of acres deemed unsafe as a result.

A few bacteria and plants have been shown to weakly degrade TNT with low efficiency; degradation of RDX is even less effective because of its chemical structure. In the last few years, however, scientists have used genetic engineering to create transgenic plants that may turn out to be very effective for phytoremediation of TNT and RDX. A transgenic strain of tobacco containing a nitroreductase gene from *Enterobacter cloacae* effectively converts TNT to less toxic chemicals (Figure 9.16). Most recently, scientists incorporated a gene called *xplA* from the bacteria *Rhodococcus rhodochrous* into *Arabidopsis thaliana*. The *xplA* gene produces an RDX-degrading enzyme called cytochrome P450, which can degrade RDS once it is absorbed into the plant (see Figure 9.16). It may now be possible to make plants that can degrade both TNT and RDX and since the poplar genome has been sequenced, bioremediation scientists working on genetically modified plants are very enthusiastic about the possibility of making transgenic poplars and other fast-growing, deep-rooted trees that can remediate TNT and RDX well below the surface soil. In the next section, we consider three of the most well-studied and highly publicized examples of bioremediation in action.

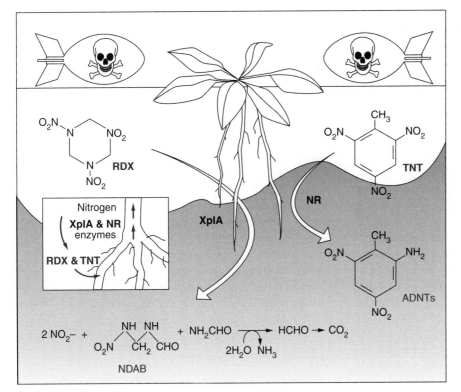

Figure 9.16 Phytoremediation of Toxic Explosives Using Transgenic Plants Transgenic plants engineered with the *XplA* or *NR* gene to degrade either RDX or TNT, respectively, absorb the explosive chemicals through their roots and then degrade them into less toxic compounds (aminodinitrotoluenes [ADNTs] or 4-nitro-2,4-diazabutanal [NDAB]) or nontoxic compounds (NH_3, CO_2). Plants engineered with the *XplA* gene use the degradation products from RDX to stimulate their growth.

9.5 Environmental Disasters: Case Studies in Bioremediation

Most bioremediation approaches involve the cleanup of contaminated areas that are relatively small, perhaps several hundred acres in size. Nevertheless, a great deal has been learned about the effectiveness of different bioremediation strategies by studying large-scale environmental disasters that have been treated by bioremediation.

Jet Fuel and Hanahan, South Carolina

In Hanahan, South Carolina, a suburb of Charleston, a leak of approximately 80,000 gallons of kerosene-based jet fuel occurred in 1975. Despite a number of cleanup measures, the spill could not be contained from soaking through the sandy soil and contaminating the water table. Ten years later contamination had reached a residential area. Removing contaminated soil was impractical because of the large area of contamination and removing contaminated groundwater would not eliminate the problem.

Scientists at the U.S. Geological Survey (USGS) determined that indigenous microbes in the soil were degrading toxic compounds in the jet fuel, and they also found that adding fertilizers to these microbes dra-

matically stimulated rates of biodegradation. This site was one of the first field applications of bioremediation. USGS scientists added nutrients to the groundwater and simultaneously removed contaminated groundwater. By the early 1990s, contamination in the groundwater supply was reduced by 75% and bioremediation had demonstrated its potential worth as an effective cleanup strategy.

The Exxon *Valdez* Oil Spill

The world relies heavily on crude oil and petroleum products that can be manufactured from oil. Petroleum products are used not just as gasoline and diesel fuel to power automobiles but also as the basis for hundreds of everyday products including plastics, paints, cosmetics, and detergents. The United States alone uses in excess of 250 billion gallons of oil each year, over 100 billion gallons of which are imported. When crude oil is transported, some amount of leakage almost always occurs. Tanker accidents spill nearly 100 million gallons of crude oil each year, and even larger volumes of oil enter our seas through naturally occurring spills and leaks.

Oil spills have had a tremendous impact on the environment, specifically on large numbers of wildlife (Figure 9.17a). Typically large spills do not have a great effect on human life directly. This is because most large spills occur in open oceans or bays far removed from swimming beaches and water supplies. Humans are more affected by small local spills, such as a leaking underground tank from a gas station that may threaten local drinking water supplies.

In 1989, the Exxon *Valdez* oil tanker ran aground on a reef in Prince William Sound off the coast of Alaska, releasing approximately 11 million gallons of crude oil and contaminating over 1,000 miles of the Alaskan shoreline. Many experimental approaches for bioremediation were implemented to clean up this spill. In many ways, Prince William Sound was a large bioremediation laboratory for trying new cleanup strategies.

As is done at most sizable oil spills, physical cleaning measures were first used to contain and remove large volumes of oil. Containment booms or skimmers—which may be surface nets, fences, or inflatable devices like buoys that float on the surface, are fixed in place, or are towed behind a boat to collect and contain oil (Figure 9.17b)—were used first. Then vacuums were used to pump oil from the surface into disposal tanks. Beaches and rocks were washed with fresh water under high pressure to disperse oil. By diluting and dissolving the oil in the sea, the oil gradually dispersed. One biotechnology cleanup application used citrus-based products to bind crude oil and allow it to be collected on absorbent pads. But after using all these physical approaches to remove the bulk of the

oil, millions of gallons of oil still remained attached to sand, rocks, and gravel both at the surface and below the surface of contaminated shorelines. This is when bioremediation went to work.

As the first step in the bioremediation process, nitrogen and phosphorus fertilizers were applied to the shoreline to stimulate oil-degrading bacteria (mostly strains of *Pseudomonas*) that were living in the sand and rocks (Figure 9.18). Indigenous bacteria immediately showed signs that they were degrading the oil. When microbes degrade petroleum products, PAHs are formed and eventually oxidized into carbon chains that can be broken down into carbon dioxide and water. Over time, chemical tests on oil from the shoreline soil showed significant changes in chemical composition, indicating that natural degradation by indigenous bacteria was working (Figure 9.19). It may take hundreds of years to clean up this oil fully, and some areas of the Alaskan environment may never return to their natural state prior to the spill.

Oil Fields of Kuwait

The deserts of Kuwait are literally a living laboratory for studying bioremediation. Ten years after the Gulf War, large areas of Kuwait's desert remain soaked in oil. During the Iraqi occupation of Kuwait from 1990 to 1991, countless oil fields were destroyed and burned, releasing an estimated 250 million gallons of

(a)

(b)

Figure 9.17 Oil Spills Pose Serious Threats to the Environment   (a) Oil spills typically have the greatest impact on wildlife. (b) Containment booms are used initially to control the spread of oil and minimize pollution of the surrounding area.

Figure 9.18 Applying Fertilizers to Stimulate Oil-Degrading Microbes Cleanup workers spray nitrogen-rich fertilizers onto an oil-soaked beach in Alaska following the Exxon *Valdez* oil spill. The fertilizers greatly accelerate the growth of natural bacteria that will degrade the oil.

Figure 9.19 Bioremediation Works The effects of bioremediation of an oil spill on a beach in Alaska are clearly evident in this photo. The portion of beach on the right was treated with fertilizers to stimulate biodegradative activities of indigenous microbes; the section of beach shown on the left was untreated.

oil into the deserts—more than 20 times the amount of oil spilled in the Exxon *Valdez* accident. Kuwaiti scientists studying these spills have found that the spilled oil has severely affected plant and animal life in many contaminated areas. Some plant species have been completely eliminated, and the long-term biological impacts will not be known for many more years.

In contrast to the *Valdez* spill, bioremediation of desert soils poses a number of different problems. Unlike the spill in Alaska, there are no waves to help

disperse and dissolve oil. Dry soil conditions of the desert also tend to harbor fewer strains of oil-metabolizing microbes, and adhesion of oil to sand and rocks slows natural degradation processes. Preliminary studies suggest that novel strains of oil-degrading bacteria are slowly working to break down oil below the surface of the sand.

The Kuwaiti government has developed a $1 billion program to address what is probably the world's largest bioremediation project. There are no previously studied sites of this size to use as a model for cleaning up arid desert environments, so bioremediation scientists from around the world are studying Kuwait's deserts with hopes of developing strategies for cleaning up these oil-drenched sands. Information that scientists learn from studying this region will certainly be of value for treating oil spills in other sandy environments. The next section provides a glimpse of how potential applications of bioremediation may be able to solve cleanup problems that have been difficult to treat.

9.6 Future Strategies and Challenges for Bioremediation

Biotechnology is a significant tool in our fight to rehabilitate areas of the environment that have been polluted through accidents, industrial manufacturing, and

Q Will the environment and wildlife of Prince William Sound ever recover from the effects of the Exxon *Valdez* spill?

A Hundreds of scientists are interested in the answer to this question. While bioremediation scientists study soil composition from contaminated areas, wildlife and fisheries biologists are carrying out many studies on marine birds, mammals, and fish. Scientists are looking at the reproductive abilities and breeding habits of certain species, migration of fish and birds, fat tissue analysis of harbor seals to identify oil-based pollutants, and the genetic effects of the spill on many species. Over 10 years after the Exxon *Valdez* spill, there are signs that the cleanup efforts have restored some areas of shoreline to pre-spill conditions, but it is also clear that many fish and wildlife species have not recovered, and the lasting effects on wildlife are still not known. Even though over $200 million and the efforts of thousands of cleanup workers have gone toward cleaning up Prince William Sound, bioremediation and time are still the critical factors that will provide long-term cleanup of the environment.

mismanagement of ecosystems. Bioremediation is a rapidly expanding science. Scientists around the world are carrying out research aimed at developing a greater understanding of the microorganisms involved in biodegradative processes, including the identification of novel genes and proteins involved in these breakdown processes. Researchers are studying microbial genetics to create genetically engineered microbes that may be able to degrade new types of chemicals, and they are developing novel biosensors to detect and monitor pollution.

Recovering Valuable Metals

The recovery of valuable metals such as copper, nickel, boron, and gold is another area of bioremediation that has yet to reach its full potential. Through oxidation reactions, many microbes can convert metal products into insoluble substances called metal oxides or ores that will accumulate in bacterial cells or attach to the bacterial cell surface. Some marine bacteria that live in deep-sea hydrothermal vents have also shown potential for precipitating precious metals. Using bacteria as a way to recover hazardous metals as part of industrial manufacturing processes is one potential application, but scientists are interested in ways these bacteria can be used to recover valuable precious metals. For instance, many manufacturing processes use silver and gold plating techniques that create waste solutions with suspended particles of silver and gold. Microbes may be used to recover some of these metals from waste solutions. Similarly, microbes may also be used to harvest gold particles from underground water supplies and cave water found in gold mines. Many bacterial strains

---

YOU DECIDE

The PCB Dilemma of the Hudson River

Winding through upstate New York, the Hudson River valley is some of the most beautiful country in the Northeast, but serious problems lurk below the glimmering surfaces of the blue water. From 1947 to 1977, the General Electric Company released over 1.2 million pounds of toxic chemicals called polychlorinated biphenyls into the river from facilities in Hudson Falls and Fort Edward, New York. PCBs were commonly used in transformer boxes, capacitors, and cooling and insulating fluids of electrical equipment manufactured before 1977. They are no longer manufactured in the United States, and their production has been banned in most of the world.

PCBs are very toxic to humans and wildlife because these fat-soluble chemicals gradually accumulate in fatty tissues. Hudson River fish are contaminated with PCBs at concentrations in excess of safe levels. Consumption of fish from most areas of the Hudson River is banned or restricted, but some people ignore these publicized restrictions because the water looks so clear and the fish do not *look* polluted. PCBs present serious health threats to humans, and prolonged exposure to PCBs is known to cause cancer, reproductive problems, and other medical conditions affecting the immune system and thyroid gland. Children are particularly susceptible to the health effects of PCBs.

High concentrations of these chemicals still lurk in the Hudson River and other environments. In the Hudson River, most PCBs are located in the sediment at the bottom of the river. To rid the river of these persistent chemicals, environmental groups and the EPA have proposed a 5- to 6-year timetable for the dredging and removal of over 2.5 million cubic yards of sediment from the different segments of the 175-mile-long river, which would remove more than 100,000 pounds of PCBs. Critics of the dredging plan suggest that dredging operations will release more PCBs into the water by stirring up the sediment. Instead of dredging, many believe that leaving the sediment in place and letting natural current flows disperse the chemicals, combined with bioremediation through bacterial degradation of PCBs, is the best way to reduce the load of PCBs in the long run.

PCB-degrading anaerobes have been detected in Hudson River sediments. Some anaerobic bacteria are involved in the first step of breaking down PCBs by cleaving off chlorine and hydrogen groups. Aerobic bacteria in the water can further modify PCBs, and others can then convert them into water, carbon dioxide, and chloride. Some predictions suggest that even after dredging, PCB levels in fish will not drop to levels acceptable for human consumption until after 2070. Even if dredging is a good plan, where will the dredged sediments go? How will these chemicals be cleaned up? Some people believe that placing PCB-laden sediments in a sealed landfill, a completely anaerobic environment, will slow down the degradative processes. Should these chemicals be left in the mud to be broken down slowly over time through natural bioremediation, or should humans intervene in an effort to speed up nature's cleanup effort? What are the pros and cons of taking action or doing nothing? You decide.

that live in such environments are actively being studied for these purposes. Plants may also provide a way to harvest metals from the environment. The wild mustard *Thlaspi goesingense*, native to the Austrian Alps, is known to accumulate metals in storage vacuoles.

Bioremediation of Radioactive Wastes

Another area of active research involves developing bioremediation approaches to remove radioactive materials from the environment. Uranium, plutonium, and other radioactive compounds are found in water from mines where naturally occurring uranium is processed. Radioactive wastes from nuclear power plants also present a disposal problem worldwide. The U.S. Department of Energy (DOE) has identified over 100 sites in 30 states that have been contaminated by weapons production and nuclear reactor development. Radioactive waste sites often have a complex mix of radioactive elements such as plutonium, cesium, and uranium along with mixtures of heavy metals and organic pollutants such as toluene. Although most radioactive materials kill a majority of microbes, some strains of bacteria have demonstrated a potential for degrading radioactive chemicals. For example, researchers at the University of Massachusetts have discovered that some species of *Geobacter* can reduce soluble uranium in groundwater into insoluble uranium effectively, immobilizing the radioactivity. However, to date, no bacterium has been discovered that can completely metabolize radioactive elements into harmless products.

One strain in particular called *Deinococcus radiodurans* is particularly fascinating. Its name literally means "strange berry that withstands radiation." Named the world's toughest bacterium by *The Guinness Book of World Records*, *D. radiodurans* was first identified and isolated about 50 years ago from a can of ground beef that had spoiled even though it had been sterilized with radiation. Subsequently, scientists discovered that *D. radiodurans* can endure doses of radiation over 3,000 times higher than other organisms including humans. High doses of radiation create double-stranded breaks in DNA structure and cause mutations, yet *D. radiodurans* shows incredible resistance to the effects of radiation. Although its resistance mechanisms are not entirely known, this microbe clearly possesses elaborate systems for folding its genome to minimize damage from radiation, and it also uses novel DNA repair mechanisms to replace damaged copies of its genome. Recent completion of the *D. radiodurans* genome map is expected to provide valuable insight into its unique DNA-repair genes.

The DOE is very interested in using *D. radiodurans* and another bacterium, *Desulfovibrio desulfuricans*, for bioremediation of radioactive sites. Recently, researchers at the DOE and the University of Minnesota created a recombinant strain of this microbe by joining a *D. radiodurans* gene promoter sequence to a gene (toluene dioxygenase) involved in toluene metabolism. This strain demonstrated the ability to degrade toluene in a high-radiation environment. In an effort to immobilize radioactive elements, scientists hope to use this same strategy to equip *D. radiodurans* with genes from other microbes that are known to encode metal-binding proteins.

Lastly, it is critical that we learn from experience. Before any potential waste site is established, scientists

---

CAREER PROFILE

Validation Technician

Bioremediation is a relatively new discipline of biotechnology, and many smaller companies are involved in varying aspects of the field. Bioremediation is a multidisciplinary industry that requires the collective efforts of many different types of scientists, including microbiologists, chemists, soil scientists, environmental scientists, hydrologists, and engineers. Microbiologists and molecular biologists work hand in hand to carry out research designed to understand the mechanisms that microorganisms use to degrade pollutants. Biologists work together with engineers to design and implement the best treatment facilities for cleaning up the pollutants in question.

One entry-level position in bioremediation is that of the validation technician. Some validation technicians corroborate EPA-filed reports. Knowledge of applicable EPA regulations, the ability to collect pollution data from field sensors (many of which are linked by satellite to monitors), and the ability to arrange these data in an accessible database are required for this position.

Good computer skills, including familiarity with database programs and satellite geographic information systems, are required. Familiarity with organisms and natural communities is also needed. From entry-level positions, validation technicians with appropriate college coursework (or applicable degree) can advance to supervisory positions. The genetic engineers who design the bioremediation mechanisms are usually research scientists with a Ph.D., but they depend heavily on data collected in field studies by validation technicians.

must make an initial assessment of the site to characterize the organisms and ecosystems present and to assess the potential effects of pollution. Industrial processes produce hundreds of different chemicals each year. In addition, it is important that biodegradation scientists continually look ahead so they can be prepared to predict how new chemicals may affect the environment and develop technologies to clean up new pollutants. Bioremediation will not be able to rid all pollutants from the environment, but well-planned bioremediation approaches are an important component of environmental monitoring and cleanup efforts.

QUESTIONS & ACTIVITIES

Answers can be found in Appendix 1.

1. Describe three specific examples of a bioremediation application. Include in your description the organisms likely to be involved in the process, and discuss potential advantages and disadvantages of each application.

2. What is the purpose of adding fertilizers and oxygen to a contaminated site that is undergoing bioremediation?

3. Suppose you learned that a plot of land, which was once used as a chemical dump, has been cleaned up by bioremediation and is now a proposed site for a new residential housing development. Would you want to live in a house that was built on a former bioremediation site? Consider potential problems associated with this scenario.

4. Lead is a very toxic metal that is not easily degraded in the environment. Lead pipes, paint containing lead, car batteries, and even lead fishing sinkers are all potential sources for releasing lead into the environment. Propose a strategy for identifying bacteria that may play a role in degrading lead.

5. Suppose your town was interested in building a chemical waste storage facility in your neighborhood. Consider "lessons learned" from past failures of waste storage, dumping, environmental contamination, and successful applications of bioremediation, and then develop a list of priorities town officials should consider before the site is built and questions that should be addressed if the site is used.

6. Access some of the websites listed on the Companion Website. Use these sites to search for information about the most prevalent pollutants found in the environment. Make a list of five pollutants

and describe where they are found, how they enter the environment, their short-term and long-term health effects (in humans or other organisms), their environmental effects, and what types of microbes may degrade these pollutants.

7. Visit the Scorecard website listed on the Companion Website. Scorecard provides pollution data reports for each state in the United States. Enter your zip code and check on the latest reports for the areas closest to where you live.

8. Visit the EPA Superfund site and search for environmental cleanup violations reported in your state. Who was responsible for these violations? What fines (if any) were levied? What efforts are planned to remediate the site(s)?

9. In November 2002, the oil tanker *Prestige* split in half and sank off the northwestern coast of Spain. At least 30,000 tons of fuel oil were spilled. Conduct a Web search to see if you can find any information about bioremediation at this site.

10. Bioremediation is yet another rapidly changing area of biotechnology. Visit Google Scholar (http://scholar.google.com/) and do a search using "bioremediation" (or search for a specific topic in bioremediation such as "phytoremediation") to find recent publications in bioremediation.

References and Further Reading

Meagher, R. B. (2006). Plants Tackle Explosive Contamination. *Nature Biotechnology*, 24:161–163.

Parales, R. E., Bruce, N.C., Schmid, A., et al. (2002). Biodegradation, Biotransformation, and Biocatalysis (B3). *Applied and Environmental Microbiology*, 68: 4699–4709.

Peterson, C. H., Rice, S. D., Short, J. W., et al. (2003). Long-Term Ecosystem Responses to the Exxon *Valdez* Oil Spill. *Science*, 302: 2082–2086.

Seshadri, R., Adrian, L., Fouts, D. E., et al. (2005). Genome Sequence of the PCE-Dechlorinating Bacterium *Dehalococcoides ethenogenes*. *Science*, 307:105–108.

Wall, J. D., and Krumholz, L. R. (2006). Uranium Reduction. *Annual Review of Microbiology*, 60: 149–166.

Visit www.pearsonhighered.com/biotechnology to download learning objectives, chapter summary, "Keeping Current" web links, glossary, flashcards, and jpegs of figures from this chapter.

Chapter 10

Aquatic Biotechnology

After completing this chapter you should be able to:

- Discuss important goals and benefits of aquaculture, and recognize and appreciate the worldwide impact of fish farming.

- Discuss the controversies surrounding aquaculture and describe its limitations.

- Describe examples of commonly used fish farming practices.

- Understand how the identification of novel genes from aquatic species may be beneficial to the biotechnology industry.

- Provide examples of transgenic finfish and their uses.

- Discuss how triploid species can be created.

- Provide examples of medical applications of aquatic biotechnology.

- Describe nonmedical products of aquatic biotechnology and their applications.

- Define biofilming and explain how scientists are looking to marine organisms as a natural way to minimize biofilming.

- Describe how marine organisms may be used for biodetection and remediation of environmental pollutants.

Horseshoe crabs *(Limulus polyphemus)* are a source of blood cells for an important test used to ensure that foods and medical products are free of harmful contaminants.

Given that water covers nearly 75% of the earth's surface, it should not surprise you to learn that aquatic environments are a rich source of biotechnology applications and potential solutions to a range of problems. Aquatic organisms exist in a range of extreme conditions, such as frigid polar seas, extraordinarily high pressure at great depths, high salinity, exceedingly high temperatures, and low light conditions. As a result, aquatic organisms have evolved a fascinating number of metabolic pathways, reproductive mechanisms, and sensory adaptations. They harbor a wealth of unique genetic information and potential applications. In this chapter, we consider many fascinating aspects of **aquatic biotechnology** by exploring how both marine and freshwater organisms can be used for biotechnology applications.

10.1 Introduction to Aquatic Biotechnology

Oceans have been a source of food and natural resources for millennia. However, human population growth, overharvesting of fish and other marine species, and declining environmental conditions have caused the collapse of some fisheries, which puts pressure on many species and strains ocean resources. Although scientists have learned a great deal about ocean biology, the vast majority of marine organisms—particularly microorganisms—have yet to be identified. It has been estimated that greater than 80% of the earth's organisms live in aquatic ecosystems.

The obvious need to utilize the potential wealth of the majority of the earth's surface combined with increasing populations, human medical needs, and environmental concerns about our planet makes aquatic science an emerging frontier for biotechnology research. Aquatic ecosystems may contain the answers to a variety of global problems. In the United States, less than $50 million is spent annually for research and development in aquatic biotechnology. In contrast, Japan spends between $900 million and $1 billion annually. The successful research of Asian countries that have invested in basic science research on aquatic biotechnology and the financial success of their products have encouraged other countries also to invest a significant amount of time and resources. When a company is awarded a patent for an aquatic biotechnology application, the average increase in stock market value has been estimated as approximately $800,000.

In the United States, the National Science Foundation and the National Sea Grant Program—which directs coastal programs involving universities and colleges, scientists, educators, and students to support marine research—have developed many initiatives to support aquatic biotechnology. Specifically, several research priorities have been identified to explore the seemingly endless possibilities of utilizing aquatic organisms. These include the following:

- Increasing the world's food supply
- Restoring and protecting marine ecosystems
- Identifying novel compounds for the benefit of human health and medical treatments
- Improving seafood safety and quality
- Discovering and developing new products with applications in the chemical industry
- Seeking new approaches to monitor and treat disease
- Increasing knowledge of biological and geochemical processes in the world's oceans

From fish farming to isolating new medical products from marine organisms, these aspects of aquatic biotechnology are among the range of issues we consider in this chapter. The next section discusses aquaculture, a large and rapidly expanding agricultural application of aquatic biotechnology.

10.2 Aquaculture: Increasing the World's Food Supply Through Biotechnology

Two fishermen strain to hoist a net overflowing with catfish that are the ideal size and health for human consumption. Every fish in the net is a keeper. When prepared for market, these catfish possess a highly desired consistency, smell, color, and taste. This scenario does not take place on a fishing dock or a commercial fishing boat but instead describes fish farmers clearing nets from a fish-rearing tank (Figure 10.1). The cultivation of aquatic animals, such as finfish and shellfish, and aquatic plants for recreational or commercial purposes is known as **aquaculture.** Specifically, marine aquaculture is called **mariculture.** Although aquaculture can be considered a type of agricultural biotechnology, it is typically considered a form of aquatic biotechnology. In this section, we primarily discuss farming of both marine and freshwater species of finfish and shellfish.

The Economics of Aquaculture

Worldwide demand for aquaculture products is expected to grow by 70% during the next 30 years. In fact, it has been estimated that, in the near future, worldwide

Figure 10.1 A Net Full of Farm-Raised Catfish Ready for Market Sale Aquaculture is an effective way of raising large quantities of fish or shellfish that are ready for market consumption in a relatively short period of time. Aquaculture also allows scientists and farmers to grow salmon selectively bred for desirable market characteristics according to what consumers prefer. The catfish shown here are market-size USDA 103 catfish, which grow significantly faster than other catfish.

demand for seafood of all kinds will exceed what wild stocks can provide by approximately 50 million to 80 million tons. Overharvesting, loss of habitat, and depressed commercial fishing industries are all contributing to the decline in production of wild seafood stocks. If demand continues to rise and wild catches continue to decline, we will see a deficit of consumable fish and shellfish. Aquaculture together with better resource management practices will in part overcome this problem.

According to the Food and Agriculture Organization of the United Nations, in 1950, aquaculture production was estimated at 1 million metric tons. Worldwide aquaculture production in 2000 was approximately 33 million metric tons, more than doubling since 1984. Recent estimates suggest that close to 30% of all fish that humans consume worldwide are produced by aquaculture. By 2010, farmed fish are expected to overtake beef products as food. The market for fish captured by conventional commercial fishing methods is approximately 92 million metric tons. Total world fisheries, aquaculture plus capture fisheries is over 125 million metric tons. Some sources estimate that aquaculture has already surpassed cattle ranching as a food source. China is the single largest fish producer, accounting for approximately a third of the world's total.

Aquaculture in the United States is big business—it is a greater than $36 billion industry providing over 20% of the world's seafood supply. Aquaculture production in the United States has nearly doubled over the last 10 years and produces more than 100 different species of aquatic plants and animals. This increase is expected to continue, and similar increases in aquaculture are occurring globally. For example, over half of the salmon sold worldwide is produced by aquaculture, with a nearly 50-fold increase in production of fish-farmed salmon over the past two decades. Aquaculture in the United States became a major industry in the 1950s when catfish farming was established in the Southeast, and aquaculture facilities now exist in every state. Some of the most successful examples of the business potential of aquaculture in the United States include the Alabama and Mississippi Delta catfish industry, salmon farming in Maine and Washington, trout farming in Idaho and West Virginia, and crawfish farming in Louisiana. Similarly, Florida, Massachusetts, and other states have established successful shellfish farms that have benefited struggling commercial fishers. For instance, Cape Cod, which houses approximately 75% of the Massachusetts aquaculture industry, has successfully raised a number of different shellfish species, including quahogs, soft-shelled clams, blue mussels, and oysters.

Many other countries are actively engaged in aquaculture practices. Chile is the second largest exporter worldwide, generating approximately $1 billion in revenue each year. Ecuador, Colombia, and Peru have rapidly growing industries. Greek farms are the leading producers of farmed sea bass in the world, producing 100,000 tons each year, which account for over 90% of aquaculture production in Europe. Norway is a leading producer of salmon. Canada produces more than 70,000 tons of Atlantic and Pacific salmon with a yearly crop value of approximately

$450 million. The largest production province in Canada is British Columbia with more than 100 salmon farms. Over the next decade it has been projected that the number of farms in Canada will increase from 120 to over 600, creating 20,000 new jobs and contributing more than $900 million to the economy. The rapid growth and success of the Chilean aquaculture industry has many countries asking, "If Chile can do it, why can't we?" As a result, expanding markets are underway in Algeria, Argentina, Scotland, Iceland, the Faroe Islands, Ireland, Russia, Indonesia, New Zealand, Thailand, the Philippines, India, and many other nations.

Many of the countries most actively engaged in developing aquaculture industries are doing so because local waters have been overfished to the point where natural stocks of finfish and shellfish have been severely depleted. Through aquaculture, markets can be created in areas where the natural resources have been lost. Aquaculture also provides the benefit of creating seafood industries in areas of the world that, because of geographic location, are not normally known for their fisheries. For instance, a successful fish farming industry has been developed in the deserts of Arizona where recycled water from aquaculture is also being used to irrigate crop fields—fish are even being raised in irrigation ditches. Similar efforts are underway in the arid areas of Australia.

In theory, increased productivity of raising fish should lead to decreased retail prices for consumers. Some aspects of raising fish are economically cheaper than animal farming or commercial fishing. For example, it takes approximately 7 pounds of grain to raise 1 pound of beef, but less than 2 pounds of fishmeal are needed to raise approximately 1 pound of most fish. As another example, farm-raised catfish grow nearly 20% faster in fish farms compared with catfish in the

TOOLS OF THE TRADE

Fishing for Fish Genes

Molecular biology techniques that enable scientists to identify genes are helping advance our understanding of human health and disease; the Human Genome Project is one example of this. Similar studies are underway to learn about the genomes of many different species of commercially valuable fish that are desirable food sources. From tuna to salmon and a variety of shellfish, scientists are attempting to identify genes that contribute to properties such as growth rates, taste, color, and disease resistance. Selective breeding approaches have been popular for producing fish with the desired characteristics. But molecular biology techniques are now being used as "tools" to identify specific genes that scientists can use to produce transgenic finfish and shellfish with enhanced properties that will make certain species most attractive to seafood lovers. So how do scientists "fish" for these genes? In some cases, whole genome sequencing studies are underway, but one common approach to this question involves a technique called **differential display PCR.**

This technique is based on comparing DNA or RNA from two different tissue samples or organisms using primers that bind randomly to sequences within a genome. It allows scientists to identify categories of genes that are expressed "differentially"—that is, genes produced in one tissue (or organism) and not another. For example, suppose scientists were trying to identify genes that might contribute to rapid growth rates in shrimp. You might hypothesize that larger shrimp have different genes or greater levels of gene expression for one or more genes involved in growth rate. To discover these genes, you would isolate mRNA from small shrimp and large shrimp, reverse transcribe the RNA to make complementary DNA, and then amplify these samples by differential display PCR. The PCR products are separated by gel electrophoresis. If large shrimp expressed mRNA for unique genes that are involved in rapid growth, then PCR products for these genes would be detected in samples from large shrimp but not small shrimp. Scientists can then determine the DNA sequence of these PCR products to figure out what genes they have identified. Some of these genes may then be used to create transgenic animals with enhanced growth capabilities.

This tool is widely used not just by aquatic biotechnologists but also by molecular biologists in many areas of research. For instance, cancer researchers use this approach to identify genes expressed in cancer cells that are not expressed in normal cells. Several groups of aquatic biotechnologists throughout the world are also using similar techniques to identify and characterize genes that bacterial and viral pathogens use to cause disease in finfish and shellfish. Such studies will not only lead to treatments for diseases that create substantial economic losses to the aquaculture industry, but they will also help scientists manage and protect native stocks of finfish and shellfish from disease.

wild, and they are ready for market sale in approximately two years. For fish species that are fed genetically engineered feed at around 10 cents/pound, the return is often 70 to 80 cents/pound on the raised fish—a good return on an investment. But as you will learn in this chapter, the economics of aquaculture are not always so favorable.

Although worldwide increases in aquaculture are likely to continue, it is unlikely that fish farming will fully replace wild catches in the near future. For one, many species are not amenable to fish farming. Because open-water species such as tuna and swordfish roam over large areas of water and require lots of food, they cannot be raised in fish farms. We discuss other barriers to aquaculture later in this chapter. The purpose of presenting the statistics in this section is not to provide you with a remedy for insomnia but rather to help you develop a perspective on the current value and importance of aquaculture and its future potential. Visit the website of the Food and Agriculture Organization of the United Nations listed on the Companion Website for a comprehensive report on world fisheries and aquaculture.

Fish Farming Practices

In many ways, aquaculture is simply an extension of the conventional land-based agricultural techniques that have been practiced for decades. Culturing organisms for human consumption is only one purpose of aquaculture. Organisms are raised for many other reasons, such as providing bait fish for commercial and recreational fishing. Small fish such as anchovies, herring, and sardines are harvested to make fishmeal and oils used in animal feed for poultry, cattle, swine, and other fish. Growing pearls, culturing species to isolate pharmaceutical agents, breeding ornamental fish including goldfish and rare tropical fish, and propagating fish to stock recreational areas are among the many applications of aquaculture.

For example, in New Jersey, fisheries biologists breed hatchery-raised trout to create a "put-and-take" fishery for trout. Each spring the state stocks over 700,000 rainbow trout, brown trout, and brook trout in streams, rivers, ponds, and lakes throughout the state. Many streams and rivers will not sustain trout year round because they get too warm in the summer or because they are of poor water quality for trout, but through stocking ("put"), anglers are encouraged to keep what they catch ("take") for table fare. Many other states have similar stocking programs for a number of warm water and cold water species, including panfish, bass, catfish, muskellunge, and hybrid striped bass—created by breeding freshwater white bass and saltwater striped bass. Such programs also provide angling opportunities for people who live in urban areas with relatively little access to rural fisheries.

As shown in Table 10.1, a variety of fish and marine organisms have been cultured successfully. In many countries this list will soon be expanded. For example, marine species such as lobsters and crabs have been farmed on a limited basis but are not yet widely available commercially.

Aquaculture practices vary widely depending on factors such as the species being farmed, life cycles of the species, environmental requirements for growth, and the length of time needed to achieve maturity for market sale. For some fish such as salmon and trout, eggs (roe) and milt (sperm) are manually harvested from breeder stocks of adult fish. In an attempt to control the quality and health of the fish produced, parents used for breeder stocks are often fish that display desired growth characteristics and physical appearances (Figure 10.2). For example, fish growth rate, health, quality, and color of the meat are among the many features considered when selecting breeder fish.

Table 10.1 IMPORTANT AQUACULTURE ORGANISMS IN THE UNITED STATES

Freshwater Organisms[*]	Marine Organisms[*]
Catfish	Oysters
Crawfish	Clams (hard and soft shell)
Trout	Marine shrimp
Atlantic salmon	Flatfishes (turbot, flounder)
Pacific salmon	Sea bass
Bait fish (fathead minnows, golden shiners, mullet, sardines)	Mussels
Tilapia	Abalone
Hybrid striped bass	Quahogs
Sturgeon	
Carp (silver, grass)	
Yellowtail	
Bream	
Goldfish	

[*]Species are ranked approximately according to total production in metric tons from highest to lowest quantity. Production of freshwater organisms exceeds that for saltwater organisms; therefore, species listed on the same row do not indicate equal production.

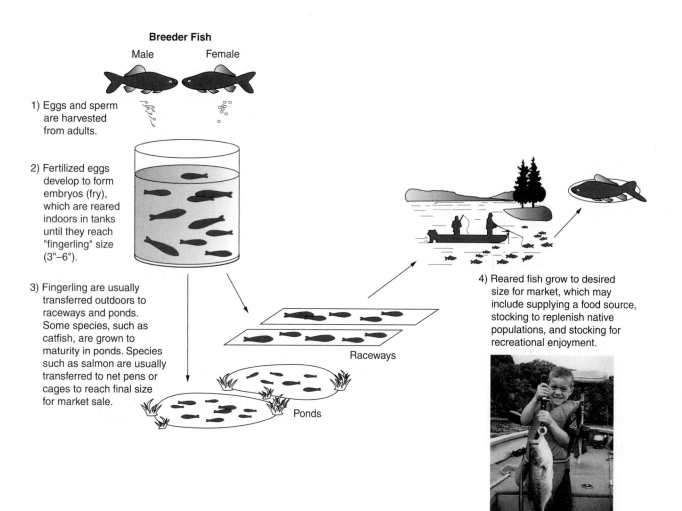

Breeder Fish

Male   Female

1) Eggs and sperm are harvested from adults.

2) Fertilized eggs develop to form embryos (fry), which are reared indoors in tanks until they reach "fingerling" size (3"–6").

3) Fingerling are usually transferred outdoors to raceways and ponds. Some species, such as catfish, are grown to maturity in ponds. Species such as salmon are usually transferred to net pens or cages to reach final size for market sale.

Raceways

Ponds

4) Reared fish grow to desired size for market, which may include supplying a food source, stocking to replenish native populations, and stocking for recreational enjoyment.

Figure 10.2 An Overview of Fish Farming

Eggs are fertilized in small tanks or containers and after a period of time, hatched embryos called *fry* are transferred to larger aquarium tanks with flowing water. Most of this culturing initially occurs indoors in a "nursery." Fish typically leave the nursery when they reach fingerling size (several inches in length, about the length of a human finger). Fingerlings, sometimes called *smolts,* are usually moved into concrete tanks called *raceways,* which may be indoors or outdoors (Figures 10.2 and 10.3). Raceways are often designed as a series of interconnected tanks that allow for the continuous flow of water through the tanks. This is particularly important for species like salmon and trout that are accustomed to battling the current to survive in streams and rivers. By mimicking environmental conditions, aquaculturists attempt to allow these species to develop physical characteristics similar to those of wild fish, such as the development of musculature and survival instincts.

Some fish species are raised in raceways until they reach market size, or they may be transferred to small shallow ponds for further growth before they are harvested. For certain species including salmon, after they reach a modest size, they can be raised to maturity in net pens, cages, and other enclosures that are placed in lakes, ponds, or estuaries. Shellfish farmers often employ rack or cage systems that are "seeded" with large numbers of tiny nursery-raised immature shellfish such as oysters or clams that adhere to the racks. Racks are then placed in estuarine environments to allow the shellfish to grow to maturity in a natural marine environment. Experts have noted that oyster grounds, which may yield approximately 10 kilograms of oysters per hectare, can yield nearly 127,600 pounds of oysters per hectare in hanging racks. Larger yields have also been reported for other species such as mussels. Most aquaculturists are constantly modifying culturing techniques in an effort to

Figure 10.3 Raceways, Ponds, Net Pens, and Oyster Racks   A wide variety of approaches are used for farming fish and shellfish including (a) raceways, (b) ponds, (c) net pens for raising salmon, and (d) rack systems for oysters and other shellfish. Many of these approaches are often used in sequence as fish grow from small fingerlings to market size.

increase the yield of a species in a given area and to minimize the expense-to-profit ratio.

Salmon farming is an excellent example of the fish-rearing process. Eggs and sperm milked from adults bred to grow fast are used for fertilization, and resulting embryos are raised for 12 to 18 months in tanks until they reach the fingerling stage. Fingerlings are usually transferred to mesh-enclosed pens placed in oceans or bays close to the shoreline. Throughout this process, the salmon diet often consists of pellets made from ground-up herring, anchovies, fish oil, and animal and poultry by-products (bone, excess meat, and the like). Food may also contain antibiotics and additive dyes to make the salmon flesh pink, and fish may be placed in pesticide baths to remove sea lice and other parasites. Throughout the culturing process, aquaculturists can change the taste of farm-raised species by experimenting with food sources such as vegetable-based foods versus foods derived from fish

products. Ultimately, unblemished fish of the appropriate size are packaged for market sale. Figure 10.2 provides an overview of fish farming.

Innovations in fish farming

Many innovative approaches for raising fish are under development. For example, in West Virginia, fisheries biologists are taking advantage of an abundance of abandoned coal mines in the mountain state. The cool mountain groundwater and water from freshwater springs seep into and fill many mines, providing high-quality environments for rearing cold water species such as rainbow trout and arctic char. This approach also provides an important use of abandoned mines that would typically serve no purpose otherwise. It is estimated that West Virginia has enough quality mine water areas to increase yearly fish production from its current rate of approximately 400,000 pounds to over 10 million pounds per year.

Some companies have experimented with **polyculture**—also called *integrated aquaculture*—raising more than one species in the same controlled environment. Raising species with different nutrient requirements and feeding habits is one way to optimize water resources. Polyculture can involve different fish species co-cultured, fish and shellfish co-raised, as well as animal and plant polyculture. For example, raising carp in the presence of plants such as lettuce is an effective approach. Roots from the lettuce absorb nutrients and fish waste products (such as nitrogen) from the water as fertilizer to support lettuce growth. Another relatively new aquaculture approach involves the use of **hydroponic systems.** These systems are small-volume water-flowing systems in which vegetables (like tomatoes and broccoli) or herbs (like basil and chives) are cultured in racks through which wastewater from fish tanks can flow. Polyculture and hydroponic systems are frequently used together when raising fish such as catfish, carp, and tilapia (Figure 10.4).

Improving Strains for Aquaculture

Many fish farming efforts involve activities designed to improve certain qualities, such as growth rate, fat content, taste, texture, and color of the finfish or shellfish being raised. As we discuss in Section 10.3, some of

Figure 10.4 Polyculture of Lettuce and Tilapia in a Hydroponic System   Hydroponic system of lettuce growing in racks above a tank containing tilapia (inset). Water from the tilapia tank, which contains nitrogenous wastes from the tilapia, is cycled through the racks of lettuce. Lettuce will use wastes in the water as fertilizer, thus creating efficient use of the water resource. By minimizing the cost of artificial filtration and removal of wastes, polyculture limits the amount of wastewater effluent for discharge into the surrounding environment.

the strategies used to improve strains for aquaculture involve molecular alterations in fish genetics. Scientists are studying mechanisms of gene expression necessary for fish reproduction, growth, and development; techniques for the **cryopreservation,** storage of tissue samples and cells at ultralow temperature usually between −20°C and −150°C; and delivery systems for hormones designed to induce spawning and improve growth of fish. New knowledge gained in these and other areas will be of great help to the aquaculture industry.

In this section, we consider a few examples of how strains of finfish or shellfish can be improved for aquaculture by discussing how selective breeding may be used to raise species with desirable characteristics. Fish farmers and scientists work together to identify individuals of a given species with desirable characteristics. For instance, in a population of catfish, not all individuals grow at the same rate or have the same body characteristics, such as muscle mass compared with the percentage of body fat. Scientists can use a number of techniques to identify fish that show fast growth rates and heavy muscle mass. For example, they can use ultrasound machines to measure muscle mass as a means to estimate fillet yield. Fish that show the highest yield are then bred to produce new generations with increased muscle mass.

Scientists can also use a sophisticated technique called **bioimpedance** to track movement of low-voltage electrical current through tissue as a way to measure lean muscle mass and body fat content. Tissue with a high muscle content and low fat content is denser than high-fat tissue. As a result, dense tissue interferes with (impedes) movement of electrical current through the tissue. These approaches are being used to develop the best-tasting, low-fat catfish, as well as a number of other fish species, on the market. Similar approaches have been used for decades for the selective breeding of cattle, swine, and many other farm animals.

Scientists at the U.S. Department of Agriculture Catfish Genetics Research Unit in Mississippi have selectively bred a new variety of catfish called USDA 103 (a name worthy of a government-issue catfish!) that grows 10 to 20% faster in ponds, trimming the time from birth to market to between 18 and 24 months (refer to Figure 10.1). It has been reported that some USDA 103 catfish can reach sexual maturity a full year faster—by 2 years of age—than other species. On the negative side, USDA 103 catfish consume more feed than other varieties of catfish.

In Arizona, researchers have used selective breeding to identify shrimp that are acclimated to growing in low-salinity water. Starting with larval shrimp grown in fresh water with added salt, scientists culture these shrimp through tanks of water with progressively lower salt concentrations. This allows for the farming of high-quality shrimp without the need for salt water, and low-salt water can then be recycled to irrigate fields of fruits and vegetables. Thus shrimp can be raised in the desert of Arizona hundreds of miles from natural sources of salt water.

Enhancing Seafood Quality and Safety

Aquaculture can be combined with cutting-edge techniques in molecular biology to create finfish and shellfish species of the color, taste, and texture consumers want. In addition scientists are working to make seafood safe so it is free of pathogens and chemical contaminants.

Igene Biotech of Columbia, Maryland, has used gene-cloning techniques to mass-produce **astaxanthin,** the pigment that gives shrimp their pink color. By including recombinant astaxanthin in fish feed, scientists can create salmon and trout with a rainbow of hues in flesh color. Astaxanthin is also thought to have potential value as an antioxidant to be used in nutritional supplements for humans. Most people like salmon with a pink color. In fact, consumer polls suggest that many people believe the redder the salmon, the higher its quality. Dark-red-colored salmon is highly prized in the best sushi bars in Japan. In reality, the pigment astaxanthin has no effect on the taste of salmon, but farmers want to grow what consumers prefer. To help fish farmers produce the fish their consumers demand, Swiss drug company Roche Holding AG, a leading producer of astaxanthin, distributes a "salmofan," which resembles a paint color chart. Aquaculturists can pick shades ranging from light pink to dark crimson and then purchase food with the concentration of astaxanthin that will produce fish with the flesh color they desire.

A variety of approaches to enhance seafood quality and safety are currently being used, and many more are in development. From detecting contaminated seafood to enhancing the taste and shelf life of seafood, aquatic biotechnologists are working to apply innovative technologies. Marine scientists are using molecular biology techniques to identify and learn more about genes encoding toxins produced by marine organisms to help them understand how these toxins can cause disease and to minimize the negative health impacts of toxin exposure. Many molecular probes and PCR-based assays are being developed for detecting bacteria, viruses, and a host of parasites that infect finfish and shellfish.

A handheld antibody test kit has been developed to detect *Vibrio cholerae* in oysters. It uses antibodies to detect proteins from *V. cholerae*, the bacteria that cause

cholera, a very serious illness characterized by severe diarrhea that can lead to dehydration and death in severe cases. Cholera is particularly a problem in developing nations with contaminated water supplies that often result from inadequate sewage treatment facilities. This kit has been widely and successfully used in South America to detect contaminated seafood and minimize cholera infections. Similar approaches have been used in Hawaii to develop a dipstick monoclonal antibody test for ciguatoxin, which affects finfish from tropical areas that are sold in the aquarium industry.

Gene probes have been developed for detecting several viral diseases of shrimp. A number of companies are developing gene probes for detecting and assessing the effects of environmental stresses on fish and shellfish. Some of these strategies are similar to those that are used to screen human single nucleotide polymorphisms to predict an individual's risk for a genetic disease, as described in Chapter 11.

Aquatic biotechnologists are interested in developing detection kits or vaccines for a number of pathogens that pose serious threats to fish raised by aquaculture. Several aquaculture companies are working to develop a vaccine for infectious salmon anemia, a deadly condition that causes internal bleeding and destroys internal organs in salmon and leads to the death of hundreds of thousands of salmon worldwide each year. Similar work is being carried out to battle against sea lice such as *Caligus elongatus*. This tiny parasite attaches to salmon, feeding on salmon blood and creating exposed lesions that render salmon susceptible to deadly infections.

Scientists in California, Idaho, Oregon, and Washington have developed a subunit vaccine against the infectious hematopoietic necrosis (IHN) virus, which is responsible for the loss of large numbers of commercial trout and salmon each year. The vaccine was developed from an IHN protein expressed in bacteria. This vaccine is injected into fish to stimulate them to produce antibodies against the IHN virus, which protect fish against infection by IHN.

Barriers and Limitations to Aquaculture

Although aquaculture is firmly established as a worldwide application of aquatic biotechnology with enormous potential to provide food in unique ways, there are concerns and obstacles. Not all species are ideally suited for aquaculture. This is particularly true for many highly prized marine species. Water quality issues can be a problem. For some species, it is difficult to maintain water with the proper flow rate to deliver adequate concentration of nutrients and to remove rapidly accumulating waste products properly. This is especially true for marine species that require large areas of ocean for roaming and do not survive when confined to small living spaces. In addition, marine organisms often have long complicated life cycles involving a series of larval stages, each of which may have different food requirements, before marketable size can be achieved. To raise adults of these species, the loss of young fingerlings can be very high and thus cost prohibitive. Aquaculture is easiest with species that have few larval stages.

Aquaculturists are constantly faced with minimizing the effects of disease on their fish populations. Given the dense, crowded conditions where many fish are raised, farmed fish are often more susceptible than native stocks to stress and disease caused by bacterial and viral pathogens, and lice. Because there is less genetic diversity and disease resistance in farmed fish, infections can spread rapidly. Most fish in a farmed population will have the same resistance and susceptibility to disease; therefore, a supply of fish can be wiped out quickly if disease is not controlled.

Some biologists are concerned that the aquaculture industry may be consuming excessive amounts of wild bait fish such as anchovies and herring. Although some bait fish are raised by aquaculture, in many areas of the world the bait fish used to make fishmeal are still netted from wild stocks. For every pound of salmon, several pounds of other typically wild-caught fish such as anchovies, herring, and mackerel are used to raise the salmon. Growing 1 pound of salmon can require 3 to 5 pounds of wild-caught fish. Scientists are looking to cure some farmed fish of their carnivorous feeding habits by changing to vegetable-based diets, but many aquaculturists are concerned about the quality of these fish. Supporters of vegetable-based diets claim that some people may like salmon that tastes less fishy.

Just as runoff of animal feces and wastes from traditional land farming is a problem that can have significant impacts in many areas of the country, environmentalists are concerned about pollution from fish farming industries. The waste-laden effluent water from aquaculture operations, which contains untreated feces, urine, and uneaten food, is typically discharged into natural waterways. These wastes are rich in nitrogen and phosphorus and can lead to algal booms and other problems. Dying algae rob water of oxygen, which can lead to fish kills. Fish wastes also harbor strains of bacterial pathogens such as *Salmonella* and *Pseudomonas*, which can affect human health; however, the likelihood of disease transmission from fish wastes to humans has not been well established. At present, it is generally thought that waste production from aquaculture has a small overall effect on

water quality, but this may change as aquaculture efforts increase worldwide.

Extermination of "pest" species at aquaculture facilities is another concern. A number of fish-eating birds (such as cormorants, pelicans, herons, egrets, and gulls) and mammals (such as seals and sea lions) can be subjected to authorized and unauthorized capture and extermination. However, many facilities employ nonlethal methods, including visual harassment such as lights, reflectors, scarecrows, human presence, and auditory devices such as predator distress calls, sirens, and electronic noisemakers to deal with these "pest" species. Underwater acoustic and explosive devices may also be used along with perimeter fencing and protective netting to deter predators from feeding at fish farms.

Concern has also been raised about the discharge of chemicals commonly used in many aquaculture operations. These include antibiotics, pesticides, herbicides (plant killers), algaecides (algae killers), and chemicals used to control parasites. In addition, residual chemicals from other antifouling compounds (such as those used to reduce growth of barnacles, algae, and other organisms) and anticorrosants may be absorbed by fish and shellfish and passed to other marine organisms and to humans. A comprehensive survey of farmed salmon has found that they contain significantly higher levels of polychlorinated biphenyls (PCBs) and other similar compounds correlated with increased risk of cancer than their wild relatives. This is a major concern for the industry and fish farmers are working on ways to reduce pollutants in farmed salmon.

A number of federal laws regulate aquaculture and its potential effect on the environment, including the Clean Water Act, which regulates discharges including aquaculture effluent; the Migratory Bird Treaty Act, which protects birds that may pose a threat to aquaculture; the Marine Mammal Protection Act, which prohibits killing of marine mammals that may be predators at aquaculture facilities; the Federal Insecticide, Fungicide, and Rodenticide Act, which governs use of these substances on crops, including fish; and the Food, Drug, and Cosmetic Act, which oversees approvals for drug use in fish farming and governs seafood safety.

The visual effects of aquaculture on the landscape are also problematic. For instance, in some areas of Scotland, there are concerns that the abundance of shellfish cages along the Scottish coastline is damaging the natural aesthetic appeal of coastal landscapes. A very serious issue raised by both aquaculturists and naturalists is the potential threat to native species by farm-raised fish that accidentally escape into the wild (refer to the "You Decide" box).

YOU DECIDE

The Risks of Fish Pollution and Genetic Pollution: Controversies of Aquaculture and Genetically Engineered Species

The term *fish pollution* describes the escape of aquaculture species that can harm natural ecosystems by altering or reducing biodiversity; introducing new parasites; competing with native species for food, habitat, and spawning grounds; interbreeding with native stocks; and destroying habitats. Farm-raised fish do escape. Many examples of fish pollution and its effects have been documented. For example, in some waterways in Florida, tilapia—a rapidly growing aquaculture panfish with a voracious appetite—have outcompeted native species such as bream for spawning areas and food. As a result, some native fish in tilapia-polluted areas have virtually disappeared. Similarly, non-native species of Pacific white shrimp farmed along the Gulf coast of Texas and the Atlantic coast of South Carolina have escaped and are free swimming in these areas.

Evidence suggests that farm-raised Atlantic salmon are breeding in waters of the Pacific Northwest. Twenty-six stocks of Pacific salmon and steelhead are currently listed as endangered or threatened. Many scientists believe the loss of these native species is due to the spread of lice and other parasites from farmed fish. In Norway, escaped farmed salmon comprise approximately 30% of the salmon in local rivers, and they are thought to outnumber resident salmon in local streams. Estimates suggest that 345,000 salmon escaped from salmon farms in British Columbia from 1991 to 1999.

A December 2000 northeaster in Maine—where salmon farming is the second-largest fishery behind lobstering—caused the uprooting of steel cages containing farm-raised salmon and released over 100,000 fish into Machias Bay. This accident is thought to be the largest known escape of aquaculture fish in the eastern United States. Many are concerned that this spill will weaken the genetic potential of future generations of wild salmon in Maine rivers. The U.S. government has already listed wild salmon in eight Maine rivers as endangered as a result of concerns about the impact of aquaculture in the area. In some spawning rivers in New Brunswick and Maine, farmed escapee salmon greatly outnumber wild salmon.

The escape of farmed fish has caused environmentalists and other groups to rally for a moratorium on new fish farms in many areas of the world and for regulation of the aquaculture industry. They see fish pollution as a problem because farmed fish can threaten native fish species and affect natural aquatic ecosystems. Many aquaculture species are similar to domestic farm animals in that they have become reliant on humans and are poorly adapted to life in the wild. Escaped farmed fish can affect the gene pool of wild stocks by competing with native stocks during breeding and interbreeding with native species. Evidence also indicates that farmed salmon produce smaller and perhaps less fertile eggs than wild stocks.

Even stronger concerns have been raised about the escape of genetically modified species such as transgenics and triploids. Although most of these species cannot reproduce, no technique ensures that all modified fish will be sterile. Once in the wild, oceans cannot be drained to retrieve them. Transgenic stocks that escape and grow in areas outside of their intended growing range often do not grow as well as they do in their intended environment. As a result, if they breed with native species, mixed stocks often lose biological fitness (the ability to reproduce) and grow slower. By introducing transgenic species, it is possible to erode adaptations that have occurred over thousands of years and reduce the fitness of native stocks. Fish with enhanced growth capabilities may be able to outcompete native fish for food and spawning sites. In 2001, Maryland became the first state to ban raising genetically modified fish in ponds and lakes connected to other waterways.

Critics also cite historical examples of the spread of non-native species as reason for concern. For example, European zebra mussels—thought to have entered the United States in ballast water of oceangoing ships entering the Great Lakes during the 1980s—have created significant problems. Colonies of these prolific shellfish have smothered habitat for other species, and through filter feeding they have caused the decline of native plankton in the Great Lakes, blocking pipes and growing on hulls of ships. Costs associated with controlling zebra mussels in the Great Lakes region alone are estimated to exceed $400 million annually.

Even though the FDA is involved in regulating the safety of transgenic fish as a food source, no clear policies exist for regulating the release of aquaculture species (including transgenics) in the United States. This must change if some of the concerns just described are to be addressed. The U.S. Department of Agriculture is looking closely at ways to assess the risk of bioengineered species and the safety of introducing genetically modified organisms into the environment in both marine and nonmarine environments. The North American Salmon Convention (NASCO), an intergovernmental organization, has rallied the salmon aquaculture industry to develop guidelines that will prevent rearing genetically engineered salmon in open waters in an effort to prevent irreversible genetic change and unforeseeable consequences of the escape of genetically engineered species into natural ecosystems.

Most scientists believe that aquaculture and genetic engineering are necessary for producing enough food to satisfy an increasing human population, but are the risks of these technologies greater than the risks of doing nothing? You decide.

The Future of Aquaculture

What lies ahead for the aquaculture industry? Much of the future direction will involve overcoming some of the barriers and limitations described here. To reduce concerns about overfishing bait-fish populations for use in fishmeal, many aquaculturists are exploring ways to grow species on different food sources that require little or no wild fish meals and oils. Advances in fish farming will likely minimize effluent discharges from aquaculture facilities and reduce the use of antibiotics, pesticides, and other chemicals used to treat farmed species. It is also likely that polyculture approaches will become more popular as aquaculturists attempt to maximize water resources.

As we discuss in the next section, molecular genetic approaches will have a powerful impact on aquaculture in the future. Using many of the strategies presented in this chapter along with recombinant DNA technology, aquatic biotechnologists will be working to increase growth and productivity of farmed species while improving disease resistance and the genetic composition of important food species. Molecular biology will have a strong impact on aquaculture, from identifying genes that control reproduction and spawning of fish and shellfish to creating genetically modified species with desired characteristics.

Work on understanding conditions that affect the mortality of marine organisms during critical times in early development will lead to new rearing practices that can raise the productivity of farmed species. Many countries are actively involved in research related to the farming of other species that are difficult to raise by aquaculture. For instance, the South Carolina Department of Natural Resources is experimenting with aquaculture techniques for raising cobia, a very popular game fish of high commercial value. Cobia roam deep open-water areas of the ocean, making them difficult to raise under confined conditions.

So far we have taken a comprehensive look at the aquaculture industry—one of the oldest applications of aquatic biotechnology. In this section, we have seen that aquaculture is not unlike many industries; it benefits society but also poses some problems. In the next section, we explore an area that is literally in its infancy: the molecular genetics of aquatic organisms.

10.3 Molecular Genetics of Aquatic Organisms

An understanding of the molecular complexities of aquatic organisms is central to aquatic biotechnology. Basic knowledge of gene expression and regulation in aquatic organisms and an understanding of genes involved in processes such as reproduction, growth, development, and survival at extreme environmental conditions will be critical for applications such as maintaining populations of endangered marine species, limiting populations of harmful species, and advancing genetic manipulations of aquatic organisms.

In addition, pathogens that affect finfish and shellfish result in large financial losses each year. Scientists are working to improve survival growth of aquaculture species, and molecular biology is being used to learn more about the immune systems of aquatic species and their susceptibility and resistance to disease-causing organisms, including disease transmission and the life cycles of the pathogens themselves.

As you will soon learn, one of the ultimate applications of our molecular genetic knowledge of aquatic organisms involves manipulating the genetic composition of marine species.

Discovery and Cloning of Novel Genes

The gene discovery process in aquatic organisms covers many interesting areas. For example, a great deal of research is dedicated to identifying new genes, learning about the environmental factors that control gene expression such as the effects of extreme temperature and deep ocean pressures, identifying the molecular basis of unique adaptations, and studying genetic and molecular factors such as hormones that control the reproduction, growth, and development of aquatic organisms. In addition to identifying novel genes, many research groups are involved in identifying mutations associated with diseases in fish. Eventually, such information will be used in the development of disease-free breeder stocks of fish with selected characteristics for U.S. hatcheries. By identifying deleterious genes that may have negative influences on fish growth, health, and longevity, scientists anticipate that many of these genes can be modified or removed to produce improved species for aquaculture. Discovering genes that can be used as probes for the identification of microscopic marine organisms such as phytoplankton and zooplankton—important food sources for many marine species—will help scientists look at environmental effects on the genetics of these microorganisms. This is a critically important issue for understanding how these microscopic organisms affect organisms higher in the food chain.

Genes are being identified as DNA markers that can be used to distinguish wild stocks of fish from hatchery-reared stocks. Biologists are interested in using these markers to identify strains that are more resistant to disease and receptive to fish farming. Such markers will also serve important roles as scientists attempt to determine the effects of farm-raised escapees on native stocks. Many of these marker genes are species specific, and they have enabled fish and wildlife officers to apply PCR techniques when investigating cases of illegal harvest and retail of fish that are subject to quotas. For example, in the northeastern United States, blackfish are prized table fare. But overharvesting of blackfish has resulted in strict size limits and restricted seasonal dates for legally harvesting blackfish. If wildlife officials encounter someone they suspect is a poacher harvesting blackfish during the closed season but the poacher only has fillets in the boat and not intact carcasses, how can officials determine what species the fillets came from? Using blackfish-specific primers, PCR assays can be run on DNA isolated from fillets to determine if the fillets are from blackfish or not.

Cloning the gene for **growth hormone (GH)** is an excellent example of how the gene discovery process can lead to advances in marine biotechnology. In Chapter 3, we discussed how cloning of human GH led to treatments for dwarfism. Recall that GH is a hormone produced by the pituitary gland that stimulates the growth of bone and muscle cells during adolescence. Cloning the salmon growth hormone gene has led to the development of **transgenic** species of salmon that demonstrate greatly accelerated growth rates compared with native strains. We consider the GH example in more detail in the next section. As another example, University of Alabama-Birmingham researchers have cloned the gene for molt-inhibiting hormone (MIH) from blue crabs. Molting (shedding) is triggered by a decrease in MIH, leading to soft-shelled crabs that can be eaten whole. Current research is underway to block the release of MIH as a way to produce soft crabs on demand for the seafood industry.

Antifreeze proteins

One of the most successful examples of the identification of novel genes with a range of very promising applications is the identification of genes for **antifreeze proteins (AFPs).** A/F Protein, Inc., of Waltham, Massachusetts, is a leader in the production of antifreeze proteins. Many of the first AFPs were isolated from bottom-dwelling fish species such as northern cod that live off the coast of eastern Canada and Antarctic fish called teleosts that live in extremely cold ice-laden waters—some of the most severe environments on earth. Subsequently, AFPs have been isolated from a number of other cold water species, including winter flounder (*Pleuronectes americanus*), sculpin (*Myoxocephalus scorpius*), ocean pout (*Macrozoarces americanus*), smelt (*Osmerus mordax*), and herring (*Clupea harengus*). Interestingly, similar proteins have been discovered in mealworm beetles.

Several types of AFPs have been discovered. Structurally, most AFPs have extensive alpha helices and are held together by large numbers of disulfide bridges. AFPs function to lower the freezing temperature of fish blood and extracellular fluids to protect fish from freezing in frigid marine waters. Seawater freezes at approximately −1.8°C. AFPs typically lower the freezing point of fish body fluids by approximately 2 to 3°C. Currently, the majority of AFPs are isolated from fish blood. To meet large quantity demands estimated for applications of AFPs, scientists are working on recombinant AFP production in bacterial and mammalian cells to accommodate anticipated future needs for AFPs worldwide.

AFPs protect living organisms from freezing in a variety of ways. They can bind to the surface of ice crystals to modify or block ice crystal formation, lower the freezing temperature of biological fluids, and protect cell membranes from cold damage. Because of these unique abilities, a number of innovative applications for these cryoprotective proteins are being developed. AFPs are being used to create transgenic fish and plants with enhanced resistance to cold temperatures and freezing. For instance, salmon cannot produce antifreeze molecules, and thus they die when exposed to near-freezing water. For example, waters off the coast of eastern Canada, too cold for wild species of salmon, are being considered as potential aquaculture habitat for freeze-resistant species of transgenic salmon containing AFP genes.

AFP gene promoter sequences are also being used in recombinant DNA experiments to stimulate expression of transgenes, including growth hormone in salmon. Transcription from AFP promoter sequences is stimulated by cold temperatures. By ligating genes of interest to the 3′ end of AFP promoter sequences, AFP promoters can be used to stimulate transcription of these "downstream" genes under cold water conditions (Figure 10.5). Thus AFP promoter gene constructs can serve as very effective transcription vectors to transcribe foreign genes, leading to increased production of protein. Such constructs may be very effective for the genetic engineering of fish, as we discuss in the next section.

AFPs have been introduced into plants to produce cold-hardy transgenic strains such as tomatoes, but these plants are not widely available yet. Few commercial crop plants produce cryoprotective proteins as effective as AFPs from fish. For many popular crops such as wheat, coffee, fruit (e.g., citrus, apples, pears, cherries, and peaches), soybean, corn, and potatoes, crop damage as a result of cold temperatures is a problem worldwide. In 1998, freezing conditions in California caused an estimated $600 million in damage to citrus crops alone.

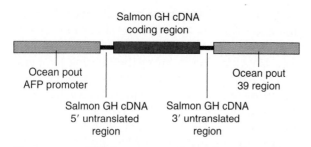

Figure 10.5 AFP Promoters Stimulate Gene Expression in Transgenic Fish A recombinant DNA plasmid construct containing an antifreeze protein gene promoter from ocean pout attached to the cDNA for salmon growth hormone can be used to produce rapidly growing fish for aquaculture. Transcription from the AFP promoter is stimulated by cold conditions; therefore transgenic fish containing this construct synthesize large amounts of growth hormone when they are raised in cool water. Increased production of growth hormones causes transgenic fish to grow faster than native nontransgenic strains.

Cryoprotection of human cells, tissues, and organs is a promising medical application of AFPs. As shown in Figure 10.6, cold storage of oocytes used for *in vitro* fertilization is one potential application. AFPs may prove useful for the storage of a number of human tissues including blood and for the development of new protocols for cryogenic storage of human organs such

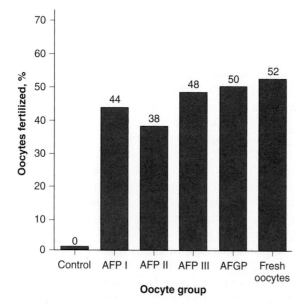

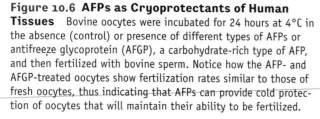

Figure 10.6 AFPs as Cryoprotectants of Human Tissues Bovine oocytes were incubated for 24 hours at 4°C in the absence (control) or presence of different types of AFPs or antifreeze glycoprotein (AFGP), a carbohydrate-rich type of AFP, and then fertilized with bovine sperm. Notice how the AFP- and AFGP-treated oocytes show fertilization rates similar to those of fresh oocytes, thus indicating that AFPs can provide cold protection of oocytes that will maintain their ability to be fertilized.

as the heart and liver prior to their use in transplantation surgery.

Finally, scientists are investigating ways that AFPs may be used to improve the shelf life and quality of frozen foods, including ice cream. Changes in the quality of frozen foods often occur during thawing and refreezing, similar to when you bring home food from the store and put it in your home freezer. It is possible that AFPs can be used to alter the ice crystallization properties of frozen foods to improve their overall quality. Scientists have even proposed using AFPs to control ice formation on aircraft and roadways.

As you can see, AFPs may be useful for a number of interesting applications. Undoubtedly the ocean harbors many species that contain many other novel genes that may be exploited to benefit biotechnology in the future.

"Green genes"

An outstanding example of a research application involving a novel gene from an aquatic organism involves the bioluminescent jellyfish *Aequorea victoria*. *A. victoria* can fluoresce and glow in the dark because of a gene that codes for a protein called **green fluorescent protein (GFP).** GFP produces a bright green glow when exposed to ultraviolet light. It has been estimated that nearly three fourths of all marine organisms have bioluminescent abilities. Bioluminescence is often used as a way for fish and other organisms to find each other in dark environments of the ocean during mating activities. Recently genes for red, orange, and yellow fluorescent proteins have been cloned from sea anemones, expanding the color palette of proteins available for researchers. In Chapter 5, we mentioned that the fluorescence of some marine species is created by bioluminescent bacteria such as *Vibrio harveyi* and *Vibrio fischeri*. We discuss this in more detail later in this chapter.

In the case of GFP, scientists have taken advantage of the fluorescent properties of this protein to create unique **reporter gene** constructs. Reporter genes allow researchers to detect expression of genes of interest in a test tube, cell, or even a whole organ. As shown in Figure 10.7, reporter gene plasmids are created by ligating the GFP gene to a gene of interest and then introducing the reporter plasmid into a cell type of choice. Once inside cells, these plasmids are transcribed and translated to produce a fusion protein that fluoresces when exposed to ultraviolet light. Only cells producing the GFP fusion protein fluoresce. In this manner, these plasmids serve to detect or "report" where the gene of interest is being expressed.

This approach has been widely used to study basic processes of gene expression and regulation. Developmental biologists have used GFP-expressing embryonic

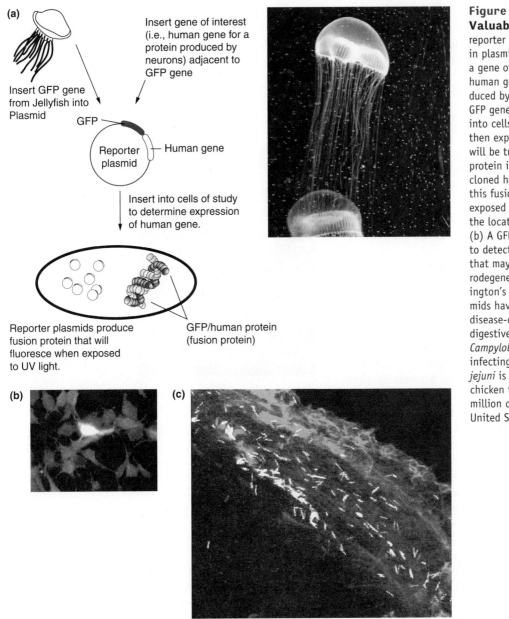

(a)

Insert GFP gene from Jellyfish into Plasmid

Insert gene of interest (i.e., human gene for a protein produced by neurons) adjacent to GFP gene

GFP

Reporter plasmid — Human gene

Insert into cells of study to determine expression of human gene.

Reporter plasmids produce fusion protein that will fluoresce when exposed to UV light.

GFP/human protein (fusion protein)

(b)

(c)

Figure 10.7 The GFP Gene Is a Valuable Reporter Gene (a) GFP reporter gene constructs can be created in plasmids by ligating the GFP gene to a gene of interest. In this example, a human gene encoding a protein produced by neurons is attached to the GFP gene and the plasmid is introduced into cells in culture. These cells will then express mRNA molecules, which will be translated to produce a fusion protein in which GFP is attached to the cloned human protein. Cells producing this fusion protein will fluoresce when exposed to ultraviolet light, "reporting" the location of the expressed protein. (b) A GFP reporter gene is being used to detect a protein in human neurons that may be responsible for the neurodegenerative condition called Huntington's disease. (c) GFP reporter plasmids have also been used to detect disease-causing bacteria in the human digestive tract such as the *Campylobacter jejuni* species shown here infecting human intestinal cells. *C. jejuni* is a common contaminant of chicken that causes approximately 2.4 million cases of food poisoning in the United States each year.

stem cells to track their differentiation into different cell types during development of embryos. Scientists have also used GFP reporter gene constructs in many innovative ways that promise to advance medical diagnostics and disease treatment (Figure 10.7b). For example, GFP genes have been used to pinpoint tumor formation in mice, to follow the progress of bacterial infections in the intestines, to monitor the death of bacteria following antibiotic treatment, and to study the presence of food-contaminating microorganisms in the human digestive tract (Figure 10.7c).

Cloning the genomes of marine pathogens

Marine biologists are interested in cloning the genomes of a variety of marine pathogens that affect wild and farmed species as a way to learn about genes these organisms use to reproduce and cause disease. In 2001, Chilean scientists deciphered the genome of the bacterium *Piscirikettsia salmonis*. *P. salmonis* infects salmon and causes a disease called rickettsial syndrome, which affects the liver of infected salmon, leading to death. Combating this syndrome is a worldwide problem for salmon farmers. Currently, no effective treatment exists. In Chile alone, this syndrome causes an estimated $100 million per year in financial losses to the salmon industry.

Armed with genome information about pathogens, scientists will be looking to develop strategies for bolstering the immune system of farmed species to improve resistance to pathogens. For example, it may be possible to inject genes or proteins from *P. salmonis* into salmon as vaccines to stimulate the immune system to defend against infection by these bacteria.

Similar genome studies are underway to understand the genetics of *Pfiesteria piscicida*, a toxic dinoflagellate that some scientists believe is responsible for major fish kills and shellfish disease in North Carolina estuaries. *P. piscicida* has also caused fish kills and disease in aquaculture facilities from the mid-Atlantic to the Gulf Coast. In fish, *P. piscicida* causes lesions, hemorrhaging, and other symptoms that can lead to death of infected fish. One reason this pathogen has attracted so much attention, bordering on hysteria in some coastal communities affected by large fish kills, is because evidence indicates that *P. piscicida* toxins can have serious health effects on humans.

Scientists are working on ways to use molecular approaches in the battle against diseases and parasites that have essentially eliminated commercial fishing for oysters in several areas of the United States. Across the country, oysters have been under siege. As a result of overharvesting, pollution, habitat destruction, and parasitic and viral diseases, oyster stocks are nonexistent in many areas formerly known as rich sources of oysters for human consumption. Protozoans have caused substantial damage to oyster populations in the eastern United States. The parasites Dermo (*Perkinsus marinus*), SSO (*Haplosporidium costale*), and MSX (*Haplosporidium nelsoni*) have devastated eastern oyster (*Crassostrea virginica*) populations in many areas of the country such as the Chesapeake Bay in Maryland and the Delaware Bay in New Jersey. Similar problems have occurred along the Gulf Coast and the coast of California.

Until recently, part of the difficulty in combating these diseases was a lack of sufficient numbers of parasites to study. Cell culture techniques have been used to overcome this problem, and researchers are now learning a great deal about the molecular biology of Dermo, SSO, and MSX. Molecular techniques have been used to learn about the life cycle of these parasites, and molecular probes are now available for their detection in the field and in aquaculture facilities. For instance, PCR-based approaches have proven very effective for the rapid and sensitive detection of Dermo, SSO, and MSX, allowing aquaculturists to screen and identify diseased oysters before widespread infections occur (Figure 10.8).

A greater understanding of genes involved in the oyster's immune system has also led to new strategies for transferring genes for disease resistance from species such as the Pacific oyster (*Crassostrea gigas*) into the more susceptible species of eastern oysters such as *C. virginica*. In the future it may be possible to create transgenic species of oysters with genes that will provide resistance to damage by these and other parasites.

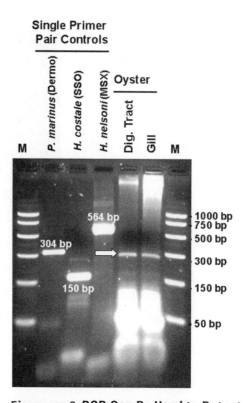

Figure 10.8 PCR Can Be Used to Detect DNA from Oyster Pathogens Dermo, SSO, and MSX are significant causes of mortality of oysters. PCR can be used to detect DNA from these pathogens in oyster tissues, helping scientists determine when these parasites are present in an effort to control oyster infection that can decimate populations of these shellfish. In this example, notice that DNA samples from the digestive tract and gill of an oyster show the presence of DNA for Dermo (indicated by *arrow*). M, DNA size markers

Genomics and aquatic organisms

Not surprisingly, genomics approaches are being used to analyze genomes for a number of aquatic species. In 2006, an international team of scientists announced that a near-complete draft of the sea urchin genome had been completed. Sea urchins are a type of echinoderm, a group of marine animals that include starfish and sea cucumbers. Urchins are believed to have originated over 540 million years ago. Their genome has been of great interest to molecular biologists because urchins share a common ancestor with humans. From this ancestor a superphylum of animals called the Deuterostomes arose; included in this phyla are echinoderms, humans, and other vertebrates. Because Deuterostomes are more closely related to each other than to animals not in this superphylum, genome scientists expect that sea urchins will be genetically similar to humans. If this is true, urchins may become very valuable experimental model organisms for molecular biologists.

Genetic Manipulations of Finfish and Shellfish

Biologists throughout the world are using genetic techniques to create breeds of finfish and shellfish with such desired characteristics as improved growth rates and disease resistance. Table 10.2 shows examples of different species that have been genetically engineered for a variety of purposes.

Genetically engineered species: transgenics and triploids

In previous chapters, we discussed how *transgenic* organisms (transgenics) can be created. Transgenic fish, like other transgenics, contain DNA from other species, and

Table 10.2 GENETICALLY ENGINEERED FISH AND SHELLFISH	
Fish Species	**Shellfish**
Atlantic salmon	Abalone
Bluntnose bream	Clams
Channel catfish	Oysters
Chinook salmon	
Coho salmon	
Common carp	
Gilthead bream	
Goldfish	
Killifish	
Largemouth bass	
Loach	
Medaka	
Mud carp	
Mummichog	
Northern pike	
Penaeid shrimp	
Rainbow trout	
Sea bream	
Striped bass	
Tilapia	
Walleye	
Zebrafish	

Source: Adapted from Goldberg, R., and Triplett, T. (2000): *Something Fishy*. Environmental Defense, www.environmentaldefense.org.

interest in transgenic fish has increased along with the aquaculture industry. Aquaculturists are interested in using recombinant DNA techniques to genetically engineer fish designed to grow faster and healthier.

As shown in Table 10.3, many species have been genetically modified for potential applications in the aquaculture industry. Foreign genes have been inserted into finfish and shellfish to accelerate growth, increase cold tolerance and disease resistance, and alter flesh qualities to improve table quality. Although transgenic strains of corn, soybean, and tomatoes have been in use in the United States for years, no transgenic fish have been approved by the FDA for human consumption. A/F Protein, Inc., has requested approval from the agency to market transgenic fish containing AFP genes. Many other companies are also seeking FDA approval for the sale of transgenic fish. Cuba is the most progressive country in the world when it comes to the use of genetically modified foods, particularly seafood. Genetically modified tilapia are already sold for human consumption in Cuba.

Aqua Bounty Farms (Waltham, Massachusetts) has produced transgenic Atlantic salmon containing a growth hormone gene from Chinook salmon—a species that demonstrates more rapid growth and larger adult-size capabilities than most Atlantic salmon. These transgenics grow nearly 400 to 600% faster than nontransgenic salmon! (Figure 10.9, on page 249) Such fish reach market-size in 18 months instead of the traditional 30 months. Aqua Bounty has been trying for more than 10 years to get FDA approval for these salmon despite data that indicates no difference in nutritional value, appearance, or taste of these fish compared to wild salmon.

Similarly, Norwegian companies have developed genetically modified farm-raised tilapia that grow nearly twice as fast as wild tilapia. Controversy as to whether genetic modification does indeed enhance the growth of farmed trout has arisen. Several conflicting studies have been reported; however, transgenic strains such as salmon and trout raised by aquaculture generally show increased growth rates compared with nontransgenic domestic strains.

In 2004, Yorktown Industries of Austin, Texas, made popular news headlines when they announced they had created the GloFish, a transgenic strain of zebrafish containing the red fluorescent protein gene from sea anemones. Advertised as the first genetically modified pet sold in the United States, when ultraviolet light illuminates GloFish, they fluoresce bright pink. A number of antibiotechnology groups voiced their protests about this novelty use of genetic engineering.

Although transgenic species are the most common type of genetically modified marine species, a number of **polyploid** species have been created. Polyploids are

Table 10.3 GENETICALLY MODIFIED SPECIES BEING TESTED FOR USE IN AQUACULTURE

Species	Foreign Gene	Desired Effect and Comments	Country
Atlantic salmon	AFP	Cold tolerance	United States, Canada
	AFP salmon GH	Increased growth and feed efficiency	United States, Canada
Coho salmon	Chinook salmon GH + AFP	After 1 year, 10- to 30-fold growth increase	Canada
Chinook salmon	AFP salmon GH	Increased growth and feed efficiency	New Zealand
Rainbow trout	AFP salmon GH	Increased growth and feed efficiency	United States, Canada
Cutthroat trout	Chinook salmon GH + AFP	Increased growth	Canada
Tilapia	AFP salmon GH	Increased growth and feed efficiency; stable inheritance	Canada, United Kingdom
	Tilapia GH	Increased growth and stable inheritance	Cuba
	Modified tilapia insulin-producing gene	Production of human insulin for diabetics	Canada
Salmon	Rainbow trout lysosome gene and flounder pleurocidin gene	Disease resistance; still in development	United States, Canada
Striped bass	Insect genes	Disease resistance; still in early stages of research	United States
Channel catfish	GH	33% growth improvement in culture conditions	United States
Common carp	Salmon and human GH	150% growth improvement in culture conditions; improved disease resistance; tolerance of low oxygen levels	China, United States
Goldfish	GH AFP	Increased growth	China
Abalone	Coho salmon GH + various promoters	Increased growth	United States
Oysters	Coho salmon GH + various promoters	Increased growth	United States
Rabbit	Salmon calcitonin-producing gene	Calcitonin production to control calcium loss from bones	United Kingdom
Strawberry and potatoes	AFP	Increased cold tolerance	United Kingdom, Canada

Note: The development of transgenic organisms requires the insertion of the gene of interest and a promoter, which is the switch that controls expression of the gene. AFP, antifreeze protein gene (Arctic flatfish); GH, growth hormone gene.

organisms with an increased number of *complete sets* of chromosomes. As we have already discussed, most animal and plant species are **diploid** (abbreviated *2n*, where *n* = number of chromosomes), meaning that they have two sets of chromosomes in their somatic cells and a single, or **haploid** number (*n*), of chromo-

Figure 10.9 Transgenic Salmon Transgenic salmon, which overexpress growth hormone genes, show greatly accelerated rates of growth compared with wild strains and nontransgenic domestic strains of salmon. On average, these transgenic strains of salmon weigh nearly 10 times more than nontransgenic salmon.

somes in their gametes. In humans, the haploid number of chromosomes is 23; therefore, the diploid number of chromosomes in human somatic cells is 46 ($2n = 46$). Figure 10.10 shows a simple representation of the differences among haploid, diploid, and polyploid species.

The majority of polyploid marine species created to date are **triploid** species. Triploid organisms contain three sets of chromosomes ($3n$). A number of different techniques can be used to create triploids. Triploids are usually derived by subjecting fish eggs

to a temperature change or chemical treatment to interfere with egg cell division. Eggs treated in this way mature with an extra set of chromosomes.

For instance, treating eggs with **colchicine**, a chemical derived from the crocus flower (*Colchicum autumale*), is one common approach to creating polyploids (Figure 10.11a). Colchicine blocks cell division by interfering with the formation of microtubules that are necessary for cell division. As a result, the chromosomes replicate in treated egg cells, but these cells are incapable of dividing. Therefore, these eggs have a diploid number of chromosomes, twice as many chromosomes as normal. Fertilization of such an egg cell by a normal haploid sperm cell results in a triploid organism. Another approach for producing triploids involves fertilizing a normal haploid egg with two spermatozoa (Figure 10.11b). The resulting embryo is a triploid containing one set of chromosomes from the egg and one set from each of two different sperm.

Polyploidy usually influences the growth traits of an organism. For instance, triploids typically grow more rapidly, in most species 30 to 50% faster, and larger than their normal diploid cousins. But most triploids are sterile because they produce gametes with an abnormal number of chromosomes, or in some cases triploids may not produce any gametes. One of the first widely used triploid strains of fish was the triploid grass carp (*Ctenopharyngodon idella*). Grass carp have a voracious appetite for many types of aquatic vegetation. In the early 1960s, the U.S. Fish and Wildlife Service imported diploid grass carp, which are

Normal choromosome complement in a somatic cell

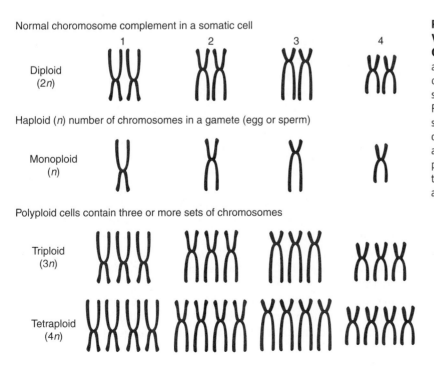

Figure 10.10 Polyploid Species Have Variations in Complete Sets of Chromosomes Somatic cells from many animal and plant cells have a diploid number of chromosomes, whereas gametes have a single set, or haploid number, of chromosomes. Polyploids contain three or more sets of chromosomes. Chromosomes from an organism with diploid number ($2n$) of eight chromosomes are shown. Although tetraploid ($4n$) and pentaploid ($5n$) organisms can be created, the vast majority of marine polyploid species are triploids ($3n$).

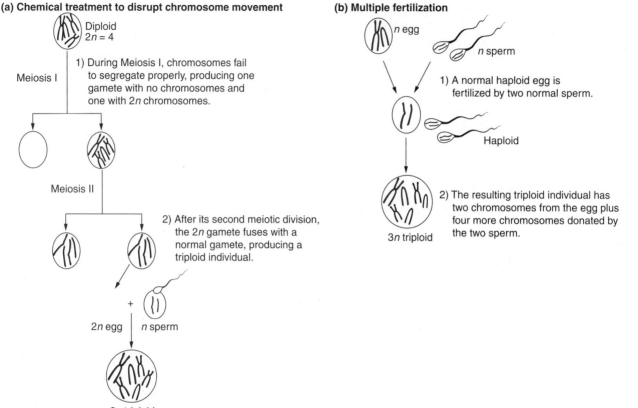

(a) Chemical treatment to disrupt chromosome movement

Diploid
$2n = 4$

Meiosis I

1) During Meiosis I, chromosomes fail to segregate properly, producing one gamete with no chromosomes and one with $2n$ chromosomes.

Meiosis II

2) After its second meiotic division, the $2n$ gamete fuses with a normal gamete, producing a triploid individual.

$2n$ egg + n sperm

$3n$ triploid

(b) Multiple fertilization

n egg

n sperm

1) A normal haploid egg is fertilized by two normal sperm.

Haploid

$3n$ triploid

2) The resulting triploid individual has two chromosomes from the egg plus four more chromosomes donated by the two sperm.

Figure 10.11 Producing a Triploid (a) Egg cells can be chemically treated to block chromosome movement during cell division to produce diploid eggs. (b) When these eggs are fertilized by a normal haploid sperm cell, triploid offspring are produced. A triploid can also be created by the fertilization of a normal haploid egg by two normal haploid sperm.

native to Malaysia. In the early 1980s, triploid grass carp were developed by temperature shocking eggs to create diploid eggs and fertilizing these eggs with normal haploid sperm. Triploids grow rapidly and can exceed 25 pounds. Because of their ability to consume large amounts of vegetation, grass carp became very popular for controlling the growth of aquatic weeds in freshwater ponds and lakes throughout the United States. Triploid grass carp became an instant favorite among pond and lake managers, who could stock these fish in lakes as a "natural" way to control weed growth, minimizing the need to use large doses of chemical herbicides.

However, not everyone has been thrilled with triploid grass carp. In some waterways, because of grass carp overpopulation (these grass carp have few natural predators), they have been so effective at eating their way through aquatic vegetation that water quality has changed substantially. Dramatic decreases in weeds used as habitat by many sport fish such as trout and bass have led to a decline in fishing in previously productive fishing waters. Also, many triploids

have escaped into waters adjacent to their original stocking, leading to vegetation loss in areas such as marshes and natural lakes not originally targeted for weed control. Finally, reproduction from these presumably sterile fish has also been documented.

Another polyploidy success story involves the triploid oyster, credited with reviving a diminishing oyster industry on the West Coast. Oyster production is typically seasonal because it is strongly influenced by weather, breeding patterns, and habitat. Not only has the development of a triploid strain of the Pacific oyster provided for year-round harvesting, but the triploid oysters also grow substantially larger than their diploid cousins (Figure 10.12). Diploids normally store sugars in their tissue and then become lean in the summer while expending energy to spawn. Lean oysters are undesirable for market sale. Because triploids do not spawn, they grow fat all summer, making them market ready.

In the next section we introduce you to some of the valuable ways in which aquatic organisms are being used for important medical applications.

(a)

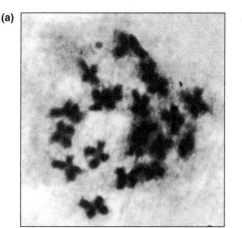

(b)

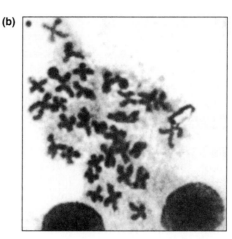

Figure 10.12 Triploid Oysters   (a) The wild Pacific oyster (*Crassostrea gigas*) is a diploid organism with two sets of 10 chromosomes. (b) Triploid strains of *Crassostrea* have three sets (3*n*) of 10 chromosomes.

10.4 Medical Applications of Aquatic Biotechnology

Relatively few products derived from aquatic organisms have medical applications, but this is rapidly changing. Many scientists believe that oceans and freshwater habitats possess near limitless opportunities for the identification of medical products. In the future, it is anticipated that new and important classes of drugs will be derived from aquatic organisms and used for human benefit, and marine organisms may be used as biomedical models to understand, diagnose, and treat human diseases.

Drugs and Medicines from the Sea

A wide number of marine species contain or are suspected to contain compounds of biomedical interest,

Q How do scientists discover bioactive aquatic products that can be used for medical treatments?

A This process often begins by a simple observation. Someone studying a marine species notes an unexplained behavior or an interesting observation. As an example, older salmon do not seem to be affected severely by osteoporosis. Does this mean they produce hormones that affect bone density? Classic biochemical approaches are often used in an attempt to address a question like this by isolating active molecules. Tissues are homogenized and separated by biochemical approaches and then often tested using *in vitro* cell culture to assay for biological activity. If active molecules are found, the gene(s) responsible for producing these molecules may be cloned (as described in Chapter 3).

including antibiotics, antiviral molecules, anticancer compounds, and insecticides. These species include sea sponges, members of the phylum Cnidaria (hydras, jellyfish, sea anemones, and a variety of corals), members of the phylum Mollusca (snails, oysters, clams, octopuses, and squids), and sharks, among many others. The wealth of marine organisms under study is broad and impressive. As research progresses, we have many reasons to be optimistic that the waters of the world will yield novel medical treatments. The future looks very promising for applications of "drugs from the deep." Let's consider a few potential medical applications of aquatic products.

Osteoporosis, a condition characterized by a progressive loss of bone mass, creates porous and brittle bones that can lead to fractures of the hip, legs, and joints, which severely hinder an individual's lifestyle. Over 90% of the roughly 25 million Americans affected by osteoporosis are women. A common treatment for osteoporosis is estrogen therapy. This medication is ineffective for many women, and the long-term health effects of estrogen are a concern. Other individuals are treated with human recombinant **calcitonin,** a thyroid hormone that stimulates calcium uptake and bone calcification and inhibits bone-digesting cells called osteoclasts. Recently, researchers have discovered that some species of salmon produce a form of calcitonin with a bioactivity that is 20 times higher than that of human calcitonin. Cloned forms of salmon calcitonin are now available for delivery as an injection form and a nasal spray.

The exquisite patterns of coral reefs around the world are created by the skeletons of corals. These structures partially consist of **hydroxyapatite (HA),** an important component of the matrix that constitutes bone and cartilage in animals including humans. The biotechnology company Interpore International has developed technology that allows HA implants to be cut into small boxes and used to fill gaps in fractured bones. These boxes are ultimately invaded by local connective tissue cells that speed repair. As a result, patients avoid needing bone grafts from other parts of their body. These implants may also serve to fill bone material lost around the root of a tooth.

Similarly, a number of adhesives have been identified in glue-like resins produced by mussels and other shellfish. A favorite cuisine in many seafood restaurants, mussels (*Mytilus edulis*) are hinged shelled mollusks that live in harsh, physically demanding environments. They typically adhere to rocks or pilings at the edges of oceans and bays. Day after day, these creatures are pounded by waves. They dry out during low tides, then get submerged and pounded by waves again as the tide rises. How do they maintain their contacts to rocks and other structures without being crushed or pulled off the rocks? The answer lies in a unique form of protein-rich superadhesive called **byssal fibers** (Figure 10.13).

Byssal fibers are several times tougher and more extensible than human tendons, which themselves are tougher than steel. Adhesive and elastic properties of byssal fibers absorb energy and stretch as waves tug away at the mussels. Although it would be cost prohibitive

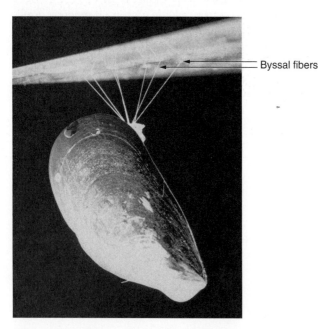

Byssal fibers

Figure 10.13 Mussels Produce Unique Adhesives Mussels and other shellfish cling tightly to rocks and other surfaces by producing a unique type of adhesive called byssal fibers. Byssal fibers are remarkably strong. These fibers can withstand tremendous physical forces such as those created by waves in the water where mussels live. Because of their strength and ability to stretch, byssal fibers are being investigated for their possible use in a variety of interesting applications for medicine and industry.

to isolate byssal fibers from mussels directly (nearly 10,000 mussels would be required for 1 gram of adhesive), scientists are using recombinant DNA techniques to express the byssal fiber genes in bacteria and yeast to produce these adhesive proteins on a large scale.

Although still several years from development, byssal fiber proteins are being considered for a wide variety of diverse applications from automobile tires to shoes and from bone and teeth repair strategies to soft body armor for soldiers. Other potential uses include surgical sutures and artificial tendon and ligament grafts.

One category of drugs from the sea that appears to provide promising medical applications of aquatic biotechnology is the identification of anti-inflammatory, analgesic (pain-killing), and anticancer compounds that may be used to treat humans. For example, researchers have identified a Pacific sponge that produces a non-steroidal compound called manoalide. This substance possesses anti-inflammatory and analgesic properties and is currently being investigated in clinical trials in humans. Over a dozen different anticancer compounds have been isolated from marine invertebrates, particularly sea sponges, tunicates, and mollusks. Many of these compounds are in various stages of clinical trials that will ideally lead to new and effective drugs on the market. Several groups of researchers are studying venomous marine creatures with the hope of identifying substances that may be used to treat nervous system disorders.

Marine cone snails, a potentially lethal species, produce conotoxins, molecules that can target specific neurotransmitter receptors in the nervous system. In 2004, the FDA approved the drug Prialt, a peptide conotoxin purified from the marine cone snail *Conus magus* by Elan Corporation of Ireland. Conotoxins such as Prialt represent a promising new source of neurotoxins with the ability to act as strong painkillers by blocking neural pathways that relay pain messages to the brain. Prialt has been successfully used to treat chronic, severe forms of pain such as back pain.

Researchers are also examining anti-inflammatory compounds found in coral extracts. Such compounds may lead to new treatment strategies for skin irritations and inflammatory diseases such as asthma and arthritis. Table 10.4 lists examples of medical compounds isolated from aquatic organisms.

Other researchers are developing culturing systems to provide adequate supplies of marine organisms such as single-celled plankton called dinoflagellates, which contain antitumor and cancer-treating abilities. Recently, a marine invertebrate called *Bagula neritina* was shown to contain minute amounts of a compound that is active against certain

Table 10.4 EXAMPLES OF MEDICAL COMPOUNDS FROM AQUATIC ORGANISMS

Organism	Medical Product	Application
African Clawed Frog (*Xenopus laevis*)	Magainins	Antimicrobial peptides first discovered in frog skin. Used to treat bacterial infections.
Coho Salmon (*Oncorhynchus kisutch*)	Calcitonin	Hormone that stimulates calcium uptake by bone cells. Used to treat osteoporosis.
Hammerhead Shark (*Sphyrna lewini*)	Neovastat	Antiangiogenic compound (blocks blood vessel formation). Used to treat cancer.
Leech (*Hirudo medicinalis*)	Hirudin	Saliva peptide from leeches used as an anticoagulant to thin blood.
Marine Cone Snail (*Conus magus*)	Prialt (ziconotide)	Synthetic peptide toxin used as pain reliever to treat chronic, severe pain and arthritis.

types of leukemia. Because large amounts of this compound will be needed for human studies, molecular cloning techniques will be important if this compound is to be produced in mass quantities.

Similarly, the Japanese pufferfish, or blowfish (*Fugu rubripes*), has been getting a lot of attention lately. *Fugu* is famous for its ability to swallow water and "puff up" when threatened and to produce a potent nerve cell toxin called tetrodotoxin (TTX). TTX is one of the most toxic poisons ever discovered (nearly 10,000 times more lethal than cyanide). Hollywood has often depicted *Fugu* in movies, showcasing its ability to produce its deadly toxin. In Japan, *Fugu* is a prized and very expensive delicacy for many sushi lovers who enjoy the food quality of this tasty fish despite the risk (eating *Fugu* kills nearly 100 people, mostly in Japan, each year).

Scientists have used TTX to develop a greater understanding of how proteins called sodium channels help neurons produce electrical impulses. TTX is a deadly poison because it blocks sodium channels and prevents nerve impulse transmission. An understanding of how TTX affects sodium channels has led to the development of new drugs that are being tested not only as anesthetics to treat patients with different types of chronic pain but also as anticancer agents in humans (Figure 10.14). Researchers are also working on sequencing the pufferfish genome, which contains nearly the same number of genes as humans but in a much smaller genome. *Fugu* also contains far less noncoding DNA (introns) than humans, so it is considered an ideal model organism for studying the importance of introns.

A steroid called squalamine, first identified in dogfish sharks (*Squalus acanthias*), appears to be a potent antifungal agent that may be used to treat life-threatening fungal infections that can fatally affect patients

Figure 10.14 Pufferfish are Helping Scientists Discover New Ways to Treat Cancer and Chronic Pain in Humans

with conditions such as AIDS and cancer. Sharks rarely develop cancer, and shark cartilage has been proposed to be a rich source of anticancer agents. Although no compounds from shark cartilage have demonstrated effectiveness in controlled clinical trials, shark cartilage extracts possess antiangiogenic compounds. **Angiogenesis** is the formation of blood vessels, a process that is often required for growth and development of many types of tumors. By blocking blood vessel formation, antiangiogenic compounds derived from marine species show promise for inhibiting the growth of certain tumors. In addition, because many aquatic organisms live in harsh environments, scientists are optimistic that we can learn from the adaptations these organisms have developed. For example, researchers are currently studying marine organisms that show tolerance to ultraviolet light as a potential source of natural sunscreens.

Even discarded crab shells from the commercial crabbing industry play a role in medical applications of aquatic biotechnology. The outer skeleton or exoskeleton of members of the phylum Arthropoda—which includes crabs, lobsters, shrimp, insects, and spiders—is a rich source of **chitin** and **chitosan.** These complex carbohydrates are structurally similar to cellulose, which forms the tough outer layer of the cell wall in plants. Cellulose is widely known as a source of dietary fiber. Similarly, chitin and chitosan are also sources of fiber. Eating vegetables and fruits to get fiber is much gentler on your digestive tract than eating crab shells. Nonetheless, ground-up extracts of crab shell can be purchased as a powder in many nutrition stores. Many skin creams and contact lenses also contain chitin, and chitin has been used to create nonallergenic dissolvable stitches that appear to stimulate healing when used in humans.

To date, few drugs from the sea have widespread use in the medical market; however, in the future, recombinant DNA technologies will lead to enhanced abilities to produce bulk quantities of bioactive compounds typically found in very low concentrations in aquatic organisms. As we discuss in the next section, one of the most successful medical applications of aquatic biotechnology has been the use of aquatic organisms for monitoring health and disease.

Monitoring Health and Human Disease

During early spring and summer along many beaches on the East Coast of the United States, a common scene has repeated itself for decades. Large numbers of horseshoe crabs (*Limulus polyphemus*) invade shallow bays to mate (see chapter-opening photo). Horseshoe crabs preceded dinosaurs on the earth. In some ways, these "living fossils" have changed very little from their initial appearance over 300 million years ago.

L. polyphemus was one of the first marine organisms to be used successfully for medical applications. In the early 1950s, it was discovered that horseshoe crab blood contains cells that kill invading bacteria. From this simple observation, a very powerful medical application of marine biotechnology was developed. The **limulus amoebocyte lysate (LAL) test** is an extract of blood cells (amebocytes) from the horseshoe crab that is used to detect bacterial **endotoxins.** Endotoxins, also called lipopolysaccharides, are part of the outer cell wall of many bacteria such as *E. coli* and *Salmonella*. Endotoxins are a type of cytotoxins, molecules that are toxic to cells. Endotoxins can cause instant death to many cells grown in culture. In humans, exposure to endotoxins from certain bacteria can result in mild symptoms such as joint pain, inflammation, and fever to more severe conditions such as a stroke. Certain endotoxins can be lethal.

Researchers discovered that horseshoe crab blood would clot when exposed to whole *E. coli* or purified endotoxins. They later determined that amebocytes—which are similar to human white blood cells—in horseshoe crab blood could be lysed, centrifuged, and freeze-dried to create a lysate that can be used in an LAL test. The LAL test, a rapid and very effective assay for endotoxins in human blood and fluid samples, is also used to ensure that cytotoxins are not present in biotechnology drugs such as recombinant therapeutic proteins. It is also used to detect bacteria in raw milk and beef. In addition, many medical companies and hospitals use the LAL test to make sure that surgical instruments, needles used for drawing blood and cerebrospinal fluid, and implanted devices such as pacemakers are endotoxin free. The LAL test is an outstanding example of the power of marine biotechnology for human benefit.

A number of assays similar to the LAL test are under development. Several of the medical products we have discussed in this section were identified while studying aquatic organisms in order to develop nonmedical products. In the next section, we consider some of these products.

10.5 Nonmedical Products

Throughout this chapter, we have examined a wide range of aquatic biotechnology applications. To further appreciate the potential of our world's waters, we now take a look at a number of aquatic products, from research reagents to food supplements, that have had an impact in the biotechnology industry.

A Potpourri of Products

In Chapter 3, we discussed the importance of *Taq* polymerase, isolated from hot-springs Archae *Thermus aquaticus*, which allowed for the development of the PCR as a powerful tool in molecular biology. The ocean also has proved to be an excellent source of enzymes and other products that have played an important role in basic and applied research. For example, bacteria living near hydrothermal vents (hot water geysers on the ocean floor) have yielded a second generation of heat-stable enzymes for use in PCR and DNA-modifying enzymes, including ligases and restriction enzymes.

Other enzymes produced by marine bacteria possess a variety of interesting properties that may result in important applications in the future. For example, some enzymes are salt resistant, which renders them ideal for industrial scale-up procedures involving high-salt solutions. In Chapter 5, we discussed the role of the bioluminescent bacterium *Vibrio harveyi* in detecting environmental pollution. Researchers have discovered marine species of *Vibrio* that produce a number of proteases, including several unique proteases that are resistant to detergents used in many manufacturing processes. As a result, these detergent-resistant proteases may have potential applications for degrading proteins in cleaning processes, including their use in laundry detergent for removing protein stains in clothes.

Vibrio is also a good source of **collagenase,** a protease used in tissue culturing. When scientists are looking to grow cells, such as liver cells in culture, they can use collagenase to digest the connective tissues holding cells together so the individual cells can be dispersed into cell culture dishes.

Another product of the sea is **carrageenan,** listed as an ingredient in many preserved foods, toothpaste, and cosmetics. This sulfate-rich polysaccharide, extracted from red seaweeds, has been used in many products for over 50 years. There is a large family of carrageenan polysaccharides. They have the ability to form gels of varying densities at different temperatures. As a result, carrageenans have been used as thickening agents and for improving how foods "feel" in our mouths when we eat them. Some of the most common applications of carrageenans include their use as stabilizing and bulking agents in chewing gum, chocolate milk, beers and wine, salad dressing, syrups, sauces, processed lunch meats, adhesives, textiles, polishes, and hundreds of other products. Marine algal products have a long history of applications, and wild seaweeds have been harvested since the beginning of humankind.

The Philippines is the world's largest producer of red seaweeds (Rhodophyta) from which many carrageenans are derived. Red algae are also prized for their use in nori, a paper-thin seaweed product used to wrap sushi. Improvements in farming seaweeds have allowed for increased production of other algal polymers including agar and alginic acids, which are important research materials used, respectively, to make agar gels for plating bacteria and for creating agarose gels for DNA electrophoresis, as discussed in Chapter 3.

Biomass and Bioprocessing

One newly emerging area of marine biotechnology involves the exploration of marine biomass. A mat of aquatic weeds of algae represents biomass, as does a school of fish. As you know, plants are responsible for the production of oxygen through the process of photosynthesis in which carbon dioxide and water are converted into carbohydrates and oxygen. Marine plants (including seaweeds, grasses, and planktons) use photosynthesis to capture and convert a tremendous amount of energy (nearly 30% of all energy production worldwide) from the sun into chemical energy. Can chemical energy from such biomass be harvested? Scientists are examining ways in which algae and plants may be used to produce alternative fuels. For example, it may be possible to take advantage of the rapid biosynthetic capabilities of marine algae with their ability to mass-produce hydrocarbons and lipids in extraordinary quantities to provide alternate sources of materials that are normally cost prohibitive to produce or isolate from natural materials. Similarly, it may be possible to convert marine biomass into fuels such as ethanol.

The U.S. Naval Research Lab has investigated potential ways to use plankton as underwater "fuel cells." Plankton at the water's surface release energy as they undergo photosynthesis, whereas plankton closer to the sediment at the bottom of the ocean (where there is less oxygen) use other reactions to generate energy. As a result, scientists have found that these plankton create a natural voltage gradient from the surface to the ocean floor that can be harvested to produce an indefinite source of electricity. In the future, the ocean may turn out to be a valuable resource for providing energy. Lastly, scientists are exploring ways in which biomass of marine algae may be used to increase absorption of carbon dioxide and decrease greenhouse effects on the earth.

Related to applications of biomass, marine scientists are exploring ways in which **bioprocessing** may involve marine products. Bioprocessing is a general term that describes engineering approaches to produce a biological product such as a recombinant protein. Algae may potentially be very valuable for expressing recombinant proteins. Researchers have found that they can make an abundance of proteins, such as antibodies, in marine algae because they can be grown on a very large scale. Marine biologists are exploring how marine organisms may be used to synthesize a variety of polymers and other biomaterials, which may be used for industrial manufacturing processes. For instance, oyster shell proteins are being considered as additives in detergents and other solvents as nontoxic, biodegradable alternatives to currently used materials.

We conclude this chapter by discussing how aquatic organisms are being considered for use in applications to clean up the environment.

10.6 Environmental Applications of Aquatic Biotechnology

Unfortunately the world's oceans have long served as dumping grounds for the wastes of humanity and industrialization, with too little thought given to the effect of pollution on fish stocks, marine organisms, and the environment. Clearly, oceans do not have an infinite ability to accept waste products without consequences, and we are seeing the results of this as wetlands and other estuarine environments—which are critical habitats for the spawning of many marine species and the growth of young marine organisms—are showing signs of severe decline due to pollution and human impact.

In Chapters 5 and 9, we discussed how bacteria can be used to detect and degrade environmental pollutants. Many microorganisms have been used to clean up aquatic ecosystems. In this section, we consider how aquatic organisms can be used to help clean up and control pollution in the environments in which they live.

Antifouling Agents

Biofilming, also called biofouling, refers to the attachment of organisms to surfaces. These surfaces could be manufactured surfaces such as the hulls of ships, inner lining of pipes, cement walls, and pilings used around piers, bridges, and buildings. Biofilming also occurs on the surface of marine organisms, especially shellfish. If you have spent any time around boat docks and piers located in marine environments, you have undoubtedly seen the effects of biofilming on structures that are in the water.

In marine environments, barnacles, algae, mussels, clams, and bacteria are among the most common organisms responsible for biofilming (Figure 10.15). As the term *biofilming* suggests, these organisms literally grow to create a layer, or usually multiple layers, that create a "film" over the surfaces to which they adhere. However, biofilming is not exclusively a marine problem. For example, similar biofilming occurs in the plumbing of your home, on contact lenses, and in your mouth. Bacteria that coat your teeth and bacteria that adhere to implanted surgical devices and prostheses are examples of biofilming. Although brushing your teeth regularly minimizes biofilming in your mouth, such simple remedies are not available for deterring biofilming in marine environments.

Biofilming can create significant problems. For instance, attachment of marine organisms to the hull of a ship can greatly increase the resistance of the ship as it moves through the water, slowing its travel time and reducing its fuel efficiency. A progressive accumulation of biofilms can clog pipes,

(a)

(b)

Figure 10.15 Unwanted Invaders Create Biofilms That Have Serious Economic Impacts   (a) Uncontrolled growth of unwanted species such as the zebra mussels presents many problems. (b) Mussels, barnacles, and other organisms can create hard biofilms.

block water intake and filtration systems for ships, and corrode metal surfaces. Furthermore, colonization of surfaces with invertebrates and mollusks creates "hard" biofouling that is difficult and costly to remove (Figure 10.15).

Traditionally, most antifouling agents have employed toxic chemicals such as copper-rich and mercury-rich paints to minimize the growth of fouling organisms. However, the metals that leach from these paints can contribute to environmental pollution. As a result, researchers are investigating the natural mechanisms that many organisms use to prevent biofouling on their own surface. If biofilming is a problem for both manufactured surfaces and the surfaces of marine organisms, how do clams, mussels, and even turtles minimize biofilming and thus prevent their shells from being completely closed by biofilming organisms? Scientists are using molecular techniques to find the answer. Some organisms are thought to produce repelling substances; other organisms appear to produce molecules that block adhesion of biofilming organisms (Figure 10.16). Many marine sea grasses, such as the eelgrass *Zostra marina*, and algae produce compounds that prevent or neutralize bacteria, fungi, and algae from adhering. In the near future, these compounds may be used to produce protective coatings for covering ship hulls, aquaculture equipment, and other surfaces susceptible to biofilming.

The fight against biofilms involves controlling the growth of marine organisms, and marine biotechnologists are also very interested in understanding how marine organisms can be used to detect and monitor sources of chemical pollution in the environment, as the initial step to developing methods to control and clean up pollution.

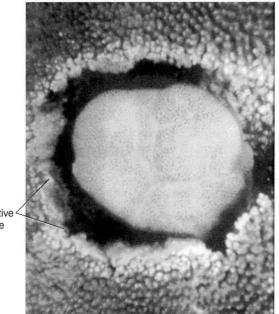

protective zone

Figure 10.16 Many Marine Organisms Have Natural Antifouling Mechanisms Many stationary marine organisms release defensive chemicals to create a protective zone (indicated by the dark-colored band around this marine coral) around them. These chemicals may deter fouling microorganisms and protect the organism from predators.

Biosensors

Many groups are working to explore the use of aquatic organisms as biosensors to detect low concentrations of pollutants and toxins in waterways. In Chapter 5, we discussed how bioluminescent strains of bacteria can be used as biosensors. Some species use bioluminescence to illuminate their environment; others use it to find mates in the dark depths of the ocean. Most bioluminescent deep-sea fish are involved in symbiotic relationships with bioluminescent bacteria such as *Vibrio fischeri*.

V. fischeri and other bioluminescent strains such as *V. harveyi* express *Lux* genes, which code for the light-emitting enzyme luciferase. In response to changing environmental conditions, the intensity of light emitted by *Vibrio* can change. Because of this ability, *Vibrio* has been used as a biosensor to detect pollutants such as organic chemicals and nitrogen-containing compounds in marine environments.

Not only are some marine organisms useful for detecting environmental pollutions, but many marine species are also believed to possess metabolic pathways for degrading a range of substances both natural and manufactured. Much research is dedicated to characterizing biochemical pathways involved in degradative processes to determine how marine organisms may be used to remediate, or clean up, the environment of a variety of hazardous substances that enter marine environments.

Environmental Remediation

Recall our discussion from Chapters 5 and 9 of how native microorganisms or genetically engineered strains have been used to degrade chemicals. In much the same way, marine organisms possess unique mechanisms for breaking down substances including toxic organic chemicals such as phenols and toluene, oil products found in harbors and adjacent to oil rigs, and toxic metals.

One of the earliest techniques used in marine remediation involved increasing the quantity of shellfish in polluted areas. Scientists introduced racks of shellfish into polluted waters to take advantage of the natural feeding method of bivalves such as clams, oysters, and mussels. Because these organisms strain the water, they act as a form of estuarine filters to remove wastes such as nitrogen compounds and organic chemicals. These chemicals, in turn, are absorbed in tissues of the shellfish. After periods of time, these shellfish can be harvested, thus removing wastes from the water.

Heavy metal contamination of marine waters is the result of many industrial manufacturing processes. As a solution to this problem, scientists have isolated

CAREER PROFILE

Careers in Aquaculture

Aquaculture offers a varied range of exciting career opportunities. Approximately 36 million people worldwide are employed in aquaculture and related areas. Those who hold entry-level or technician positions maintain appropriate environmental conditions for the species that are being cultured, feed or supply nutrients in some form, and dispose of the fecal material or uneaten food. Although computer monitoring of many systems is gaining acceptance, visual inspection of farmed organisms (the "crop") cannot be underestimated. For example, changes in growth patterns, feeding, or behavior can foreshadow problems with the crop that must be addressed. Nevertheless, performing manual processes is usually an important part of the job. Even the facility foreperson and manager will likely be covered with water or waste materials from time to time—this is part of the business.

There are many potential career paths in aquaculture. Engineers design aquaculture production systems. Biologists and food scientists develop feeds for optimum growth. Plant and animal geneticists breed new hybrids of species presently under culture that may grow faster or have characteristics that are more suitable for sales or hardiness. Veterinarians must attend to sick fish and nurse them back to health. There are also opportunities for sales positions in aquaculture because eventually someone has to sell the end product. If the cultured product is not sold at a profit, considerable effort has been wasted.

Those entering into the aquaculture field at the technician level should have a background in the sciences or animal husbandry. There are limited jobs for those with a high school education, but most employers would look for a motivated person with at least an associate's degree. A B.S. degree in biology or animal science and experience in an aquaculture facility is preferred, and those going into research will need at least a master's degree, if not a Ph.D. Experience is a wonderful teacher, and those who have an interest in fish or plant husbandry should make every attempt to garner experience wherever possible. In addition, if there is one thing that every aquaculturist becomes familiar with, it is plumbing! If a plumbing course ever presents itself, take it.

Knowledge of both air and water pumps and basic hydrodynamics is critical because it is integral to aquaculture.

Fish and plants are living things. If they are ignored, they will die. Therefore, the water crop must be monitored every day, several times a day. Fish must be fed, waste material must be removed, oxygen and nitrogen levels must be monitored and adjusted, predators must be addressed, and the overall condition of the system must be evaluated. Without a person to monitor and maintain a crop, the process will fail, very quickly in most cases.

Working conditions vary because aquaculture does not take place in one specific type of place. Fish and shellfish can be grown in cold or very warm climates. Species can be grown outside or inside, and plants are usually grown in greenhouses. The working conditions may be wet, but they are usually fairly comfortable, except for net pen cultures of salmon, which might place a person on a floating net pen in cold inclement weather chipping ice off the structure.

Salaries in aquaculture, as in most agricultural sciences, lag behind those in other areas of biotechnology. Those who go into the aquaculture business for themselves will face the same wage challenges as farmers who must wait two to three years before they can sell their crops. However, the interaction with the crop and freedom to work in often beautiful outdoor situations are appealing. People take more than a passing interest in the fish or shellfish that are being grown, and a sense of ownership and pride is attached. The same is true of the ornamental plants that are grown.

The job market outlook in aquaculture is strong because people worldwide are eating more fish. If the world's oceans can produce only a finite amount of fish and seafood, then aquaculture will need to step up and meet this demand. This area of agriculture and biotechnology will definitely expand in the future, as will all the jobs associated with it. Be sure to visit the Marine Careers section of the Sea Grant website listed at the Companion Website, an outstanding resource for career information on aquaculture and many other areas of aquatic biotechnology.

Contributed by Gef Flimlin, Rutgers Cooperative Extension, Toms River, New Jersey.

marine bacteria that oxidize metals such as iron, manganese, nickel, and cobalt. Some of these bacteria can also be used to extract important metals from low-grade ores. Additionally, some marine bacteria and single-cell algae express metallothioneins, a family of metal-binding proteins. These species thrive in water contaminated with cadmium and other heavy metals, where they literally mop up cadmium from the surrounding environment and then degrade toxic metals into harmless by-products. Scientists are looking at ways to use these organisms to extract, recover, and recycle important and expensive metals such as gold and silver from manufacturing processes.

Many marine organisms produce substances that are valuable for degrading and processing a variety of waste materials. By using aquatic organisms or their products, it is possible to stimulate waste degradation in natural environments or in bioreactors seeded with these organisms in much the same way that sewage treatment plants rely on bacteria to degrade fecal wastes. For example, microbiologists at the USDA have experimented with growing nitrogen-metabolizing algae on large mats called scrubbers so they can be used as natural filters. These scrubbers work like charcoal filters in an aquarium in that they bind nitrogenous wastes. Water contaminated with farm animal wastes is passed over the scrubbers, and the algae absorb and metabolize the wastes. These wastes provide nutrients for the algae, which grow into thick mats. The algae are periodically harvested by cutting them back, but they are allowed to regrow like grass. Water cleaned in this way has been used for irrigation, and some of the harvested algae have even been used as livestock food. Similar experiments have been done in Florida using water hyacinths to clean water rich in phosphorus, nitrogen, and other nutrients.

QUESTIONS & ACTIVITIES

Answers can be found in Appendix 1.

1. Describe the differences between transgenic fish and triploids, and discuss how each type of genetically engineered fish can be created.

2. Suggest ways to assess the risk of genetically engineered marine species. Consider this from the standpoint of risks associated with consumption of genetically engineered species as well as risks associated with introducing these species into natural environments.

3. Create a list of benefits and problems associated with aquaculture.

4. Describe the most interesting aspect of aquatic biotechnology you learned about in this chapter. What topic did you choose? Why did you find it interesting?

5. Name some properties of an aquatic organism that might be attractive to aquatic biotechnologists interested in identifying novel compounds that might be valuable for a biotechnology application.

6. Provide examples of finfish and shellfish species that are important for aquaculture.

7. In this chapter, we have primarily focused on applications of aquatic biotechnology that benefit humans. Many aquatic biotechnologists are interested in using biotechnology to improve native populations of aquatic organisms (finfish or shellfish) around the world. Suggest at least three ways in which biotechnology may be used to restore fish stocks.

8. The GloFish has been marketed as the world's first genetically modified pet. Even though the GloFish was not designed for human consumption, many different groups (including scientists) have voiced strong concerns against this use of transgenic animals. According to Yorktown Technologies, the company that produced the GloFish, the Environmental Protection Agency (EPA), Department of Agriculture (USDA), U.S. Fish and Wildlife Service, and the Food and Drug Administration (FDA) all claimed there was no need for any of these agencies to regulate or approve sale of the GloFish. Why do you think federal agencies decided it was not necessary to regulate the GloFish? Describe possible reasons why the GloFish is so controversial. What might some of the primary concerns be about this fish?

9. What are biofilms? Give examples of biofilms in aquatic environments as well as humans. How may aquatic organisms be used to help scientists combat biofilming?

10. What is the limulus amoebocyte lysate (LAL) test? Briefly describe how it works and its uses.

References and Further Reading

Devlin, R. H., Biagi, C. A., Yesaki, T. Y., et al. (2001). Growth of Domesticated Transgenic Fish. *Nature*, 409: 781–782.

Hew, C. L., and Fletcher, G. L. (2001). The Role of Aquatic Biotechnology in Aquaculture. *Aquaculture*, 197: 191–204.

Hites, R. A., Foran, J. A., Carpenter, D. O., et al. (2004). Global Assessment of Organic Contaminants in Farmed Salmon. *Science*, 303: 226–229.

Stix, G. A. (2005). Toxin Against Pain. *Scientific American*, 292: 88–93.

Visit www.pearsonhighered.com/biotechnology to download learning objectives, chapter summary, "Keeping Current" web links, glossary, flashcards, and jpegs of figures from this chapter.

Chapter **11**

Medical Biotechnology

After completing this chapter you should be able to:

- Provide examples of model organisms and explain why they are important.

- Describe different karyotyping techniques that can detect chromosome abnormalities and molecular techniques for genetic testing.

- Provide examples of why pharmacogenomics can change how many genetic disease conditions may be treated in the future.

- Discuss how monoclonal antibodies may be used for treating disease.

- Understand the purpose of gene therapy, compare and contrast different gene therapy strategies, and recognize limitations of gene therapy.

- Define regenerative medicine and provide examples of how cell and tissue transplantation and organ engineering can be used.

- Understand what stem cells are and describe how they can be isolated. Provide examples of possible therapies that may be developed from stem cells.

- Compare and contrast therapeutic cloning and reproductive cloning.

- Briefly explain the importance of the Human Genome Project for learning about and treating genetic diseases.

Customized medicine is a goal of medical biotechnology.

Thus far, we have provided examples of current applications of biotechnology that already impact our lives every day. We have also considered potential new areas of biotechnology. There is perhaps no topic in biotechnology that provokes greater optimism and debate than **medical biotechnology.** Applications of medical biotechnology have existed for decades. For instance, 100 years ago, leeches were commonly used to treat illness by so-called bloodletting. Some doctors believed that by using leeches to suck blood out of a patient, diseases were being removed from the body. It is an exciting time to learn about medical biotechnology because sophisticated advances in this field are occurring at a mind-boggling rate. In fact, even the leech is getting attention again—although not for bloodletting, but for enzymes found in its saliva that can dissolve blood clots and possibly be used to treat strokes and heart attacks.

Medical biotechnology incorporates many of the topics that we have already discussed. From developing new drugs to the prospects of using stem cells and cloning, the possibilities of medical biotechnology are incredible but also incredibly alarming to many people, including scientists. Applications of medical biotechnology will affect the type of health care treatments you receive in the future.

In this chapter, we consider a range of different applications of medical biotechnology and discuss many potential impacts of this very exciting area. We begin by providing an overview of how molecular biology techniques can be used to detect and diagnose disease and by considering innovative medical products developed through biotechnology. We then present an introduction to applications and examples of gene therapy before discussing regenerative medicine. The chapter concludes by relating the Human Genome Project to the discovery of human disease genes.

11.1 The Power of Molecular Biology: Detecting and Diagnosing Human Disease Conditions

The year 2003 marked the 50-year anniversary since Nobel Prize winners James Watson and Francis Crick revealed the structure of DNA as a double-helical molecule. Since then, molecular biology has advanced at an astonishing pace, providing molecular techniques that give scientists and medical doctors very powerful tools for combating human diseases.

Models of Human Disease

Many of the applications you will learn about in this chapter are possible because of important **model organisms.** As you learned in Chapter 7, in particular, mice, rats, worms, and flies have played critical roles in helping scientists study human disease conditions. We think of ourselves as unmatched by other species for our ability to communicate through speech and writing as well as walking upright, creating music, making a good pizza, and exploring distant planets. But we are not really unique at the genetic level. From yeast and worms to mice, we share large numbers of genes with these organisms. A number of human genetic diseases also occur in model organisms. Therefore, scientists can use model organisms to identify disease genes and test gene therapy and drug-based therapeutic approaches to

Q What are clinical trials?

A Before any therapy—such as a new drug treatment—is approved by the U.S. Food and Drug Administration for wide use in humans, most drugs or therapy strategies are tested through a series of rigorous steps called clinical trials. Most clinical trials are classified into three different phases. For instance, when testing a new drug for treating cancer, clinical studies begin with Phase I trials, which involve initial studies of a drug in a small number of healthy volunteer patients. Phase I trials are commonly called safety studies because they are used to determine if a new treatment is safe for humans, to determine safe doses of a drug, and to determine the best route of administration (oral versus injection into bloodstream, etc.). These studies do not usually test how effective the drug may be for treating an illness. Model animals are an essential first step prior to initiating Phase I trials. If a drug proves to be too toxic in animals, it is not likely to be approved for Phase I trials.

In Phase II trials, the safety of the drug is further evaluated on larger groups of people, generally a few hundred participants, and the effectiveness of the drug is carefully studied, usually for several years. Lastly, Phase III trials study the new drug for its effectiveness compared with other drugs considered to be the standard or most effective current treatment. Phase III trials typically involve thousands of patients often with different backgrounds and stages of illness throughout the country. About 25% of all drug applications pass Phase III. Approximately 80% of Phase III drugs are ultimately approved by the FDA.

check their effectiveness and safety in preclinical studies before using them for **clinical trials** in humans.

Model organisms are critically important to scientists because we cannot manipulate human genetics for experimental purposes. It is, of course, unethical and illegal to force humans to breed or to remove their genes to learn how they function. However, these approaches are widely used to study genes in model organisms. Mice, rats, chicks, yeast, fruit flies, worms, frogs, and even the zebrafish, a common fish in home aquariums, have all played important roles in our understanding of human genetics. Many important genes are highly conserved from species to species. If we identify important genes in model organisms, we can form hypotheses and make predictions about how these genes may function in humans.

Many genes identified in different model species have been shown to be related to human genes based on DNA sequence similarity. Such related genes are called **homologs.** For instance, a gene thought to play a role in human illness can be eliminated in model organisms through gene knockout, as we discussed in Chapter 7. The effects on the organism can be studied to learn about the functions of the gene. For instance, several years ago scientists discovered that mice can become obese if they lack a single gene called *Ob* (Figure 11.1). The *Ob* gene codes for a protein hormone called leptin, which travels through the bloodstream to the brain to regulate hunger. The subsequent discovery of a human homolog for leptin has led

to a new area of research with great promise for providing insight on fat metabolism in humans and the genetics that may influence weight disorders. Some childhood diseases of obesity are affected by mutation of the *Ob* gene. Extremely obese children in England have been treated with leptin and have responded very well in preliminary studies.

Historically, significant scientific discoveries in almost all fields of biology, including anatomy and physiology, biochemistry, cell biology, developmental biology, genetics, and molecular biology, were first made in model organisms and then related to humans. For instance, in developing embryos, some cells must die to make room for others. How does the body know where to develop certain organs and determine which cells must die to make room for others? Studies of the roundworm *Caenorhabditis elegans* have greatly advanced the answers to these important developmental questions. Maps of *C. elegans*, which has 959 cells, have been created that allow scientists to determine the fate or lineage of all cells in the embryonic worm that develop to form the nervous system, intestine, and other tissues of the worm. Of these cells, 131 are destined to die in a form of cell suicide known as programmed cell death, or **apoptosis.** During development of a human embryo, sheets of skin cells create webs between the fingers and toes; apoptosis is responsible for the degeneration of these webs prior to birth. But apoptosis is significant in other ways. We know that apoptosis is involved in neurodegenerative diseases such as Alzheimer disease, Huntington disease, amyotrophic lateral sclerosis (Lou Gehrig disease), and Parkinson disease, as well as arthritis and forms of infertility. Model organisms will help us better understand the genes involved and slow or stop these degenerative processes.

We have already seen that one of the goals of the Human Genome Project was to develop a better understanding of genetic similarities and differences between humans and other species, particularly other mammals. The Human Genome Project and comparative genomics studies have clearly demonstrated that we share a large number of genes with other organisms. Hundreds of human genes show relatedness to bacterial genes, providing solid evidence for the evolution of genes from bacteria to higher organisms. Can you believe that you share approximately 50% of your genes with the pesky fruit flies you bring home on fruit from the grocery store? It may seem hard to believe that it takes only about twice as many genes to make a human as it does a fruit fly. And plants, such as rice, have even more genes than we do. We share nearly 40% of our genes with roundworms and 31% of our genes with yeast—the same yeast we use to help make bread rise and ferment alcoholic beverages.

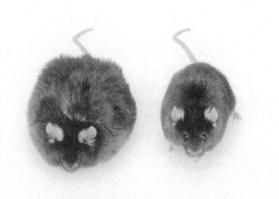

Figure 11.1 Obese Mice   Model organisms are very valuable for helping scientists learn about human disease genes. The mouse on the left has been genetically engineered by gene knockout to lack the *obese* (*Ob*) gene, which produces a protein hormone called leptin, from the Greek word *leptos* meaning "thin." Leptin is produced by fat cells and travels through the bloodstream to the brain, essentially telling the brain when the body is full. The *Ob* knockout mouse weighs almost five times as much as its normal sister (right).

We share even more genes with mice; approximately 90% of our genes are similar in structure and function.

Many genes that determine our body plan, organ development, and eventually our aging and death are virtually identical to genes in fruit flies. Moreover, mutated genes that are known to give rise to disease in humans also cause disease in fruit flies. Approximately 61% of genes mutated in 289 human disease conditions are found in the fruit fly. This group includes the genes involved in prostate cancer, pancreatic cancer, cystic fibrosis, leukemia, and many other human genetic disorders. Heart disease is another example of a condition that scientists are studying in model organisms. Nearly 1 million people die each year in the United States from heart disease. Researchers are using gene knockouts to develop so-called heart attack mice that are deficient in genes required for cholesterol metabolism. They hope that these mice will show elevated blood cholesterol levels similar to those that occur in atherosclerosis—hardening of the arteries—so they can test therapies to combat heart disease.

Lastly, researchers have been hunting for a cure for AIDS for 20 years. Creating a small animal model for AIDS is a high priority for many AIDS scientists. In addition to humans, HIV and related viruses cause disease in primates such as chimpanzees and rhesus macaque monkeys, but these animals are very expensive. Each animal can cost over $50,000, and they are only available in limited numbers. Researchers are making slow but steady progress toward developing rodent models infected with HIV. Even if such animals are created, there is no guarantee their disease will adequately mimic human AIDS. In humans, HIV infects and destroys human T lymphocytes (T cells). A main impediment to a rodent model is that HIV does not recognize and bind to receptor proteins on mouse T cells. Nevertheless, scientists might be able to express human T-cell proteins in mice to trick the virus into infecting mice.

Biomarkers for Disease Detection

In theory, with the right diagnostic tools it may be possible to detect most every disease at an early stage. For many diseases, such as cancer, early detection is critical for providing the best treatment and improving the odds of survival. One detection approach is to look for **biomarkers** as indicators of disease. Biomarkers are typically proteins produced by diseased tissue or proteins whose production is increased when a tissue is diseased. Many biomarkers are released into body fluids such as blood and urine as a product of cell damage—released by dead and dying cells such as cells undergoing apoptosis. For example, a protein called **prostate-specific antigen (PSA)** is released into the bloodstream when the prostate gland is inflamed, and elevated levels can be a marker for prostate inflammation and even prostate cancer. Detecting individual genes or gene expression patterns also provides scientists with biomarkers for disease (see Figure 11.8 later in this chapter), and many biotechnology companies are actively involved in searching for better biomarkers that can be used for early detection and disease diagnosis.

Detecting Genetic Diseases

Techniques in molecular biology have proven to be extremely valuable for detecting many different genetic diseases.

Testing for chromosome abnormalities and defective genes

Until relatively recently, most genetic testing occurred on fetuses for the purpose of identifying the sex of a child or to detect a small number of genetic diseases. Most of these procedures involved testing for genetic conditions that occur as a result of alterations in chromosome number. If there are problems with chromosome separation during the formation of sperm or egg cells, a fetus may contain abnormal numbers of chromosomes. One of the most best understood examples of a disorder created by an alteration in chromosome number is **Down syndrome.** Most individuals with Down syndrome have three copies of chromosome 21 (trisomy 21). Affected individuals show a number of symptoms, including mental retardation, short stature, and broadened facial features. Recently, scientists developed a strain of mice with almost a complete copy of human chromosome 21. These mice show characteristics of Down syndrome and may turn out to be a very valuable model for understanding the genetics of this condition.

Fetal testing for Down syndrome is fairly common, particularly in pregnant women older than 40 years, because the incidence of Down syndrome is related to the age of eggs produced by the woman. Trisomy 21 and other abnormalities in chromosome number can be tested in a fetus to provide parents with information that may be used to determine if they want the pregnancy to continue. If a defect is detected, genetic tests also provide information that can be used to treat fetuses during pregnancy and after the child is born.

So how is a developing fetus tested for Down syndrome? Two different techniques can be used: **amniocentesis** and **chorionic villus sampling.** Amniocentesis is performed when the developing fetus is around 16 weeks of age. A needle is inserted through the mother's abdomen into the pocket of amniotic fluid surrounding and cushioning the fetus

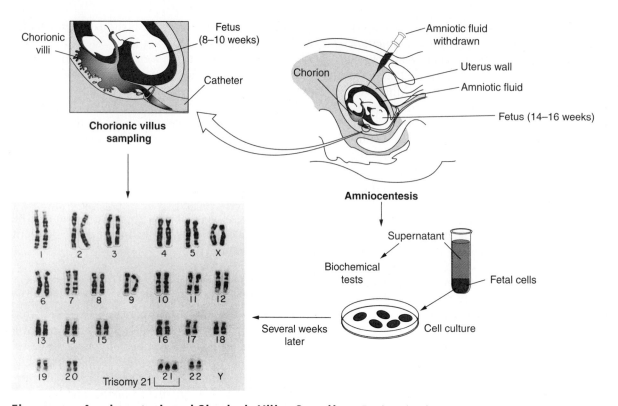

Figure 11.2 Amniocentesis and Chorionic Villus Sampling Fetal testing for chromosomal abnormalities is most commonly achieved through either amniocentesis or chorionic villus sampling. This karyotype from a person with Down syndrome shows three copies of chromosome 21 (trisomy 21).

(Figure 11.2). This fluid contains cells shed from the fetus such as skin cells. Isolated cells are cultured for a few days to increase cell number, and then the cells are treated to release their chromosomes, which are spread onto a glass slide. The chromosomes are stained with different dyes that bind to proteins attached to the DNA, creating patterns of alternating light and dark bands on each chromosome. Based on the size of each chromosome and its banding pattern, it is relatively easy to align all chromosomes into pairs and count chromosome number. Recall from Chapter 2 that this technique is called a **karyotype.** Karyotypes are also used to determine the sex of a child by the presence of the sex chromosomes (X and Y).

Chorionic villus sampling (CVS) can also be used for fetal testing (Figure 11.2). During this procedure, a suction tube is used to remove a small portion of a layer of cells called the chorionic villi, fetal tissue that helps form the placenta. An advantage of CVS over amniocentesis is that enough cells are obtained so the sample can immediately be used for karyotyping. Another advantage of CVS is that the procedure can be done earlier in the pregnancy, around 8 to 10 weeks. But because the fetus is so small, this procedure carries

a higher risk for disturbing the fetus and causing a miscarriage than amniocentesis.

Karyotyping is easily carried out on adults to check for chromosome abnormalities. Typically, blood is drawn from an adult, and the white blood cells are used for karyotyping. A relatively new technique for karyotyping in both fetuses and adults is **fluorescence *in situ* hybridization (FISH).** In FISH, a chromosome spread is prepared on a slide and then fluorescent probes are hybridized to each chromosome. Each probe is specific for certain "marker" sequences on each chromosome. In some cases, FISH can be performed with probes that fluoresce different colors. This produces a **spectral karyotype.** FISH is very useful for identifying missing chromosomes and extra chromosomes, but in particular FISH makes it much easier than conventional karyotyping to detect defective chromosomes. A number of human genetic diseases created by chromosomal abnormalities occur when a portion of a chromosome is deleted or a piece of chromosome is swapped from one chromosome to another because of problems in chromosome replication. For instance, in a type of leukemia (a cancer of the white blood cells) called chronic myelogenous leukemia, DNA is exchanged between chromosomes 9 and 22 so genes from 9 are swapped onto 22 and vice versa (Figure 11.3). This exchange can be detected by FISH using different colored fluorescent probes for each chromosome.

Most genetic diseases result from mutations in specific genes instead of abnormalities in chromosome number or defects in chromosome structure. Because more sophisticated techniques are being developed, scientists can detect *individual* diseased genes in both fetuses and adults. This capability will become increasingly common as we learn more about disease genes as a result of the Human Genome Project.

Some genetic diseases can be detected in both embryos and adults from either amniotic cells or blood cells, respectively, using **restriction fragment length polymorphism (RFLP) analysis** (pronounced "rifflips"). The basic idea behind RFLP analysis is that defective gene sequences may be cut differently by restriction enzymes than their normal complements because nucleotide changes in the mutant genes can affect

restriction enzyme cutting sites to create more or fewer. Recall from Chapter 8 that RFLP analysis is used for DNA fingerprinting. As an example, if DNA from a healthy individual and DNA from an individual with sickle-cell disease are both cut with restriction enzymes, they will be of different sizes because of how restriction enzymes cut each gene. This can be clearly observed when the DNA fragments are subjected to Southern blot analysis with a probe for the β-globin gene, the gene affected in sickle-cell disease (Figure 11.4). Hence we have the term *r*estriction *f*ragment *l*ength *p*olymorphisms—fragments of different lengths or forms ("poly" means many and "morphism" refers to the form or appearance of something) created by restriction enzymes.

Sickle-cell disease occurs when a person has two mutant versions of the β-globin gene. The mutant copies of β-globin protein produce an abnormal form of hemoglobin that affects the size and shape of red blood cells, giving them a characteristic "sickled" appearance (see also Figure 2.20).

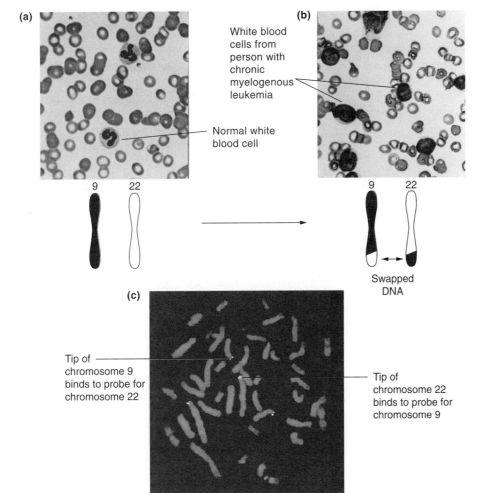

(a)

White blood cells from person with chronic myelogenous leukemia

Normal white blood cell

9   22

(b)

9   22

Swapped DNA

(c)

Tip of chromosome 9 binds to probe for chromosome 22

Tip of chromosome 22 binds to probe for chromosome 9

Figure 11.3 FISH Can Be Used to Detect Chromosome Defects Chronic myelogenous leukemia, a cancer of white blood cells created when genes on chromosome 9 and 22 are swapped, can be detected by using specific probes for each chromosome that fluoresce different colors.

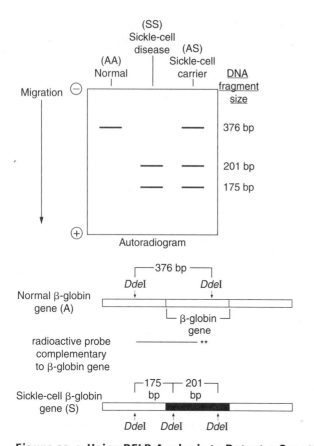

Figure 11.4 Using RFLP Analysis to Detect a Genetic Disease   Human DNA cut with a restriction enzyme (*Dde*I) was subjected to agarose gel electrophoresis followed by Southern blotting to transfer the DNA to a nylon membrane and hybridization with a radioactive probe specific for the β-globin gene. The normal β-globin gene (A) contains two cutting sites for the enzyme *Dde*I; mutations in the sickle-cell (mutated) β-globin gene (S) create three cutting sites for *Dde*I. Differences in DNA fragment sizes (polymorphisms) are detected by autoradiography depending on where the probe binds to complementary sequences in the β-globin gene. A healthy person with two copies of the normal β-globin gene (AA) shows a single band at 376 bp; a person with sickle-cell disease (SS) would have two copies of the mutant β-globin gene and show bands at 201 and 175 bp. A person considered a "carrier" would have one normal and one defective β-globin gene (AS) but would not have sickle-cell disease because there is one functioning copy of the β-globin gene; bands show at 376, 201, and 175 bp.

One disadvantage of RFLP analysis is that it can only be used to analyze gene defects in which a mutation changes a restriction enzyme recognition sequence in a gene. An approach called **allele-specific oligonucleotide (ASO) analysis** allows for the detection of a single nucleotide change in any gene even if the mutation does not change a restriction site. In this technique, DNA is isolated from human cells, usually white blood cells, and then amplified by the polymerase chain reaction (PCR) using primers that flank a

disease gene of interest. Amplified DNA is then blotted onto nylon filters and hybridized separately to two different ASOs as probes. ASOs are small single-stranded oligonucleotide sequences, usually around 20 nucleotides in length. An ASO that will hybridize to a normal gene and an ASO for the mutant gene are used. Figure 11.5 shows an example of how ASO analysis can be used to test for the sickle-cell gene. PCR-based tests such as this are becoming increasingly more valuable for detecting diseased genes. One major advantage of PCR is its high sensitivity for detecting defects in small amounts of DNA. Consequently, PCR and ASO analysis as well as FISH are being used to screen for gene defects in single cells from 8- to 32-cell-stage embryos created by *in vitro* fertilization. Such **preimplantation genetic testing** allows individuals to select a healthy embryo prior to implantation.

Currently, several dozen defective genes can be tested for in adults or fetuses using many of the techniques we have described (Table 11.1). Refer to "Tools of the Trade" in Section 11.5 for several excellent websites where you can learn more about human disease genes.

Single nucleotide polymorphisms

One of the many intriguing findings of the Human Genome Project is the discovery that subtle alterations in

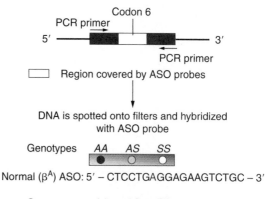

Figure 11.5 Using PCR and ASOs to Test for the Sickle-Cell Gene   ASO tests are very valuable for detecting single nucleotide mutations such as the one that causes sickle-cell disease (see also Figure 2.20). In this example DNA from white blood cells is amplified by PCR, blotted onto filter paper, hybridized to ASO probes, and visualized by autoradiography. The ASO for the normal hemoglobin gene (β^A) will only bind to DNA on the filters if the normal gene is present; the ASO for the mutant hemoglobin gene (β^S) will only bind to the defective gene if present on the filter. DNA from individuals with two copies of the sickle-cell hemoglobin gene (SS) will only bind to the β^S ASO and not the normal β^A ASO.

Table 11.1 GENETIC DISEASE TESTING

Genetic Disease Condition	Genetic Basis for Disease and Symptoms
Cancers (brain tumors; urinary bladder, prostate, ovarian, breast, brain, lung, and colorectal cancers)	A variety of different mutant genes can serve as markers for genetic testing.
Cystic fibrosis	Large number of mutations in the CFTR (cystic fibrosis transmembrane conductance regulator) gene on chromosome 7. Causes lung infections and problems with pancreatic, digestive, and pulmonary functions.
Duchenne muscular dystrophy	Defective gene (dystrophin) on the X chromosome causes muscle weakness and muscle degeneration.
Familial hypercholesterolemia	Mutant gene on chromosome 19 causes extremely high levels of blood cholesterol.
Hemophilia	Defective gene on the X chromosome makes it difficult for blood to clot when bleeding.
Huntington disease	Mutation in gene on chromosome 4 causes neurodegenerative disease in adults.
Phenylketonuria (PKU)	Mutation in gene required for converting the amino acid phenylalanine into the amino acid tyrosine. Causes severe neurological damage, including mental retardation.
Severe combined immunodeficiency (SCID)	Immune system disorder caused by mutation of the adenosine deaminase gene.
Sickle-cell disease	Mutation in β-globin gene on chromosome 11 affects hemoglobin structure and shape of red blood cells, which disrupts oxygen transport in blood and causes joint pain.
Tay-Sachs disease	Rare mutation of a gene on chromosome 5 causes certain types of lipids to accumulate in the brain. Causes paralysis, blindness, retardation, and respiratory infections.

DNA called **single nucleotide polymorphisms** (**SNPs;** pronounced "snips") represent one of the most common forms of genetic variation among humans. SNPs are single-nucleotide changes in DNA sequences that vary from individual to individual (see Figure 1.11).

SNPs have been found on all human chromosomes. It has been estimated that one SNP occurs approximately every 1,000 to 3,000 base pairs (bp) in the human genome. Over 1.4 million SNPs have been identified to date. If a segment of chromosomal DNA from two different people is compared by DNA sequencing, approximately 99.9% of the DNA sequence will be exactly the same. A majority (greater than 80%) of the 0.1% variation in bases between these two people will be SNPs. Most SNPs have no effect on a cell because they occur in nonprotein coding regions (introns) of the genome. But when a SNP occurs in a gene sequence, it may cause a change in protein structure that produces disease or influences traits in a variety of ways, including conferring susceptibility for some types of disease conditions.

SNPs represent variations in DNA sequences that ultimately influence how we respond to stress and disease. The first SNP discovered to be associated with a disease condition was sickle-cell disease. Because SNPs occur frequently throughout the genome, they may serve as valuable genetic markers for identifying disease-related genes. Some SNPs might be used to predict susceptibilities to diseases such as stroke, diabetes, cancer, heart disease, behavioral and emotional illnesses, and a host of other disorders that may have a genetic basis.

SNPs are thought to be so promising that pharmaceutical companies have invested millions of dollars in a collaborative partnership called the HapMap Project. Many SNPs on the same chromosome are clustered in groups called haplotypes. "Hap" is an abbreviation for haplotype. HapMap is an international effort among companies, academic institutions, and private foundations with an established goal of identifying and cataloging the chromosomal locations (loci) of the more than 1.4 million SNPs that are present in the 3 billion bp of the human genome and to understand the roles of SNPs in disease diagnosis and treatment.

Identifying sets of disease genes by microarray analysis

Another relatively new technique for studying genomes will also play an important role in the detection of genetic disease. **DNA microarrays,** also called gene chips, are glass microscope slides spotted with genes (see Chapter 3). A single microarray can contain thousands of genes. Researchers can use microarrays to screen a patient for a pattern of genes that might be expressed in a particular disease condition. As shown in Figure 11.6, microarray data can then be used to figure out the patient's risk of developing disease based on the number of expressed genes for the disease the patient shows.

For instance, microarrays created with known diseased genes or certain SNPs will be valuable for

studying expressed genes in a patient. To do this, DNA or RNA is isolated from a patient's tissue sample—a blood sample or even a scraping of cells lining the cheeks. The patient's DNA is tagged with fluorescent dyes and then hybridized to the chip. Spots on the microarray where the patient's DNA bound are revealed by fluorescence (refer to Figure 3.17). Binding of a patient's DNA to a gene sequence on the chip indicates that the person's DNA has a particular mutation or SNP. There are even companies working on handheld chip devices that doctors can use to get nearly instant information about a patient's genetics. Microarrays are currently being used to identify genetic differences in patients with various types of cancer. Treatment strategies to combat cancer are being designed based on subtle differences in the expression of cancer-causing genes.

Throughout this chapter, we consider how biotechnology may help develop new treatments for age-related diseases such as Parkinson disease, Alzheimer disease, heart disease, cancer, stroke, and arthritis. We know that more than 50,000 Americans are over 100 years old and that genes play a major role in the longevity of human aging. In fact, it has been estimated that as many as 1,000 genes may control lifespan. To this end, scientists are using microarrays to zero in on genes that may influence aging. Studies of siblings in their 90s and 100s suggest that at least one gene on chromosome 4, and probably many others, provide clues as to why some people live longer than others.

Protein microarrays are another new option for disease diagnosis. They are used in much the same way as DNA chips. For instance, these chips can contain hundreds or thousands of antibodies spotted on a chip. By applying blood proteins from a patient, researchers have been able to detect illness indicated by the presence of proteins from disease-causing organisms.

In the next section we consider how biotechnology will result in new products that can be used to treat human disease.

11.2 Medical Products and Applications of Biotechnology

Identifying novel drugs and developing new ways to treat disease are major areas of medical biotechnology. In Chapters 4 and 5, we discussed how different proteins, for example, insulin and human growth hormone, produced by recombinant DNA technology have been used to treat human disease conditions. Here we consider a few examples of important products of medical biotechnology that you will hear more about in the future.

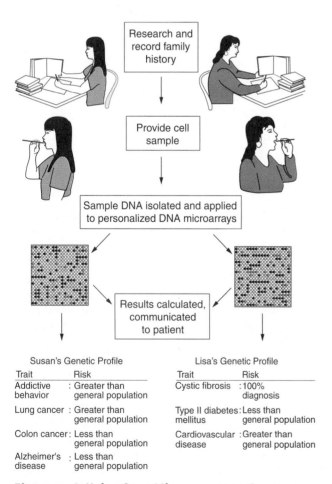

Figure 11.6 Using Gene Microarrays to Create a Genetic Profile

Genetic Testing: Issues to Consider

Since genetic testing became available, a number of concerns have been raised about how genetic information could be used and what it may be used for beyond personal and private decisions. Consider some of these issues:

- Should we test unborn children or adults for genetic conditions for which there is currently no treatment or cure?

- What are acceptable consequences if parents learn their unborn child has a genetic defect?

- What are the psychological effects of a false result, which may indicate erroneously that a healthy person has a disease gene or a gene defect that goes undetected in a person with a genetic disorder?

- How do we ensure privacy and confidentiality of genetic information and avoid genetic discrimination? Who should have access to your genetic information? How could your genetic background be used to discriminate against you? How could your health or life insurance company's access to your genetic information affect your premiums? Could your premiums be raised based on "genetic" risk in the same way that premiums are raised

based on other risks, such as how old you are and the car you drive?

- What are your obligations to inform others such as a potential spouse or employer of your knowledge about a possible genetic disorder?

- If genes are discovered for undesirable human behaviors, how would these genes be perceived in legal courts if accused criminals use genetics as their basis for a not guilty by reason of genetics plea?

- Would society implement mechanisms to prevent or dissuade individuals with genetic defects from having children?

- Errors in genetic testing can have tragic consequences. Currently there are no federal standards for quality control of genetic testing in the United States. Should mandatory proficiency testing be a requirement to minimize errors?

As you can see, genetic testing is certainly not without its controversies and limitations, and there are few easy answers to these issues. Visit the "Your Genes, Your Choices" website listed on the Companion Website for a thought-provoking series of ethical dilemmas created by genetic testing and genetic technology. What would you do if you had to face the scenarios presented at this site? You decide.

The Search for New Medicines and Drugs

It is estimated that cancer may soon surpass cardiovascular disease as the leading cause of death in the United States. But many promising breakthroughs on the horizon may empower doctors with new strategies for treating different types of cancer. Scientists are investigating many of the genes involved in the growth of cancer cells, including genes called **oncogenes.** Oncogenes produce proteins that may function as transcription factors and receptors for hormones and growth factors, as well as serve as enzymes involved in a wide variety of ways to change growth properties of cells causing cancer. Scientists are also actively studying tumor suppressor genes that produce proteins that can keep cancer formation in check.

Oncogenes and tumor suppressor genes are getting so much attention because researchers are working on ways to identify proteins made by these genes as targets for small molecule inhibitors—drugs that can bind to protein and block their function. Similarly, researchers are working on drugs that can act as

"activators" to bind to and stimulate important proteins that may be used to fight disease. In addition to small molecule drugs, there is an incredible amount of research designed to personalize medicine and improve drug delivery.

Pharmacogenomics for personalized medicine

The discovery of SNPs is partially responsible for a newly emerging field called **pharmacogenomics:** customized medicine. It involves designing the most effective drug therapy and treatment strategies based on the specific genetic profile of a patient. Pharmacogenomics is based on the idea that individuals can react differently to the same drugs, which can have varying degrees of effectiveness and side effects in part because of genetic polymorphisms (Figure 11.7). It is unclear whether pharmacogenomics will be a cost-effective approach to medicine or not; nonetheless, this area of medical biotechnology holds great potential.

Many drugs currently used in **chemotherapy** may be effective against cancerous cells but also affect

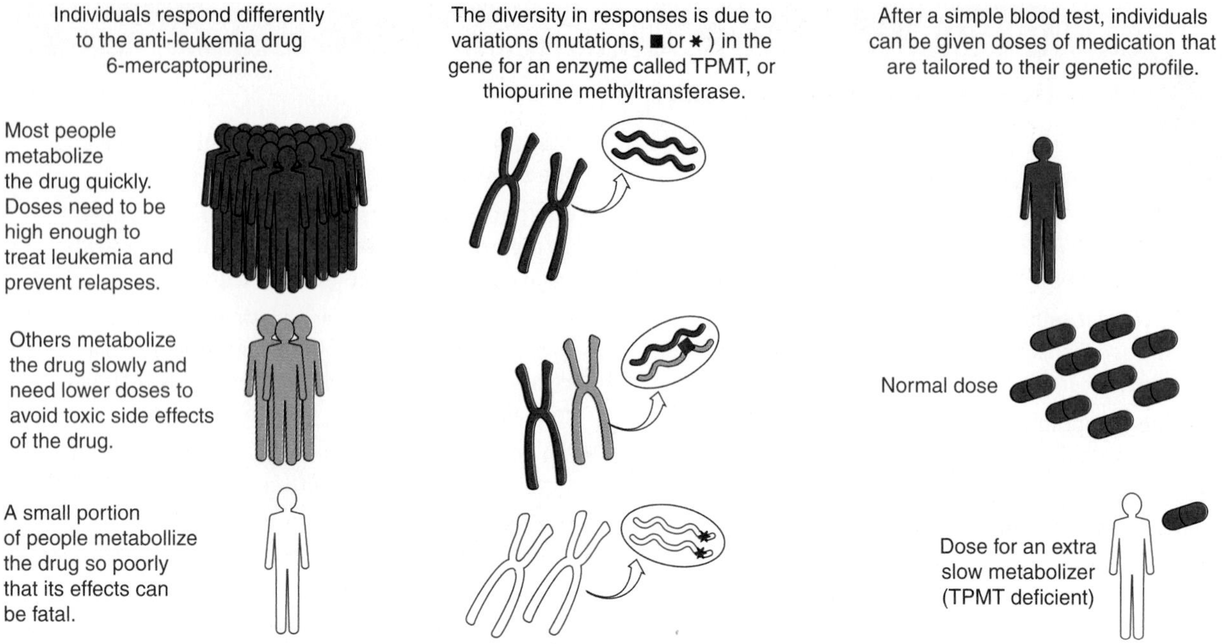

Individuals respond differently to the anti-leukemia drug 6-mercaptopurine.

The diversity in responses is due to variations (mutations, ■ or ✱) in the gene for an enzyme called TPMT, or thiopurine methyltransferase.

After a simple blood test, individuals can be given doses of medication that are tailored to their genetic profile.

Most people metabolize the drug quickly. Doses need to be high enough to treat leukemia and prevent relapses.

Others metabolize the drug slowly and need lower doses to avoid toxic side effects of the drug.

A small portion of people metabollize the drug so poorly that its effects can be fatal.

Normal dose

Dose for an extra slow metabolizer (TPMT deficient)

Figure 11.7 Pharmacogenomics   Different individuals with the same disease often respond differently to a drug treatment because of subtle differences in gene expression. The dose that works for one person may be toxic for another—a basic problem of conventional medicine. This example shows patient response to a chemotherapy compound called 6-mercaptopurine (6-MP), which has long been used to treat children with a form of blood cancer called acute lymphocytic anemia (ALL). Since the discovery that different genetic variations of the gene for the 6-MP degrading enzyme thiopurine methyltransferase are important for determining how patients respond to 6-MP, physicians now routinely run genetic tests (or blood tests to measure the enzyme levels) on ALL patients before determining the proper dose of 6-MP.

normal cells. Hair loss, dry skin, changes in blood cell counts, and nausea are all related to the effects of chemotherapy on normal cells. Researchers have been looking for so-called magic bullet drugs that destroy only cancer cells without harming normal body cells. If such drugs were designed, patients might get well faster because the drugs would have little or no effects on normal cells in other tissues. Consider the following example of pharmacogenomics in action. Breast cancer is a disease that shows familial inheritance for some women. Women with defective copies of the genes called *BRCA1* or *BRCA2* may have an increased risk of developing breast cancer, but many other cases of breast cancer do not exhibit a clear mode of inheritance. Perhaps there are additional genes or non-genetic factors at work in these cases. If a woman has a breast tumor thought to be cancerous, a small piece of the tissue could be used to isolate DNA for SNP and microarray analysis, which could be used to determine which genes are involved in this particular woman's form of breast cancer. Armed with this genetic information, a physician could design a drug treatment strategy—based on the genes involved—that would be

specific and *most effective* against this woman's type of cancer. A second woman with a different genetic profile for her type of breast cancer might undergo a different treatment.

Scientists at Genentech used this strategy to develop Herceptin, a type of monoclonal antibody (monoclonals are discussed in the next section) approved by the FDA in 1998. Herceptin binds to a protein (HER-2) produced by the human epidermal growth factor receptor 2 gene, which is expressed in about 25% to 30% of breast cancer cases. Women with HER-2 positive tumors typically develop aggressive breast cancer with a greater likelihood of metastasis (spreading) and poorer prognosis for survival. Herceptin has proven to be effective in some women, but in others, tumors become resistant to the antibody. A similar problem has occurred with other pharmacogenomics drugs developed for treating other cancers.

One of the first successful examples of pharmacogenomics involved a drug called *Gleevec* used to treat chronic myelogenous leukemia (CML), the condition discussed in Figure 11.3. Gleevec targets a protein called the BCR-ABL fusion protein, which is created by the swapping of DNA that occurs in CML and has generally proven to be a relatively effective way to treat the disease. Visit the Howard Hughes Medical Institute website listed on the Companion Website and check out the animations link for a good way to observe the action of Gleevec.

Increasingly, gene expression data from DNA microarrays will be used to diagnose patients based on the genes they express to then provide him or her with a pharmacogenomics treatment based on those genes. Figure 11.8 shows an example of microarray data for individuals diagnosed with different forms of leukemia. Notice how the patients can be grouped into different genetic categories of leukemias based on the clusters of genes most actively expressed. Because each group of patients expressed large numbers of different genes and proteins, there is no reason to think they will all respond well to the same chemotherapy. Knowing this, different chemotherapy approaches can be customized for each category of patients. As a result, this has greatly increased patient survival rates. A similar approach has been used for breast cancer patients, and it is expected that microarray data and pharmacogenomics approaches will increasingly become routine aspects of disease diagnosis and treatment.

Improving techniques for drug delivery

In addition to developing new drugs to battle disease, many companies are working to develop innovative ways to deliver drugs to maximize their effectiveness. Sometimes even a well-designed drug is not as effective as desired because of delivery problems—getting the drug to where it needs to function. For instance, if a drug to treat knee arthritis is taken as an oral pill, only a small amount of the drug will be absorbed by the body and transported via blood to tissues of the knee joint. Other factors that influence drug effectiveness are drug solubility (its ability to dissolve in body fluids), drug breakdown by body organs, and drug elimination by the liver and kidneys.

Microspheres, tiny particles that can be filled or coated with drugs, may be one way to improve drug effectiveness. These particles are often made out of materials that closely resemble the lipids (fats) in cell membranes. Delivery of microspheres as a mist sprayed into the airways through the nose and mouth has been used successfully for treating lung cancer and other respiratory illnesses such as asthma, emphysema, tuberculosis, and flu (Figure 11.9). Refer to Section 11.3 for a discussion of how microspheres called liposomes are used in gene therapy. Researchers are also investigating

Patient Groups

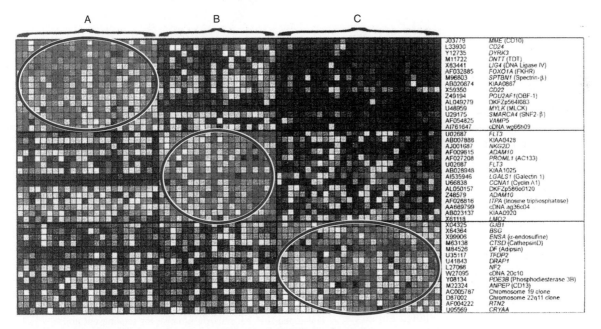

Figure 11.8 Microarrays for Gene Expression Analysis and Pharmacogenomics Shown here are microarray results indicating gene expression profiles in leukemia patients. Each column represents data for one patient, and each row is a different gene (gene symbols and names are on the far right side of the figure). Because this black and white image was reproduced from a color image, gray spots shown here represent highly expressed, active genes. Red ovals highlight clusters of actively expressed genes (gray and white spots) that can be used to categorize patients into different groupings based on expressed and relatively inactive (dark spots) genes.

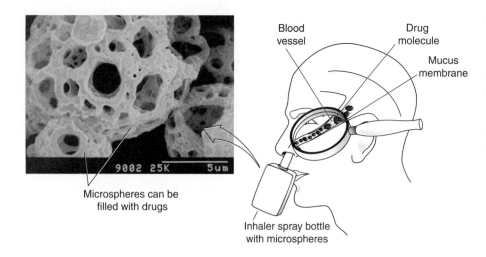

Microspheres can be
filled with drugs

Blood
vessel

Drug
molecule

Mucus
membrane

Inhaler spray bottle
with microspheres

Figure 11.9 Microspheres for Drug Delivery Microspheres can deliver drugs to specific locations in the body or be distributed throughout the body depending on how they are used. In this example, drug-containing microspheres are sprayed into the nose where they will enter blood vessels and rapidly enter the bloodstream to travel throughout the body.

ways to package anticancer drugs into microspheres for implantation in the body adjacent to growing tumors and anesthetics for pain management.

In 2006, the FDA approved Exubera, an inhalable version of insulin produced by Nektar Therapeutics of San Francisco and sold by pharmaceutical giant Pfizer. Exubera is a recombinant form of insulin delivered as an inhalable powder and offers diabetics the first alternative to needle-based delivery of insulin. But after barely a year, Pfizer stopped selling Exubera because of slow sales. Among the reasons that have been given for Exubera's failure are the unwillingness of physicians and patients to try something new, a bulky inhaler used to deliver the particles, and a poor marketing strategy.

Nanotechnology and nanomedicine: Biotechnology at the nanoscale

Nanotechnology is an area of science involved in designing, building, and manipulating structures at the nanometer scale. A nanometer (nm) is one billionth of a meter, and it's the size scale of molecules. For reference, a human hair is approximately 200,000 nm in diameter, DNA is about 2 nm in diameter, and bonds between many atoms are around 0.15 nm long. Nanotechnology is a big business, with applications in material manufacturing, energy, electronics, and engineering, but **nanomedicine**—applications of nanotechnology for improving human health—is of particular interest for medical biotechnology. Scientists envision tiny devices in the body carrying out a myriad of medical functions, including nanodevice sensors to monitor blood pressure, blood oxygen levels, and hormone concentrations as well as nanoparticles that can unclog blocked arteries and detect and eliminate cancer cells.

Several nanotechnology-enabled drugs are making their way to the market, primarily in areas of cancer treatment, and over 150 nanotechnology cancer thera-

pies are in development. Scientists have developed so-called smart drugs using viruses or tiny nanoparticles such as gold particles that are introduced into the body to seek out and target viruses or specific cells, such as cancer cells, to deliver a cargo that would treat or destroy those that are damaged rapidly and effectively in a silent manner with few side effects (Figure 11.10).

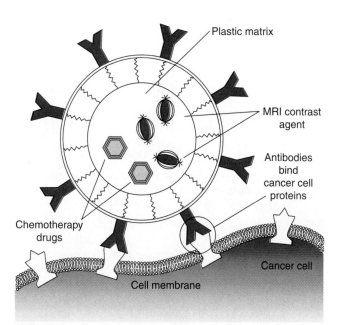

Plastic matrix

MRI contrast
agent

Antibodies
bind
cancer cell
proteins

Chemotherapy
drugs

Cell membrane

Cancer cell

Figure 11.10 Tumor-Seeking and Tumor-Killing Nanoparticles Multifunctional nanoparticles have been developed that have shown promise for seeking out and detecting tumor cells to help image the cells and to deliver tumor-killing drugs that can kill cancer cells. This example shows a plastic nanoparticle covered with antibodies against cancer cell proteins that allow the particle to bind to cancer cells. Inside the particle are contrast agents, which can be used for magnetic resonance imaging or X-ray approaches to detect the tumor, and chemotherapy drugs that can diffuse out of the particle to kill tumor cells.

Artificial Blood

Since the 1930s, blood transfusions have been performed routinely and successfully in the United States. Transfusions are often necessary for treating trauma victims, providing blood during surgeries, and treating people with blood-clotting disorders such as hemophilia. In the 1980s, the realization that HIV had contaminated many blood supplies led to new testing techniques that have made blood supplies much safer. Blood donated for transfusions is now tested for pathogens such as HIV and hepatitis viruses B and C before it is stored, but donated blood only has a shelf life of a few months and must be refrigerated. However, throughout many areas of the world, particularly in developing countries where screening procedures are not very good, there is a serious need for safe blood, free of infectious bacteria and viruses. Many of these concerns have prompted scientists to seek ways to develop artificial blood or blood substitutes.

Major advantages of artificial blood could include a disease-free alternative to real blood, a constant supply of blood in the face of blood shortages and emergency situations, and a supply of blood that can be stored for long periods of time. Also, unlike donated blood, synthetic blood would not have to be matched to the recipient's blood type to avoid rejection by the immune system. A major limitation in the development of synthetic blood to date is that artificial bloods have been designed to serve the primary task of normal red blood cells—transporting oxygen to body tissues, a role carried out by the oxygen-carrying protein hemoglobin. Red blood cells are literally hemoglobin factories and filled with this important protein. But normal red blood cells perform other functions such as providing the body with a source of iron, and hemoglobin is also important for removing carbon dioxide from the body. Researchers have yet to create blood substitutes that can perform all the functions of normal blood; nevertheless, many promising products are under development.

In 2000, South Africa became the first country to approve a blood substitute—a product called Hemopure, produced by Biopure Corporation, a Massachusetts company. So how is artificial blood made? Artificial bloods are cell-free solutions containing molecules that can bind to and transport oxygen in much the same way as normal hemoglobin. Some blood substitutes such as Hemopure are made from the hemoglobin of cattle; others are made from human hemoglobin. Cow blood is collected from food cattle at slaughterhouses

and then processed to purify the hemoglobin. Many other types of artificial blood being tested are produced using fluorocarbons, chemicals that can bind oxygen just as hemoglobin does and then release oxygen to the surrounding tissues. Ultimately, artificial blood products must provide safe alternatives to real blood transfusions. Much work remains to be done, but the potential benefit of these products has many companies investing large amounts of money and time to develop viable blood substitutes.

Vaccines and Therapeutic Antibodies

In Chapter 5 we looked at how vaccines can be used to stimulate the body's immune system to produce antibodies and provide a person with protection against infectious microbes. Certainly vaccination has been very effective for protecting us from pathogens that cause

Q What does blood matching mean?

A There are many ways to match or type blood cells. The most common scheme is called the ABO blood typing system. This system is based on surface proteins called *agglutinogens* located on the cell membrane of red blood cells. Individuals are categorized as having type A blood if their red blood cells have the A agglutinogen (surface protein). Type B blood cells have the B agglutinogen, type AB blood cells have both the A and B agglutinogens, and type O blood cells have neither the A nor the B agglutinogen. Similar matches of agglutinogens must usually be used for blood transfusions. For instance, a person with type A blood can receive a transfusion with type A blood but not type B blood. If a recipient receives a mismatched blood transfusion, antibodies in the recipient's immune system recognize the type B red blood cells as foreign and attack these cells, forming clumps of antibodies and red blood cells, which can cause life-threatening problems for the recipient.

If you know your blood type, you know you have a particular ABO type and also that your blood is indicated as "+" or "-." These designations refer to another surface protein, the Rh factor, so named because it was discovered in Rhesus monkeys. Human red blood cells that have the Rh factor are Rh positive (+); blood cells without the Rh factor are Rh negative (-). O- individuals are considered universal donor types because they can donate to recipients of any blood type; AB+ individuals are considered universal recipients.

polio, tetanus, typhoid, and dozens of others. As we discussed in Chapter 5, development of vaccines against some of the most deadly pathogens such as HIV and hepatitis is a very active and important area of research.

Many scientists hope that vaccination may be useful against conditions such as Alzheimer disease and many different types of cancers, but vaccination for these purposes is still mostly unproven in humans. Cancer vaccines are being experimented with as therapeutic treatments that are not preventative but designed to treat a person who already has cancer. In this approach, a person is injected with cancer cell antigens in an effort to stimulate the patient's immune system to attack existing cancer cells. In fact, there is considerable excitement about new types of "naked DNA" vaccines in which plasmid DNA encoding genes that produce antigens are injected directly into tissue where cells take up the plasmid and express the antigens that stimulate antibody production by the body. In other DNA vaccines, an immune response is mounted against the DNA itself. Recall from Chapter 5 that a DNA vaccine for West Nile virus has been effective in horses. Vaccine developers hope that this breakthrough will lead to new improvements in human vaccine development.

As we discussed in Chapter 5, the primary purpose of vaccination is to stimulate antibody produc-

tion by the immune system to help ward off foreign materials. However, antibodies themselves might be used to treat an existing condition as opposed to preventing infectious microbes from causing disease. Using antibodies in some types of therapy makes good sense because antibodies are very specific for the molecules or pathogens to which they are produced and can find and bind to their target with great affinity. Since their development in 1975, **monoclonal antibodies (MAbs),** purified antibodies that are very specific for certain molecules, have been considered "magic bullets" for disease treatment. To make a MAb, a mouse or rat is injected with purified antigen to which researchers are trying to make antibodies. Figure 11.11 shows production of MAbs specific for proteins from human liver cancer cells. After the mouse makes antibodies to the antigen, a process that usually takes several weeks, the animal's spleen is removed. The spleen is a rich source of antibody-producing B lymphocytes or simply B cells. In a culture dish, B cells are mixed with cancerous cells called myeloma cells that can grow and divide indefinitely. Under the right conditions a certain number of B cells and myeloma cells will fuse together to create hybrid cells called **hybridomas.**

Hybridoma cells grow rapidly in liquid culture because they contain antibody-producing genes from

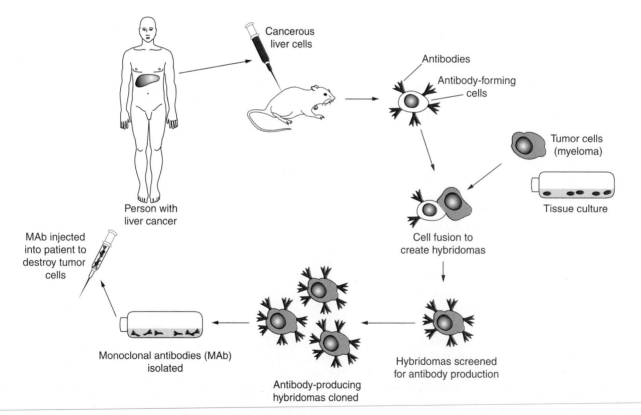

Figure 11.11 Making Monoclonal Antibodies

B cells. These cells are literally factories for making antibodies. Hybridoma cells secrete antibodies into the liquid culture medium surrounding the cells. Chemical treatment is used to select for hybridomas and discard unfused mouse and myeloma cells so that researchers have pure populations of fused, antibody-producing cells. Hybridomas, which are especially good at making large amounts of the antibody, can be transferred to other culture dishes and frozen away at ultra-low temperatures so a permanent stock of cells is always available. Antibodies can be isolated from hybridoma cultures in large batches by growing hybridoma cells in batch culture using bioreactors.

Monoclonal antibodies can be injected into patients where they seek out and target the antigens to which they were produced. The MAbs in Figure 11.11 would bind to liver cancer cells and work on destroying the tumor. In 1986, the FDA approved the first monoclonal antibody, OKT3, which was used to treat organ transplant rejection. In the 1990s, MAbs were developed to treat breast cancer (Herceptin) and lymphoma (Rituxan). There are currently over a dozen MAbs being used worldwide to treat cancer, cardiovascular disease, allergies, and other conditions. Scientists even envision attaching to MAbs chemicals or radioactive molecules that target damaged or cancerous cells and use their payload to kill these cells. Several groups have even tried attaching MAbs to enzymes from snake venom. This may sound like a very odd partnership, but snake venom proteins (that can kill mammalian cells by causing the lysis, or rupture, of cells) may be used for the selective elimination of tumor cells. If venom proteins are attached to MAbs specific for a certain cancer, the antibodies could find the tumor cells and go to work rupturing the cancer cells. Of course, much work is needed to ensure that venom proteins or other molecules attached to MAbs only attack intended cells, but this concept and other related approaches demonstrate potential applications of MAbs.

Research also indicates that therapeutic antibody strategies may be of value for treating people addicted to harmful drugs such as cocaine and nicotine in cigarettes. In the United States alone, more than 13 million people abuse drugs. Scientists believe that it may be possible to stimulate antibody production to drugs such as cocaine. The idea is to produce antibodies that bind to the drug as the antigen, trapping and preventing the drug from affecting brain cells. Monoclonal antibodies have also been used for several years in common tests for conditions such as strep throat, and most home pregnancy kits use MAbs to detect hormones produced during pregnancy. Monoclonals for disease treatment have still not lived up to their initial hype, and there have been some setbacks in the field.

For example, MAb treatment of Alzheimer patients produced severe inflammation in several people due to a human antimouse antibody response. As discussed in Chapter 4, humanizing antibodies alleviate some of the problems with MAbs. Increasingly it appears that MAbs will continue to be valuable tools for medicine in the 21st century. In the next section we consider gene therapy, a promising and controversial topic of medical biotechnology.

11.3 Gene Therapy

Gene therapy involves the delivery of therapeutic genes into the human body to correct disease conditions created by a faulty gene or genes. Think about the awesome power and potential of gene therapy—providing a person with normal genes to supplement defective genes and cure disease or even using normal genes to replace faulty ones. But how are these genes delivered? How can genes be sent to the proper tissues and organs that need treatment? Can gene therapy be effective and safe? These are the major questions surrounding gene therapy research. Here we consider these and other questions as we provide an overview of gene therapy strategies for treating and attempting to cure disease.

How Is It Done?

The two primary strategies for gene delivery are *ex vivo* **gene therapy** and *in vivo* **gene therapy** (Figure 11.12). In *ex vivo* therapy (*ex* means "out of," *vivo* is Latin for "something alive"), cells from a person with a disease condition are removed from the patient, treated in the laboratory using techniques similar to bacterial transformation, and then reintroduced into the patient. Technically speaking, introducing DNA into animal or plant cells is called **transfection.** For instance, liver cells from a patient suffering from a liver disorder would be surgically removed and cultured. Appropriate therapeutic genes would then be delivered into these cells using vectors and other approaches that we discuss in the next section. These genetically altered liver cells would then be transplanted back into the patient without fear of rejection of the tissue transplant because these cells came from the patient initially (Figure 11.12).

In vivo gene therapy strategies involve introducing genes directly into tissues and organs in the body without removing body cells (Figure 11.12). One challenge of *in vivo* gene therapy is delivering genes only to the intended tissues and not tissues throughout the body. Scientists have primarily relied on using viruses

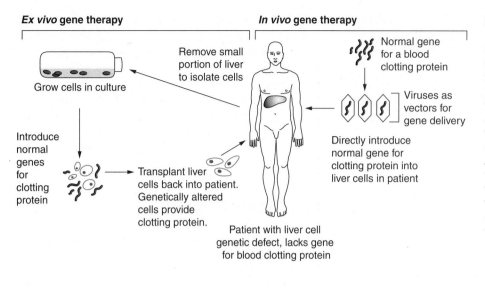

Ex vivo gene therapy

In vivo gene therapy

Remove small portion of liver to isolate cells

Grow cells in culture

Normal gene for a blood clotting protein

Viruses as vectors for gene delivery

Introduce normal genes for clotting protein

Transplant liver cells back into patient. Genetically altered cells provide clotting protein.

Directly introduce normal gene for clotting protein into liver cells in patient

Patient with liver cell genetic defect, lacks gene for blood clotting protein

Figure 11.12 *Ex Vivo* and *in Vivo* Gene Therapy in a Patient with a Liver Disorder *Ex vivo* gene therapy involves isolating cells from the patient, introducing normal genes for the clotting protein into these cells, and then transplanting cells back into the body where these cells will produce the required clotting protein. *In vivo* gene therapy involves directly introducing DNA into cells while in the patient. In either mode of gene therapy, genes may be introduced into cells as DNA packaged into viruses as vectors or as naked DNA.

as vectors for gene delivery, but in some cases genes have been directly injected into some tissues. So far, *ex vivo* strategies have generally proven to be more effective than *in vivo* approaches.

Delivering the payload: Vectors for gene delivery

A major challenge that must be overcome if gene therapy is to become a reliable tool for treating disease is achieving a safe and effective delivery of therapeutic genes—the payload. Depending on the genetic condition to be treated, some therapeutic strategies may require long-term expression of a corrective gene, whereas others may require rapid expression for shorter periods of time. How can the genes be delivered and expressed to produce adequate amounts of protein to correct disease? A majority of gene delivery strategies, both *ex vivo* and *in vivo*, rely on viruses as vectors to introduce therapeutic genes into cells.

A viral vector would use a viral genome to carry a therapeutic gene or genes to "infect" human body cells, thereby introducing the therapeutic gene. Scientists have considered various viruses such as **adenovirus,** which causes the common cold; influenza viruses, which cause the flu; and herpes viruses, which can cause cold sores and some cause sexually transmitted diseases, as potential **vectors** for gene delivery. Even human immunodeficiency virus-1 has been considered as a gene therapy vector. For any viral vector to work, scientists must be sure these vectors have been genetically engineered and inactivated so they neither produce disease nor spread throughout the body and infect other tissues.

Most viruses infect human body cells by binding to and entering cells and then releasing their genetic material into the nucleus or cytoplasm of the human cell. This is usually DNA, but some viruses contain an RNA genome. The infected human cell then serves as a host for reproducing the viral genome and producing viral RNA and proteins. Viral proteins ultimately assemble to create more viral particles that break out of the host cells so they are free to infect other cells and repeat the life cycle.

It may seem strange that viruses would even be considered for carrying genes to cure human diseases. However, scientists have realized that many harmful viruses are very effective at introducing their genome into cells. They reasoned that if these viruses could be genetically altered to deliver therapeutic genes safely, then we could use viruses for beneficial purposes. In many ways viruses are perfectly designed as gene therapy vehicles or vectors for gene delivery. For instance, adenovirus—which approximately 80% to 90% of the population has been infected with by childhood because it causes the common cold—can infect many types of body cells fairly efficiently. Retroviruses such as lentivirus and even HIV are of interest as vectors because on entering a host cell they copy their RNA genome into DNA and then randomly insert their DNA into the genome of the host cell where it remains permanently, a process called **integration.** One advantage of using retroviruses for gene delivery is that they can integrate therapeutic genes into the DNA of human host cells, allowing permanent insertion of genes into the chromosomes of a patient's cells as a way to provide lasting gene therapy.

Viruses have also been widely studied as gene therapy vectors because some viruses only infect certain body cells. This might allow for *targeted* gene therapy—the ability to deliver genes only to the tissues infected by a certain virus. For instance, a strain of herpes virus

(HSV-1) primarily infects cells of the central nervous system. This strain is a candidate for targeted gene delivery to cells of the nervous system, which may be an effective way to treat genetic disorders of the brain such as Alzheimer disease and Parkinson disease. Researchers are also investigating ways that the genetically altered herpes viruses can be used to destroy brain tumors. Preliminary trials of this approach in humans have shown some promise.

Most human cells do not take up DNA easily. If they did, it could then be possible to transfect cells by simply mixing them with DNA in a tube in much the same way that transformation is achieved with bacterial cells. However, some success has been demonstrated for both *in vivo* and *ex vivo* strategies using "naked" DNA. Naked DNA is simply DNA by itself, without a viral vector, that is injected directly into body tissues. Small plasmids containing therapeutic genes are often used for this approach. Cells of certain tissues will take in some of the naked DNA and express genes delivered this way. Naked DNA delivery techniques have been most effective in organs such as the liver and muscle (Figure 11.13). One of the major problems with transfecting human cells *in vivo* is that because a relatively small number of cells take up the injected DNA, there may not be enough cells expressing the therapeutic gene for gene therapy to have any effect on the tissue. Scientists are working on ways to overcome these problems and deliver naked DNA more effectively. For example, electroporation (recall that in Chapter 5 we discussed how electrical stimulation is used to move plasmids into bacteria cells) can be used to stimulate movement of DNA into cells.

One approach that may be a promising way to deliver DNA without viral vectors involves structures called liposomes. **Liposomes** are small-diameter hollow microspheres made of lipid molecules, similar to fat molecules present in cell membranes. Liposomes are packaged with genes and then injected into tissues or sprayed (in the next section we consider how this approach has been used to treat cystic fibrosis). A similar technique involves coating tiny gold particles with DNA and then shooting these particles into cells using a DNA gun. A DNA gun is really a pressurized air gun that delivers gold or liposome particles through cell membranes without killing most cells. Biodegradable gelatin particles are also being studied as gene-carrying vectors.

Another technology that scientists are experimenting with for delivering therapeutic genes involves engineering and constructing so-called artificial chromosomes. This technique involves making small chromosomes that consist primarily of nonprotein-coding DNA, which contains a therapeutic gene. The key to this approach is that the artificial chromosome contains structures similar to normal human chromo-

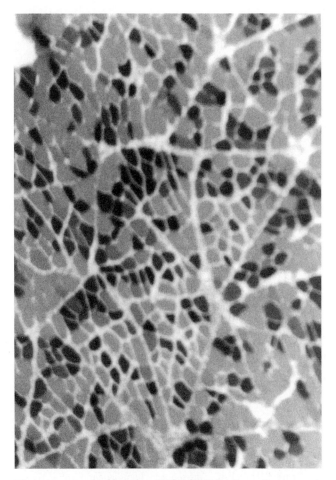

Figure 11.13 Delivering Naked DNA Some tissues, such as the muscle tissue shown in this figure, can take up naked DNA that has been injected into the tissue. The dark-colored cells have taken in and expressed a gene delivered in a plasmid. Light-colored cells have not taken in the gene. Scientists are working to overcome this obstacle. Because many cells do not take up naked DNA, this technique is often not very efficient.

somes that allow for permanent incorporation into cells and replication.

Antisense RNA technology and RNA interference for gene therapy

Recall that in Chapter 3 we discussed the use of **antisense RNA technology** as a way to block translation of mRNA molecules to silence gene expression. This approach was used to create the Flavr Savr tomato (see Figure 6.7). The basic concept of antisense RNA technology is to design an RNA molecule that will serve as a complementary base pair to the mRNA you want to inhibit, thus blocking it from being translated into a protein (Figure 11.14). This approach for shutting off a gene is frequently called **RNA** or **gene silencing.** Since the development of antisense RNA technologies in the 1970s, scientists have thought that RNA silencing approaches would

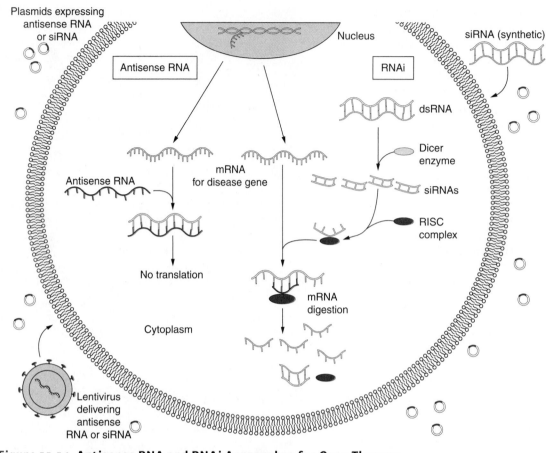

Plasmids expressing antisense RNA or siRNA

Nucleus

Antisense RNA

RNAi

siRNA (synthetic)

dsRNA

Dicer enzyme

siRNAs

Antisense RNA

mRNA for disease gene

RISC complex

No translation

mRNA digestion

Cytoplasm

Lentivirus delivering antisense RNA or siRNA

Figure 11.14 Antisense RNA and RNAi Approaches for Gene Therapy by Gene Silencing   Antisense technology and RNAi are two ways to silence gene expression and turn off disease genes. In antisense technology (left) an antisense RNA molecule binds to the mRNA for the disease gene and prevents it from being translated into a protein. With RNAi technology, double-stranded RNA (dsRNA) molecules are delivered to the cell. The dsRNAs are then cleaved by Dicer into short interfering RNAs (siRNAs), which are escorted by the RISC protein complex to bind to their target mRNA causing its degradation and thus preventing translation.

be promising ways to turn off disease genes as a gene therapy approach.

Antisense RNAs have been effectively used for gene silencing in cultured cells, but this technology has yet to live up to its promise as a treatment for disease. The recent emergence of **RNA interference (RNAi)** has reinvigorated gene therapy approaches by gene silencing (refer to Figure 3.20). With RNAi, double-stranded RNA molecules are delivered into cells where the enzyme Dicer chops them into 21-nt-long pieces called **small interfering RNAs (siRNAs;** Figure 11.14). The siRNAs then join with an enzyme complex called the **RNA-induced silencing complex (RISC),** which shuttles the siRNAs to their target mRNA where they bind by complementary base pairing. The RISC complex degrades the siRNA-bound mRNAs so they cannot be translated into protein.

The main challenge to RNAi based-therapeutics so far has been *in vivo* delivery of the antisense RNA,

dsRNA, or siRNA. RNAs degrade quickly in the body. It is also hard to get them to penetrate cells and to target the right tissue. Two common delivery approaches are to inject the antisense RNA or siRNA directly or as a plasmid vector that is taken in by cells to be transcribed to make antisense or RNAi molecules (Figure 11.14). Liposome, lentivirus delivery mechanisms, and attachment of siRNAs to cholesterol and fatty acids are also used to deliver silencing RNAs. Another problem is that most complex diseases are not caused by just one gene. For example, as mentioned earlier in this chapter, the gene *BCL-2* is overexpressed in about 50% of breast cancer cases, and antisense RNA technology can silence this gene *in vitro*, but breast cancer is a multigene disease and it is not yet possible to silence all of the genes involved in breast cancer.

A large number of antisense and RNAi clinical trials are underway in the United States for blindness. One form, called macular degeneration, targets a gene called *VEGF*. The protein encoded by *VEGF* promotes blood vessel growth. Overexpression of this gene leading to excessive production of blood vessels in the retina leads to impaired vision and eventually blindness. This disease is one that many feel will soon become the first condition to be treated by siRNA therapy. Others include several different cancers, influenza (scientists have had good success blocking flu infections in the lungs of mice), diabetes, multiple sclerosis, arthritis, and neurodegenerative diseases.

Curing Genetic Diseases: Targets for Gene Therapy

Most gene therapy researchers are focusing on genetic disorders created by single gene mutations or deficiencies—such as sickle-cell disease—because in theory these conditions may be easier to cure by gene therapy than genetic diseases involving multiple genes that must interact in complex ways. Current estimates indicate there may be more than 3,000 human genetic disease conditions caused by single genes. Table 11.1 on page 267 shows many of the diseases that are potential candidates for treatment by gene therapy. Recent trials for treating deafness, arthritis, and blindness have shown great promise. Here we consider a few examples of gene therapy in action.

The first human gene therapy

The first human gene therapy was carried out in 1990 by a group of researchers and physicians at the National Institutes of Health in Bethesda, Maryland, led by W. French Anderson, R. Michael Blaese, and Kenneth Culver. The patient, 4-year-old Ashanti DaSilva, had a genetic disorder called **severe combined immunodeficiency (SCID).** Patients with SCID lack a functioning immune system because they have a defect in a gene called adenosine deaminase (*ADA*). *ADA* produces an enzyme involved in metabolism of the nucleotide deoxyadenosine triphosphate (dATP). A mutation in the *ADA* gene results in the accumulation of dATP, which, at high concentration, is toxic to certain types of T cells, resulting in a near-complete loss of these cells in the patient with SCID. This condition is appropriately called "severe combined" immunodeficiency because mutations in the ADA gene deliver a knockout blow to the immune system's ability to make antibodies and fight off disease. Without functioning T cells, B cells cannot recognize antigen and make antibodies. Prior to gene therapy strategies, most patients with SCID did not live past their teens because their immune systems simply could not fight off infections from microbes.

To treat Ashanti, the normal gene for ADA was cloned into a vector that was then introduced into a retrovirus that had been inactivated to become non-pathogenic. An *ex vivo* gene therapy approach was used in which a small number of T cells were isolated from Ashanti's blood and cultured in the lab. Her T cells were infected with the ADA-containing retrovirus, and the infected T cells were further cultured. Because retroviruses integrate their genome into the genome of host cells, the retrovirus was integrating the normal ADA gene into the chromosomes of Ashanti's T cells during this culturing. After a short period of culturing, these ADA-containing T cells were reintroduced back into Ashanti (Figure 11.15).

Ashanti received multiple treatments. Within a few months after gene therapy, the T cell numbers in Ashanti began to increase. After two years, ADA enzyme activity in Ashanti was relatively high, and she was showing near normal T-cell counts with about 20% to 25% of her T cells showing the added ADA gene. Ashanti is currently enjoying a healthy life. Since Ashanti, gene therapy has successfully restored the immune systems of over two dozen children with SCID.

Treating cystic fibrosis

Cystic fibrosis (CF) is one of the most common genetic diseases. Approximately 1,000 children with CF are born in the United States each year, and currently more than 30,000 people in the United States have been diagnosed with CF. This disease occurs when a person has two defective copies of a gene encoding a protein called the **cystic fibrosis transmembrane conductance regulator (CFTR).** The normal CFTR protein serves as a pump at the cell membrane to move electrically charged chloride atoms (ions) out of cells. Chloride ions enter cells a number of ways and are required for many cellular reactions. The CFTR is important for maintaining the proper balance of chloride inside cells. Mutations in the *CFTR* gene, which may cause the total absence of the protein or result in defective protein, are responsible for CF. More than 500 mutations in the *CFTR* gene that can cause CF have been detected.

The CFTR is made by cells in many areas of the body, including the skin, pancreas, liver, digestive tract, male reproductive tract, and respiratory tract (trachea and bronchi). Abnormally functioning or absent CFTR causes an imbalance in chloride ions inside the cells because the defective CFTR does not pump out these ions (Figure 11.16). In organs such as the trachea, an accumulation of chloride ions in these cells leads to extremely thick sticky mucus that clogs the airways. This occurs because water moves into chloride-rich cells in an effort to balance chloride concentrations inside the cells. Normally, mucus in the trachea helps sweep dust and particles out of the airways to keep

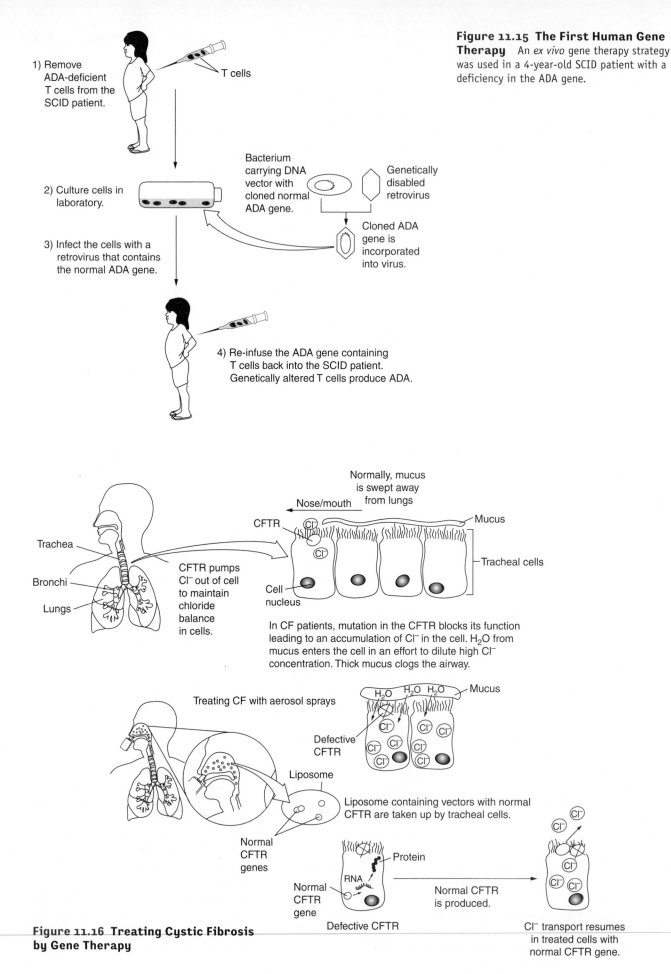

1) Remove ADA-deficient T cells from the SCID patient.

T cells

2) Culture cells in laboratory.

3) Infect the cells with a retrovirus that contains the normal ADA gene.

4) Re-infuse the ADA gene containing T cells back into the SCID patient. Genetically altered T cells produce ADA.

Bacterium carrying DNA vector with cloned normal ADA gene.

Genetically disabled retrovirus

Cloned ADA gene is incorporated into virus.

Figure 11.15 The First Human Gene Therapy An *ex vivo* gene therapy strategy was used in a 4-year-old SCID patient with a deficiency in the ADA gene.

Trachea

Bronchi

Lungs

CFTR pumps Cl⁻ out of cell to maintain chloride balance in cells.

Normally, mucus is swept away from lungs

Nose/mouth

CFTR

Mucus

Tracheal cells

Cell nucleus

In CF patients, mutation in the CFTR blocks its function leading to an accumulation of Cl⁻ in the cell. H_2O from mucus enters the cell in an effort to dilute high Cl⁻ concentration. Thick mucus clogs the airway.

Treating CF with aerosol sprays

H_2O H_2O H_2O Mucus

Defective CFTR

Liposome

Liposome containing vectors with normal CFTR are taken up by tracheal cells.

Normal CFTR genes

Protein

RNA

Normal CFTR gene

Defective CFTR

Normal CFTR is produced.

Cl⁻ transport resumes in treated cells with normal CFTR gene.

Figure 11.16 Treating Cystic Fibrosis by Gene Therapy

these materials from reaching the lungs. But when water enters trachea cells, the mucus becomes extremely thick. In addition to clogging the airways, the thick mucus provides an ideal environment for microbes to grow; as a result, patients with CF experience infections from bacteria such as *Pseudomonas*.

Infections of the airways and lungs can lead to pneumonia and respiratory failure, the leading cause of death among patients with CF. There are similar effects in many other organs of the body. Males experience infertility owing to problems related to ion transport in male reproductive organs, and both males and females have extremely salty sweat owing to abnormalities in ion transport in sweat glands. In fact, prior to the discovery of the *CFTR* gene, a sweat test was the standard for diagnosing CF and is still used today.

Treatments for patients with CF vary from back clapping—holding the patient almost upside down and slapping the back—and moving body positions to drain lung mucus, to using drugs to thin the mucus and antibiotic treatment to fight infections. However, there is no cure for CF. This disease often causes death as a result of lung malfunctions and respiratory infections early in life, with nearly half of all patients with CF dying before age 31. Recently, many new forms of treatment have enabled these patients to live to well into adulthood. One form of gene therapy has helped some patients. In 1989, the *CFTR* gene was discovered, and shortly afterward scientists began gene therapy trials by introducing the normal *CFTR* gene into liposomes and spraying these liposomes into the nose and mouth of CF patients as an aerosol or administering liposomes into the airways via a hose (Figure 11.16). Liposomes fuse with lipids in cell membranes of tracheal cells, releasing the normal *CFTR* gene into the cytoplasm of cells. The normal *CFTR* gene is copied into mRNA, and the normal protein translated. The normal CFTR protein enters cell membranes and starts to transport chloride ions out of cells, thereby thinning the mucus and alleviating CF symptoms.

Gene therapy for CF is not a reliable cure yet. It is expensive, requiring multiple reapplications because DNA delivered via liposomes does not integrate into chromosomes. Each time tracheal cells divide, and they divide rapidly, delivered genes are lost, and more spraying is required. Also, *CFTR*-containing liposomes are not taken up by all cells. Even cells containing the delivered gene may not produce enough CFTR protein to allow for adequate transport of chloride ions. And there have been problems with the expressed CFTR protein being toxic to cells. Although a cure for CF through gene therapy is not yet available, scientists are aggressively moving forward on strategies that may eventually lead to the permanent introduction of the normal gene or the correction of defective *CFTR* in an effort to improve dramatically the lives of people with CF.

Challenges Facing Gene Therapy

Scientists have always been concerned about the potential risks associated with gene therapy and the safety of these procedures. Discussions about the safety of gene therapy have greatly intensified since an 18-year-old volunteer named Jesse Gelsinger died during a gene therapy clinical trial at the University of Pennsylvania in 1999. Jesse's death was directly attributed to complications related to the adenovirus vector used to deliver therapeutic genes to treat him for a genetic disorder (ornithine transcarbamylase deficiency), which affected his ability to break down dietary amino acids.

Over 500 gene therapy clinical trials have been carried out around the world, a majority of these in the United States, and more than 600 trials are ongoing in 20 countries. Jesse Gelsinger was the first person in a gene therapy clinical trial to die as a result of his treatment. His death raised more questions about using viral vectors, placed greater emphasis on the development of nonviral vectors, and called for greater scrutiny of gene therapy and tighter restrictions on gene therapy trials.

In 2002, concern about gene therapy approaches involving retroviruses were further elevated as a result of gene therapy trials in France for treating X-linked severe combined immunodeficiency syndrome (SCID-X). In this trial, three of the 11 children treated developed leukemia because of the therapy, and one of these children died in 2004. They had received injections of *ex vivo* treated bone marrow cells that were treated with a retrovirus-delivered therapeutic gene. The children were 1 and 3 months old at the time of treatment, and they had returned home to a normal life seemingly cured until they developed a leukemia-like cancer about 2½ years after gene therapy. Their cancer was caused by retrovirus vectors randomly integrating the therapeutic gene into a critical location of the genome containing the promoter region for a gene called *LM02*, which encodes a transcription factor required for normal formation of white blood cells. This integration led to aberrant transcription of the *LM02* gene and overexpression of *LM02* that triggered uncontrolled division of mature T cells.

This tragedy resulted in a temporary cessation of a large number of gene therapy trials and the U.S. FDA completely stopped most retroviral studies. Trials eventually resumed but with greater patient monitoring.

Currently there are more barriers to gene therapy than solutions to medical problems:

- How can expression of the therapeutic gene be controlled in the patient? For many cells, normal function requires that correct amounts of key proteins are made at the right time. What happens if therapeutic genes are overexpressed or if a gene shuts off shortly after it has been introduced?

- How can scientists safely and efficiently target only the cells that require the therapeutic gene without affecting other cells in the body where the gene is not needed?

- How can gene therapy be targeted to specific regions of the genome to prevent the random integration problems encountered in the French trials or instability and movement of the inserted DNA?

- How can gene therapy provide lasting, permanent treatment without frequent administration of the therapeutic gene?

- How can rejection of the therapeutic gene be avoided? Whenever gene therapy is used, it is not always known if the recipient's immune system will reject the protein produced by therapeutic genes or reject genetically altered cells containing therapeutic genes.

- How many cells must express the therapeutic gene to treat the condition effectively? This will vary depending on the disease condition, but it remains to be determined if a majority of diseased cells must be affected by the therapeutic gene or if a disease can be treated by only correcting a small number of cells.

These and other barriers must be overcome before gene therapy becomes a safe and reliable treatment approach, but scientists are making excellent progress in this field. They are also making incredibly rapid advances in another hot area of medical biotechnology called regenerative medicine, the topic of the next section of this chapter.

11.4 The Potential of Regenerative Medicine

Currently, doctors treating human illness are primarily limited to approaches such as surgical techniques, radiation treatment, and drug therapy. Although these approaches all have a place in medicine, pills, for instance, are generally not the answer to *curing* human disease, and they certainly don't offer the ability to regenerate tissue and restore functions of damaged organs. For example, when a person has a condition such as a heart attack or stroke, tissue damage often

results. When an organ is damaged, the only way to restore functions of the organ fully is to replace the damaged tissue with new cells, something that often does not occur naturally in heart and brain tissue.

When organ development occurs in the embryo, changes in expression of many genes must occur in an ordered sequence. Changes occur in cells that even drugs cannot fix, and no one drug can stimulate the growth and repair of new tissue when an organ is severely damaged. Even with new knowledge gained from the Human Genome Project, it is highly unlikely that any one drug or even a few drugs could be used to stimulate hundreds of changes in gene expression with the proper timing required for tissue regeneration and restored organ function.

Regenerative medicine, growing cells and tissues that can be used to replace or repair defective tissues and organs, is an exciting new field of biotechnology that holds the promise and potential for radically changing medicine and the delivery of health care as we know it. Most researchers in the field agree that the goal of regenerative medicine is not to extend human lifespan and achieve immortality but to improve the healthy quality of life.

Cell and Tissue Transplantation

Organ transplantation is not a new idea, but applications that involve transplanting specific cells and tissues to replace or repair damaged tissues are relatively new aspects of medical biotechnology research.

Fetal tissue grafts

Neurodegenerative diseases occur gradually, leading to progressive loss of brain functions over time. Alzheimer disease and Parkinson disease are perhaps the two most well-known examples. These diseases rank first and second, respectively, as the most common neurodegenerative disorders. For Parkinson disease alone, approximately 50,000 cases are diagnosed yearly, and an estimated 500,000 Americans currently have the disease. Parkinson disease is created by loss of cells in an area deep inside the brain called the *substantia nigra*. Neurons in this region produce a chemical called *dopamine*, which is a neurotransmitter—a chemical used by neurons (nerve cells) to signal one another. Loss of these dopamine-producing cells causes tremors, weakness, poor balance, loss of dexterity, muscle rigidity, reduced sense of smell, inability to swallow, and speech problems, among other effects. Most treatments involve drugs that increase dopamine production or accumulation in the brain; however, after about 4 to 10 years of drug treatment, the disease progresses, and the effectiveness of these drugs diminishes, leading to a poor quality of life for the patient, who typically dies of complications related to the disease.

Unlike fetal neurons, which can divide, most adult neurons will not repair themselves when damaged, and most neurons do not undergo cell division. Scientists have long been interested in using fetal neuron transplants as a way to treat Parkinson disease and other neurological conditions. The basic idea is to introduce fetal neurons with the hopes that these cells can establish connections with other neurons in an effort to replace the damaged brain cells and restore brain function. After demonstrated success in rodents, fetal tissue transplants have been used since the late 1980s, and well over 100 patients have received such transplants. In this procedure, physicians typically drill a hole in the skull and surgically inject neurons into areas of the brain most affected in Parkinson patients. Most human fetal tissue comes from embryos or fetuses obtained from accident victims and legally induced aborted embryos. Patients receiving fetal transplants have shown varying degrees of improvement, including relief of Parkinson symptoms by over 40% and in some cases almost eliminating most symptoms even several years later. Fetal transplants have not provided full recovery.

As of 2001, over 250,000 individuals have been paralyzed by trauma to the spinal cord, and nearly 2 million people worldwide are living with spinal cord injuries. Each year approximately 85,000 people suffer spinal cord injuries, including roughly 10,000 people in the United States. Damage can occur when the cord is crushed or the nerve fibers are severed. Incomplete or complete severing may result in paraplegia, paralysis of the lower body, or quadriplegia, paralysis of the body from the neck down, depending on where the injury occurred in the cord. Many strategies have been used in an attempt to repair spinal cord injuries. One approach has been to graft nerve fibers from fetal or adult neurons into the damaged area of the spinal cord in an attempt to bridge the severed cord (Figure 11.17). These bridge implants have shown promise in dogs and rats. As scientists learn more about inflammatory chemicals that hinder nerve growth and the factors that stimulate nerve growth, it may be possible to use such molecules to minimize scar tissue formation, to reduce damage caused by scar tissue from supporting cells called glial cells, block growth inhibitor molecules, and stimulate neuron regeneration at the same time.

Organ transplantation

Organ transplantation can and does save lives. Approximately 8 million surgeries related to tissue damage and organ failure are performed in the United States each

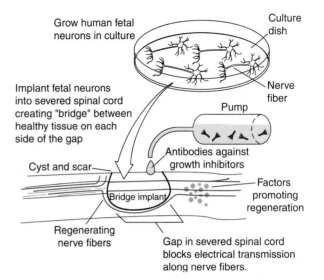

Figure 11.17 Bridging the Gap In many types of spinal cord injuries, a gap occurs when the spinal cord is severed. This blocks electrical transmission of nerve impulses and can result in paralysis and the loss of body sensations such as temperature, pain, and touch. One experimental strategy for repairing spinal cord injuries involves implanting fetal neurons into the severed area of the spinal cord to create a bridge between normal neurons. Scientists are using this strategy in combination with antibodies that may block chemicals that inhibit spinal cord repair and growth factors that can stimulate neuron regeneration in an effort to bridge the gap in spinal cord injuries.

year, but about 4,000 people die each year while waiting for an organ transplant. At least 100,000 people die each year without ever qualifying to be on the waiting list. Well over $400 billion is spent on organ failure and tissue-related health care costs in the United States. This number is nearly half of the nation's health care bill.

Autografts, transplanting a patient's own tissue from one region of the body to another, can alleviate some transplantation problems. For example, coronary bypass operations involve removing segments of a vein from the leg and connecting this blood vessel surgically to arteries in the heart as a bypass around obstructed vessels. But if a patient needs a heart or a liver transplant, a human who can donate an organ for the recipient must be found. Even when a human donor who appears to be a match is found, organ rejection is a major problem. Rejection typically occurs when the recipient's immune system recognizes that the donor organ is foreign. Matching organs for transplantation involves tissue typing to check if a donor organ is compatible for a recipient. Tissue typing is based on marker proteins found on the cell surface (membrane) of every cell in the body. Tissue typing proteins are part of a large group of over 70 genes called the **major histocompatibility complex (MHC),** aptly named because *histo* means "tissues," and MHC molecules must be matched between donor and recipient to have a compatible organ transplantation.

There are many different types of MHC proteins. One common group, called the human leukocyte antigens (HLAs, named because they were first discovered on white blood cells, leukocytes), are found on virtually all body cells. Immune system cells such as B and T cells recognize HLAs on all body cells present since birth as "self" (belonging to the same individual), whereas any other cells are "nonself" or foreign cells that may be attacked by the immune system and destroyed. Some common HLAs are found on most human tissues, and others are unique to a given individual. To have a successful transplantation of an organ from one human to another requires a close match of several types of HLAs between the donor organ and the recipient's cells; otherwise, the recipient will reject the transplanted organ.

Since the first human liver transplant in 1963, transplant surgeons have been using immunosuppressive drugs to weaken the immune system and minimize organ rejection. Most transplant recipients must use immunosuppressive drugs for the rest of their lives. One obvious problem with this approach is that patients on immunosuppressive drugs can and do develop infections, which can be life threatening, because of their weakened immune system. The lack of sufficient human organ donors and organ rejection are major reasons why scientists are looking at other ways to provide donor organs.

Xenotransplantation—the transfer of organs from different species—may one day become a viable alternative to human-to-human organ donation, thus helping relieve the tremendous need for human donor organs. Baboons were once considered the animal of choice for providing organs to human recipients. The first animal-to-human organ transplant in a child, carried out in 1984 by doctors at Loma Linda University Medical Center in California, involved transplanting a baboon heart into Baby Fae, a 12-day-old girl. Baby Fae lived with the baboon heart for three weeks before she died from complications related to organ rejection. Similar transplants have been performed without great success. Although baboons and other primates may still be candidates for providing organs, many groups are choosing to investigate the potential for using pigs as an organ source. Pigs may be a good choice because they are plentiful, easy to breed, and relatively inexpensive. Many pig organs are also similar in function and size to the types of human organs. Progress on using pig organs for transplantation in humans has been slowed by concerns that viruses may be transmitted from pigs to humans, causing the transplanted organs to be rejected and creating other health problems.

Transplantation scientists have recently combined molecular techniques and transplantation technologies to produce cloned pigs that may help overcome current fears of organ rejection and viral disease transmission. Researchers at the University of Missouri have created cloned piglets that lack a gene that causes pig organs to be rejected by the immune system of humans. Using gene knockout techniques, researchers have eliminated a key gene called *GGTA1* (β-1, 3-galactosyltransferase). *GGTA1* produces a sugar on the surface of pig tissues, which if present when transplanted into a human, would be recognized as a foreign antigen leading to antibody production and immune rejection of the organ. The *GGTA1* knockout pigs were cloned using nuclear transfer cloning techniques discussed in Chapter 7. Creating *GGTA1* knockout pigs may be a way to generate pigs that could produce organs for transplantation the human immune system may not recognize as foreign (Figure 11.18).

Xenotransplantation does not always have to involve the transfer of a whole organ, as you will learn in the next section. Scientists are working hard to develop ways to deliver small clusters of cells as a technique for cell and tissue transplantation.

Cellular therapeutics

Cellular therapeutics involves using cells to replace defective tissues or to deliver important biological molecules. One alternative for avoiding organ rejection for transplants involving human tissue and xenotransplants is to use living cells that have been encapsulated into tiny plastic beads or tubes called **biocapsules** or microcapsules. Biocapsules may also contain genetically

Figure 11.18 Pigs Could Potentially Save the Lives of Patients Waiting for a Transplant These piglets have been engineered to lack a sugar-producing gene that causes human bodies to reject pig organs, potentially providing a source of rejection-free pig organs.

engineered cells designed to produce therapeutic molecules such as recombinant proteins.

Biocapsules have tiny holes in their walls, making them permeable to nutrient exchange and allowing molecules produced by the encapsulated cells to escape from the capsule and enter the bloodstream or surrounding tissues (Figure 11.19). For instance, capsules containing insulin-producing cells (beta cells) from the pancreas, implanted in patients with type I diabetes, would produce insulin that could travel out of the capsule into the bloodstream of the patient to all body organs that require insulin. Another important feature of biocapsules is that they protect cells from being attacked by the recipient's immune system by hiding them within capsules so immune cells and antibodies cannot enter the capsule to destroy these cells. Although not a permanent cure, biocapsules can provide lasting release of molecules into the body. This approach would likely require that biocapsules be changed every few months; however, in the case of diabetes, this might be a better alternative than daily injections of insulin.

Tissue Engineering

Whether you pay $30,000 or $300 for your first car, one thing is certain, over time parts will wear out and break. A trip to the local mechanic can repair and replace some car parts, but eventually the car wears out to the point of no return, and it is time for another car. Wear and tear of human body parts also takes its toll. Over time, organs do not work as well as they

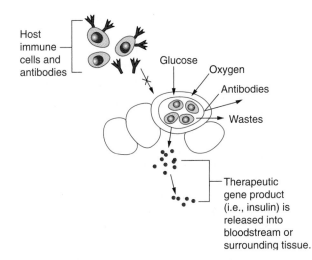

Figure 11.19 Biocapsules Encapsulated cells may provide valuable ways to deliver therapeutic molecules. Cells in the biocapsule are protected from attack by immune cells and antibodies of the host. At the same time, biocapsules allow molecules produced by the cells to leave the capsule and provide therapeutic benefits for the host. This figure illustrates how insulin-producing cells could be used to provide a patient with diabetes with a source of insulin.

should; in some cases, an organ may stop functioning altogether. Even if a person lives a relatively healthy life, the wear and tear of aging or a sudden event such as a stroke or heart attack will lead to a decline in organ function and perhaps organ failure. But our bodies are not like cars: We cannot go to a warehouse of body parts for replacements.

In the future, however, the emerging science of **tissue engineering** may someday provide tissues and organs that can be used to replace damaged or diseased tissues. This small but growing industry—there are over 60 biotechnology companies involved in tissue engineering in the United States alone and experts predict the field will grow by 50% in the next decade—is actively involved in research to produce human tissues and organs that can be used to replace damaged tissues or possibly even replace worn tissues. For instance, engineered sheets of human skin have proven very useful for treating severe burn victims who have lost substantial portions of their skin, a very painful and life-threatening condition.

Tissue engineering scientists often begin by designing and constructing a framework or scaffold made of biological substances such as calcium, collagen, a polysaccharide called alginate, or biodegradable materials. The scaffold is usually shaped as a mold of the organ to be replaced, and its purpose is to create a three-dimensional framework onto which cells are placed. Growing human cells on the scaffolding is called *seeding* because the cells literally act as "seeds" to create more cells that will grow over the scaffold. Scaffolds seeded with cells are bathed in a nutrient-rich media, and over time cell layers build up over the scaffolding material to assume the shape of the scaffold. So far, sheets of skin grafts have proven to be the most successful organs grown by tissue engineering; however, engineered bone structures for healing bone fractures have also worked very well, and scaffolding to engineer teeth has shown promise.

In the 1990s, tissue engineering pioneer Dr. Charles Vacanti and colleagues made headlines around the world when they revealed a mouse with an engineered ear growing on its back (Figure 11.20). In this example, a biodegradable scaffold in the shape of an outer ear was attached to a mouse and then seeded with cartilage cells from cows. The cartilage cells infiltrated the scaffolding and produced cartilage as they grew, and then as the scaffolding degraded, the cartilage developed enough strength to support itself. The human ear-looking tissue was never transplanted onto a human. Because it was made of cow cells it would have been rejected by the human immune system anyway. Also, this tissue was just the outer ear without the inner ear structures that actually detect sound. But it provided strong evidence that tissue engineering could work.

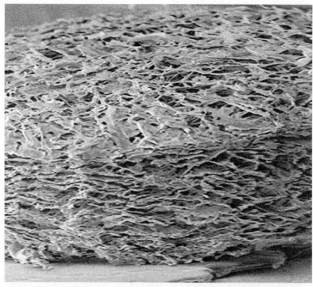

Figure 11.20 Tissue Engineering (a) An example of collagen scaffolding similar to the kind used to engineer an outer ear on the back of a mouse (b).

Creating large and complex organs such as the liver and kidney has proven to be much more difficult, although fetal tissue has been used to grow a rudimentary kidney in rats that was able to produce a urine-like fluid. At least two Phase II clinical trials are underway in which human bladders were created using tissue biopsies from patient's bladders to seed scaffolding and produce new bladders. The field of tissue engineering is really in its infancy, but progress is advancing at an incredibly rapid rate, and there is every reason to expect that engineered organs will be a reality. Tissue engineering will also benefit from knowledge about genes discovered through the Human Genome Project. As an example, we will consider how applications of the telomerase gene may be used to advance the tissue engineering field.

The telomere story

In normal human cells, the ends of chromosomes contain sequences of DNA nucleotides called **telomeres.** Telomeres are usually 8,000 to 12,000 bp units of the repeating sequence 5'-TTAGGG-3'. They can be thought of as the plastic tabs at the end of shoelaces that prevent the laces from unraveling, a sort of chromosome "cap." Normal cells have a limited ability to proliferate. Most human body cells can divide a maximum of 50–90 times before they show signs of aging—a process called **senescence**—which eventually leads to cell death. A cell's lifespan is affected in part by telomeres.

Each time a cell divides, telomeres shorten slightly. This occurs because of a basic flaw that prevents DNA polymerase from completely copying the ends of both strands of a DNA molecule. In many ways, telomeres serve as a biological clock for counting down cell divisions leading to senescence and cell death. Telomeres shorten, and senescence occurs until the cell can no longer divide. If there are multiple copies of the TTAGGG repeats at the ends of chromosomes, cells can lose this DNA without the loss of precious gene sequences. Eventually loss of repeat sequences produces a critical loss of DNA so cells no longer divide (Figure 11.21).

Scientists have long known that many cancer cells can divide indefinitely—a property called *immortality*. One way that cancer cells achieve immortality is through the actions of an enzyme called **telomerase.** Telomerase repairs telomere length at the ends of chromosomes by adding DNA nucleotides to cap the telomere after each round of cell division. Telomerase is not active in normal cells but is active in over 90% of human cancers. By preventing telomere shortening, telomerase activity is a major reason why cancer cells can divide indefinitely. In fact, biologists call cancer cells immortal cells because of their ability to avoid senescence indefinitely.

Many studies indicate that telomere shortening is involved in the aging process—the aches and pains, wrinkles, arthritis, gray hair, and other symptoms humans experience as we age. Telomerase is not a "fountain of youth" cure for the effects of old age. Aging and cancer are far more complex than just one enzyme. Although high levels of telomerase are found in almost every human cancer cell, telomerase itself does not *cause* cancer. Telomerase in combination with genetic mutations in genes that control cell division can create immortal cells that avoid senescence. Overproduction of telomerase often correlates with the aggressive growth of tumors. Researchers are working on treatment strategies including the use of telomerase as an anticancer vac-

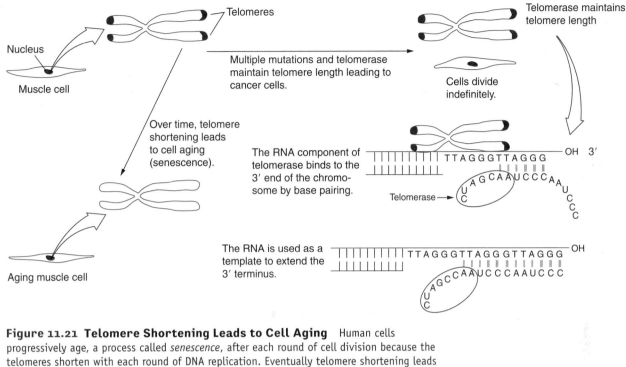

Figure 11.21 Telomere Shortening Leads to Cell Aging Human cells progressively age, a process called *senescence*, after each round of cell division because the telomeres shorten with each round of DNA replication. Eventually telomere shortening leads to cell death. Through mutations that affect the functions of different genes and expression of the enzyme telomerase, cancer cells can divide indefinitely and avoid senescence. Telomerase plays a role in this process by continually filling in telomere sequences to prevent them from shortening during cell division.

cine target to inhibit telomerase and halt cancer cells from dividing.

From a tissue engineering perspective, scientists are investigating how introducing telomerase genes into cultured human cells can allow them to produce normal human cells that display immortality. If successful, immortal human cells could be very valuable for treating individuals with age-related disorders from arthritis to neurodegenerative diseases. Such cells could also be used for myriad functions, such as providing skin cells for healing bedsores and ulcers and treating patients with late-onset blindness, muscular dystrophy, and even brain disorders. In the future, the technology might even be available to remove aging cells from a patient, introduce telomerase genes to these cells to extend their lifespan, and then return the cells to the patient. Some of this work could be done in combination with **stem cells,** which are perhaps the hottest and most controversial topic in medical biotechnology today.

Stem Cell Technologies

The CDC's National Center for Human Statistics indicates that approximately 3,000 Americans die every day from diseases that may one day potentially be treated by stem cell technologies. In the future, stem cell research may affect the lives of millions of people throughout the world. The tremendous promise and controversy surrounding stem cells has made stem cell research and related topics regular front-page news items and TV headlines.

What are stem cells?

To understand what stem cells are, we need to look briefly at the development of the human embryo. We do this by considering how *in vitro* fertilization (IVF) is carried out. IVF first gained public attention in 1978 when Louise Brown, the first test tube baby, was born. To create a child by IVF, sperm and egg from donor parents are mixed together in a culture dish to produce an embryo. After several days of division, the embryo is surgically implanted in the uterus of a woman, usually the egg donor, who has been treated with hormones to prepare the uterus for implantation of an embryo. When a couple agrees to undergo IVF, several embryos are usually created, but often only one is implanted during each procedure. The remaining embryos are frozen for future use as needed. Potentially, the leftover embryos can be a source of human embryonic stem cells, but they are also the source of a great deal of controversy.

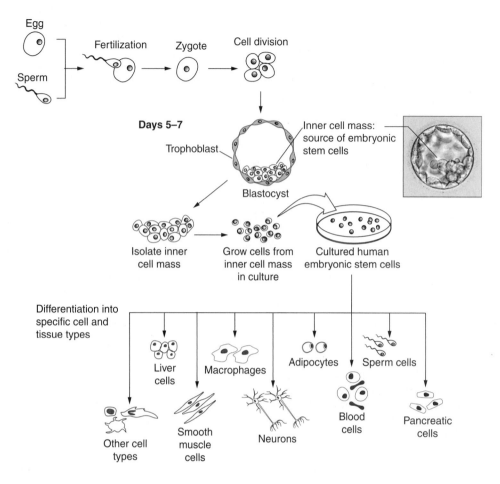

Figure 11.22 Isolating and Culturing Human Embryonic Stem Cells Cells isolated from the inner cell mass of human embryos can be grown in culture as a source of human ES cells. Under the proper growth conditions, ES cells can be stimulated to differentiate into virtually all cell types in the body.

An embryo goes through a predictable series of developmental stages (Figure 11.22). Following fertilization of a sperm and an egg cell, the fertilized egg is called a **zygote.** The zygote divides rapidly and after three to five days first forms a compact ball of about 12 cells called a **morula,** meaning "little mulberry." In around five to seven days after fertilization, the dividing cells create an embryo consisting of a small hollow cluster of approximately 100 cells called a **blastocyst.** The blastocyst is approximately a seventh of a millimeter in diameter. The blastocyst contains an outer row of single cells called the trophoblast; this layer develops to form part of the placenta that nourishes the developing embryo. The area of cells of primary interest to stem cell biologists is a small cluster of around 30 cells tucked inside the blastocyst that form a structure known as the **inner cell mass** (Figure 11.22). These cells are the source of **human embryonic stem cells (hESCs).**

During embryonic development, cells of the inner cell mass develop to form the embryo itself. Stem cells in the inner cell mass have the ability to undergo **differentiation,** a maturation process in which cells develop specialized functions. Differentiation is a com-

plex process involving many genes that must be activated and silenced, and differentiating cells rely on chemical signals such as growth factors and hormones from other cells to help them change. Stem cells are so special because they can eventually differentiate to form all of the more than 200 cell types that make up the human body. Stem cells are called **pluripotent** because they have the potential to develop into a variety of different cell types.

Successful isolation and culturing of the first hESCs from a human blastocyst was reported in 1998 by James Thomson of the University of Wisconsin at Madison who had cultured hESCs from rhesus monkeys two years earlier. Also in 1998, John Gearhart and colleagues at Johns Hopkins University isolated embryonic germ cells, primitive cells that form the gametes—sperm and egg cells—from human fetal tissue and demonstrated that these cells can develop into different cell types (Figure 11.22). These discoveries followed the work of other scientists who had isolated stem cells in species such as mice, pigs, cows, rabbits, and sheep. In fact, stem cell researchers credit much of what is now known about isolating hESCs from pioneering work initiated in mice in the 1980s—another outstanding example of how model organisms contribute greatly to advancing science.

Human ES cells are unspecialized cells with two major properties that have scientists very excited.

- ES cells can self-renew indefinitely to produce more stem cells.

- Under the proper growth conditions, hESCs can differentiate into a variety of mature cells with specialized functions.

Human ES cells avoid senescence and show no signs of aging, in part because they express high levels of telomerase. Several groups have maintained stem cells for over three years and over 600 rounds of division without apparent problems. Cultured cells such as these that can be maintained and grown successively are called **cell lines.** Stem cells also grow rapidly and can be frozen for long periods of time and still retain their properties. Under the right conditions, when stimulated with different molecules that include hormones and substances called **growth factors** (scientists are working hard to identify many of these factors), stem cell lines can be coaxed to differentiate into different types of cells. Stem cells typically need to be cultured over a layer of other cells, usually derived from mice, called "feeder cells," which provide hESCs with many unknown components that help them multiply and prevent them from differentiating. Several labs have recently reported success in growing stem cells without feeder layers, however, by adding growth factors such as fibroblast growth factor to hESCs. This is a major advantage because feeder cells can often mix in and contaminate stem cells when they are stimulated to differentiate.

In the laboratory under the proper culturing conditions, embryonic stem cells from humans, mice, rats, primates, and other species differentiate into a myriad of cells including skin cells; brain cells (both neurons and glial cells, which support, nourish, and protect neurons); cartilage (chondrocytes); spermatozoa; osteoblasts (bone-forming cells); liver cells (hepatocytes); muscle cells including smooth muscle, which forms the walls of blood vessels; skeletal muscle cells, which form the muscles that attach to and move the skeleton; and cardiac muscle cells (myocytes), which form the muscular walls of the heart.

How do researchers obtain hESCs? Human ES cells for research purposes are derived from the blastocysts of embryos that are no longer needed by couples for IVF or from human embryos created by IVF from sperm and egg cells donated for the purpose of providing embryos for research materials. Typically leftover blastocysts would either be destroyed or frozen indefinitely, but with the consent of the couple, they can be used to derive embryonic stem cells as described in Figure 11.22. In U.S. fertility clinics alone, an estimated 400,000 frozen unused embryos and several thousand eggs are discarded annually.

A reproductive quirk of sorts may enable researchers to harvest stem cells without the need for human embryos. Prior to fertilization, human eggs actually have a full number of chromosomes (46). Upon fertilization, one set of these (23) will be merged with 23 chromosomes from the sperm. The other 23 chromosomes from the egg are ejected into a remnant cell called the *polar body*, which does not develop further. It may be possible to treat human eggs with chemicals so they retain their entire set of 46 chromosomes and then induce embryonic development from the egg. This process of creating an embryo without fertilization is called **parthenogenesis,** which occurs naturally in some animals such as salamanders, some reptiles, birds such as chickens, and certain insects, as a normal means of reproduction. Creating human embryos by parthenogenesis may be another way to derive ES cells. Stem cells from parthenogenic primate eggs have already been isolated, and recently hESCs have been isolated from a human embryo created by parthenogenesis (we discuss this discovery further in the next section). Stem cells derived by parthenogenesis still may not eliminate concerns about using embryos for research because people may not be comfortable with a parthenogenic embryo's unclear status as a life-form because it was not derived by fertilization. So far, human parthenogenic eggs do not appear to be capable of developing to form full-term infants.

Can adult stem cells do everything embryonic stem cells can do?

Research on hESCs is very controversial because of their source—an early embryo. During the last few years, scientists have discovered **adult-derived stem cells (ASCs),** cells that reside in mature adult tissue and could be cultured and differentiated to produce other cell types. ASCs appear in very small numbers, and although they have been isolated from the brain, intestine, hair, skin, pancreas, bone marrow, fat, mammary glands, teeth, muscle, and blood, they have not yet been discovered in all adult tissues.

Some opponents of hESC research claim that ASCs are a more acceptable alternative than using hESCs because isolating ASCs does not require the destruction of an embryo. ASCs can be harvested from people by fine-needle biopsy: A thin-diameter needle is inserted into muscle or bone tissue. It may even be possible to isolate ASCs from cadavers. We also know that ASCs are present in fat (adipose) tissue, which could potentially be an outstanding source of stem cells especially if you consider that over 500,000 L of fat tissue collected by liposuction and other cosmetic surgery techniques are discarded in the United States each year.

Experiments have shown that ASCs from one tissue can differentiate into another different specialized cell type. For instance, an ASC isolated from muscle tissue could be used to develop into a blood cell. But other studies have demonstrated that ASCs may not be as pluripotent as hESCs. Much more research is required to determine if ASCs will be as valuable as hESCs may be.

Can stem cells be generated without destroying an embryo?

Soon there may be several ways to readily produce cells with hESC properties without destroying an embryo, and many labs around the world are working hard to accomplish this. One approach involves creating stem cell lines from a single cell (called a *blastomere*) from a pre-blastocyst-stage embryo called a *morula* that contains 8 to 10 cells. Removing a single cell, which is done for preimplantation genetic testing of *in vitro*–produced embryos, does not destroy the embryo, which can further develop to the blastocyst stage and beyond. In another approach, scientists at the Massachusetts Institute of Technology turned off a gene called *CDX2* in a skin cell and then fused the cell with an oocyte to create a blastocyst that gave rise to hESCs, but the altered blastocyst could not implant in the uterus so it was incapable of developing into a full organism. Whether this disabled embryo would be acceptable to those who oppose embryonic stem cell research remains to be seen, but this approach has many people working on understanding genes that control hESC growth and development.

In early 2007, researchers from Wake Forest University published work demonstrating they had isolated stems cells from human amniotic fluid. In the lab, these so-called **amniotic fluid-derived stem (AFS) cells** were coaxed to become neurons, muscle cells, adipocytes, bone, blood vessels, and liver cells. It is not entirely clear if these cells are truly different from hESCs or ASCs, but if so they may be a key breakthrough in stem cell technologies.

Creating stem cells by nuclear reprogramming of somatic cells

One of the most promising new approaches for isolating stem cells without creating an embryo involves a technique called **nuclear reprogramming of somatic cells.** The basic concept of this approach is to use genes involved in cell development to push a somatic cell back to an earlier stage of development and affect gene expression, to reprogram the somatic cell genetically to return to a pluripotent state charac-

teristic of the stem cells from which it was derived. In 2005, one of the first successful reprogramming techniques involved fusing hESCs with skin cells called *fibroblasts*. The hybrid cells generated displayed several properties of hESCs *in vitro* and *in vivo*.

In the past two years, several laboratories have achieved success with nuclear reprogramming of mouse and human cells, and this approach is being heralded as a revolution in stem cell biology research. One approach has involved using retroviruses to deliver four transgenes *Oct3/4*, *Sox2*, *c-myc*, and *Klf4* into fibroblasts (Figure 11.23). Expression of these four genes, which encode transcription factors involved in cell development, "reprograms" the fibroblasts back to an earlier stage of differentiation. Such reprogrammed cells are called **induced pluripotent stem (iPS) cells.** iPS cells demonstrate many properties of hESCs, such as self-renewal and pluripotency, and appear to be indistinguishable from hESCs.

These iPS cells also express genes such as *Nanog* and *Oct* that are characteristic markers known to be expressed in undifferentiated hESCs. Other experiments also demonstrated that these iPS cells could differentiate into other cell types including neural cells and cardiac muscle cells.

In late 2007, two different research groups demonstrated they had successfully derived iPS cells by reprogramming human skin cells taken directly from a volunteer. One group reprogrammed skin cells taken from the face of a 36-year-old woman and connective tissue cells from the joints of a 69-year-old man, and another group used skin cells from the foreskin of a newborn boy. iPS cells were also recently produced without using the *c-myc* gene. This is a potentially important advance because *c-myc* is a known cancer-causing gene (oncogene). These promising results demonstrate that nuclear reprogramming may be a potentially viable way to generate donor person-specific stem cells without the need for an embryo. Watch for exciting new developments in the next few years involving nuclear reprogramming and iPS cells.

Potential applications of stem cells

There are many potential applications for stem cells—from growing healthy tissues, to studying them to understand and treat birth defects, to genetic manipulation for delivering genes in gene therapy approaches, to creating whole tissues in the laboratory using tissue engineering. Many scientists believe that stem cell technologies will play key roles in developing treatments for diseases such as stroke, heart disease, Parkinson disease, Alzheimer disease, Lou Gehrig disease, diabetes, and other conditions (refer to Table 11.2).

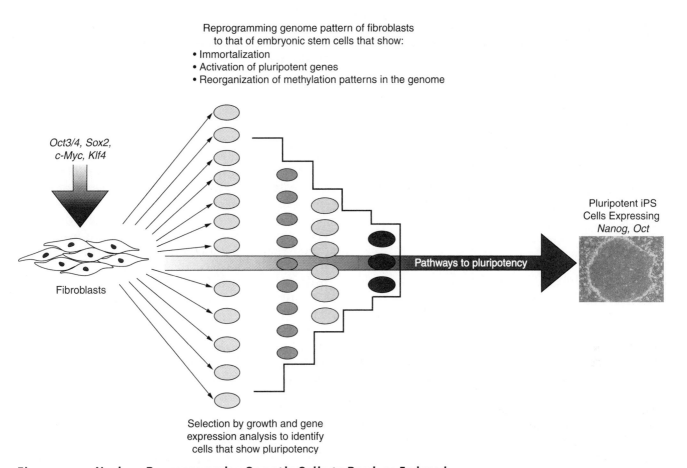

Figure 11.23 Nuclear Reprogramming Somatic Cells to Produce Induced Pluripotent Stem Cells Introducing four transcription factor genes (*Oct3/4, Sox2, c-myc, Klf4*) into mouse fibroblasts induces formation of induced pluripotent stem (iPS) cells. During this process a number of cells are defective in reprogramming and must be selected out during culture and by selecting them for expression of endogenous marker genes that are characteristic marker genes (such as *Nanog* and *Oct*) known to be expressed in pluripotent stem cells.

Potential and *promise* are frequently used words when discussing stem cell applications, but use of these cells for treating disease is still largely unproven. Here we consider some of the most promising examples of stem cell applications to date. Patients with leukemia, a cancer that causes white blood cells to divide abnormally producing immature cells, frequently require chemotherapy or radiation treatment to destroy the defective white blood cells. As a result, the patient's immune system is greatly weakened. Leukemia treatments may also involve blood transfusions to replace white blood cells and red blood cells damaged by chemotherapy. Using stem cells to make white blood cells is becoming an effective way to treat leukemia. Stem cells from umbilical cord blood have also been used to provide red blood cells for sickle-cell patients and individuals with other blood deficiencies. Isolating stem cells from cord blood is becoming so popular that, in many U.S. states, parents can opt to pay to have cord blood stem cells frozen indefinitely should their child need them in the future.

So far there have been a number of promising results in animal models and in human clinical trials using stem cells for tissue repair. For example, stem cells from fat have been used to form bone tissue in the human skull. Repair of heart tissue has shown strong potential. Stem cells might be used to replace dead and dying cells following trauma such as a heart attack. Heart attacks are a leading cause of death in the United States killing nearly half a million people each year. Death of cardiac muscle cells weakens the heart and can prevent it from beating with the proper strength to maintain normal blood flow. Adult cardiac muscle cells do not repair themselves well. Researchers at New York Medical College and the National Human Genome Research Institute have successfully injected adult stem cells from mouse bone marrow into damaged areas of the mouse heart

Table 11.2 STEM CELL–BASED THERAPIES MAY POTENTIALLY BENEFIT MILLIONS OF PEOPLE

Disease Condition	Number of Patients in the United States
Cardiovascular disease	58 million
Autoimmune diseases	30 million
Diabetes	16 million
Osteoporosis	10 million
Cancers (urinary bladder, prostate, ovarian, breast, brain, lung, and colorectal cancers; brain tumors)	8.2 million
Degenerative retinal disease	5.5 million
Phenylketonuria (PKU)	5.5 million
Severe combined immunodeficiency (SCID)	0.3 million
Sickle-cell disease	0.25 million
Neurodegenerative diseases (Alzheimer and Parkinson diseases)	0.15 million

Source: Adapted from Stem Cells and the Future of Regenerative Medicine, www.nap.edu/catalog/10195.html.

(Figure 11.24). These stem cells can develop into cardiac muscle cells, form electrical connections with healthy muscle cells, and improve heart function by over 35%. A similar approach was used to transplant human ES cells into the ventricular wall of damaged hearts in pigs and in humans involved in clinical trials. In both examples transplanted stem cells differentiated to form cardiac muscle cells that restored a significant percentage of electrical activity and contractility to the damaged areas. One study used stem cells from human umbilical cord blood to improve cardiac function in rats.

Scientists are optimistic that this approach may someday work in humans. Consider this: In the future, a surgeon may order a few grams of cardiac muscle cells from a regenerative medicine lab to transplant into a heart attack patient in much the same way that surgeons routinely order blood from a blood bank for a transfusion during a surgical procedure.

In late 2007, a promising and innovative treatment using iPS cells to correct sickle cell anemia in mice was reported. iPS cells were produced from skin cells of transgenic mice that express a mutated version of the human sickle cell hemoglobin gene and display

sickle cell disease. The iPS cells were genetically engineered to correct the hemoglobin gene mutation. Blood stem cells were then made from the corrected iPS cells and transferred into donor sickle cell mice which produced functional red blood cells that corrected the disease condition. This is an incredibly exciting result combining aspects of both stem cell technologies and gene therapy.

Earlier in this chapter, we discussed problems faced by patients who suffer from spinal cord injuries. In the last few years, researchers have disproved a long-standing belief that the human brain and spinal cord cannot grow new neurons. Adult stem cells have been isolated from the brain and used to make neurons in culture, and scientists have already demonstrated that ES cells can be differentiated to form neurons that can be injected into mice and rats to improve neural function in animals with spinal cord injuries. Researchers at Johns Hopkins University have demonstrated that human stem cell transplants can enable mice with paralyzed hind limbs to walk. These studies were carried out on mice that were paralyzed after they were infected with a virus similar to the poliomyelitis virus that causes polio. There is still much work to be done in the spinal cord repair and regeneration field, but researchers are optimistic that neural stem cell transplants may be ready for human clinical trials in the next three to five years, offering hope to the many individuals who are affected by spinal cord injuries.

Many fundamental questions about both adult and embryonic stem cells, however, must be answered before stem cell technologies are viable treatment strategies. For instance, how can stem cell differentiation be controlled once the cells are placed in the body? One problem with injecting stem cells instead of differentiated adult tissues is that scientists cannot fully control the spread of stem cells to other places in the body; nor can they control differentiation of stem cells into other tissues than their intended use. Injected ES cells have formed tumors, including types of tumors called *teratomas* that contain mixes of differentiated tissues such as teeth, bone, and hair, all in one tumor. Another problem is avoiding chromosome abnormalities that are known to occur when stem cells differentiate. For example, alterations in chromosome number (trisomy 12 and trisomy 17 and others) can be a common problem with stem cells. Other important questions to be answered are these:

■ Is there an "ultimate" adult stem cell that could turn into every tissue in the body? Some scientists are optimistic; others are very doubtful.

■ Why do stem cells self-renew and maintain an undifferentiated state?

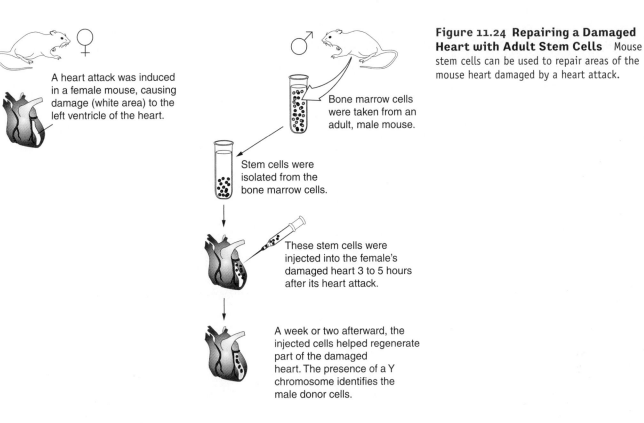

Figure 11.24 Repairing a Damaged Heart with Adult Stem Cells Mouse stem cells can be used to repair areas of the mouse heart damaged by a heart attack.

A heart attack was induced in a female mouse, causing damage (white area) to the left ventricle of the heart.

Bone marrow cells were taken from an adult, male mouse.

Stem cells were isolated from the bone marrow cells.

These stem cells were injected into the female's damaged heart 3 to 5 hours after its heart attack.

A week or two afterward, the injected cells helped regenerate part of the damaged heart. The presence of a Y chromosome identifies the male donor cells.

- What factors trigger division of stem cells?
- What are the growth signals (chemical, genetic, environmental) that influence the differentiation of stem cells?
- What factors affect the integration of new tissues and cells into existing organs?
- Can nuclear reprogramming of somatic cells, or other approaches that do not require an embryo, become a reliable technique for producing pluripotent stem cells with the properties of hESCs?
- Will it be possible to reprogram human cells without using oncogenes or retroviruses, and can the efficiency of the reprogramming process be improved?

Answers to these and many other questions will help scientists and physicians in their quest to develop stem cell–based applications for treating human disease.

Cloning

We have discussed many types of cloning in this book. Remember that *cloning* refers to making a copy of something—a gene, a cell, or an entire organism. Recombinant DNA technology is used for gene cloning. When bacteria or cultured cells divide in a petri dish or bioreactor, clones of cells are being produced. But clearly no aspect of cloning is as controversial as animal cloning. With the announcement of the cloning of Dolly the sheep on February 24, 1997, the world was immediately faced with the prospect that biotechnology could result in human cloning. Dolly's creation generated a great increase in public awareness and additional discussion about cloning. In this section, we consider scientific implications of human cloning applications.

Therapeutic cloning and reproductive cloning

There are really two approaches to cloning: **reproductive cloning** and **therapeutic cloning** (Table 11.3). The intent of reproductive cloning is to create a baby. Dolly was the first of many other mammals to be produced by reproductive cloning. Unlike reproductive cloning, therapeutic cloning provides stem cells that are a genetic match to a patient who requires a transplant. In therapeutic cloning, the chromosomes from a patient's cell (for instance, skin cells) are injected into an enucleated egg—an egg that has had its nucleus removed—that is then stimulated to divide in culture to create an embryo (Figure 11.25). The embryo produced will not be used to produce a child; instead, the embryo will be grown for several days until it reaches the blastocyst stage so it can be used to harvest stem cells. Stem cells isolated from this embryo can be grown in culture and then introduced into the donor patient (Figure 11.25).

Table 11.3 COMPARISON OF STEM CELLS, THERAPEUTIC CLONING, AND REPRODUCTIVE CLONING

	Embryonic Stem Cells	Adult Stem Cells	Therapeutic Cloning (Somatic Cell Nuclear Transfer)	Reproductive Cloning
Final or "end" product	Undifferentiated stem cells (isolated from fetal or embryonic tissue such as an embryo at the blastocyst stage) growing in culture	Undifferentiated stem cells (isolated from adult tissue such as bone marrow cells) growing in a culture dish	Undifferentiated stem cells growing in a culture dish (obtained from the person who will also serve as the recipient of these cells)	"Cloned" human
Purpose/ application	Source of stem cells for research and for treating human disease conditions such as replacing diseased or injured tissue	Source of stem cells for research and for treating human disease conditions such as replacing diseased or injured tissue	Source of stem cells that are genetically matched to recipient for treating human disease conditions such as replacing diseased or injured tissue	Create, duplicate, or replace a human by producing an embryo for implantation, leading to the birth of a child
Surrogate mother required	No	No	No	Yes
Human created	No	No	No	Yes
Time frame	A few weeks of growth in culture	A few weeks of growth in culture	A few weeks of growth in culture	9 months, the duration of a normal biological pregnancy (after growth of the embryo in culture)

Source: Adapted from Vogelstein, B., Alberts, B., and Shine, K. (2002). Genetics. Please Don't Call It Cloning! *Science*, 295: 1237, and Stem Cells and the Future of Regenerative Medicine, www.nap.edu/catalog/10195.html.

Therapeutic cloning may be a very valuable way to provide patient-specific stem cells that could be used to treat disease without fear of immune rejection by the recipient because he or she was the original source of the cells. In theory, stem cells from a patient can also be used to create cell lines from humans with genetic diseases to provide scientists with unprecedented potential to study and learn more about human disease conditions. We say "in theory" because it has not been proven that therapeutic cloning in humans can work.

Many scientists do not like the term *therapeutic cloning* because it implies creating a human clone. Creating stem cells to treat human diseases is not the same as cloning a human being. Most stem cell researchers prefer the term **somatic cell nuclear transfer** because nuclear transfer truly describes the biological processes taking place. Although not an efficient process, somatic cell nuclear transfer works because enucleated eggs are essentially already programmed for development. Even after the nucleus has been removed, enucleated eggs have the proteins and organelles necessary for rapid cell division. When scientists carry out nuclear transfer experiments, they use a fine-diameter pipette that places a small amount of suction pressure on the egg to hold it in place (Figure 11.25). Then a very thin-diameter glass needle is inserted into the egg to remove the nucleus. This enucleated cell serves as the host cell during nuclear transfer. (Many cells die during this process of removing the nucleus.) Scientists then inject into the egg the nucleus, or in some cases a whole cell, from a donor cell, which serves as the new source of genetic material for the egg (Figure 11.25).

The egg is incubated in growth media in a culture dish containing specialized growth molecules to stimulate rapid division leading to the formation of an embryo within several days. For therapeutic cloning, the blastocyst would be used as a source of stem cells and then destroyed. For reproductive cloning, the blastocyst would be implanted in a woman's uterus to allow it to grow and develop to form a baby, which would take the normal nine months. Recall from our discussion in Chapter 7 that reproductive cloning by

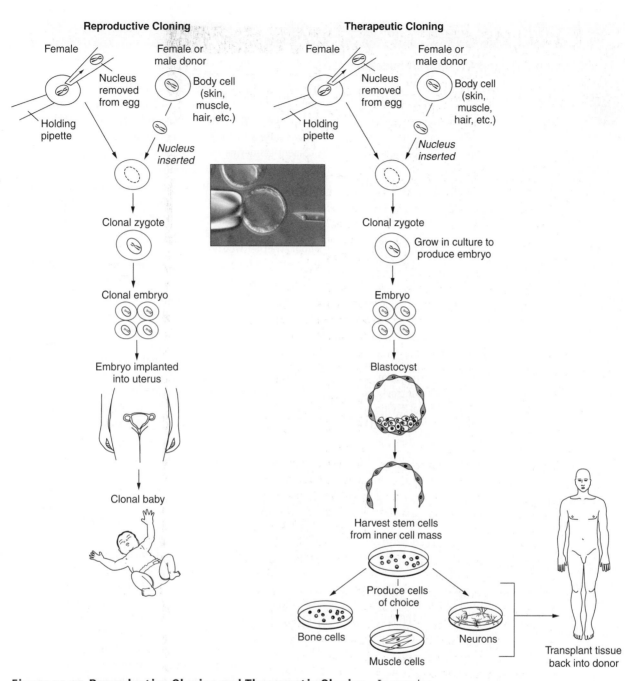

Figure 11.25 Reproductive Cloning and Therapeutic Cloning In reproductive cloning, the goal is to produce a cloned baby. In therapeutic cloning, stem cells that are genetically identical to the cells taken from a patient are produced to provide patient-specific stem cell therapy. Photo shows a holding pipette (left side of egg) holding an egg while the nucleus is being removed with a glass micropipette (right).

nuclear transfer is a very inefficient process at best. In the case of Dolly the sheep, it took 277 nuclear fusion attempts to produce only one successfully implanted embryo that developed completely and formed Dolly.

Many scientists think that reproductive cloning of humans is unethical, immoral, and scientifically unsafe. There have been several well-publicized announce-

ments from a few private groups about their ongoing plans to produce humans by reproductive cloning. Generally, these claims have generated much skepticism and have been widely disapproved and strongly denounced by most of the scientific community. In Chapter 12, we discuss some of the policies and regulations that affect cloning researchers throughout the

YOU DECIDE

Stem Cell and Cloning Debates

Frequently, when science and medicine produce innovative discoveries, society is often not prepared for the consequences of new technology. In 1850, the development of anesthesia was considered very controversial. Many worried about unanticipated adverse reactions from anesthesia, and religious groups protested over "painless" childbirth based on scripture, suggesting that Eve was to go forth from the Garden of Eden to deliver children in pain. Over 150 years later, few people dispute anesthesia as an important tool for complex surgeries and even routine procedures such as having a wisdom tooth extracted.

When recombinant DNA technology was first developed, there was great fear and speculation about what would result from such experiments (recall from Chapter 3 the Asilomar Conference to discuss the dangers of this work). To date, recombinant DNA technology has resulted in many innovative and safe products that have been used to treat more than 250 million people worldwide. Not unlike the anesthesia and recombinant DNA controversies, clergy, politicians, researchers, and the general public currently debate the merits of stem cells and cloning. At the root of these debates is the source of human stem cells, in particular human ES cells and their potential uses and abuses. In large part, human ES cells are controversial because of their source—the early human embryo. Knowing that ES cells may have enormous potential for treating and curing many devastating diseases and providing people with an opportunity for healthier, longer lives, what do you think about their use? Is it fair that medical progress for disease treatment could be delayed by allowing federal funds to be used only for research on existing ES cell lines? Embryonic stem cells are perhaps the most controversial scientific issue ever debated by the public and by politicians. The range of questions surrounding stem cells and cloning is seemingly endless.

- Is it acceptable to produce a human embryo for the sole purpose of destroying it for other uses?

- Some fear that stem cells and cloning technologies will cause a great need for human eggs to support research. Is it acceptable to pay women to collect their eggs surgically?

- What is the moral status of early embryos created by therapeutic cloning?

Some people believe that a person is formed at the moment an egg is fertilized, and so they consider therapeutic cloning equivalent to killing a child deliberately for the benefit of another person. Others believe the early embryo is a cluster of living cells with the *potential* for forming a person, but the early embryo itself is not a human being. Also, how can we justify destroying embryos that are developed through *in vitro* fertilization approaches? Why not use these embryos in an attempt to reduce pain and suffering in other humans?

Scientists define life in many ways. Biologists agree that the cluster of cells called the blastocyst is alive at the cellular level. Although all life-forms warrant respect, the blastocyst is not a person because it does not have limbs, a nervous system, organs, or other physical features of a human individual. So a major source of debate continues about whether we should assign moral status to human embryos and if so at what stage. Does the moral value of an embryo increase as it develops? Or is its moral value equal to that of a baby or adult? If an early embryo is deemed a living person, then it has all rights of other living persons. Consequently, intentionally destroying an embryo is immoral. Tax-paying citizens must decide how their money will be spent and what they believe is ethical, responsible, and safe research. The scientific establishment must share its knowledge to ensure that citizens make well-informed decisions on such topics as ES cells and cloning.

Human cloning is banned in the United States as is the use of public funds to produce embryos for harvesting stem cells, but many countries have less restrictive policies than the United States. There is concern that severe restrictions on stem cell and cloning research in the United States could create a "brain drain" in which top scientists in these fields will move to countries where cloning is legal. The biotechnology industry in the United States could suffer as a result. Even if therapeutic cloning is ever approved in the United States, will its acceptance eventually make it more likely that people will accept reproductive cloning? Probably not. Therapeutic cloning, which is intended to be used to treat illness, is a very different issue than creating a new human. If creating iPS cells turns out to be a viable way to produce stem cells, scientific and ethical debates about therapeutic cloning and the use of hESCs may become irrelevant because stem cells could be generated without an egg or embryo.

Generally, Americans expect the highest level of health care in the world. If scientists in another country use therapeutic cloning to produce treatments for Parkinson disease, Alzheimer disease, and others, how will the American public feel about not having access to such technologies in the United States? Some have even argued that reproductive cloning is a fundamental right of living in the United States. How would you feel if you were part of an infertile couple who could not have a biologically related child any way other than through reproductive cloning? Are there proper and ethically acceptable applications of using early embryos and their cells? Can the same be said for cloning? You decide.

world. At the time this book was printed, no definitive proof existed that a human has ever been cloned.

In November 2001, a private biotechnology company in Massachusetts, Advanced Cell Technology (ACT), announced a breakthrough in cloning when they reported they had cloned the first human embryos by somatic cell nuclear transfer. ACT is the same company that recently derived stem cells from a blastomere, as discussed in the previous section. U.S. legislation prevents federal funding of cloning research, but ACT is a private company working without federal funds. ACT claims that its experiments were designed to test a technique for producing human embryos that could be used in transplantation therapy.

ACT tried several techniques. First, it began with 19 human eggs from 7 consenting female donors identified from egg-donor volunteer ads ACT placed in local papers. ACT scientists removed the nucleus from each egg cell and injected these enucleated eggs with the nuclei from cells in the ovary called *cumulus cells* that surround the outer surface of egg cells and help nourish developing egg cells. In this case, enucleated egg cells from one set of women served as host cells for injecting nuclei from cumulus cells from other women. Of eight egg cells injected, two divided to become four-cell embryos and only one developed to the six-cell stage, but growth did not continue beyond this stage.

ACT also tried a parthenogenesis approach to create human embryos, and of 22 eggs, 6 developed to form what ACT claimed to resemble blastocyst-stage embryos; however, none of these embryos contained an inner cell mass that could be used to isolate stem cells. Many scientists have dismissed the ACT claims as very premature. Skeptics argued that because ACT's embryos did not progress past the six-cell stage, the embryos were developmentally defective and nuclear fusion did not occur properly. Additionally, the ACT embryos did not double the number of cells with each division as would occur in normal embryos; furthermore, ACT did not produce stem cells from the embryos they created.

In Chapter 13 we briefly discuss controversies surrounding the work of Dr. Woo Suk Hwang of Seoul National University (see "You Decide" on page 334). In 2004, Dr. Hwang's research group reported they had cloned a human embryo by somatic cell nuclear transfer cloning and then subsequently derived an embryonic stem cell line from this embryo. These two discoveries were heralded as major breakthroughs because it meant that therapeutic cloning was no longer a theory but a reality. In 2005, Hwang's team published work indicating they had successfully cloned more blastocysts from humans—about 10 times more effectively than their previous work. Further, they announced

they had created stem cell lines that were identical matches to 11 different skin cell donors with different diseases, thus establishing that patient-specific cloned embryos and stem cells for therapeutic cloning could be produced. However by the end of 2005, it became clear that aspects of Hwang's data and publication had been falsified. This was a setback for stem cell research, and the negative publicity surrounding this work further increased concern among skeptical groups that oppose embryonic stem cell research. Although recently, researchers at Harvard demonstrated that Hwang's group had unwittingly developed the first hESC line derived by parthenogenesis. Their genetic analysis of stem cells that Hwang's lab claimed were derived from an embryo created by SCNT revealed the embryo was in fact generated by parthenogenesis of an egg that had retained its nucleus during the SCNT procedure.

In early 2008, the California biotechnology company Stemagen announced that they had used DNA from skin cells to create human embryos by somatic cell nuclear transfer. These embryos were grown to a point where they were ready for implantation but were not implanted and stem cells were not isolated from the embryos. Stemagen offered these results as proof that it is feasible to generate donor-specific human embryos for therapeutic cloning. Some stem cell scientists have expressed mixed optimism and skepticism given the fraudulent claims of the Hwang lab until further data from Stemagen is available to confirm their results.

Embryonic Stem Cell and Therapeutic Cloning Regulations in the United States

Regulatory issues surrounding the use of stem cells and cloning are discussed in Chapter 12. In particular, refer to "You Decide" on pages 314–315. Although there are international guidelines on ethical use of stem cells, currently no international policy governs ES cells and therapeutic or reproductive cloning. In the United States, since August 2001, there has been a ban on using federal funds to create an embryo for the purpose of isolating embryonic stem cells. This ban did provide for the use of federal funds for research on 78 cell lines that had already been established. However, many stem cell researchers feel these cell lines are far less valuable than initially believed. It has been widely reported that many of these lines have abnormalities in chromosome number and contamination by mouse feeder cells and only about a dozen of them appear to be free of contamination and readily available to researchers.

Many scientists and stem cell advocacy groups are rallying hard for lifting the ban on federal funding for creating new ES cell lines. Bill HR 810 proposed expanding current policy to include the use of federal funding for research on embryos created for fertility treatments or donated from IVF clinics and establishing new federal guidelines for working with stem cells. It passed the House in 2005 but was short the votes needed to override a presidential veto. After HR 810 passed the Senate in 2006, it was vetoed by President Bush. A new bill, HR 3, the Stem Cell Research Enhancement Act, passed the House in early 2007 with more votes than HR 810, but it was still 37 votes short of those needed for a veto override. HR 3 was passed by the Senate (63 to 34), but it fell short of the two-thirds majority needed to override a veto, and it too was subsequently vetoed by the President.

Many private foundations are providing hundreds of millions of dollars for stem cell researchers to work on ES cells, and several have enacted legislation to create stem cell research institutes and to provide state-funded support for ES research. Some of the more active states include California (which in 2004 passed Proposition 71 approving a budget of nearly $300 million in bonds and over $3 billion overall), New Jersey, Connecticut, Illinois, Maryland, and Wisconsin. Around the world, regulations vary on the production of new ES cell lines and policies on therapeutic cloning. For instance, production of new lines and therapeutic cloning is legal in the United Kingdom, Israel, South Korea, China, and Singapore. Therapeutic cloning is banned in Brazil, Australia, and the European Union (although ES cells can be derived from unused IVF embryos where legal in member nations).

11.5 The Human Genome Project Has Revealed Disease Genes on All Human Chromosomes

We have discussed the Human Genome Project in several sections of this book. In Chapter 3 we looked at how recombinant DNA techniques are used to identify and clone genes (you may want to review Chapter 3 prior to reading this section), and in this chapter we considered a few examples of how molecular biology can be used for genetic testing. Many of the disease genes that can currently be tested for were discovered through the Human Genome Project, and scientists have been developing complex "maps" showing the locations of normal and diseased genes on all human chromosomes. We conclude this chapter by taking a brief look at how scientists pieced together the human genome puzzle.

Piecing Together the Human Genome Puzzle

Because a single chromosome contains millions of base pairs, sequencing and assembling millions of base pairs to construct one complete chromosome is a challenging task. One way to sequence an entire chromosome involves a random cloning process called shotgun cloning and sequencing (refer to Chapter 3, Figure 3.20). In this process, chromosomal DNA is cut into smaller pieces with several different restriction enzymes.

Depending on the size of each DNA fragment generated by digesting with restriction enzymes, large pieces are usually cloned into yeast artificial chromosomes (YACs) or bacterial artificial chromosomes (BACs), and smaller pieces of each large fragment are cloned into cosmid or plasmid vectors prior to sequencing (Figure 11.26). The idea behind this approach is to generate and sequence pieces of DNA fragments randomly with the hope of sequencing short overlapping pieces. High-powered computers are then put to work to look for and align overlapping sections of DNA from individual pieces cut with the two restriction enzymes. By completing this genetic puzzle, it is possible to reconstruct a chromosome by aligning these overlapping fragments and assembling a stretch of continuously overlapping sequences (Figure 11.26). Data that result from this type of study produce a **restriction map** as a physical map of overlapping pieces and restriction enzyme–cutting sites on a chromosome.

Many doubted that shotgun strategies could be used effectively to sequence larger genomes, but development of this technique and novel sequencing strategies rapidly accelerated the progress of the Human Genome Project. Although shotgun approaches are effective for sequencing and mapping segments of a chromosome, such approaches are not practical for identifying expressed genes because only a very small percentage of human DNA consists of genes. Much of our genome contains non–protein-coding DNA. In some ways, using a shotgun approach to identify gene sequences in the genome is like searching for a needle in a haystack. By working with mRNA, genome scientists are studying expressed genes in a tissue and not non–protein-coding DNA. The mRNA can be copied into DNA. Recall that this copied DNA is called **complementary DNA (cDNA)** because it is identical (complementary) to a sequence of mRNA. Pieces of cDNA can be sequenced to produce fragments called **expressed-sequence tags (ESTs).** ESTs represent small pieces of DNA sequenced from genes that are expressed in a cell. Rarely do ESTs span an entire gene, but these pieces can be used as tags to determine the

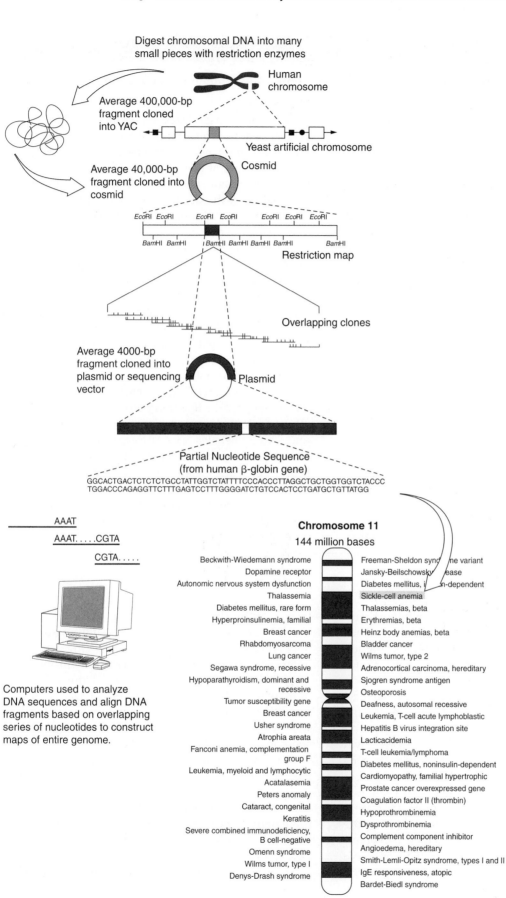

Digest chromosomal DNA into many small pieces with restriction enzymes

Human chromosome

Average 400,000-bp fragment cloned into YAC

Yeast artificial chromosome

Average 40,000-bp fragment cloned into cosmid

Cosmid

EcoRI EcoRI   EcoRI EcoRI   EcoRI EcoRI EcoRI

BamHI BamHI   BamHI BamHI BamHI BamHI   BamHI

Restriction map

Overlapping clones

Average 4000-bp fragment cloned into plasmid or sequencing vector

Plasmid

Partial Nucleotide Sequence
(from human β-globin gene)

GGCACTGACTCTCTCTGCCTATTGGTCTATTTTCCCACCCTTAGGCTGCTGGTGGTCTACCC
TGGACCCAGAGGTTCTTTGAGTCCTTTGGGGATCTGTCCACTCCTGATGCTGTTATGG

AAAT

AAAT.....CGTA

CGTA.....

Computers used to analyze DNA sequences and align DNA fragments based on overlapping series of nucleotides to construct maps of entire genome.

Chromosome 11
144 million bases

Beckwith-Wiedemann syndrome
Dopamine receptor
Autonomic nervous system dysfunction
Thalassemia
Diabetes mellitus, rare form
Hyperproinsulinemia, familial
Breast cancer
Rhabdomyosarcoma
Lung cancer
Segawa syndrome, recessive
Hypoparathyroidism, dominant and recessive
Tumor susceptibility gene
Breast cancer
Usher syndrome
Atrophia areata
Fanconi anemia, complementation group F
Leukemia, myeloid and lymphocytic
Acatalasemia
Peters anomaly
Cataract, congenital
Keratitis
Severe combined immunodeficiency, B cell-negative
Omenn syndrome
Wilms tumor, type I
Denys-Drash syndrome

Freeman-Sheldon syndrome variant
Jansky-Beilschowsky disease
Diabetes mellitus, insulin-dependent
Sickle-cell anemia
Thalassemias, beta
Erythremias, beta
Heinz body anemias, beta
Bladder cancer
Wilms tumor, type 2
Adrenocortical carcinoma, hereditary
Sjogren syndrome antigen
Osteoporosis
Deafness, autosomal recessive
Leukemia, T-cell acute lymphoblastic
Hepatitis B virus integration site
Lacticacidemia
T-cell leukemia/lymphoma
Diabetes mellitus, noninsulin-dependent
Cardiomyopathy, familial hypertrophic
Prostate cancer overexpressed gene
Coagulation factor II (thrombin)
Hypoprothrombinemia
Dysprothrombinemia
Complement component inhibitor
Angioedema, hereditary
Smith-Lemli-Opitz syndrome, types I and II
IgE responsiveness, atopic
Bardet-Biedl syndrome

Figure 11.26 Cloning and Sequencing Pieces of a Human Chromosome to Create Chromosome Maps By cutting chromosomes into small pieces, genome scientists can clone these pieces into vectors such as plasmids, cosmids, BACs, or YACs. The fragments can then be sequenced, and the overlapping pieces can be strung together to make a continuous map. Using computers to analyze the sequence of these fragments, genes (such as globin) involved in genetic disease conditions can be identified as shown on this map of chromosome 11. Note: Only a partial map of disease genes located on chromosome 11 is shown.

TOOLS OF THE TRADE

Using the Internet to Learn About Human Chromosomes and Genes

We have discussed the importance of bioinformatics for analyzing genetic information and creating databases as tools that scientists around the world can use to compile, share, and compare DNA sequence information. The wealth of freely available databases that catalog information about human chromosomes and genes resulting in part from the Human Genome Project is a great example. An excellent way to learn about what the Human Genome Project has revealed is to review some of the chromosome maps available on the Web. For instance, if you are interested in the Y chromosome, you could review maps and descriptions of the genes found on it to find out why this chromosome is partially responsible for making male humans.

We encourage you to use the websites cited here to follow a chromosome or gene of interest for a few months to see what kinds of information you can uncover. These sites are great resources. You can, for example, use them to learn more about a rare disease gene related to a disease condition affecting someone close to you. These sites present up-to-date information that cannot be found in even the most recent books. If you cannot find a gene of interest in these sites, it probably has not been identified yet! The sites described here are among the best sites for learning about human disease genes and chromosome maps.

The Department of Energy Human Genome Program Information site provides excellent chromosome maps of identified genes. Visit www.ornl.gov/hgmis/posters/chromosome for basic maps of human genes. The Online Mendelian Inheritance in Man site (OMIM; www.ncbi.nlm.nih.gov/entrez/query.fcgi?db=OMIM) is a great database of human genes and genetic disorders. In the keyword box, type in the name of a gene or disease that interests you. For example, type in "breast cancer" and then click the search button. When the next page appears, you will see a list of genes implicated in breast cancer along with corresponding access numbers highlighted in blue as links. Clicking on one of the links will take you to a wealth of information about that gene including background information, links to scientific papers about the gene, gene maps, and even nucleotide and protein sequence data (when available). You might also want to search this site to see if a gene has been found for a particular behavioral condition (for example, alcoholism or depression).

The National Center for Biotechnology Information (NCBI) sponsors The Human Genome website (www.ncbi.nlm.nih.gov/genome/guide/human), which allows you to access detailed maps of chromosomes and disease genes using a feature called Map Viewer. The NCBI Genes & Disease site (http://www.ncbi.nlm.nih.gov/books/bv.fcgi?rid=gnd&ref=sidebar) also provides a student-friendly set of pages on human disease genes that have been mapped to chromosomes.

---

sequence of an entire gene by piecing together overlapping ESTs, as shown in Figure 11.26. ESTs have played an important role in the identification of human genes. Visit the NOVA Online "Sequence for Yourself" site for informative animations on cloning and sequencing DNA and assembling cloned DNA fragments to create a physical map of a chromosome as was done in the Human Genome Project.

Progress on the Human Genome Project was greatly accelerated by the development of fast-paced computer-automated sequencing instruments (refer to Figure 3.13) that worked around the clock to generate large amounts of sequence data. After DNA was sequenced, bioinformatics was used to catalog the sequence information, interpret the sequence to determine if it contained protein-coding instructions, and compare to databases of known sequences to find out if this sequence has already been determined or if it represented a novel piece of chromosome sequence.

Figure 11.26 shows a big-picture representation of how DNA sequence can be determined from restriction fragments of DNA and how overlapping restriction fragments can be assembled to create a chromosome map. This figure also shows that a sequence of the human globin gene on chromosome 11 was identified. Recall that mutation of the globin gene is the cause of sickle-cell disease. Also notice the number of disease genes located on chromosome 11.

Chromosome maps are now available that pinpoint the locations of normal and disease genes of interest (refer to Figure 1.2). Figure 11.27 shows simplified maps for each chromosome, highlighting one or two prominent genes on each that are known to be involved in human genetic disease. The Human Genome Project has led to the development of follow-up projects such as **The Cancer Genome Atlas Project (TCGA),** a comprehensive effort to identify genomic changes involved in a variety of different cancers.

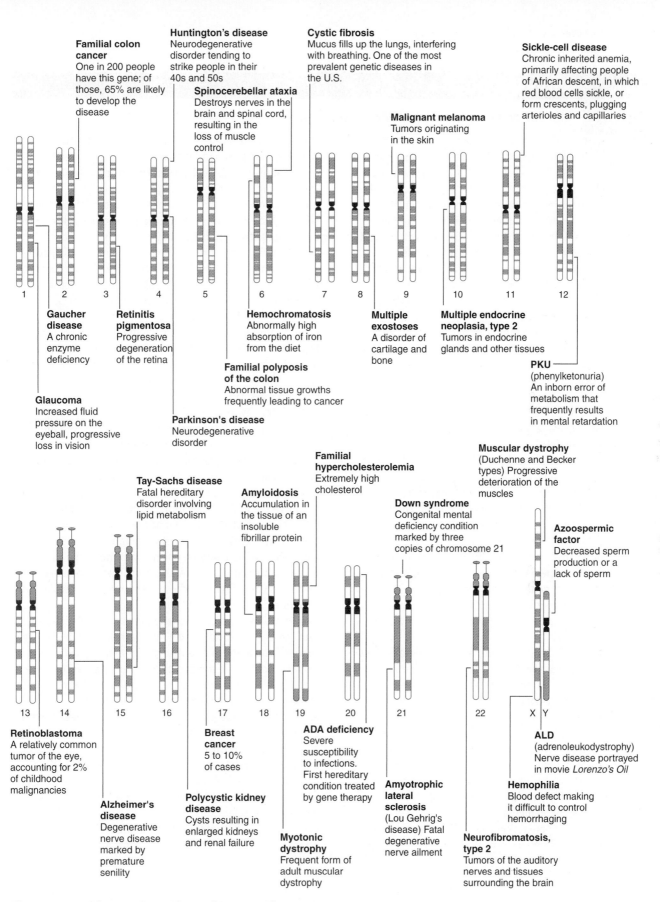

Familial colon cancer
One in 200 people have this gene; of those, 65% are likely to develop the disease

Huntington's disease
Neurodegenerative disorder tending to strike people in their 40s and 50s

Spinocerebellar ataxia
Destroys nerves in the brain and spinal cord, resulting in the loss of muscle control

Cystic fibrosis
Mucus fills up the lungs, interfering with breathing. One of the most prevalent genetic diseases in the U.S.

Malignant melanoma
Tumors originating in the skin

Sickle-cell disease
Chronic inherited anemia, primarily affecting people of African descent, in which red blood cells sickle, or form crescents, plugging arterioles and capillaries

Gaucher disease
A chronic enzyme deficiency

Retinitis pigmentosa
Progressive degeneration of the retina

Hemochromatosis
Abnormally high absorption of iron from the diet

Multiple exostoses
A disorder of cartilage and bone

Multiple endocrine neoplasia, type 2
Tumors in endocrine glands and other tissues

Familial polyposis of the colon
Abnormal tissue growths frequently leading to cancer

PKU (phenylketonuria)
An inborn error of metabolism that frequently results in mental retardation

Glaucoma
Increased fluid pressure on the eyeball, progressive loss in vision

Parkinson's disease
Neurodegenerative disorder

1 2 3 4 5 6 7 8 9 10 11 12

Tay-Sachs disease
Fatal hereditary disorder involving lipid metabolism

Amyloidosis
Accumulation in the tissue of an insoluble fibrillar protein

Familial hypercholesterolemia
Extremely high cholesterol

Down syndrome
Congenital mental deficiency condition marked by three copies of chromosome 21

Muscular dystrophy
(Duchenne and Becker types) Progressive deterioration of the muscles

Azoospermic factor
Decreased sperm production or a lack of sperm

Retinoblastoma
A relatively common tumor of the eye, accounting for 2% of childhood malignancies

Alzheimer's disease
Degenerative nerve disease marked by premature senility

Breast cancer
5 to 10% of cases

Polycystic kidney disease
Cysts resulting in enlarged kidneys and renal failure

ADA deficiency
Severe susceptibility to infections. First hereditary condition treated by gene therapy

Myotonic dystrophy
Frequent form of adult muscular dystrophy

Amyotrophic lateral sclerosis
(Lou Gehrig's disease) Fatal degenerative nerve ailment

ALD (adrenoleukodystrophy)
Nerve disease portrayed in movie *Lorenzo's Oil*

Hemophilia
Blood defect making it difficult to control hemorrhaging

Neurofibromatosis, type 2
Tumors of the auditory nerves and tissues surrounding the brain

13 14 15 16 17 18 19 20 21 22 X Y

Figure 11.27 Disease Gene Maps of Human Chromosomes Maps show one or two genes on each human chromosome that are involved in a genetic condition. Many more genes than are shown in this figure are located on each chromosome. Note: Chromosomes are not drawn to scale.

CAREER PROFILE

Career Options in Medical Biotechnology

Medical biotechnology offers an exciting range of potential career choices, primarily in biotechnology and pharmaceutical companies. Industry experts suggest that students interested in a career in medical biotechnology identify an area of interest and then gain life and work experience that helps them fit a company's needs. Courses in chemistry, cell biology, molecular biology, biochemistry, and bioinformatics are particularly valuable. In addition to appropriate coursework, an internship or summer research experience at a local chemical, pharmaceutical, or biotechnology company is highly recommended.

If possible, get involved in a research project with a professor while you are an undergraduate student. Undergraduate research can provide invaluable opportunities for learning how to plan research projects, execute and troubleshoot experiments, and interpret data. Such research may even lead to a presentation at regional or national meetings—a great way to network and interact with other scientists—or even a publication. Your research professor can also provide an essential reference to help you land your first job.

Hiring experts strongly recommend that students study companies of different sizes to get a feel for the type of work and size of company that appeals to them. Students should study the companies that intrigue them. Are you interested in drug discovery and development, cell or gene therapies, cancer research, aging research, surgical materials, genomics, or curing childhood diseases? Would you prefer a large company to a smaller, more personalized company or even a biotechnology start-up company?

Consider what you want to do, and then identify companies whose needs match your skills. With the wealth of information available online, this is easier than ever. Most biotechnology companies have websites, and excellent links to these sites are presented in Chapter 1. On an interview, demonstrating a little background knowledge and a true interest in a company can be impressive.

There are many entry-level job opportunities in medical biotechnology for people with associate's or bachelor's degrees. Many people start as laboratory technicians. In some companies, this position involves routine procedures such as preparing solution and setting up

materials for experiments, but individuals who show initiative are often given latitude to assume greater responsibilities and decision-making responsibilities in research projects. Application scientists develop new products and procedures, often working directly at the lab bench conducting research. In some companies, however, application scientists may give on-site product demonstrations and present technical seminars to potential customers. Application scientists are also frequently involved in customer relations issues, teaching customers how to troubleshoot a product, interpret data, and the like. Clinical scientists work together with physicians and research scientists to help carry out clinical trials for testing a new drug or medical product. Medical biotechnology companies also employ people who want to combine an interest in science with business skills through marketing and product sales positions.

The interview process at most biotechnology companies is very rigorous. Companies want people who are very organized, systematic, and attentive to detail. Medical biotechnology research is a team effort. Good writing and communication skills are essential because you will be working with other people routinely and interacting in verbal and written form. In addition, virtually every biotechnology and pharmaceutical company today seeks people with computer skills who can work together with teams of software engineers and information technology professionals to create new software to manage and store data, test hypotheses, and model molecular structures.

Industry insiders consistently point out that enthusiasm, an ability to take novel approaches to problem solving, and a commitment to professional growth are valuable tools that companies look for in potential employees. People who are eager and excited about the challenge of applying their skills in teamwork approaches to problem solving are highly desired. Even if a person's educational background is not an exact match for a particular position, as is often the case, enthusiasm and a willingness to learn can determine whether the applicant is hired. Biotechnology companies seek people who want to make a difference—if you have a burning desire to use science to help improve human health, a career in medical biotechnology might be for you.

Scientists are particularly interested in genetic changes that trigger normal cells to become cancerous cells in the brain, ovaries, and lungs because cancers of these organs affect large numbers of Americans. Eventually

scientists expect that new information from TCGA will be used to better diagnose and treat cancer.

What and who we are is clearly much more than the total number of our genes. Being human is far more complicated than our genome—we are infinitely more complex than just the sum of our genes and body parts. What defines us genetically is the complexity of how our genes are used and how proteins interact with one another to provide the myriad functions and unique characteristics of the creatures we call humans. The Human Genome Project has provided us with unprecedented knowledge about human chromosomes and the genes they contain. However, much more work lies ahead to determine the functions of a majority of human genes and the complex ways proteins encoded by our genes function to play important roles in normal cellular activities as well as diseased conditions.

QUESTIONS & ACTIVITIES

Answers can be found in Appendix 1.

1. How can molecular biology techniques be used to identify genetic disease conditions in fetal or adult humans? Provide two examples.

2. What is gene therapy? Explain the differences between *ex vivo* and *in vivo* gene therapy, and give an example of a human genetic disease treated by each approach. Provide two examples of how therapeutic genes can be delivered into cells, and discuss the challenges scientists must solve to make gene therapy an effective technique for treating human genetic disease conditions.

3. Compare and contrast embryonic and adult stem cells. Include an explanation of where each type of stem cell comes from and how each type can be isolated. Give two examples of how stem cells may be used to help treat human disease conditions.

4. Compare and contrast therapeutic cloning and reproductive cloning by preparing a table listing the pros and cons of each technology.

5. Define pharmacogenomics, and explain how it may change health care in the future.

6. Briefly describe how the Human Genome Project will result in advances in medical biotechnology.

7. Visit Clinical Trials.gov and search for MAb treatments currently under clinical trials. Visit the American Society for Gene Therapy site to search for current gene therapy clinical trials that are in progress.

8. Search the U.S. House of Representatives website (http://www. house. gov/ internet) for HR 810, a stem cell bill that passed the House and Senate but the President vetoed. What was the purpose of this bill? Would you support this bill? Did your state representatives vote in favor or against this bill? Search the site for information on bill HR 3 to learn about the most current bill passed by the House in 2007.

9. Describe two gene-silencing techniques and how they may be used for gene therapy.

10. What is regenerative medicine?

11. What are SNPs? How can SNPs be used to diagnose human genetic disease conditions?

References and Further Reading

Colavito, M. C. (2007). *Gene Therapy*. M. A. Palladino, Ed., San Francisco, CA: Benjamin Cummings.

Collins, F. S., and Barker, A. D. (2007). Mapping the Cancer Genome. *Scientific American*, 296: 50–57.

Cowan, C. A., Atienza, J., Melton, D. A., et al. (2005). Nuclear Reprogramming of Somatic Cells After Fusion with Human Embryonic Stem Cells. *Science*, 309: 1369–1373.

De Coppi, P., Bartsch, G., Siddiqui, M. M, et al. (2007). Isolation of Amniotic Stem Cell Lines with Potential for Therapy. *Nature Biotechnology*, 25: 100–106.

Dykxhoorn, D. M., and Lieberman, J. (2006). Knocking Down Disease with siRNAs. *Cell*, 126: 231–235.

Friend, S. H., and Stoughton, R. B. (2002). The Magic of Microarrays. *Scientific American*, 286: 44–53.

Gilbert, D. M. (2004). The Future of Human Embryonic Stem Cell Research: Addressing Ethical Conflict with Responsible Scientific Research. *Medical Science Monitor*, 10: RA99–103.

Hacein-Bey-Abina, S., Von Kalle, C., Schmidt, M., et al. (2003). *LM02*-Associated Clonal T Cell Proliferation in Two Patients After Gene Therapy for SCID-X1. *Science*, 302: 415–419.

Hanna, J., Wernig, M., Markoulaki, S., et al. (2007). Treatment of Sickle Cell Anemia Mouse Model with iPS Cells Generated from Autologous Skin. *Science*, 318: 1920–1923.

Langer, R. (2003). Where a Pill Won't Reach. *Scientific American*, 288: 50–57.

Lau, N. C., and Bartel, D. P. (2003). Censors of the Genome. *Scientific American* 289: 34–41.

Meissner, A., Wernig, M., and Jaenisch, R. (2007). Direct Reprogramming of Genetically Unmodified Fibroblasts into Pluripotent Stem Cells. *Nature Biotechnology*, 25: 1177–1181.

Nettelbeck, D. M., and Curiel, D. T. (2003). Tumor-Busting. *Scientific American*, 289: 68–75.

Palladino, M. A. (2006). *Understanding the Human Genome Project*, 2/e. San Francisco, CA: Benjamin Cummings.

Seeman, N. D. (2004). Nanotechnology and the Double Helix. *Scientific American*, 290: 64–75.

Shamblott, M. J., Axelman, J., Wang, S., et al. (1998). Derivation of Pluripotent Stem Cells from Cultured Human Primordial Germ Cells. *Proceedings of the National Academy of Sciences U.S.A.*, 95: 13726–13731.

Takahashi, K., Tanabe, K., Ohnuki, M., et al. (2007). Induction of Pluripotent Stem Cells from Adult Human Fibroblasts by Defined Factors. *Cell*, 131: 861–872.

Thomson, J. A., Itskovitz-Eldor, J., Shapiro, S. S., et al. (1998). Embryonic Stem Cell Lines Derived from Human Blastocysts. *Science*, 282: 1145–1147.

Tribut, O., Lessard, Y., Reymann, J-M., et al. (2002). Pharmacogenomics. *Medical Science Monitor*, 8: RA152–163.

Yu, J., Vodyanik, M.A., Smuga-Otto, K., et al. (2007). Induced Pluripotent Stem Cell Lines from Human Somatic Cells. *Science*, 318: 1917–1920.

Visit www.pearsonhighered.com/biotechnology to download learning objectives, chapter summary, "Keeping Current" web links, glossary, flashcards, and jpegs of figures from this chapter.

Chapter 12

Biotechnology Regulations

After completing this chapter you should be able to:

■ Describe the APHIS (USDA) permitting process, including the precautions that must be taken to prevent accidental release of bioengineered plants into the environment.

■ List the six criteria that must be met before a plant is eligible for "notification" under APHIS guidelines.

■ Describe the role of the EPA in regulating biotechnology products.

■ Describe the FDA's role in regulating food and food additives produced using biotechnology.

■ Describe the FDA's role in regulating pharmaceutical products, including phase testing.

■ Cite examples of the regulatory agencies' ability to respond expeditiously to emerging situations.

■ Describe the functions of patents in science, and explain how patents encourage discovery.

■ Explain why DNA sequences are considered patentable.

FDA Inspector at work. The FDA is one of the government agencies responsible for evaluating the safety of biotechnology products.

The use of a mislabeled drug or consumption of a contaminated food can cause death. The public looks to the government to enact and enforce laws that provide protection from unsafe or ineffective food and drugs. Medical products and foods are regulated by the government, and these regulations have profound effects on the biotechnology workplace, especially in the United States. It usually takes years to test a new drug to ensure its safety and effectiveness. In the meantime, lives are lost that the new drugs might have saved. There is a cost if marginal food must be discarded, just as there is a cost if agricultural productivity is diminished because potentially toxic herbicides are not used; clearly, there is a conflict between ensuring safety and reducing costs. New and innovative products such as those made by methods of biotechnology promise some benefits and pose some risks. When former president Bill Clinton and U.K. prime minister Tony Blair pledged to limit gene sequence patents by, among other things, sharing information attained by researchers working on the Human Genome Project, panic spread in the biotech industry, which led to the widespread sale of biotech stocks. Many applaud a restriction on the patenting of gene sequences because they argue that such patenting is morally unacceptable. At the heart of this debate is your understanding of what it means for something to be "patentable." Your opinion may change after reading about protections and patents in this chapter.

12.1 The Regulatory Framework

During the 19th century in America, peddlers freely promoted their patented medicines (see an amusing example in Figure 12.1), but their main concern often was skipping out of town before the local authorities caught on to the fact that most of these so-called medicines were, in fact, scams. At the same time, many large companies were poisoning the environment—often without any real understanding of the long-term results of their actions. When disasters occurred or public opinion was aroused, the government could no longer ignore major public safety events, and as a result, it established protective agencies. The new federal agencies were charged with overseeing the safety of food and medicine and protecting the environment from polluters, and these agencies regulate these same industries today.

Unlike the traditional early regulatory protocols that were developed after the fact, when biotechnology emerged as a field of research, a regulatory system to oversee the industry was already in place: The 1974 Asilomar Conference was a voluntary meeting to

Don't Be Too Fat

Don't ruin your stomach with a lot of useless drugs and patent medicines. Send to Prof. F. J. Kellogg, 1366 W. Main St., Battle Creek, Michigan, for a free trial package of a treatment that will reduce your weight to normal without diet or drugs. The treatment is perfectly safe, natural and scientific. It takes off the big stomach, gives the heart freedom, enables the lungs to expand naturally, and you will feel a hundred times better the first day you try this wonderful home treatment.

Figure 12.1 Patented Medicines The public looks to the federal government to enact and enforce laws that provide protection from unsafe or ineffective food and drugs. The history of new regulations has followed tragedies caused by unregulated products (like the one advertised above).

establish guidelines by researchers themselves when some of those involved in gene transfer saw the potential for spread of disease genes if safeguards were not in place. The National Institutes of Health (NIH) was the first federal agency to assume regulatory responsibility over biotechnology. Among other things, the NIH published research guidelines for recombinant DNA techniques. Recall from Chapter 3 that the Recombinant DNA Advisory Committee of the NIH sets the guidelines for working with recombinant DNA and recombinant organisms in the laboratory. The NIH continued to monitor and review all DNA research until 1984. At that point, the government published the "Coordinated Framework for Regulation of Biotechnology," which made biotechnology research a joint responsibility of the NIH, the USDA, and the EPA.

This groundbreaking report proposed that biotechnology products would be regulated much like traditional products. For example, bioengineered plants would be monitored by the same agencies that regulate other plants, and bioengineered drugs would fall under the same rules that apply to all other drugs. In essence, this report proposed that biotechnology products would not pose regulatory and scientific issues that are substantially different from those posed by traditional products. This concept is central to understanding the regulation of biotechnology in the United States. Established as a formal policy in 1986, the "Coordinated Framework for Regulation of Biotechnology" describes the federal system for evaluating products developed using modern biotechnology. The framework is based on health and safety laws developed to address specific product classes. Three agencies are responsible for most of the products that are the result of biotechnology: the U.S. Department of Agriculture (USDA), the Environmental Protection Agency (EPA), and the Food and Drug Administration (FDA). The USDA makes sure a given organism is safe to grow, the EPA makes certain it is safe for the environment, and the FDA determines whether it is safe to consume (see Table 12.1). These three agencies cooperate to achieve their goals, and all are often involved in the oversight and regulation of a single biotechnology product. Visit the Case Studies website (listed on the Companion Website) for detailed illustrations of teamwork among these agencies. Even though bureaucracies are often depicted as slow to change and slow to act, it is important to understand that the regulatory process is constantly being refined to ensure that products, both bioengineered and traditional, are safe. In this chapter, we note specific examples that demonstrate the responsiveness of the system to emergent technologies, risks, and needs.

We begin our survey of biotechnology regulations by looking at the responsibilities of each of the three major regulatory agencies. We also consider the role that federal legislation plays in creating new regulations for the industry. Next, we learn about patents and their role in promoting biotechnology research. Finally, we consider the global marketplace as a context for the development of biotechnology in the United States.

Table 12.1 PRIMARY FEDERAL REGULATORY AGENCIES IN THE UNITED STATES

Regulatory Oversight of Biotechnology Products

Agency	Product Regulated
U.S. Department of Agriculture	Plants, plant pests (including microorganisms), animal vaccines
Environmental Protection Agency	Microbial/plant pesticides, other toxic substances, microorganisms, animals producing toxic substances
Food and Drug Administration	Food, animal feeds, food additives, human and animal drugs, human vaccines, medical devices, transgenic animals, cosmetics

Major Laws that Empower Federal Agencies to Regulate Biotechnology

Law	Agency
The Plant Protection Act	USDA
The Meat Inspection Act	USDA
The Poultry Products Inspection Act	USDA
The Eggs Products Inspection Act	USDA
The Virus Serum Toxin Act	USDA
The Federal Insecticide, Fungicide, and Rodenticide Act	EPA
The Toxic Substances Control Act	EPA
The Food, Drug, and Cosmetics Act	FDA, EPA
The Public Health Service Act	FDA
The Dietary Supplement Health and Education Act	FDA
The National Environmental Protection Act	USDA, EPA, FDA

Source: www.fda.gov.

12.2 U.S. Department of Agriculture

Created in 1862, the USDA has many functions related to the advancement and regulation of agriculture, including regulating plant pests, plants, and veterinary biologics. A **biologic** is broadly defined as any medical preparation made from living organisms or their products. Insulin and vaccines are examples of biologics.

Animal and Plant Health Inspection Service

The **Animal and Plant Health Inspection Service (APHIS)** is the branch of the USDA responsible for protecting U.S. agriculture from pests and diseases. Because genetically engineered insects and plants are potentially invasive, they are treated as plant pests and regulated by the agency under the requirements of the Federal Plant Pest Act. APHIS provides permits for developing and field testing genetically engineered plants. If an experimental organism poses a potential threat to preexisting agriculture, the agency makes certain that safeguards are in place.

Permitting Process

After a new plant has been engineered in a laboratory, APHIS requires several years of **field trials** to investigate everything about the plants, including disease resistance, drought tolerance, and reproductive rates. APHIS will not approve a field trial application unless precautions are taken to prevent accidental cross-pollination. This requirement can be a significant undertaking. For example, trial planting sites for transgenic corn must be at least 1 mile away from fields where seed corn is grown and harvested. A 25-foot perimeter of fallow (unplanted, tilled) land must surround the entire trial site field. In some cases, the tassels at the ends of the ears of corn must be bagged to confine the pollen (as seen in Figure 12.2), and the tassels themselves must later be removed. Furthermore, no plants are allowed to escape the field test site, which means all "volunteer" plants that naturally reseed must be destroyed and no plants, or even parts of plants, can be removed unless they are to be destroyed. As an extra precaution, the entire test field is usually kept fallow for the season following the harvest of the trial crop. APHIS employees inspect the sites before, during, and after trials to make sure all tests are conducted in accordance with the requirements. An example of this protection can be illustrated by the fact that in 2002, two growers in Iowa and Nebraska were required to destroy all the seed corn produced from fields where they had pre-

Figure 12.2 APHIS Regulations in Action   Corn growing in a field with its tassels bagged to prevent pollination of unintended plants in nearby fields in compliance with regulations from the Animal and Plant Health Inspection Service of the USDA.

viously grown transgenic soybeans for protein (human antibody) extraction but did not remove a few "volunteer" plants that could have contaminated the subsequent seed corn crop (as required by the USDA). They were also fined $250,000 each for this infraction. The case illustrates that the regulators are watching, and the regulations are being enforced.

The APHIS Investigative Process

Of course, the ultimate objective for a grower is to harvest a marketable product. The first step in that process is to petition APHIS for **deregulated** (or nonregulated) **status.** Genetically engineered plants with this status are monitored in the same way as traditional plants. Before granting this status, APHIS reviews field-test reports, scientific literature, and any other pertinent records before it determines whether the plant is as safe to grow as traditional varieties.

APHIS considers three broad areas while evaluating the petition for deregulation:

1. *Plant pest consequences.* APHIS examines the biology of the plant to evaluate the possible threat to other plants. This is known as the "plant pest" risks. The agency investigates the possibility that the new genetic material might lead to a new plant disease. It also considers whether crossbreeding with native plants could create new "superweeds."

2. *Risks to other organisms.* The agency investigates the risk to wildlife and desirable insects that might feed on the crop or be exposed to the pollen.

3. *Weed consequences.* Finally, the agency considers the possibility that the new plant will be unwelcome and invasive—in other words, a weed. The plant's reproductive strategies are an especially important consideration. A plant that reseeds itself easily, has a great mechanism for spreading its seeds, and is resistant to cold and drought presents a real problem in the future. For instance, if the dandelion were presented as a new biotechnology plant, it would be rejected on the basis of its amazing success in propagation.

The Notification Process

An alternative system is in place to fast-track some new agricultural products. This alternative is called **notification.** Six criteria must be met before notification is an option.

1. The new agricultural product must be one of only a limited number of eligible plant species. These species, including corn, cotton, potatoes, soybeans, tobacco, and tomatoes, have all been thoroughly studied in past trials, and most of their characteristics are well understood.

2. The new genetic material must be confined to the nucleus of the new plant; it cannot be floating in the cell on plasmids or viral vectors (nuclear genes tend to remain with the new engineered plant). See Chapter 3 for a refresher on vectors used in the genetic engineering of plants.

3. The function of the genes being introduced must be known, and *no* plant disease can be caused by the protein being expressed.

4. If the new agricultural product will be used for food, the new genes cannot cause the production of a toxin, an infectious disease, or any substance used medically. (See the previous example in the "Permitting Process" section.)

5. If the gene is derived from a plant virus, it cannot have the potential to create a new virus.

6. The new genetic material must not be derived from animal or human viruses (to prevent spread as a new human disease).

If a bioengineered plant meets all six of these criteria, the agency approves the field trial within 30 days. Under notification, the plant developer is still required to meet the standards demanded under the permitting process. The same precautions must be taken to prevent accidental spread of the new plant, and detailed records of the trial must be maintained.

After the safety of the new plant is determined, a grower can cultivate, test, or use the plant for cross-breeding purposes without monitoring and approval by APHIS. The new biotechnology product can then be brought to market. But these requirements may not be all that is needed, and the EPA regulates all new products introduced into the environment.

12.3 The Environmental Protection Agency

Established in 1970, the EPA has responsibilities ranging from protecting endangered species to establishing emission standards for cars. Another major duty is regulating pesticides and herbicides. As a natural extension of that responsibility, the EPA regulates any plant that is genetically engineered to express proteins that provide pest control. For example, the EPA monitors the production of a recently approved engineered food potato resistant to potato beetle and the potato roll virus produced through transfer of the Bt plasmid, as discussed in Chapter 6. The EPA also supervises the use of herbicide-tolerant plants. In this case, the EPA is interested not only in the plant but also in the herbicides that will be used in the fields where the herbicide-tolerant plants are grown. Of particular importance is the pesticide residue remaining on the crop. The EPA monitors these levels to make certain the crop itself is safe for the public to consume. Both USDA (APHIS) and EPA approval are needed for examples like the ones shown in Figure 12.3.

Experimental Use Permits

Plant developers who plan to create a plant that expresses pesticidal proteins must first contact the EPA. Field experiments involving 10 acres or more of land or 1 acre or more of water cannot be conducted without an **experimental use permit (EUP)** issued by the agency. EUPs require careful record keeping because the plant cannot be registered for use as a pesticide without clear data that there will not be "unreasonable adverse effects" as defined by U.S. pesticide law. In other words, the plants themselves must meet the same standards as chemical pesticides used in the fields.

The First Experimental Use Permit

In July 1985, Advanced Genetic Sciences applied for the first experimental use permit involving genetically modified organisms. The company developed two

(a)

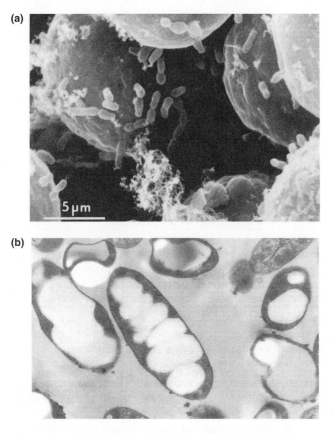

(b)

Figure 12.3 Environmental Release of Genetically Modified Organisms Concerns that genetically modified organisms may have adverse effects on crops or the environment are recognized and regulated by the USDA (APHIS) and the EPA. (a) Plant cell infected by *Bacillus thuringiensis*. (b) A cell producing polyhydroxyalkanoates (a bioplastic).

Figure 12.4 "Ice-Minus" Sprayers Protected by EPA Requirements In the first case of an environmental test of modified organisms, "moon suits" like the ones illustrated here were used to protect the sprayers against unknown biohazards when applying "ice-minus" bacteria to strawberry plants.

genetically altered strains of naturally occurring bacteria that could potentially protect crops from frost damage. Scientists had long known that natural bacteria produce a protein that helps seed (start growth of) frost crystals. The modified bacteria did not produce this protein. Crops sprayed with these harmless bacteria could theoretically survive subfreezing temperatures and still produce fruit. Recall from Chapter 6 that field studies with these "ice-minus" bacteria were very controversial. Somewhat surprisingly, these bacteria fell under the auspices of the EPA. Frost-causing bacteria were considered pests, so the new bacteria were classified as "pesticides." The EPA approved the field tests on the condition that researchers take enormous precautions to prevent the spread of the bacteria. The bacteria were applied to a 0.2-acre plot surrounded by a 15-meter-wide bare soil buffer zone. A total of 2,400 strawberry plants were included in the test. The individuals who applied the bacteria wore protective clothing (as illustrated in Figure 12.4) applied the bacteria using low-pressure handheld

sprayers. Numerous monitoring samples were taken inside and outside the testing plot at the time of the application and for a year after the application. The field trials were a success, and the bacteria eventually entered the market in 1992. This case set the stage for future agricultural applications of biotechnology.

Deregulation and Commercialization

The next phase of the process is commercialization, during which the product is approved for market. However, before anything is available for sale, the EPA spends about a year reviewing the data collected during the experiments. This review concentrates on four general areas of concern. First, the EPA considers the source of the gene, how it is expressed, and the nature of the pesticide-protein produced. Second, the health effects of the bioengineered plant are studied in depth. Experimenters use mice to measure the impacts of the protein on living organisms, including how easily the protein is digested and whether it can trigger allergies. Third, because the EPA is responsible for protecting the environment at large, the agency also must investigate the "environmental fate" of the pesticide protein. One concern is the rate of degradation of the pesticide and whether it lingers in the soil or water. Another issue is the possibility that the pesticide protein might escape through cross-pollination with weed plants to create a superweed. Fourth, the EPA is also interested in the effects on nontarget species. These nontarget species can include helpful insects such as honeybees and ladybugs, as well as other animals including fish, birds, and rodents.

Like the USDA, the EPA can grant deregulated status to any plant that meets the requirements of all

Figure 12.5 Monarch Butterfly Contamination Questioned Bt corn was thought to be toxic to a nontarget species, the monarch butterfly, but studies showed that Bt corn planted in the area presented no measurable risk to the butterflies.

of these tests. Any engineered plant with deregulated status is registered with the EPA and can be sold or distributed like any other plant. But even in these cases, the mission of the EPA is not complete. The agency must be prepared to respond to new information, and it has the power to amend or revoke existing regulations whenever required. You may recall the concern that Bt corn was toxic to a nontarget species, the monarch butterfly (Chapter 6). As soon as suspicions were raised, the agency acted quickly to suspend the commercial production of the plant. Anyone who has witnessed the spectacle of tree branches weighed down by migrating colonies of monarchs or seen them fluttering in high mountain meadows can appreciate the EPA's expedient investigation into the potential threat. As it turned out, studies proved that Bt corn presented no measurable risk to the butterflies (shown in Figure 12.5); consequently, production of the corn was allowed to continue. This example demonstrates an important feature of the bureaucratic regulatory framework in the United States: It is responsive to emerging public concerns and proceeds with caution. Perhaps most important, however, the regulatory bodies depend on good scientific evidence to come to their conclusions. The public is similarly protected by regulations relating to the release of food and drugs.

12.4 Food and Drug Administration

The FDA is charged with making certain that the foods we eat and the medicines we use are safe and effective. Because both food and drug products are hot sectors of biotechnology development, the FDA is busy monitoring new developments in the industry. The fundamental issue is that biotechnology products should be as safe as their conventional counterparts (see www.fda.gov/cder/handbook/develop.htm,

a searchable site of FDA regulatory centers and their responsibilities).

Food and Food Additives

When a new food product or additive is being developed, the FDA serves as a consultant to biotechnology developers (companies, university researchers, etc.) and advises them on testing practices. Studies usually focus on whether the new product has any unexpected and undesirable effects. In addition, the protein is evaluated to see if it is substantially the same as naturally occurring proteins in food. Also, because many common foods, including milk, shellfish, peanuts, and eggs, are known to contain allergy triggers, any protein derived from such sources must be considered a potential allergen and labeled accordingly. Note that companies are not required to consult with the FDA when developing biotechnology food products. However, most companies in the United States have taken advantage of the FDA's expertise before bringing a food product to market. If a food product or additive poses no foreseeable threat, the FDA can grant **generally-recognized-as-safe (GRAS)** status. An example of a product with GRAS status is genetically engineered chymosin, which is used in 80% of the cheese manufactured in the United States, as described in Chapter 4. The Flavr Savr tomato (discussed in Chapter 6) is an example of a food product that won FDA approval as the first commercial plant product to hit the U.S. market. Sometimes a product that has been deemed safe by the FDA will still have a tough time in the marketplace. An example is the use of bovine somatotropin (BST), also known as bovine growth hormone (BGH). This substance can increase the productivity of milk in dairy cattle. Although it occurs naturally in the animals, many consumers have concerns that the hormone will show up in the milk. Although 30% of dairy animals may be treated with BST, the debate continues whether there might be negative effects on the cattle or the humans who consume the milk they produce. No hard data suggesting that BST presents any danger have been uncovered yet, as evidenced by an eight-year study by the FDA (see report in reference list). However, if this or any other product ever proves to be unsafe, the FDA has the responsibility and power to remove it from the market.

The Drug Approval Process

In addition to regulating food and food additives, the FDA is responsible for regulating new pharmaceutical products. Biotechnology drugs must meet the same exacting standards as any other new drugs. Imagine for a moment that a new biotechnological treatment for cystic fibrosis has had promising results during animal trials. The next step is contacting the FDA and filing an **investigational new drug (IND) application** with the **Center for Drug Evaluation and Research.** Before approving the substance for further testing on humans, the FDA considers the results of previous experiments, the nature of the substance itself, and the plans for additional testing. To provide effective regulatory review of biological products, the Center for Biologics Evaluation and Research conducts active mission-related research programs. This research greatly expands knowledge of fundamental biological processes and provides a strong scientific base for regulatory review (see http://www.fda.gov/cber/research.htm).

Good Laboratory (GLP), Clinical (GCP), and Manufacturing (GMP) Practices

In 1975, FDA inspection of a few pharmaceutical testing laboratories revealed poorly conceived and carelessly executed experiments, inaccurate record keeping, poorly maintained animal facilities, and a variety of other problems. These deficiencies led the FDA to institute the **good laboratory practice (GLP)** and **good manufacturing practice (GMP)** regulations to govern animal studies of pharmaceutical products. GLPs require that testing laboratories and manufacturers follow written protocols (standard operation procedures, or SOPs), have adequate facilities, provide proper animal care, record data properly, and conduct valid toxicity tests. A similar set of standards, **good clinical practices (GCPs),** protect the rights and safety of human research participants and ensure the scientific quality of clinical experiments.

Development of new products from cells and tissues for therapeutic use, isolation and identification of genes, and introduction of genes into human cells and tissues, microbes, plants, and animals are all current and expanding biotechnologies. However, laboratory workers who handle gene vectors, recombinant DNA, and biological organisms containing recombinant DNA, bacteria, and fungi risk infections as a result of these practices. Biotechnology companies are aware of the potential liabilities and financial implications related to employee infection and have made implementing the appropriate biosafety practices a high priority, as illustrated in Figure 12.6.

Phase Testing of Drugs

If the FDA is satisfied that the new drug warrants further investigation, does not present obvious risks, and will be tested in scientifically sound methods, the drug wins IND status. The next step is phase testing,

Figure 12.6 Lab Technician Uses a Micropipette Under Biosafety Level 1 Conditions There are three levels of protection against biological hazards: A Class 1 cabinet protects the operator from airborne material generated at the work surface, where the air flows over the work surface and is filtered before being vented back into the room. A Class IIB cabinet draws high-efficiency particulate air (HEPA)-filtered air across the work surface, refilters the air before venting into the room, and is suitable for bacterial and tissue cultures. A Class III cabinet isolates the material completely in a gas-tight area, which is only required when working with the most hazardous agents.

which is different (more stringent) for proposed drugs than for foods. First, during Phase I (safety), between 20 and 80 healthy volunteers take the medicine to see if there are any unexpected side effects and to establish the dosage levels. Phase II (efficacy) begins the testing of the new treatment on 100 to 300 patients who actually have the illness the drug is designed to treat. If no detrimental side effects are noted and the drug seems to have some positive effect, it is ready to go to Phase III (comparative benefit to other current drugs) testing. This phase involves between 1,000 and 3,000 patients in double-blinded (to remove potential bias from affecting results, neither patient nor clinician knows whether placebo or drug is being administered) tests and lasts for 3½ years. Generally, a drug can be marketed only after its benefits and long-term safety have been established in Phase III studies. Most drugs never make it this far. Only 20% of drugs that go through Phase I testing make it all the way to Phase III and on to FDA approval as an **NDA (new drug authorization).** Even after a new drug enters the market, it continues to be monitored by the FDA indefinitely. No therapeutic drug product can be marketed in the United States until an NDA has been issued by the FDA. The number of new drugs that the FDA has approved has fallen by half since 1996, with only 20 approved in 2005. Phase III trials consume 70% of clinical development costs, yet 40% of compounds now fail at this final hurdle largely due to the inability to demonstrate superiority over placebo (see Figure 12.7). Similar to an NDA, a biotech company will file for a **Biological License Agreement (BLA)** if they are seeking approval of a biologically derived product such as a viral therapy, blood compound, vaccine, or protein derived from animals. The FDA reviews information that goes on a drug's professional labeling (information on how to use the drug) and the FDA inspects the facilities where the drug will be manufactured as part of the approval process.

An important recent exception to this testing procedure must be noted. The FDA does permit the approval of drugs and vaccines intended to counter biological, chemical, and nuclear terrorism without first proving their safety and worth in Phase II and III trials. This is also true for drugs designated "orphan drugs," as designated by the FDA Office of Orphan Products Development (OOPD): drugs with small numbers of beneficiaries but with great benefit, such as human botulism immune globulin (BIG), an antibody for botulism poisoning. Of course, it would be unethical to expose human beings deliberately to smallpox or nerve gas to test the value of treatments. The OOPD administers the major provisions of the Orphan Drug Act (ODA), which provide incentives for sponsors to develop products for rare diseases. The ODA has been very successful—more than 200 drugs and biological products for rare diseases have been brought to market since 1983. In contrast, the decade prior to 1983 saw fewer than 10 such products come to market. New drugs of this kind

will still undergo Phase I testing because that part of the process requires exposure to the drug only, not to the dangerous substance it is meant to counteract—and so the FDA can determine that the drug itself will not cause unwanted side effects. For example, the FDA expedited approval of the new drug Cipro, the antibiotic used to treat anthrax. This is more evidence of the bureaucracy's ability to respond to emergent situations.

Faster Drug Approval versus Public Safety

Biotechnology companies spend about $1.2 billion to bring a biological product to market, and it takes about 15 years to receive FDA approval for marketing. Because of financial concerns, biotechnology companies would like to speed up this process; however, the FDA would prefer to keep the process at its present pace to guarantee adequate testing. A compromise may be needed to provide products that the public needs (for example, gene therapy for diseases that will eventually kill patients) and safety that is acceptable to regulatory agencies like the FDA.

For instance, there are currently five suppliers of human growth hormone. Each company was required to get individual approval from the FDA, each doing its own animal studies (Phases I and II) and clinical studies (Phase III). The U.S. Pharmacopeia Convention (USP) has proposed an alternate method for standardizing the measurement of activity *in vitro* (in glass or lab culture). This change would mean that each of these companies could accelerate the testing and proceed to the human safety trials a little earlier. From the point of view of the pharmaceutical companies, this change in methodology

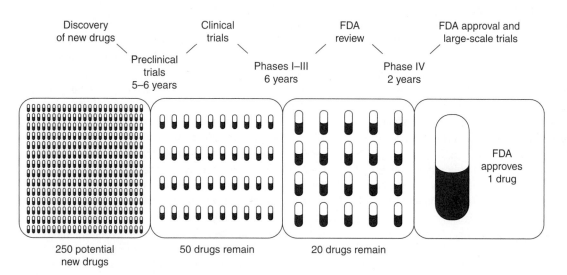

Figure 12.7 The Drug Pipeline Only 20% of the drugs that enter Phase I will make it to Phase III, and 40% of these will fail due to the inability to show superiority over placebo.

would address the backlog of products being submitted for approval by using standardized methods without compromising safety. Whether this change will be approved can only be determined by time. The possible requirement to label all foods produced by biotechnology is another issue that will be decided by time.

12.5 Legislation and Regulation: The Ongoing Role of Government

The regulation of biotechnology, as other industries, is a matter of politics as well as science. The regulatory agencies are government organizations, and they exist to enforce laws. Ideally, those laws rest on good scientific practice, but they also reflect philosophical values. It is not uncommon for value systems to clash, and when they do, a long process of negotiations and compromises may be needed to reach a consensus.

It is not surprising that biotechnology has spawned controversy. Consider the implications of using human embryonic stem (ES) cells. Early research has been extremely promising, and some researchers believe we are on the verge of discovering cures for deadly diseases like Parkinson disease, cancer, and diabetes. But scientific promise is only one side of the controversy. Many Americans find the notion of marketing ES cells ethically repugnant. From their perspective, the practice is as inhumane as harvesting an organ from an unwilling donor. Legislation that calls for either a complete ban or a two-year moratorium on research has been promoted. As the debate progresses, both sides will do their best to shape policy that will regulate this aspect of biotechnical research in the future (see Chapter 13 for further discussion of ethical issues surrounding ES cells).

YOU DECIDE

Should Reproductive Cloning of Humans be Banned?

A majority of the stem cell work described in Chapter 11 has been conducted by private companies or academic institutions using funding from private organizations. This has occurred because of a ban on the use of federal funds, established in 1996, for research that would destroy embryos. One of the challenges facing President George W. Bush when he took office in January 2001 was to consider whether to completely lift the ban on using federal funds for stem cell research, modify the U.S. government's position on funding this work, or impose new restrictions.

Although scientists are concerned about the ethical treatment of human embryos, many stem cell researchers in the United States were also concerned that the ban on federal funding of human embryo research could cause the United States to fall behind other countries actively engaged in embryonic stem cell research. Some feared a "brain drain" would occur whereby many of the best stem cell researchers in this country would leave to work in other countries. At the other end of the spectrum were politicians and lay public groups concerned about the values of human life and the ethics of using stem cells derived from embryos.

During his first few months in office, President Bush sought input from many scientific, public, political, and religious factions before deciding if the United States would fund ES cell research. Some opponents of embryo research expressed concern that use of ES cells might lead to an increased demand for embryos, resulting in problems such as additional abortions. Some critics claim that an embryo has the moral status of a person and any action that destroys an embryo is a desecration of human life.

In a televised speech on August 9, 2001, President Bush addressed the nation to discuss research involving stem cells in what the president described as "a complex and difficult issue, an issue that is one of the most profound of our time." In this speech President Bush presented his administration's positions on the use of federal funds to support scientific research on stem cells derived from human embryos. After discussing the potential of stem cells for treating a range of human diseases, ethical issues of the beginnings of life, and emphasizing his opposition to human cloning, President Bush declared that federal funds will be allowed to be used for research on approximately 60 to 70 existing stem cell lines produced by private research, on the basis that they were created from embryos that had already been destroyed. But federal funds could not be used to *harvest* embryos for isolating stem cells.

President Bush's decision also made available about $200 million in federal funds for stem cell researchers, but a majority of pioneering work is already being

YOU DECIDE

funded through partnerships with industry. For instance, James Thomson, the University of Wisconsin-Madison researcher who isolated the first human ES cells, was funded by Geron Corporation, the largest U.S. biotechnology company involved in human ES cell research. Because private companies have carried out or funded a majority of stem cell work in the United States, some of these companies have closely protected their results for proprietary reasons to gain exclusive commercialization rights to cell lines and technologies they have developed. Although it is generally acknowledged that federal regulation of stem cell research is needed, university researchers and companies don't want to lose their competitive ability to gain rights to patent novel discoveries and potential applications of stem cells.

Many of the top stem cell researchers in the United States consider the Bush announcement a bittersweet decision for several reasons. For instance, cultured cell lines derived from ES cells may not show the full potential to differentiate that freshly isolated ES cells have. It is known that many cultured cells accumulate mutations over time and behave differently, so they may not reflect the true abilities of freshly isolated ES cells. Existing cell lines may have a limited ability to develop into all cells of the body. Lastly, some companies may be reluctant to freely share these limited stem cell lines that they generated using their own funds. Time will tell how President Bush's decision will affect U.S. stem cell researchers relying on federal funds and restricted to doing research on existing cell lines. Meanwhile, private companies and privately funded researchers continue to pursue ES cell research independently of federal restrictions on using embryos for deriving ES cells.

In 1979, after the first successful birth of a child created by *in vitro* fertilization (IVF), a U.S. Ethics Advisory Board recommended that federal funding of human embryo research for the purpose of IVF was acceptable. No action was really taken on the report produced by this board until 1994, when an NIH advisory board endorsed the use of unused embryos from IVF for research. This recommendation was rejected by President Clinton, and subsequently Congress blocked federal funding of all human embryo research.

After human ES cells were cultured by researchers at the University of Wisconsin in 1998 using private funds, in 1999 the National Bioethics Advisory Committee appointed by President Clinton recommended a modification to the congressional ban to fund embryo research to support federal funding for isolating ES cells from unused embryos and recommended a moratorium on human cloning. Under the direction of Harold Varmus, the NIH also proposed similar action for the use of stem cells already produced with nonfederal support. These suggestions languished until August 2001 when President Bush modified the ban on the use of federal funds for research on stem cells derived from embryos allowing federally funded scientists to work only on stem cell cultures or cell lines already established.

On July 31, 2001, the U.S. House of Representatives passed the Human Cloning Prohibition Act of 2001 by a vote of 265 to 162 to ban federal support for all human cloning research—both therapeutic cloning to derive ES cells and reproductive cloning. Currently several bills are under consideration by the Senate to oppose cloning for any purpose. In the United States, there are no laws preventing privately funded researchers from creating stem cells from embryos, and most states do not have laws preventing private groups from creating embryos by IVF or by somatic cell nuclear transfer.

So what does the rest of the world think about the prospects of stem cells and cloning? Producing early-stage human embryos for any purpose other than implantation into the uterus of the woman who produced the egg is illegal in Germany. The United Kingdom has the most tolerant rules in Europe regarding embryonic stem cells. The U.K. permits research on embryonic stem cells derived from aborted embryos and leftover embryos created for research by *in vitro* fertilization. Therapeutic cloning is also legal in Britain. In 1990 the U.K. enacted the Human Fertilisation and Embryology Act. Approved by both the House of Lords and the House of Commons, the act allows embryo research and permits scientists to create embryos for research purposes. Embryonic stem cell research to develop therapies for human diseases is also allowed. However the U.K. House of Commons will soon debate a bill that would ban implantation of a cloned embryo (reproductive cloning). Similar legislation is being considered in Holland, Israel, and Italy. Japan, Canada, France, and Australia allow scientists to derive ES cells from embryos left over from IVF, but these countries have banned reproductive and therapeutic cloning.

Now that you have the background information, should reproductive cloning of humans be banned? You decide.

Labeling Biotechnology Products

Another controversial issue is the labeling of foods that contain genetically modified organisms (GMOs). From the perspective of the FDA and the current White House administration, food labels must include information about the ingredients and any claims made or suggested by the manufacturer. The FDA also requires special labeling of foods that present known safety or usage issues. If a biotechnology food product included a protein that is not usually found in the food and is a known allergen, the FDA would require special labeling to notify allergic consumers of the risk. This standard is also applied to traditional food products. Many chocolate candies include a warning that they may contain traces of peanuts. However, because biotechnology foods are not inherently more or less dangerous than their traditional counterparts, the FDA does not require labeling to indicate the method of production (the European Union, however, does require labeling when genetic modification has occurred if the product is to be sold in Europe).

Some consumer groups are pursuing a change in labeling laws that would require information about the use of genetic engineering in food production. They argue that this disclosure is essential to their right to know and to make informed choices based on their own values and beliefs. An interesting example of this real-world debate focuses on the use of BST to increase milk production, as discussed earlier. Over the past decade, a series of comprehensive studies have indicated that milk produced using BST presents no significant health risks. For this reason, the FDA continues to hold the position that no label is required. Furthermore, when dairy producers began labeling their products as "BST Free," the FDA intervened. The agency stated that such labels are actually misleading because milk production in dairy cows depends on naturally occurring BST. This has spurred sales of milk labeled "from cows not treated with BST." As indicated in Chapter 6, some people fear the introduction of proteins into foods that will stimulate allergies or otherwise add antibiotic resistance substances to our foods. Safeguarding against these potential hazards is what the regulatory agencies have been doing for decades.

These two examples are typical of the many controversies that have attended the growth of the biotechnology industry. It is safe to say that the future is likely to present many more such debates as politics and society adjust to the role of biotechnology in our world. For this reason, one of the primary responsibilities of all regulatory agencies is to provide an avenue for public comment. This aspect of regulation is an important element of democracy in action and is shared by more than one agency, as you can see in Table 12.2. You can learn more about the role of public comment in shaping policy, or even participate yourself, when you visit the regulatory agency websites listed on the Companion Website.

Table 12.2 EXAMPLES OF SHARED RESPONSIBILITIES BY FEDERAL REGULATORY AGENCIES

New Trait/ Organism	Regulatory Review Conducted by	Reviewed for
Viral resistance in food crop	USDA	Safe to grow
	EPA	Safe for the environment
	FDA	Safe to eat
Herbicide tolerance in food crop	USDA	Safe to grow
	EPA	New use of companion herbicide
	FDA	Safe to eat
Herbicide tolerance in ornamental crop	USDA	Safe to grow
	EPA	New use of companion herbicide
Modified oil content in food crop	USDA	Safe to grow
	FDA	Safe to eat
Modified flower color in ornamental crop	USDA	Safe to grow
Modified soil bacteria that degrades pollutants	EPA	Safe for the environment

Source: www.fda.gov.

The Fluvirin Failure

Some 48 million potential doses of flu vaccine were lost by the Emeryville, California, company Chiron when British regulators from the Medicines and Healthcare Products Regulatory Agency (MHRA) found bacterial contamination in some lots of the vaccine. Companies that operate in the United Kingdom must follow GMP regulations that are as strict or stricter than those in the United States. After the MHRA closed the plant in 2004 (it has since reopened after correcting the problems), the FDA performed its own investigation and listed 20 violations in a warning letter sent to Chiron in December 2004. The company found *Serratia* bacteria in nine of its flu vaccine lots. Because it could not trace where the problem started, it was forced to destroy 91 lots of vaccine. Standard operating procedures are required for the traceability of problems like these according to Current Federal Regulations section 211.

How can other companies avoid lapses in GMP like the Chiron incident? FDA regulators recommend that firms should evaluate every aspect of production, including people, procedures, equipment, materials, record keeping, audits, and training. Current good manufacturing practices require continual updating, and companies are required to provide their employees with this training.

In an attempt to create quicker results with smaller numbers of individuals in clinical drug trials,

the FDA is encouraging drug companies to design clinical trials with flexible enrollment and dosing. Not everyone agrees this is a good idea.

In a typical clinical trial, drug dosage and the number of patients in the control and experimental groups are predetermined and unchangeable. In contrast, adaptive trials allow changes in drug dosage, patient pool sizes, and other changes in response to incoming data. In other words, the parameters are changed "on the fly" as new results are received. Considering that Phase II and III trials can easily drag on for five or more years, everyone is supportive of changes that will possibly speed results.

Bayer Healthcare used an adaptive approach in its Phase II trials for a new cancer drug. They did not know ahead of time which cancer would respond best to the new drug, so they enrolled a number of patients with different advanced cancers. Within a short time they knew the drug responded best to kidney cancer, so they "adapted" their trials to enroll and test only patients suffering from advanced kidney cancer.

The major concern is that the committee to decide on adaptive responses be completely objective. It is quite possible that drugs in trials will fail faster and cost companies more, and results could be manipulated with the right "adaptations." The FDA is releasing guidelines for evaluating multiple possible outcomes and methods to enrich trials with patients that are most likely to benefit. Will adaptive design provide faster results? Will the cost of drugs go up or down? Who will benefit the most, the patient or the drug company?

12.6 Introduction to Patents

Patents are regulated by a governmental agency, but biotechnology products are rarely the type of ideas that we think of patents protecting. A patent gives an inventor or researcher exclusive rights to a product and prohibits others from making, using, or selling the product for a certain number of years. When patent laws were first conceived, nobody could have conceived of patenting DNA or proteins. It was expected that inventors would be engineering better mousetraps, not better mice. The **U.S. Patent and Trademark Office (USPTO)** issued the first patent on a bacterium with a unique gene sequence (genetically engineered organism) in 1980. Since that time, the office has granted more than 2,000 patents for plant, animal, and human genes. Patents have also been granted for transgenic animals and plants, monoclonal antibodies and hybridomas, isolated antigens and vaccine compositions, and methods for cloning or producing proteins.

YOU DECIDE

Would We See Drug Development Without Patents?

Gene patents are here to stay, but many questions remain unresolved. For instance, what happens when a single company "owns" a biotechnology product with potentially life-saving applications? As Table 12.4 illustrates, many companies have applied for patents on gene therapies for cancer.

History shows that medical practice can be affected by patent law. For example, Myriad Genetics holds exclusive patents on *BRCA1* and *BRCA2*, the genes that largely determine a woman's risk for breast cancer. The company recently began demanding fees from any lab that ran blood tests for the genes, a move that threatened to make the test unavailable to many women. In January 2001, the National Cancer Institute (NCI) struck a deal with Myriad Genetics that allows all NCI and NIH labs to use the test at a discount. Still, some other labs no longer are able to afford the test.

Should one company be allowed to own the rights to important medical examinations or treatments? Should the government step in to make sure such procedures are available to everyone at a modest price, or would such a move trample the rights of the patent holder? You may wish to evaluate the case studies discussed in the U.S. Patent and Trademark Office website.

You decide.

These patents have translated into big profits for biotechnology companies, but they have also stirred considerable controversy. Many people believe that genes—especially human genes—should not be treated like light bulbs or mousetraps. To their way of thinking, it should not be possible to patent life or critical components of life such as DNA and proteins.

To fully understand this issue, we need to look at the functions of patents. To win a patent, a discovery must meet three basic requirements: It must be novel, it must be nonobvious, and it must have some utility. (The concept of "utility" is discussed in more detail later in this chapter.) Traditionally, patents have been awarded for inventions, but many natural products clearly meet all three requirements, as illustrated in Table 12.3. For example, a newly discovered DNA sequence can be novel, nonobvious, and potentially very useful, and the USPTO generally does not hesitate to issue patents on sequences for proteins of known function.

In short, the process of patenting genes fully fits the letter and the spirit of patent law. Still, many concerns and uncertainties remain. The challenge for the government is to find a way to protect the rights of individual researchers while promoting science as a whole, as is the case for better cancer drugs (see Table 12.4).

The Value of Patents in the Biotechnology Industry

For the biotechnology industry, strong patents mean strong businesses. Patents are the primary asset by

---

Table 12.3 PATENTS, TRADEMARKS, AND TRADE SECRETS

Patents provide rights for up to 20 years for inventions in these broad categories:

Utility patents protect useful processes, machines, articles of manufacture, and compositions of matter. Some examples are fiber optics, computer hardware, and medications.

Design patents guard the unauthorized use of new, original, and ornamental designs for articles of manufacture. The look of an athletic shoe, a bicycle helmet, and the *Star Wars* characters are all protected by design patents.

Plant patents are the way we protect invented or discovered, asexually reproduced plant varieties. Hybrid tea roses, Silver Queen corn, and Better Boy tomatoes are all types of plant patents.

Trademarks protect words, names, symbols, sounds, or colors that distinguish goods and services. Trademarks, unlike patents, can be renewed forever as long as they are being used in business. The roar of the MGM lion, the pink of the Owens-Corning Pink Panther, and the shape of a Coca-Cola bottle are familiar trademarks.

Copyrights protect works of authorship, such as writings, music, and works of art that have been tangibly expressed. The Library of Congress registers copyrights that last the life of the author plus 50 years. *Gone With The Wind* (the book and the film), Beatles recordings, and video games are all copyrighted.

Trade Secrets are information that companies keep secret to give them an advantage over their competitors. The formula for Coca-Cola is the most famous trade secret.

Source: www.uspto.gov/web/offices/ac/ahrpa/opa/museum/1intell.htm

---

Table 12.4 GENE THERAPIES BEING STUDIED IN CANCER PATIENTS THAT MAY RECEIVE PATENTS AND REGULATORY APPROVAL

Approach	Number of U.S. Trials Approved since 1988 or Awaiting Federal Approval
Antisense therapy (to block synthesis of proteins encoded by deleterious genes)	4
Chemoprotection (to add proteins to normal cells to protect them from chemotherapies)	7
Immunotherapy (to enhance the body's immune defenses against cancer)	58
Pro-drug, or suicide gene, therapy (to render cancer cells highly sensitive to selected drugs)	21
Tumor suppressor genes (to replace a lost or damaged cancer-blocking gene)	6
Antibody genes (to interfere with the activity of cancer-related proteins in tumor cells)	2
Oncogene down-regulation (to shut off genes that favor uncontrolled growth and spread of tumor cells)	2

Source: Fiattman, G. I., and Kaplan, J. M. (2001). "Patenting Expressed Sequence Tags and Single Nucleotide Polymorphisms," *Nature Biotechnology*, 19: 683.

which a successful biotechnology company will be valued during all stages of its development. Suppose a team of scientists discover a vaccine against a deadly virus. Without a patent, no pharmaceutical company would go forward with the daunting tasks of conducting costly clinical trials, obtaining FDA approval, and developing a comprehensive marketing plan. Without patent protection, a rival company could buy the drug off the shelf, duplicate it, and sell it at a discount rate without spending a dime on research.

In the United States, patents are enforced for up to 20 years from the earliest date of filing. That means other companies can market their own versions of the product after the original developers have had exclusive rights for two decades. This plan both encourages new discoveries and prevents long-term monopolies.

To obtain a U.S. patent, a company must file an application that adequately describes the product. The company must also disclose what it considers to be the best use of the product at the time that the application is filed. USPTO experts with specialty technical backgrounds examine all patent applications. Patents are awarded on a first-come, first-served basis. If two companies file applications for the same discovery at around the same time, the two sides could be headed for a legal battle. In such cases, a little preparation goes a long way.

The following guidelines protect a product that is heading for patenting:

1. *Keep good records.* Every researcher should keep detailed notes within a bound notebook or in a secure electronic format as specified in FDA regulations. Each entry should be signed, dated, and witnessed by an individual who is not directly involved in the research. The notebook should contain all conceptual ideas and supportive data. This evidence may be used to support the company's patent or to invalidate rival patents.

2. *Do your homework.* Continue to monitor the activities of potential competitors and others in the field through trade literature, published patent applications, and issued patents. The USPTO website is a valuable source of information. You can also check commercial databases such as Derwent, Medline, and Biosis.

Patenting DNA Sequences

Although the vast majority of DNA sequences may never have medical or agricultural value, some could play an important role in producing food or fighting disease. Naturally, organizations that identify those sequences want to enjoy the fruits of their efforts by obtaining protection under the patent laws. The

Q If you assisted your professor in a discovery and published your work, would it prevent the discovery from being patented?

A A chemistry professor at a college applied for a patent on a new chemical compound after three of his students had published their theses including the new compound. The claim to the patent was originally rejected by the patent examiner as lacking novelty (see the discussion of novelty in Section 12.6) because the new compound was already known from the published theses. The decision was reversed on appeal by the Federal Circuit Court (in re Cronyn), even though premature public disclosure of an invention, in written, oral, or electronic form one calendar year before application prevents patenting.

The problem with premature disclosure is particularly pervasive in academia, where the advancement of knowledge, rather than commercial gain, is the goal. In this case, the students' theses were not generally cataloged in either the main library or the chemistry library. They were not listed in the main catalogs of the libraries, nor were they assigned Library of Congress numbers. The Federal Circuit Court held that to be considered "publicly available," the theses had to be accessible to the public. This is obviously an unusual case, and companies are particularly careful to inform investigators of how easily patent rights may be compromised or lost by premature public disclosure.

genomic frontier has been mapped, and today's pioneers are eagerly and aggressively staking out what they regard as their territory. Here is a closer look at the requirements of a patent application.

Put in legal terms, patentable subject matter includes any new and useful machine, manufacturing process, or composition of matter, or any new and useful improvement. Thus, to be patentable, a product must have some utility, as discussed earlier. Under the guidelines, applicants must assert a utility for the claimed invention that is specific, substantial, and credible. To claim a **specific utility** in the case of a DNA sequence, a researcher must know exactly what the DNA sequence does. It is not enough to say it is a probe or a marker (see Chapter 3 for a refresher on how these work). The researcher must disclose what, precisely, it probes or marks (for instance, a probe for a specific human disease gene). A **substantial utility** defines a real-world use. One example would be a cloned DNA fragment encoding a protein with a known function or use. To meet the

credible utility requirement, the researcher must convince the patent office that the application is backed by sound science.

Even if a gene sequence passes all three tests, it can still be denied a patent if it is too similar to an existing patented sequence. Remember from Chapter 3 that the function of a gene can be dramatically altered by switching a single nucleotide. For this reason, patents on gene sequences are usually very specific. Put another way, a segment of DNA usually must match a patented sequence exactly in order to violate—or "infringe" on—that patent. If a company claims patent infringement on two similar sequences, a court must decide if the sequences have essentially the same function. This is known as the **doctrine of equivalents.** The requirements for protecting intellectual property rights and the sale of products with patents are not the same outside the United States and require separate applications in these countries.

According to several studies, approximately 20% of all human genes have been patented, primarily by private biotechnology companies. Not surprisingly, the most heavily patented genes are typically those implicated in human health and disease. Almost 50% of known cancer genes have been patented. In some cases, genes such as *BRCA1*, one of the genes involved in breast cancer, have patent rights issued to multiple companies resulting in disputes over commercial rights and ownership of such sequences. Some genes have up to 20 patents covering rights to

TOOLS OF THE TRADE

A Safe and Efficient Regulatory Work Environment

What can a group of well-trained scientists with excellent government funding, limited regulations, and a large supply of germ plasma produce? Chinese scientists have created the largest plant biotechnology production capacity outside of North America. They have shown that government support is essential if new products are to be developed quickly in developing countries. As a result, farmers in China are cultivating more acres of genetically modified plants than any other developing country, and they are doing this with the support and encouragement of their government. These crops are monitored and regulated by the Chinese government under regulations developed within the country.

Although China has spent the past 50 years building the most successful agricultural research system in the developing world (employing more than 70,000 scientists), research in modern plant biotechnology did not begin until the mid-1980s. In response to rising pesticide use and the emergence of a pesticide-resistant bollworm population in the late 1980s, Chinese scientists began research on genetically modified cotton. In 1997, the commercial use of genetically modified cotton was approved, after devastating pest-related crop losses. The response by China's poor farmers to the introduction of Bt cotton eliminated any doubt that genetically modified crops can play a role in developing countries. By 2000, farmers planted Bt plant varieties on 20% of China's cotton acreage and reduced an average of 13 sprayings (and a savings of $762 per hectare per season). Although yields and the price of Bt cotton and non-Bt varieties were the same, the cost savings and reduction in labor enjoyed by Bt cotton users reduced the cost of producing a kilogram of cotton by 28%.

A simplified approval process is not the only reason for their success. Chinese scientists have identified more than 50 plant species and more than 120 functional genes that scientists are using in plant genetic engineering. Of these genetic technologies, 353 were made between 1996 and 2000. China's Office of Genetic Engineering and Safety Administration approved 251 cases of genetically modified plants for field trials, environmental releases, or commercialization. Of these, regulators approved 45 genetically modified plant applications for field trials, 65 for environmental release, and 31 for commercialization. Transgenic rice, resistant to three of China's major rice pests—stem borer, plant hopper, and bacterial leaf blight—have passed at least two years of environmental release trials.

Unlike the rest of the world, in which most plant biotechnology research is financed privately, China's government funds almost all of its plant biotechnology research and planned to raise these research budgets by 400% before 2005 (although they fell somewhat short of their goal). Some U.S. companies have found this favorable regulatory environment an enticement. Monsanto (based in St. Louis, Missouri) is pursuing a joint venture with the Hebei Provincial Seed Company in China to develop other plant varieties. China is vigorously pursuing opportunities for contract research, and it has become an exporter of biotechnological research methods and commodities. The sales of genes, markers, and other tools, in addition to genetically modified varieties, may cause other countries with more restrictive policies regarding genetic modification to look at this world leader.

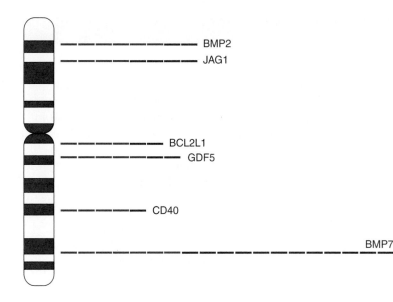

Figure 12.8 An Illustration of Patent Activity on One Human Chromosome Chromosome 20 is shown with each horizontal bar representing a patent claiming a gene sequence located in that region. The labels indicate the loci of highly patented genes.

commercial uses of the genes such as diagnostic tests for genetic testing. See Figure 12.8 for a map of patent hotspots on chromosome 20.

YOU DECIDE

Patenting of Expressed Sequence Tags

In the early days of gene sequencing in the early 1990s, researchers at the National Institutes of Health (NIH) described a new method of identifying areas of the human genome by using expressed sequence tags (ESTs). These sequences are close to important genes and could be used as marker points to determine other sequences that overlapped. Unlike fully sequenced genes with known functions, an EST indicates only the fact that the nucleic acid is expressed in this area. Some researchers (even within the NIH) sought to file patent applications claiming ESTs, and any gene or protein associated with them (logically, if the sequence is expressed, it must be important to the cell, shouldn't it?). The USPTO began to reject EST applications in 2001 based on lack of "utility." Companies with EST patent applications have appealed this decision and have lost (re Fisher 2005). The utility requirement that a genetic sequence can be patented if the protein is known that is expressed from that sequence, and if the protein has a function as a biological product, was upheld. Do you agree with this decision, or do you think it will slow research to find new sequences of importance? Or do you think ESTs should not be patentable and available to all in GENBANK (see www.ncbi.nlm.nih.gov)? You decide.

12.7 Biotechnology Products in the Global Marketplace

Biotechnology is a global enterprise. At this point, the world community is still involved in preliminary negotiations about the regulation of biotechnology products. There have been some promising beginnings. The United Nations, for example, has produced a voluntary code for the deliberate release of GMOs. Public comment is being solicited since it was proposed in 2000, and no final binding regulations have been agreed on (see reference). An agency of the UN, the Food and Agriculture Organization (FAO), is focusing its attention on biotechnology in developing nations, providing guidance on safety and risk assessment. In general, no consensus or clear set of commonly held standards for the import and export of biotechnology products exists. There is reason to be optimistic, however. Recent developments in Europe can serve as a model for international cooperation in the trade and development of biotechnology products.

In recent years, the European Union (EU) has begun to stand out as both a hotbed for innovation and as an attractive market for biotechnology products. The formation of the EU in 1993 dramatically streamlined the process for introducing new biotechnology products to the marketplace. Today, the EU stands as a textbook example of a government body committed to promoting scientific progress.

A key part of the new system is the **European Agency for the Evaluation of Medicinal Products (EMEA),** the equivalent of the FDA. The EMEA authorizes medicinal products for human and veterinary use. It also helps spread scientific knowledge by providing information on new products in all 11 official EU languages. Once a product has been approved

CAREER PROFILE

Quality Control Inspector

Quality control inspectors in biotechnology companies are responsible for inspecting and checking products to make sure they meet the specified levels of quality described on the package inserts (and that the products' specifications are filed with the appropriate regulatory agencies). This is a good career for a detail-oriented person who likes to work in the lab. This area has a continued high rate of employment. Quality control is involved at every stage of production. Some examine raw materials to verify specified requirements, others test equipment and verify accuracy, and others test the standard operating procedures established by the company (based on FDA, USDA, EPA, and related regulations).

Although a high school diploma and experience in quality control is adequate in some companies, most employers prefer an associate's degree in a technical area and a few years of experience. Quality control inspectors are often required to determine error levels in measurements, making a good math background a necessity. Good hand-eye coordination and mechanical aptitude are also required. A strong background in the regulations that govern the products of the industry is a must. Quality control inspectors with a bachelor's degree and experience can be promoted to manager. Many transfers occur within biotechnology companies from quality assurance to quality control. The Quality Assurance Department is responsible for maintaining all of the paperwork that is reviewed when any regulatory agencies show up for an inspection. A good career ladder exists in many companies from quality assurance to quality control.

by the EMEA, it can be marketed in all of the 15 countries in the EU. Thanks to this streamlining, new drugs can be brought to market throughout the EU in a year, instead of four or five, and contrasted with the eight to eleven years that the FDA usually takes to approve a new drug. Since its inception, the centralized procedure has approved more than 100 human medicines and 25 veterinary products, as shown in Table 12.5.

There is one noteworthy difference between the EMEA and the FDA. As previously discussed, the FDA is closely involved with product development from the moment a company seeks permission to conduct clinical trials. Within the EU, however, approval for clinical trials is granted by authorities in each country, not the EMEA. In many cases, the EMEA has no prior contact with the company or even knowledge of the product until it is submitted for review. According to the most recent polls (2000), 75% of pharmaceutical companies in the EU are satisfied with the process as it stands.

QUESTIONS & ACTIVITIES

Answers can be found in Appendix 1.

1. Which U.S. agency regulates food additives?

2. Which U.S. agency regulates GM crops?

3. Which U.S. agency regulates water quality?

4. Which U.S. agency regulates environmental release of genetically engineered organisms?

5. How do patents regulate drugs and devices?

6. How have new streamlined approval processes affected the European drug market?

7. Where is the greatest number of patent applications in cancer therapy?

8. In your opinion, what should have happened according to FDA regulations to prevent the Fluvirin incident?

9. In your opinion, what is wrong with patenting an EST?

10. Why are drugs often available in the EU before they are available in the United States?

Table 12.5 EUROPEAN UNION PATENT APPROVAL STATISTICS	
Medicines for Human Use	
Total Applications	228
Authorizations Granted	118
Medicines for Nonhuman Use	
Total Applications	37
Authorizations Granted	13

Source: Fox, S. (2002). GEN, "Bringing Drugs to Market in European Union," *GEN*, 20(5).

References and Further Reading

Davis, P., Kelley J. J., Caltrider, S. P., et al. (2005). ESTs Stumble at the Utility Threshold. *Nature Biotechnology*, 23(10): 1227–1229.

Fiattmann, G. I., and Kaplan, J. M. (2001). Patenting Expressed Sequence Tags and Single Nucleotide Polymorphisms. *Nature Biotechnology*, 19: 683.

Fox, S. (2002). Bringing Drugs to Market in European Union. *GEN*, 20 (5).

Huang, J., Rozelle, S., Pary, C., and Wang, Q. Plant Biotechnology in China. *Science*, 295: 674–766.

Jensen, K., and Murray, F. (2005). Intellectual Property Landscape of the Human Genome. *Science*, 310: 239–240.

Koppal, T. (2002). Inside the FDA. *Drugs Discovery*, 5(10): 32–38.

Lawrence, S. (2007). Biotech Drugs Cost $1.2 Billion. *Nature Biotechnology*, 25(1): 9.

Michaels, D. (2005). Doubt Is Their Product. *Scientific American*, 292(6): 96–101.

Stix, G. (2006). Owning the Stuff of Life. *Scientific American*, 294(2): 76–83.

Vastag, B. (2006). New Clinical Trials Policy at FDA. *Nature Biotechnology*, 24(9): 1043.

Wilkie, D. (2005, March 14). The Chiron Case. *The Scientist*, p. 40.

Yang, X., Tian X. C., Kubota, C., et al. Risk Assessment of Meat and Milk from Cloned Animals. *Nature Biotechnology*, 25(1): 77–83.

Visit www.pearsonhighered.com/biotechnology to download learning objectives, chapter summary, "Keeping Current" web links, glossary, flashcards, and jpegs of figures from this chapter.

Chapter 13

Ethics and Biotechnology

After completing this chapter you should be able to:

■ Define bioethics and explain how it relates to biotechnology.

■ Identify different approaches to ethical thought.

■ Identify potential ethical problems associated with biotechnology.

■ Pose questions and approaches that address the ethical problems identified in this chapter.

■ Identify outcomes and pitfalls associated with different ethical approaches.

■ Describe the ethical considerations for research with humans.

■ Discuss interactions among science, economics, communication, and public policy.

■ Understand and explain controversies and ethical issues surrounding genetic testing, stem cells, and cloning.

■ Describe possible pathways to careers in bioethics.

Bioethics considers topics such as human life. John and Lucinda Borden testifying at a U.S. House hearing on embryo research. John holds their twins Luke and Mark; Lucinda holds a picture of the twins as embryos. Luke and Mark were adopted frozen embryos.

13.1 What Is Ethics?

Ethics identifies a code of values for our actions, especially toward other humans. In simple terms, ethics could be considered a guide to separate right and wrong, good and evil. The area of ethics that deals with the implications of biological research and biotechnological applications, especially regarding medicine, is called **bioethics.** It considers social and moral aspects and potential outcomes of the use of biological and medical techniques.

In some ways, ethics might seem like plain old common sense. Of course you shouldn't intentionally cause cancer in a group of people without their knowledge just to test your new anti-cancer drug. But in many cases the choices are not so clear cut. Often the decisions must deal with potential trade-offs, compromises, or possibly even sacrifices. Bioethics deals with some of the most fundamental questions confronting our society. And in many respects the decisions made now can affect the future of science, of humanity, and of the world in which we live.

The intent of this chapter is not to tell you what to think about bioethics. Instead, the intent is to get you to understand *how* to think about bioethics—to encourage you to ask questions, to think about how to ask the right questions, acquire all the facts, and make decisions based on information rather than emotional reactions. It is important for you to appreciate that ethics is a *dilemma-based discipline.* Ethical dilemmas arise when an important problem or situation requires careful consideration and thought to make what one believes to be sound ethical decisions.

One fundamental question that should be asked when dealing with bioethical issues is not "Can this be done?" but "Should this be done?" And if something should be done, the question becomes "How can it be done in the right way?" Such questions are important for everyone to consider, especially in areas of biotechnology in which discoveries and their applications can have a great impact on human health and the environment. These questions get to the heart of not only society but also of science and its role in society. For example, consider the photo shown in Figure 13.1. Scientists recently implanted mouse cells into chicken embryos to demonstrate that the mouse cells could receive signals from the chicken embryo to induce development of teeth—something that does not normally occur in chickens. Chick embryos started to form teeth, but they were not allowed to develop into adults. But some people are outraged at experiments such as this because it creates images of "Frankenstein-like" mutant chickens with teeth that aren't normal. The rooster image shown in Figure 13.1 was created by photographers for shock value and headlines. So just because technology

Figure 13.1 Growing Teeth in Chickens. Should This Be Done?

is available to create a chicken embryo with teeth, should it be done in the first place? Should the embryo be allowed to grow into an adult chicken with teeth? Is this ethical? What do you think?

Approaches to Ethical Decision Making

Before we explore some of the current bioethical issues, we first examine some of the more important methods of ethical thought. The study of ethics is as old as humanity itself. Questions of our duties and responsibilities to other members of the human community have been with us for ages. Hippocrates (c. 460–361 B.C.) might be considered the first **bioethicist.** He emphasized the patient rather than the disease in his practice of medicine, viewing the worth of the individual and the sanctity of human life as of primary importance. For years, physicians have pledged to follow the central tenets of the **Hippocratic Oath—**"do not kill," "to help, or at least do no harm"—in their duty to patients and to their profession.

Ethical thought and methods to approach problems can often be divided between two main viewpoints (although there are certainly other approaches as well). One, the **utilitarian approach,** or consequential ethics, started with the Scottish philosopher Jeremy Bentham (1748–1832) and the English philosopher John Stuart Mill (1806–1873). This approach states that something is good if it is useful, and an action is moral if it maximizes pleasure among humans; put more simply, this approach aims to produce the "greatest good for the greatest number." The second main approach—the **deontological approach, Kantian approach,** or duty ethics—comes primarily from the German philosopher Immanuel Kant (1724–1804). This approach focuses on certain imperatives, or absolute principles, which we should follow out of a sense of duty and should dictate our actions.

Modern bioethics can be traced to much more recent times and is primarily the work of two ethicists in the 1970s: Joseph Fletcher and Paul Ramsey. These men refined the two primary approaches to ethical thought mentioned—Fletcher for utilitarianism (also termed "situational ethics") and Ramsey for deontology (or "objectivism").

Utilitarianism emphasizes consequences, not intentions. Another way to phrase this emphasis is that "the ends justify the means." The idea is to calculate what the consequences of an action would be and to weigh different consequences against one another. If we can analyze various courses of action to determine which will have the greatest positive effect on the greatest number of people, then we can provide an answer to the question of what we ought to do. In some ways, this is an elegant method for making decisions. All avenues can be assessed for potential benefit, and the decision becomes more quantitative. The disadvantage of this calculation is that we must assign a value to all of the things considered. How do we assign these values? Some would say that some of the most important things in life (love and family) are not easily quantified, whereas other things (material goods and lifespan) could be emphasized in the calculations because they are quantifiable. Another source of concern could also be the individual doing the calculating and assigning values. For example, the utilitarian calculation would be less than ideal if done by someone who believes males are worth more than females or that the primary consideration should be profit margin.

Deontology, or objectivism, starts from the point of view that there are at least some absolutes (definitive rules that cannot be broken), and we have a moral obligation or commitment to adhere to these absolutes. One absolute usually expressed is for the value of human life, expressed by Kant in this way: "Act in such a way that you always treat humanity, whether in your own person or in the person of any other, never simply as a means, but always at the same time as an end." Another way this is expressed is as respect for others—treating others as ends in themselves, rather than as means to an end. This approach has often been associated with religious traditions, but it is a mistaken notion to assume an objective ethical approach is solely a religious approach or even that all religious approaches to ethical issues begin with the same absolutes. Deeply held convictions are just that—personal points of reference to which an individual adheres. An advantage of objectivism is that it gives firm guidelines in many situations where ethical decisions are required, providing a clear-cut ethical formula for decision making. A disadvantage of this approach is that it may be, or at least may be perceived to be, too rigid in its decision-making process, not taking what may be important factors into account or considering possible changes in values. And there may be situations or issues where no clear conviction or absolute exists, requiring further examination of the issue to define the moral imperatives and discern the course of action.

To illustrate the difference between utilitarianism and objectivism, consider a situation in which a person is desperately hungry and has no money to buy food. Wandering by a store, the person sees a loaf of bread left out on a table. A utilitarian calculation might take into account the person's need, the available food, and the minuscule loss in value to the store and consider it would be okay to take the bread. An objectivist approach might have the absolute "it is wrong to steal" and consider it unethical to take the bread.

Note that there can be other ethical approaches and methods beyond the two main approaches mentioned here, and even these two approaches to ethical decision making can sometimes be blended. When approaching ethical decisions, a key objective is to gather information, consider the facts, and make a thoughtful and informed decision. And in debates on ethically contentious issues, it is neither wise nor polite to deride or belittle another person's decision. Be sensitive to the effects of your own conduct. Aim to understand and defend your own ethical decisions rationally, and strive to consider the decisions of others.

Ethical Exercise Warm-Up

What follows is a typical scenario regarding tough ethical decision making. Although the situation may seem a bit harsh because it deals with a life-or-death scenario, some of the biotechnologies being developed also deal with issues of life and death and have far-reaching consequences.

A family pulls up to the Grand Canyon in their car. The parents get out to check out a refreshment stand and lock the car doors, leaving three young children asleep in the back seat. But they do not set the brake well enough. After they've moved some distance away, the car slowly begins to roll toward the edge of the cliff. You are the only one who notices. A large

TOOLS OF THE TRADE

Careful Thought and an Open Mind Are Powerful Tools for Bioethicists

In contrast to the various laboratory and industrial applications of biotechnology, bioethics requires no specialized equipment. The main tools of the trade are logic and an open, inquisitive mind. Bioethicists need to be able to assess a situation carefully, considering the possibilities from many different angles. Considerations need to include medical, religious, philosophical, legal, scientific, and social concerns and outcomes. Bioethicists need to be able and willing to consider many different viewpoints, which are often conflicting. Bioethicists also need to be inquisitive, able to ask probing questions and determine what underlying reasons are behind the concerns. They may also find it necessary to do extensive literature searches, delving into the background of a topic from numerous perspectives. Language skills can be an additional asset because not all perspectives or literature may be accessible in a single language. Sorting through possible outcomes and pointing out potential pathways for resolution of concerns require reason and logic, as well as patience. In the modern world, knowledge of regulatory agencies and rules is also extremely important. This means that bioethicists must not only be able to understand the regulations and laws currently in place but be able to interact with policy makers to facilitate the creation of sound regulations and laws. Finally, good interpersonal and communication skills (both oral and written) are essential for making the concerns, questions, and outcomes clear to all involved.

YOU DECIDE

Right or Wrong?

The chief executive officer (CEO) of a small startup biotech company is involved in negotiating a multimillion dollar merger of his company with a larger biotech company as a result of a promising stem cell cloning technology developed by the startup. However, the CEO is also aware of a recent significant problem with this technology. Does the CEO reveal the problem and risk jeopardizing the merger, or does the CEO allow the merger to move ahead without discussing the problem? You decide.

13.2 Biotechnology and Nature

Humans have used biotechnology for a long time. Fermentation, for example, is an ancient biotechnology technique that uses naturally occurring microbes. Nature has been doing biotechnology experiments for much longer than humans. Bacteria routinely swap plasmids, and recombination and mutation occur, allowing the expression of new genes or new combinations of genes that were not present before. However, the game changes when humans get involved and create new genetic combinations. In those instances, we need to evaluate our directions and analyze our involvement in the overall experiment.

Developed in the 1970s, recombinant DNA technology allows great potential in the manipulation of the genetic combinations possible in nature. As we discussed in Chapter 3, because of concerns with this new technology, scientists themselves met at a conference in Asilomar, California, in 1975 and called for a **moratorium** (a temporary but complete stoppage of any research) until the safety of the technique and possible consequences could be assessed. One primary concern was that genetically engineered bacteria would escape from the laboratory into the environment, possibly creating new diseases, spreading old diseases in a more virulent manner, or creating an imbalance in the ecosystem that might lead to decimation of some species.

In the end, scientists determined that recombinant DNA technology could be controlled in a way that would preserve safety for humans and for the environment while allowing the science to continue. In particular, guidelines were developed for different levels of biosafety containment depending on the inherent dangers of the experimental system used. For example,

man is standing close to where the car is headed. You could push him in front of the car, which would stop its rolling, although he would either be crushed or pushed over the edge himself. What would you do?

First, do you have all the facts in this situation? So far we have painted the scenario so you have only been given two choices: to let the car go over the edge or to sacrifice the one man for the three children. The question seems to be simply "Can you trade his life for theirs?" A utilitarian approach might consider this is an ethical trade, one life for three, and also consider that the three are young lives; an objectivist approach might consider that it is wrong to endanger or kill any human life, no matter the eventual consequences. Are there other possible choices or alternatives?

experiments with nonpathogenic bacteria and non-pathogenic gene sequences require only minimal safety equipment, whereas experiments with known pathogens, human cells, or potentially pathogenic genes require more stringent containment procedures, and research with particularly virulent pathogens mandates the strictest containment and safety procedures.

Cells and Products

As we have discussed throughout the book, both bacteria and eukaryotic cells can be genetically engineered to express foreign genes and proteins. This has proven to be an indispensable approach for producing medically valuable products. The cells can be grown in large volumes in bioreactors, and the products can be easily harvested in the culture medium and purified. Beyond the safety questions that revolve around cultured bacteria or human cells, other issues—especially ethical concerns—need to be considered. Think about the ethical challenges involved with genetic modification of individual cells and the products produced as a result of these changes.

When a product concerns human application, there is not only the obvious issue of safety but also the issue of effectiveness, or **efficacy,** in its intended use. This is obviously important for an intended patient because the patient wants an effective treatment. But efficacy is also important for the manufacturer; if the product is not effective, there can be a tremendous economic waste. The next question then becomes whether testing can ethically be performed in either cell culture or animal studies. For any drug, an important consideration will be the dose at which the drug is effective, with minimal side effects and toxicity. Consequently, it will also be important to establish whether the drug has any carcinogenic or teratogenic hazards. These considerations prevent future problems, such as finding out that the drug cures the disease at one dose but kills the patient or causes cancer or birth defects at another dose, because they raise the ethical concern of harming rather than helping the patient and the potential problems involved in using the patients themselves as guinea pigs to test drug effectiveness.

We must be mindful of the **humane treatment** of animals in these studies. We need to determine how many experimental animals will be the minimum needed to test the drug, what types of treatments will be necessary for the tests, and whether rodents or primates must be used. The choice of species can affect the action of the drug. One of the best examples of species difference in drug action occurred with the drug **thalidomide,** which was originally designed as a mild sedative. It was tested in standard laboratory rodents and found to be safe. However, many of the pregnant women who were prescribed this sedative gave birth to babies with severe birth defects. When the drug was tested on marmosets (a type of monkey), birth defects similar to those seen in humans resulted. It turns out that drug metabolism can vary among species. This fact seems to indicate that all drugs intended for humans should be tested on humans or primates. We consider the question of human experimentation in detail later, but the possible differences in effects on humans versus various animal species must always be kept in mind.

GM Crops: Are You What You Eat?

A key advance in genetic engineering, and also one of the biggest controversies, has come from the production of **genetically modified (GM) organisms,** particularly GM crops. The aim is to produce plants that can resist pests, disease, or harsh climates, allowing better production of crops. Yet many people are opposed to GM crops and have an aversion to ingesting genetically modified food. Before you decide whether this concern is founded on sound reasoning or on emotions, consider the facts.

GM crops and other genetically modified plants present several areas of concern. The first area involves the plant itself. You need to determine if the alterations in the plant's genetics provide a benefit to the plant or at least do not produce a less vigorous plant. The project probably would not go ahead if it did not provide some benefit, such as pest resistance, but one question that might be worth answering is whether the integrity of the species (maintaining the original genetic composition of a species, without major change, such that it is still essentially the same species) is somehow preserved along with the alteration. You should seek to define for yourself whether such **species integrity** is important

YOU DECIDE

Buyer Beware?

Because of public concerns over the possible safety of GM foods, there have been proposals for conspicuous labeling of all GM foods, even if there is only a minute concentration of any GM product in the food. The cost of testing and labeling could add significantly to the price of these foods, and the actual need for labeling in terms of safety is disputed. Should public fears over GM foods require that all such foods be labeled for consumers? Is this a good marketing scheme? You decide.

or whether creating a "better" plant species is more desirable than trying to maintain an "old" species. In doing so, determine for yourself whether genetic modification of organisms, in this case plants, violates any ethical codes.

Another question, on a broader scale from the first, is the possible effect of altered plants on the ecosystem and on overall **biodiversity** (the range of different species present in an ecosystem). We must determine the effect of the introduction of a genetically modified plant on the local environment. Because we are focusing on crop plants, the desired effect will likely be not only increased growth and production from the GM crop plant but also some effect on potential pests and diseases. One example is Roundup-Ready soybeans. The soybean is genetically modified to resist Roundup herbicide, allowing a farmer to spray the crop and kill noxious weeds that would interfere with growth of the crop, without harming the soybeans. Another example is Bt corn (Figure 13.2). Recall from Chapter 6 that this GM crop is engineered to produce a toxin from the bacterium *Bacillus thuringiensis* that kills corn borer larvae and other insect pests, which destructively feed on the plant.

Some research suggested a possible toxic effect of Bt crops on monarch butterflies, even though they were not a target insect and do not feed on corn. It would thus be important to know if the toxin affects specific species or groups of insects, and whether nontarget insects can also be affected. One question that was considered was whether the toxin could be spread or was confined solely to the corn plant. Because corn is wind pollinated, the pollen might be carried to other plants and be toxic to some insects at a distance. Researchers needed to determine the likelihood of this happening. For the monarch butterflies, the study indicated that corn pollen could be spread by wind to milkweed plants (which are a food source for monarchs) located next to the GM cornfield. Monarchs feeding on the milkweed could then ingest the corn pollen (and the toxin). Scientists had to ask whether this was a likely occurrence and how much pollen and toxin it would take to kill a monarch butterfly. Think about the types of experiments you might design to test these questions. In recent years, several long-term studies have shown no adverse effects of monarch exposure to Bt crops.

There might be possible effects of cross-pollination between Bt corn and transfer of engineered genes to other noncrop species (for example, other plants that might acquire the toxin gene and kill desirable insects, such as monarch butterflies). You would need to consider the likelihood of such an occurrence. The whole question of introducing GM plants into the natural environment needs careful scrutiny to consider possible long-term changes to the ecosystem and the effects on biodiversity. Another consideration is the spread of GM plants to other areas beyond where they are cultivated. Develop your own plans for controlling the spread of GM plants through the environment.

The idea of the likelihood of an event, the **statistical probability,** is another important concept to keep in mind. As ethical decisions are considered, it is crucial to determine accurately what chance exists

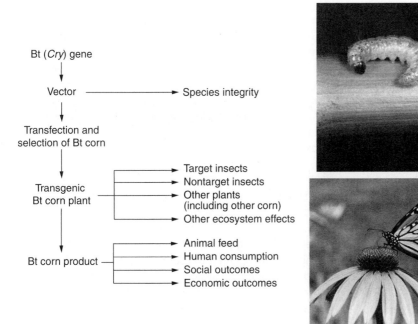

Figure 13.2 Possible Interactions and Considerations for a GM Crop, Bt Corn Bt-modified crops were designed to protect plants against pests such as the cotton bollworm (top photo) and corn borer, but soon after their use concerns were raised about possible effects on nontarget insects such as monarch butterflies (bottom photo).

for a "bad" event to happen. Another consideration might be just how negative the effect of the possible event would be. Because many consequences take time to develop, the probability of an outcome occurring must be part of the consideration.

Bioethicists and scientists often use statistical measures to determine **risk assessments.** Risk assessment considers the likelihood that something harmful or unintended will happen. Whether you realize it or not, risk assessment is part of your decision-making process every day. For instance, you know that each time you drive your car there is a risk that you might be in an accident. Assuming you are not afraid of driving, your risk assessment probably tells you that the likelihood of an accident does not outweigh the need to get where you want to go. Cell phone use, air travel, drinking alcohol, or making poor dietary choices are other good examples of risk assessment areas in your regular life.

Another question to consider regards the product of the GM crop: We need to consider how it will be used, whether it is safe to feed to animals, and whether it is safe for humans. We also need to gather all the facts and make informed evaluations. Because the toxin is directed against insects, it seems unlikely it would affect animals or humans. But making this assumption is not enough; there needs to be evidence and tests that could

verify its safety. Think of some experiments you could do to test the safety of a GM product.

We should not just focus on the particular gene in question but also ask whether other genes or products present in the GM crop might need to be considered. For example, in many cases, antibiotic-resistance genes are used as selection markers for genetically engineered cells. Are the genes still present in the GM crop and, if so, can these genes be transferred from ingested food to gut bacteria? You should again consider what the likelihood is that ingested DNA or proteins would survive digestion. Figure 13.2 shows a chart of some of the possible interactions and considerations.

We also need to determine whether the product from a GM crop should be quarantined after the crop has been cultivated. This again comes back to our question of safety, especially regarding human exposure to or consumption of the product. It is virtually impossible to differentiate nonengineered corn from Bt corn by the naked eye or even by microscopy. Several companies have now marketed test kits using antibody or DNA tests to detect GM crops. For Bt corn, some of the crop inevitably ended up in human products, including taco shells. The primary concern was the possibility of allergic reactions to the modified

YOU DECIDE

Is the Rejection of GM Foods by Food Companies a Liability?

In the recent past it has been a policy by food companies to reject GM food ingredients in their foods—and in some cases to advertise this rejection (e.g., "All Natural"). Is this the best policy for the consuming public?

Every year numerous food products are recalled from the U.S. food market because of the presence of insect parts, toxic molds, bacteria, and viruses. These contaminants are "naturally" present in the soil where farmers grow food. Toxins like fumonisin and some other mycotoxins that grow on grains are highly toxic, causing fatal diseases in livestock that eat them, and esophageal cancer in humans. The FDA is acutely aware of the dangers of mycotoxins and has established maximum fumonisin levels in food and feed products such as corn. These recommended maximums have not been entirely successful.

In 2003, the UK Food Safety Agency (FDA for the UK) tested 6 organic cornmeal products and 20 conventional cornmeal products for fumonisin contamination. All six organic cornmeals had elevated levels, from 9 to 40 times

the allowable level and were voluntarily withdrawn. In contrast, when corn is genetically modified by inserting the Bt gene, it expresses a protein that is not only toxic to corn-boring insects but reduces infection by the mold *Fusarium* and the levels of fumonisin, a mycotoxin. The expressed protein is toxic to corn-boring insects but is harmless to birds, fish, and mammals, including humans. When we consider the potential for harm to infants and adults from fumonisin in foods and allergenic products (that can be removed by gene splicing), the liability to food companies for not using GM products mounts up.

If food companies and food processors manage to avoid legal liability for their insistence on using crops that could be safer due to genetic modification, are they still "morally guilty" for the damage they knowingly are creating by using non-GM food products? Should food companies be required to use the safest possible food production methods, whether by genetic modification or other means? How does the consuming public advocate for safe food? You decide.

corn. Can you think of a procedure for testing the likelihood of such a reaction?

One consideration to keep in mind is whether the toxin would survive food preparation and still stimulate an allergic reaction. Some countries strictly limit the importation of GM crops, and there is a movement to label foods to designate their origin in terms of potential genetically modified plants. Some people believe that, based on the data, these restrictions are not valid and may unduly frighten consumers. What would be your reaction to finding a "GM" label on your pizza or cornflakes?

Besides environmental and health concerns, social and economic questions also arise from the potential use of GM crops. The ability to modify plants for better, less costly production could drastically change the agricultural industry. Potentially, more abundant food could be available at a reduced cost both to the farmer and to the consumer. These advantages may be offset by potential disadvantages, however, such as the safety concerns described previously. Other possible uses for GM crops include the production of medically useful compounds. Safety and efficacy again become key considerations in the ethical assessment for use of these compounds. Current U.S. policy requires not only the usual range of tests

YOU DECIDE

How Much Return on the Investment?

Monsanto, a company based in St. Louis, Missouri, created Roundup-Ready soybeans that are genetically engineered to withstand Monsanto's Roundup weed killer, a commonly used herbicide that kills almost all plants. Roundup-Ready seed costs several dollars per bushel more than conventional soybean seeds. Recently, Monsanto used private investigators to check out reports that farmers in New Jersey had been recycling seed harvested from Roundup-Ready soybeans planted the previous year. Using seed harvested from a previous crop is a common practice for many farmers and allows them to save money on seed for planting a new crop. Because Monsanto invested so much money in their technology, they wanted farmers to buy new Roundup-Ready seeds each year, but farmers claim they were never told their seed could not be replanted. Monsanto claims farmers are using their high-tech expensive product for free.

Should the farmers be allowed to continue their traditional practice of recycling seed? Or should biotechnology companies be able to enforce restricted uses of their products? You decide.

for safety and efficacy of medical compounds but also that growth of the GM plants be restricted. Fields must be surrounded only by other plants that should not cross-pollinate with the GM plant, all plant material must be removed at harvest, and the field cannot be used for a number of years after harvest.

Animal Husbandry or Animal Tinkering?

Genetic modification of animals raises many of the same questions posed by the genetic modification of plants. Early applications of biotechnology to animal husbandry have included antibiotic supplements in feeds and injections of growth hormone or steroids to increase growth of the animals. Scientists need to consider the application of these supplements and injections from an ethical standpoint. First, because these are agricultural animals, the effects of genetic modification on the products from the animals (milk, meat, and so on) and the safety of these products for human consumption were the main concern. One consideration was the length of time the supplements (especially hormones) would persist within the animal, that is, whether they would still be present when consumed by humans. If so, scientists also need to determine whether there would be any effects on the consumer and whether the hormone, if present, would survive any cooking or the digestive process.

Little concern has been expressed about the effect of GM agricultural animals on the environment, but there are still questions about species integrity and the health of the animal, as well as the safety of animal products for human consumers. Similar considerations need to be made regarding animals genetically engineered to produce medically useful products. These could take the form of transgenic animals modified to produce a clotting factor in their milk, transgenic animals such as pigs engineered with human genes so their organs could be transplanted into humans without being rejected, or cloned cows to be used for meat. At what point would you consider alteration of an animal to be unethical?

Seek to define for yourself whether there is an ethical boundary for manipulation of a species or for cross-species manipulations. Be sure to consider whether there is a point at which the animal might acquire enough human genes, cells, or attributes that you would consider it human and whether this would change your ethical viewpoint on using the animal for research. Refer back to Figure 13.1. Recently, English and French scientists implanted mouse cells into chicken embryos to demonstrate that mouse cells could recognize developmental signals from chicken cells to stimulate tooth formation. Developmental biologists are interested in this and other subsequent studies in part because although the genes involved in these developmental signals are not active in birds (teeth were lost in birds about 70 to 80 million years ago), it indicates these genes can be activated and can stimulate tooth development when the proper cells are present. You may ask, "So what? Who wants to make chickens with teeth?" These are good questions, and you may certainly ask if these types of experiments are ethical uses of animals. But also consider that information from these experiments can be fundamentally valuable to understand tooth development and potentially in the future to develop novel treatments for tooth formation, replacement, and regeneration in humans—the main reasons for this research.

Genetically modified wild species of animals present another set of ethical questions including environmental concerns. Additionally, the same ethical concerns mentioned previously for transgenic plants apply to transgenic animals. In an attempt to control pests or otherwise balance an ecosystem, some people have proposed that GM animals be released into the wild. List some potential ethical problems associated with these proposals, as well as ways to control or remove the ethical concerns.

The Human Question

Many of the thorniest questions regarding biotechnology research and its application, and some of the most contentious debates, revolve around humans. Even some of the simplest scientific procedures can evoke strong emotions and stir profound controversy when humans are the subjects. Why do you think the use or potential use of humans experimentally causes such strong reactions? Later we discuss how simply defining a human has become an area for debate. For now, we look at a simple example based on a potential anticancer drug generated through biotechnology. Suppose the drug has moved along to the point where it is ready for clinical trials. We must decide to whom we will administer the drug as a test. Because it is an anticancer drug, we will naturally give it to patients who have the type of cancer targeted by this drug. We must decide, however, whether to give it to all patients with that type of cancer or only patients in the most advanced stages of the cancer, those who may have exhausted all other means of treatment and for whom the new experimental drug is a last resort. The rationale is that, for this subset of patients there is no other possible treatment, and so the experimental treatment presents at least some hope. However, the drug may work better with patients who are earlier in the progression of the cancer, and thus might be more effective. Nonetheless, the patients who are in earlier stages of the disease have other alternatives that have already shown safety and efficacy (at least to some extent).

After you have picked the patient group for the clinical trial, another dilemma arises. Patients have a right to be informed fully of the potential effects of the experimental treatment, both good and bad. Only when so informed can they be willing participants in the trial. This is termed **informed consent.** Patients give their consent to proceed with the experiment, fully informed of the potential benefits and potential risks that lie ahead. The concept of informed consent is vital to any procedure involving a human. If a patient is unable to give consent on her or his own (because the individual is too young or in a coma, for example), a family member or guardian can give proxy consent. Determine for yourself why informed consent is so vital.

Placebos present one more problem as far as experimental procedures on humans are concerned. Standard scientific practice involves using an experimental group (in this case, patients who would receive the drug) and a control group (in this case, patients who would receive a placebo, a safe but noneffective

YOU DECIDE

Should Clinical Trial Data be Public, or Is It the Property of the Drug Company?

A bill that may pass the U.S. Senate calls for mandatory disclosure of the results of clinical trials in a public database. Some claim the information may help patients determine whether to participate, but some in the drug industry say it is unnecessary and will stifle innovation. In 2002, the National Institutes of Health launched Clinicaltrials.gov to help patients and physicians find information on nearby clinical trials. The current registry contains over 31,700 clinical trials in more than 139 countries. This registry has been voluntarily utilized by most biotech and drug companies because it excludes early-stage trials, where sensitive business information could be lost. The new bill would require information from Phase 1 trials, which enroll healthy patients in small numbers and according to some industry spokespersons would be of little value to patients. The advocates for the bill claim that companies could benefit from sharing early-stage trials to prevent new drugs from repeating the same mistakes that produced negative side effects. In defense of the industry, biotech companies rely on venture capital funding, and if early data are shared with other companies it is claimed that funding for innovation would dry up. Is the benefit worth the risk that it may reduce funding for drug research? Is there a better solution? You decide.

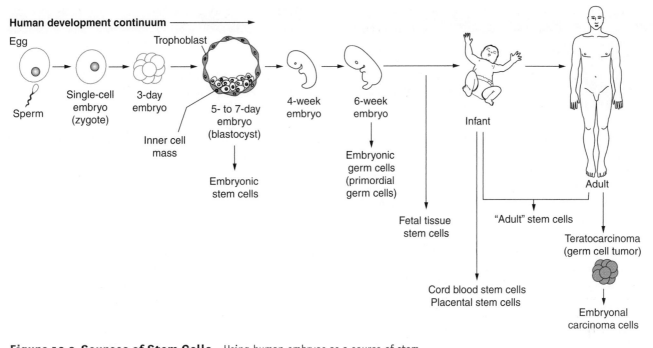

Figure 13.3 Sources of Stem Cells Using human embryos as a source of stem cells is a very controversial topic, in part because it raises the question, "When is the embryo considered to be a person?"

treatment such as a sugar pill or saline injection). In a completely randomized **double-blind trial,** neither the patients nor the doctors administering the treatment would know who received the real drug and who received the placebo. Informed consent should play a role in this type of experiment. Using placebos is part of objective science, but we need to look at whether it is ethical. Objective science may not always be the best approach or the ethical approach.

What Does It Mean to Be Human?

Many of the current ethical debates around biotechnology—in particular the debates about stem cells and cloning—revolve around the moral status of the human embryo. As we discussed in Chapter 11, stem cells may hold the potential for tremendous breakthroughs in **regenerative medicine,** allowing repair or replacement of damaged and diseased tissue in many diseases

such as heart disease, stroke, Parkinson disease, and diabetes. There are many possible sources for stem cells (Figure 13.3), including embryos, fetuses, umbilical cord blood, adult tissues, and even "tamed" tumor cells. Much of the debate has centered on the scientific question of the abilities of these different stem cells to transform into other tissue types for treatment of diseases; however, a key element in the debate has been the ethical question. As a society, we have not been able to decide whether it is ethical to destroy early-stage human embryos for research that may potentially treat thousands of patients. Unlike organ donation, where an individual may donate one of two paired organs while alive or the donation occurs after the donor has died, the process of harvesting the embryonic stem cells destroys the donor (the embryo; Figure 13.4).

We must resolve the focal question, which examines the moral or ethical status of a human embryo. Biologically the embryo is a human being, species

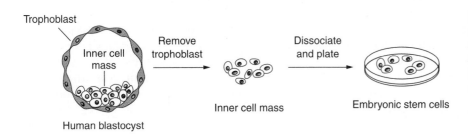

Human blastocyst

Figure 13.4 Isolation of Embryonic Stem Cells from the Early Embryo Also Destroys the Embryo Is it ethical to destroy a human embryo to harvest stem cells that may benefit many other humans?

YOU DECIDE

Celebrity, Shame, and Stem Cells

Dr. Woo Suk Hwang of Seoul National University in South Korea was thought to be a pioneer in stem cell research. Dr. Hwang was treated to free airfare to travel around the world, high fees for speaking engagements, high-society parties, regular appearances on television, and other perks not typically associated with life as a scientist. Hwang became a source of national pride in Korea primarily for claims that his research group had created the first cloned human embryo and stem cells derived from human patients. This work was published in 2004 and 2005 in highly heralded papers in the prestigious journal *Science*. But by December 2005, Woo Suk Hwang's celebrity and his career were quickly falling apart in disgrace, and he resigned from his position. Current and former members of his own research team as well as a collaborating lab revealed evidence suggesting falsification of data and allegations of unethical approaches for obtaining donated eggs. It was subsequently determined that data and figures had indeed been falsified and manipulated, and the lab did not clone an embryo or derived patient-specific stem cells as claimed. Among these unethical actions was concern about how eggs had been collected for the work (initially Hwang reported that anonymous donors provided the egg) and lies about the number of eggs used. It turned out that junior researchers in Hwang's lab were among the donors and donors were paid for their eggs. Women were also paid to take fertility drugs to provide eggs. Hwang's fall from grace was highly publicized around the world, and it served to further support the claims of groups speaking out against stem cell work that were already skeptical about the potential of stem cells lauded by researchers.

Bioethicists claimed that Hwang failed to adhere to Western society ethical standards when he paid women for their eggs and used junior researchers who worked for him to provide eggs. But the South Korean Health Ministry deemed there was no ethical problem because the eggs were donated voluntarily, with no evidence that junior researchers were coerced to give eggs. They determined that Hwang did not violate existing laws. The ministry further explained its conclusions by saying that Western-trained physicians and scientists considered donations from vulnerable junior researchers as unethical because they might feel pressure to satisfy their supervisors and donate involuntarily. The ministry further offered that although researchers in South Korea abide by ethical procedures when conducting research, they generally don't focus on ethical issues as much as Westerners. Was is ethical for Dr. Hwang to collect eggs from junior researchers? You decide.

Homo sapiens, just starting out on the developmental journey. We need to weigh the relevance of different developmental milestones for being considered human against the simple biological fact that the embryo is a member of the species. Biologically, an embryo is a member of the human species, but the question of the moral status of a human embryo goes beyond the biological and also revolves around what some have termed the status of **personhood.** This term has been used to define an entity that qualifies for protection based not on an intrinsic value but rather on certain attributes, such as self-awareness. List for yourself the advantages and disadvantages of this concept of personhood. One concern with this concept might be the question of who decides which attributes count in evaluating whether a particular human can be valued as a person. A system such as this would mean such attributes not only can be gained (as we develop physically or mentally) but also can be lost (through aging, disease, or injury).

Regarding embryo research, some have taken this attitude: "Not a person, not a problem." They reason that a human embryo is microscopic, not yet possessing a beating heart, brainwaves, arms, or legs. Of course those things develop later, but at a very early stage, the embryo lacks what we usually associate with our concept of humanity. There are obviously various views of the status of the human embryo, and no consensus. Some say it is simply a clump of cells, just like a chunk of skin. Others believe it is a form of human life deserving of profound respect—a potential person. Still others maintain that an embryo has the same moral value as any other member of the human species. Consider the question of whether *any* human cell deserves respect as a potential person. When combined, an egg cell and a sperm cell form an embryo, and any somatic cell can contribute a nucleus to form a cloned embryo. Consider then whether there is something special about having a complete human genome in an egg cell.

When considered in the context of using human embryos for their embryonic stem cells, the debate regarding the necessary destruction of the embryo is contentious. Some evidence points to the potential application of embryo research and embryonic stem

Q Regardless of whether you consider an embryo a human life or not, can you think of humans who might not meet a threshold that requires the attribute of self-awareness?

A People who are temporarily or permanently disabled may not meet this criterion. This group could include those in a coma, those with advanced Alzheimer's disease or other neurological diseases, and those who have suffered a traumatic brain injury. Depending on the definition of self-awareness, infants might not meet this threshold, nor would some people who are mentally impaired. Actually, no one meets the criterion during sleep!

YOU DECIDE

The Same or Different?

Consider the following scenario. An embryo is created (pick your favorite technique) for a couple who wants a child. However, it's known that the couple carries a genetic disease, with a 100% certainty that the embryo will have the disease. Scientists working with the couple isolate the embryonic stem cells from the embryo. In culture, they are able to repair the genetic problem in the cells. Some of the embryonic stem cells are then packaged using tetraploid embryo cells, creating an embryo that is implanted into the woman and carried to term. The infant that is born does not have the genetic disease, and neither will any of the child's offspring. In addition to the several ethical questions that can be discussed regarding this scenario, consider this one: Is the born child the same human/individual/person as the original embryo? You decide.

cells in the treatment of numerous diseases. Certainly, embryonic stem cells can theoretically be used to form any tissue, with the potential for transplants to repair or replace damaged or diseased tissue. This might suggest it would be ethical to destroy human embryos for research, if from that destruction it might be possible to perform research that potentially could lead to treatments for patients suffering from disease. However, based on current published evidence, one question that first must be asked is whether embryonic stem cells are as good for potential treatments as claimed. Another consideration that has both ethical and scientific components is whether other viable alternatives, such as adult stem cells, are just as good. Mounting evidence indicates that adult stem cells may be just as effective for treating diseases as embryonic stem cells. The ethical questions then take on a new form asking whether embryo research is necessary so science can explore all possible avenues for medical breakthroughs or whether the calculation now indicates the alternative makes ethically contentious research unnecessary. Of course, one ethical position would be that even if there are no alternatives, the ethical cost is too high to justify destruction of embryos.

Interestingly, even some people who view a human embryo as a potential person and not a realized individual oppose destruction of human embryos on ethical grounds. The concern is not directly with the embryo and its status but rather with how society views any human life. From their point of view, we are embarking on a slippery slope where the destruction of human beings for medical use or experimentation might move from the use of embryos to the use of born individuals. Once again, we return to the question of personhood, especially as a societal construct that might rank different humans based on their quality of life and on their usefulness to society.

Spare Embryos for Research Versus Creating Embryos for Research

As we consider the question regarding the moral status of the embryo, there are varying views based on the original purpose for which the embryo was created or on the method by which the embryo was created. Contrary to a common misconception that aborted fetuses are used for stem cell research, as we discussed in Chapter 11, the primary source of embryos for this research is "excess" embryos from *in vitro* fertilization (IVF). These embryos, left in frozen storage after a couple has used other embryos for implantation and ideally pregnancy and birth, may be donated (with the couple's consent) for research. Depending on the IVF clinic or the country, frozen embryos may be discarded after a certain period of time. List for yourself the necessary ethical considerations in this instance. Some say it is ethically valid to use these embryos for research if the alternative is they will be discarded, that some ethical good can be salvaged from their existence if they can contribute to a potential therapy or an increase in scientific knowledge. Others say their destruction for research crosses an ethical line inconsistent with the purpose for which they were originally created.

Another potential source of human embryos is the specific creation of embryos for research purposes. Some have argued that the use of spare embryos for research is ethically valid (reasoning this could be salvaging a potential good out of certain destruction) but that specific creation of embryos for research is not.

To determine whether this argument is ethically consistent, consider whether there is any difference in the embryos biologically, or only in a view of the intent or use of the embryos.

Cloning

Creation of embryos by cloning (somatic cell nuclear transfer) raises many of the same questions encountered with stem cell research, with the added complexity of the technique (Figure 13.5) and the potential "identity" of the clone. A cloned embryo created for transfer into a uterus and implantation faces many risks and safety factors for its own development and growth, both before and after birth. The success rate for live births and the subsequent survival of the clones is extremely low, and we know there can be a number of problems with the health of clones.

One consideration is whether creating a cloned human embryo with the intent of initiating a pregnancy and live birth should be considered another type of assisted reproductive technology similar to IVF or whether (because of the safety risks and low success rates) it should be considered unethical human research. Societal questions regarding the identity of a born human clone also must be taken into consideration. For example, if a couple decides to create a cloned child, using a donor cell from the wife, the clone will not be a genetic daughter but instead will be the wife's sister, a late-born twin, and will not be related to the husband at all. Ethical considerations related to the clone include how the lack of relatedness to one parent

might change kinship and family relationships. Given that the clone is a "copy" of one of the "parents," there may be expectations put on a clone once born to "live a better life" than the person who was cloned. Another potential concern arises if a previously existing person, now deceased, is cloned. The genetic makeup of a clone will already be known, already dictated because the process of cloning reproduces a previously existing individual. A clone may be expected to live up to that genetic legacy, with heightened expectations by the parents and others based on what was achieved by the donor of the genetic material used to create the clone.

The cloned cat "cc" (Figure 13.6) looks very similar to the cat that donated her genetic material, but her coat pattern is slightly different from the donor used to create her. These differences occur because of subtle genetic changes during development. Our genes determine many of our physical characteristics and predispose us to various diseases or behaviors, but of course after we are born there are many experiences and environments that make us who we are and who we will become. Those experiences cannot be duplicated, so the clone will grow up differently than the one who was cloned and may behave quite differently. A clone of Einstein might become an artist instead of a scientist. A great deal more affects us and our makeup than just our genetics, including our environment and experiences.

So far we have considered the creation of a cloned human embryo with the intent of producing a live-born child. However, this is not the only proposal for creation of cloned human embryos. Creating human embryos by therapeutic cloning could lead to matched embryonic

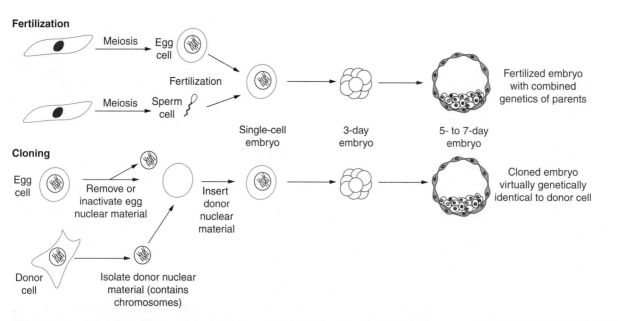

Figure 13.5 Creating Embryos by Fertilization Versus Cloning (Somatic Cell Nuclear Transfer)

Figure 13.6 The First Cloned Cat, "cc" (the Carbon Copy Cat)

cells for patients (Figure 13.7) and to valuable human research models for the study of genetic diseases and cancer. Although this might sound like a potentially valuable and ethically valid reason for creating cloned human embryos, others argue that such embryos should not be produced.

One argument against the creation of cloned human embryos is not based on the embryo's inherent value as a human or person, but on the argument that the creation of human embryos for such purposes could lead to

human commercialization, making any human life a commodity to be bought, sold, and used, cheapening life in the process. Still others have argued we should not be creating human embryos in a manner that manufactures human life, the so-called designer embryos.

One argument in favor of creating cloned human embryos relies on the assumption that if an embryo is not created by normal means (in this case they mean by fertilization), it is not human. Each of these arguments is based on different definitions of what it means to be human and what value is placed on human life. Interestingly, nearly 30 years ago, similar ethical debates about what it means to be human were debated when Louise Brown, the so-called test tube baby, was conceived by *in vitro* fertilization. And now even the most conservative groups, including many religious sects, consider *in vitro* fertilization to be a completely ethical means to conceive an embryo.

Patient Rights and Biological Materials

Consider how you would feel if the following scenario happened to you. You are being treated for a disease such as leukemia and you donate blood, bone marrow, and spleen cell samples for analysis as part of your treatment. You later learn that the physicians treating you developed cell lines from your tissues that they patented and then used to receive substantial financial compensation. This exact scenario has occurred, resulting in a number of patient lawsuits. In these cases, patients have claimed that prior to providing their informed consent to extract the cells, they were unaware physicians would conduct research on their cells and that the physicians would benefit financially. Patients and lawyers have claimed these

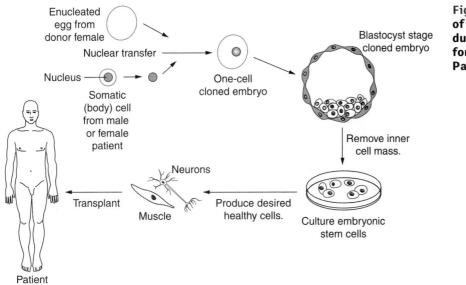

Figure 13.7 Theoretical Scheme of Therapeutic Cloning to Produce Matched Healthy Tissues for Transplantation into a Patient

examples are a breach of the doctor-patient relationship and an example of a "Principle of Unjust Enrichment"—that physicians inappropriately benefited financially from their patients' tissues. In most of these cases, patients have sought a share of financial compensation for their cells. In several high-profile cases, courts have ruled that physicians do have a duty to disclose the physician's personal interest in research and potential economic matters unrelated to patient treatment. But courts have ruled that donors of cells and other biological materials do not have ownership rights of their biological materials, and physicians and scientists who develop patents and receive financial benefits from these materials are not liable for patient compensation.

Many donors are unaware they do not own their own cells and they relinquish a large degree of control over their tissues when donated. Related to this, there have even been examples in which donated sperm samples were used to inseminate eggs to produce children without the consent of the individual who donated the sperm, thus making men fathers without their complicity. As stem cells and regenerative medicine technologies become more common, undoubtedly patients' rights regarding biological materials will become even more complex and require greater scrutiny and ethical consideration.

Regulations in Flux

Current regulations in the United States regarding cloning and stem cell research are mixed and in a state of debate. Since 1996, the U.S. Congress has prohibited the use of federal funds for the creation of human embryos by fertilization, cloning, parthenogenesis, or any other means for use in research, or the use of federal funds for research that involves the destruction of human embryos. However, note that these federal rules only apply to use of government money: At present no federal laws prohibit embryo destruction or human cloning research funded by the private sector.

President Bush's address to the nation on August 9, 2001, regarding human embryonic stem cell research only covered federal funding. His decision was that federal funds could be used for research on human embryonic stem cell lines that already existed, but they could not be used for embryo creation or destruction (in line with the current regulations by Congress). Numerous bills have been introduced in Congress to enact laws relating to stem cell research and human cloning; some would allow more embryo research and therapeutic cloning, whereas others would be more restrictive of human embryo research and prohibit all human cloning. A total ban on human cloning has passed the U.S. House twice, first in the summer of 2001 and again in the spring of 2003. However, the U.S. Senate has yet to debate the issue.

In the absence of federal laws related to embryo research and human cloning, many states have enacted their own laws. Some states have enacted laws that prohibit embryo destruction for research and/or human cloning (for example, Michigan, Iowa, Virginia, Louisiana, Pennsylvania, and Arkansas); others (such as California) have passed laws encouraging human embryonic stem cell research and human therapeutic cloning. Virtually all states have been debating the issues in their legislatures. However, at this point, there is only a patchwork of laws and regulations related to human embryo research and cloning in the United States.

The international situation is only slightly more coherent. A few countries (the United Kingdom, China, and Singapore) have enacted liberal laws allowing human embryo research and human therapeutic cloning, but most countries and international bodies have been more restrictive, at least regarding human reproductive cloning. Several countries—including Australia, Canada, France, and Germany—have enacted or will likely soon pass complete bans on human cloning for any purpose. The first three countries listed have indicated they will also pass laws only allowing restricted use of some frozen embryos for research. International bodies such as the European Union and the United Nations are also moving toward complete bans on human cloning. Debates regarding such research (and the ethical status of such research) have taken on global proportions. A recent analysis by the Ethics Working Party of a group called The International Stem Cell Forum revealed that of 50 countries studied, all either expressly prohibited human cloning or strongly discouraged it.

Your Genes, Your Self

The Human Genome Project has led to the identification of genes responsible for, or contributing to, many disease states. This knowledge has also led to strategies for treating or preventing diseases. But because the story told in our DNA could become so easily read, there is a growing concern over the privacy of that information. Our DNA sequence can be a truly unique identifier. Researchers especially must take care to guard the confidentiality of those who have donated DNA for sequencing and testing. Because a free flow of scientific information ensures rapid dissemination of ideas and speeds advances, we look next at some of the ways scientists can safeguard genetic information to assure individual research subjects or groups of subjects that their genetic privacy will be maintained. This is important not only to reassure research subjects but also for scientists to maintain the continued trust of people and their willingness to participate in research projects.

As identification of genetic traits becomes more routine in clinical settings, physicians will need to ensure genetic privacy for their patients. There are significant concerns as to how genetic information could be used negatively by employers, insurance companies, governmental agencies, or through perceptions acquired by the general public. Genetic privacy and prevention of genetic discrimination will be increasingly important in the coming years. Currently, no federal laws regarding genetic privacy and genetic discrimination exist, so individuals must rely on the good ethical practices of the people with whom they are dealing. In 2008, the U.S. House of Representatives passed bill H.R. 483, the **Genetic Information Nondiscrimination Act.** The Senate also passed an identical bill (S. 358) and shortly thereafter President Bush signed the GINA into law. This act prohibits discrimination based on genetics and the improper use of genetic information in health insurance and employment.

More or Less Human?

Genetic technologies are advancing rapidly, and after many years of research there have now been the first few successful instances of human gene therapy. As we discussed in Chapter 11, these first successes were with children suffering from severe combined immunodeficiency syndrome (SCID). Individuals with SCID are born without an effective immune system, primarily because of one defective gene in their immune cells. Successful treatments have used adult stem cells from bone marrow of the children, using genetic engineering to add the correct gene and replacing the engineered stem cells back into the patients (refer to Figure 11.15). So far the indications are that the treated children have all been cured of their immune deficiency and are able to leave their protective isolation bubbles and live essentially normal lives.

Think about some of the ethical considerations associated with gene therapy technology. Gene therapy treatments such as those used to correct SCID seem wildly successful, so one might think there should be no concerns. However, because these are experimental medical procedures, there are certainly issues of informed consent, safety, and efficacy. One safety issue that has arisen is the potential for cancer formation. As discussed in Chapter 11, an adenovirus vector used for gene therapy has led to leukemia in several children. This consequence has led to considerable concern about future genetic therapies, especially using this particular viral vector. In addition, some countries have halted all genetic therapy attempts on humans until further experiments can validate the safety of the treatment. So one ethical consideration might be the risks associated with altering the genetics of an individual, especially the specificity of the targeting for insertion of the replacement gene.

Current and proposed somatic gene therapies involve the treatment of existing genetic diseases. However, we should consider the possibility of genetic treatments for conditions where there is only a genetic *tendency* toward a particular disease, and there is no certainty the disease will occur. One possibility is breast cancer, where genes (such as mutations in *BRCA1* and *BRCA2*) have been identified that can increase the risk for development of the cancer. We need to consider the difference between a genetic disease and a genetic attribute. For a genetic disease (such as SCID), an existing genetic problem definitely causes the disease seen in the individual. However, a genetic attribute does not mean the disease condition exists and does not necessarily always lead to the disease. If we consider attributes that may not necessarily lead to a disease or may not even be associated with increased risk for disease, we extend our considerations of genetic engineering beyond the therapeutic and begin to contemplate whether we will consider other potential genetic modifications as ethically acceptable, going beyond treatment for an existing condition to alterations and even enhancements of the human genetic composition.

You may want to consider whether genetic treatment for an existing disease is equivalent to genetic engineering to prevent a disease. This may seem to be a strange question because a disease is readily identifiable compared with a normal healthy condition. But the real question relates to what might be considered normal versus abnormal. There may be a fine line for ethical distinction between treating a known genetic problem and preventing even the possibility of a potential problem.

To extend this prospect a little further, consider whether enhancing our individual genetics, perhaps to increase muscle mass or oxygen-carrying capacity of red blood cells, might also be considered medically necessary for the health of the individual, even if it were really just a personal preference. Prohibiting such genetic alterations could be considered an infringement of individual rights. Because this individual's genome would be altered to something different than the normal, we need to reconsider the question of whether the individual would still be considered human and just what we might include in that definition.

The question of what makes an organism human becomes more pressing when we look at the potential of **germline genetic engineering.** In this case, the genetic alteration is done in the sperm, egg, or early embryo. This can actually make the manipulation

YOU DECIDE

What to Treat Using Gene Therapy

In Chapter 11, we discussed some of the unresolved scientific questions surrounding gene therapy. There are also many ethical concerns about gene therapy and the risks associated with gene therapy. Do the patient and his or her family members understand the risks associated with gene therapy trials? For instance, Jesse Gelsinger (recall from Chapter 11 that Jesse died as a result of his gene therapy) had a relatively mild form of a disease that was being successfully treated with medication. Should he have been a gene therapy trial participant? Gene therapy is currently very expensive. Should everyone have access to gene therapy treatments regardless of the cost? What types of disease and conditions should be subject to gene therapy (for example, only deadly diseases)? What about treatable diseases for which gene therapy would be used so patients would not need to take daily, but effective, medications? What about cosmetic conditions such as baldness? You decide.

more effective than somatic genetic engineering because the genetic modification can affect every cell of the individual. As a result, the genetic alteration will be inheritable. Potentially this could mean the elimination of some genetic diseases because those genes could be removed from the human gene pool. Of course, any problems resulting from the genetic alteration could also be inheritable, as could any genetic enhancements. Consider whether this type of genetic modification, which affects not only the treated individual but also future generations, might be ethically acceptable. Some of the potential outcomes for removing or adding genes in the human gene pool could include the elimination of genetic diseases or disease susceptibility, enhanced human performance, increased or decreased human lifespan, or increased cancer risks. One Nobel laureate has been an advocate for this type of research because it could lead to development of a "better human being," but we might pause to consider just what constitutes "better" and whose definition should be utilized in these decisions. Another potential outcome sometimes mentioned for germline genetic alteration could be the creation of a new superior species of human—the basic concept behind **eugenics,** which Adolf Hitler attempted to achieve not through genetic engineering

but by attempting to eliminate what he perceived to be inferior individuals. One potential negative associated with this outcome could be splitting humans into different genetic social classes.

13.3 Economics, The Role of Science, and Communication

We cannot escape the fact that money can play a major role in research decisions. Private investments as well as government funds fuel research, and individuals and companies alike seek to use biotechnology for discoveries that will be profitable. Not only is scientific research expensive, but projects, companies, and careers may live or die depending on the funding for and profitability of the research. There is clearly not enough money for every scientific proposal, and each is evaluated based on whether it will increase our knowledge about fundamental aspects of science or produce a discovery or product that improves our lives. Panels of scientists reviewing the proposed research usually make these decisions, but the amount of money and even some of the decisions on funding direction also come from those providing the money, especially the U.S. Congress. Most of the decisions are based on the potential success of the science, its novelty, and its potential application to health and general knowledge. However, some of the decisions may also be affected by how well the scientists are known by the reviewers or by political pressure from members of Congress.

We should consider not only how research proposals are evaluated and their likelihood for success, but also the possibility of their contributing to the breadth of scientific knowledge (because it is difficult with science to know from where the next major breakthroughs might come and what background knowledge might lead to such breakthroughs). Additional considerations are the costs and access to any treatments that may be produced. Funding for research that leads to treatments or products that will be so expensive only the rich can afford them may be a colossal waste of money and time, and might suggest other avenues be explored instead. Because the possible sources of funding for research are limited, funding decisions may need to be weighed regarding whether funding one type of research might decrease the funding of another, potentially more successful line of research.

Intellectual property rights are also important as we consider the ethical implications of research and its funding. **Patenting** of intellectual property (isolated genes, new gene constructs, new cell types, GMOs, or even embryos) can be potentially lucrative for the discoverers but may also pose ethical and scientific problems. For example, consider the possibilities for a human gene that has been isolated and characterized—and then patented by the discoverer. The person or company holding the patent could require that anyone attempting to do research with the patented gene pay a licensing fee for its use. Should a diagnostic test or therapy result from the research, more fees and royalties may be required. This could potentially make it difficult or expensive to carry out research on some genes or limit the clinical use of the patented gene.

Some physicians have already complained they cannot afford to pass on the charges of certain genetic tests to their patients, owing to licensing restrictions. However, limiting or preventing patents for genes or genetic tools could remove the incentive for pursuing such research, especially for companies that hope to profit from their research. Compile a list of potential ethical problems associated with patenting of genes or cells, and consider possible alternatives or compromises that might be reached to allow research to continue.

Ethical questions regarding biotechnology also touch on science's role in society (Figure 13.8). Science and technology have provided discoveries and inventions that make our lives happier and healthier. But it is important to consider whether scientists should have unlimited freedom for research. It is often difficult to

know what the source of the next major breakthrough will be, and sometimes, new discoveries arise from unexpected sources and paths that many scientists may not even consider worth following. We need to determine how science can best serve society. The whole concept of regulating something as unpredictable and free flowing as scientific discovery is difficult. It is also hard to know who should decide these questions—scientists because they know how science works, policy makers because they must set the rules, or society because it is most affected by the decisions. Consider what types of regulations might best serve the needs both of science in furthering discovery and of society in furthering health care and ethical interests. Bioethics becomes critically important to the biotechnology industry because ethical discussions and debate are often the driving force for making laws in general (laws that govern the population's behavior and certainly laws that will regulate the biotechnology industry) once it becomes a general consensus of a particular population that a law is necessary.

Accurate, honest communication is vital to the success of science in general and biotechnology in particular. Scientists must be willing and able to communicate openly and candidly with other scientists but more importantly with the general community. A public that cannot understand and appreciate the importance of the contributions that science makes to their daily lives will not support its endeavors. Straightforward communication is necessary, without overstating the potential of the research or the imminence of the results and without using confusing terminology. If the public and the policy makers believe they have been misled, the outcomes might be disastrous both for science and for society. If the public believes scientists are insensitive to ethical questions in their research, scientists will have a hard time earning the public's trust. Consideration of all the facts is important. Integrity in the research is essential, but so is integrity in the communication of science.

Many of these ethical questions involve difficult decisions affecting not only yourself but other lives as well, both now and into the future. It would always be easier if the decisions, especially the tough ones, could be made for you. But that approach assumes the decision maker has all the facts or has objectives that match your own. It also involves giving up your individual freedom in making those decisions, as well as your individual responsibility. With freedom comes responsibility—if you make choices, you are responsible for those choices, and your choices affect not only you but others as well. A person who is clever knows the best means to achieve an end; a person who is wise knows which ends are worth striving for.

"I FIND IT HARDER AND HARDER TO GET ANY WORK DONE WITH ALL THE ETHICISTS HANGING AROUND."

Figure 13.8 Ethical Decision Making Is an Essential Part of Science

CAREER PROFILE

Career Paths in Bioethics

There are over 100 academic bioethics research centers around the world. With ethics issues related to the Human Genome Project, aging, cloning, and GMO, public interest in bioethics is at an all-time high, and there are many exciting career opportunities in bioethics. Traditionally, medical ethics has been a concern from the time Hippocrates swore to "First, do no harm," but bioethics only became a discipline in the 1980s. As concerns over issues such as abortion and euthanasia were raised, doctors consulted with religious scholars, priests, rabbis, and philosophers to develop the field of bioethics. In the late 1980s, there was increasing consultation and involvement from lawyers and scientists. All these pathways (medical, religious, philosophical, legal, scientific) have contributed to the development of the discipline of bioethics, and there is still no very clear path to follow for a career in bioethics.

Many people working in bioethics train in science or medicine initially and then pursue a master's degree in bioethics. Some begin in law school first, some start with a Ph.D. in philosophy, and now there are a few Ph.D. programs in bioethics. Other starting points may be in medical anthropology, medical sociology, history of medicine, or nursing. There is still much controversy over the best training to develop proper credentials for bioethics. One nearly universal agreement is that training should be interdisciplinary: It should include a broad knowledge base, language skills, and the ability to communicate across such multiple disciplines as religion, science, and medicine.

Choosing the right path of training depends to a great extent on a person's interests. Start by assessing your own interests and skills as they relate to bioethics. You should also investigate what possible jobs are available so you can anticipate the necessary training. A good place to start is the NIH "Careers in Bioethics" website (see the Keeping Current list of websites on the Companion Website).

Ethical skills can convince employers of your potential. When biotechnology employers look at potential employees, they do not simply look at academic credentials and work experience. Meeting the specifications in the job description is essential, but possessing some of the intangible skills that can make prospective employers call you for an interview is often more important.

Employers often look for well-rounded employees who not only will work well in teams but also are diplomatic, resourceful, and able to build networks among their peers. Let employers know you possess these skills, even if you have acquired them over the years in other work experiences than those called for in the application. Communication skills are important in knowledge industries like biotechnology, and being fluent in more than one language can give you an advantage. Being able to make convincing presentations is also important, and familiarity with the media to make it effective is a necessity.

Adapting to changing environments is another important skill because biotechnology companies are constantly changing with the technology. Provide adequate examples of your ability to adapt, think on your feet, and understand what the customer (shareholder) is saying.

Show you have put a lot of pride and personal effort in your work and understand how people perceive you. Try to highlight various ways in which your skills, knowledge, and experience have led to concrete achievements. Be prepared to defend any so-called soft skill you possess with clear examples, but don't be afraid to sell yourself or your ethical abilities.

Adapted from P. Watson (2003), Transferable Skills for a Competitive Edge, *Nature Biotechnology*, 21: 211.

QUESTIONS & ACTIVITIES

Answers can be found in Appendix 1.

1. What are the two leading approaches to ethical thought, and how do they differ?

2. "Last September a little girl from California named Molly received a lifesaving transplant of umbilical cord blood from her newborn brother, Adam. Molly who was then eight years old, suffered from a potentially fatal genetic blood disorder known as Fanconi anemia. But what made the procedure particularly unusual was that Adam might not have been born had his sister not been sick. He was conceived through *in vitro* fertilization, and physicians specifically selected his embryo from a group of others for implantation into his mother's womb after tests showed he would not have the disease and he would be the best tissue match for Molly." Was this ethical? Adapted from R. M. Kline (2001), Whose Blood Is It, Anyway? *Scientific American*, 284(4): 42–49.

3. Make a chart with a list in one column of all the possibilities you can think of for interactions a GM plant might have with the ecosystem. Then in the second column list the possible outcomes, good or bad, for each of the listed interactions.

4. If a GM crop was engineered to produce growth hormone, what are some of the possible ethical outcomes and questions that need to be anticipated?

5. Is there a difference between a human being and a person? Why or why not?

6. There is never enough money available, even from the government, to support every scientific research proposal that is submitted to funding agencies. How should decisions be made about which of these proposals are a waste of money and which will yield discoveries that may change human life forever? When competing scientists may be part of the decision making, how should the decision be made on how tax money is spent for research? Should ethical concerns be a part of such funding decisions?

7. Suppose that gene therapy could be used to lower cholesterol and saturated fat levels in the blood and allow humans to consume fat-rich fried foods with little risk of heart disease. However, for some individuals such gene therapy could unknowingly be ineffective and result in premature death due to heart disease. Should this gene therapy be approved for use in humans? Answer this question by briefly describing how a bioethicist who believes in a utilitarian approach might view this scenario, and contrast this perspective with a deontological viewpoint of this dilemma.

8. When drug companies find out that a drug target is not responding to their new drug, do they have a legal obligation to inform other drug companies?

9. If a food allergen can be removed from food products by using a recombinant plant food source, and a consumer is harmed by the unwillingness of a producer to utilize the GM source, could the company be found liable for the harm created?

10. If a single cell is removed from an embryo *in vitro* and kept to be used for future organ regeneration but the embryo is not harmed and can continue to develop normally, is this an ethical use of human embryonic tissue?

References and Further Reading

Bouchie, A. (2006). Clinical Trial Data, to Disclose or Not to Disclose. *Nature Biotechnology*, 24(9): 1058–1060.

Fukuyama, F. (2002). *Our Posthuman Future: Consequences of the Biotechnology Revolution*. New York: Farrar, Straus & Giroux.

Gilbert, S. F., Tyler, A. L., and Zackin, B. J. (2005). *Bioethics and the New Embryology: Springboards for Debate*. Sunderland, MA: Sinauer Associates.

Greely, H. T. (2005). Banning Genetic Discrimination. *New England Journal of Medicine*, 353(9): 865–867.

Kilner, J. F., Hook, C. C., and Uustal, D. B., eds. (2002). *Cutting-Edge Bioethics*. Grand Rapids, MI: William B. Eerdmans.

Levine, C. (2006). *Taking Sides: Clashing Views on Controversial Bioethical Issues*, 11/e. Dubuque, IA: McGraw-Hill/Duskin.

MacQueen, B. D. (2002). The Moral and Ethical Quandary of Embryonic Stem Cell Research. *Medical Science Monitor*, 8(5): ED1–4.

Miller, H. (2006). Why Spurning Food Biotech Has Become a Liability. *Nature Biotechnology*, 23(9): 1075–1077.

Morrison, A. R. (2003). Ethical Principles Guiding the Use of Animals in Research. *The American Biology Teacher*, 65(2): 105–108.

Stock, G. (2002). *Redesigning Humans: Our Inevitable Genetic Future*. New York: Houghton Mifflin.

Vastag, B. (2006). New Clinical Trials at FDA. *Nature Biotechnology*, 24(9): 1043.

Visit www.pearsonhighered.com/biotechnology to download learning objectives, chapter summary, "Keeping Current" web links, glossary, flashcards, and jpegs of figures from this chapter.

Appendix 1

Answers to Questions & Activities

Chapter 1

1–2. Open ended; answers variable.

3. Bioremediation

4. Pharmacogenomics involves prescribing a treatment strategy based on the genetic profile of a patient—fitting the drug to the patient's genetics. An RNA or DNA sample from a patient with a medical condition can be analyzed to determine whether the genes that person expresses match the genetic profile of a particular disease process. If they do, a specific treatment approach can be designed according to the best known treatment strategy based on the genetic profile of the patient.

5. No product can leave a biotechnology company until it passes rigorous tests by the internal unit called quality control (QC). Quality assurance (QA) is responsible for monitoring everything that enters and leaves a company to ensure product safety and quality and tracing sources of problems identified by complaints as specified by regulatory agencies such as the Food and Drug Administration. QA and QC measures are important for just about every step of research and development and product development.

6–10. Open ended; answers variable.

Chapter 2

1. Genes are sequences of DNA nucleotides—usually from 1,000 to around 4,000 nucleotides long—that provide the instructions (code) for the synthesis of RNA. Most genes produce mRNA molecules encoding proteins but some genes produce RNAs that do not code for proteins. Genes are contained in chromosomes, tightly packed coils of DNA and protein. Chromosomes enable cells to separate DNA evenly during cell division. Chromosomes contain multiple genes, and the number of genes in a chromosome can vary depending on its size.

2. The complementary strand will be an antiparallel strand with the sequence 3'-TCGGGGCTGAGATAAG-5'.

3. Biologists use the phrase "gene expression" to talk about the production of mRNA from a particular gene. From mRNA, cells translate proteins that are responsible for many aspects of cell structure and function.

4. Chargaff's rules describe that the percentage of adenines in a genome is approximately equal to the percentage of thymines in a genome; the percentages of guanines and cytosines are also roughly equal. This is true because DNA consists of complementary base pairs. Adenines in one strand form hydrogen bonds with thymines in the opposite strand; guanines form hydrogen bonds with cytosines. Applying this knowledge, if DNA contains approximately 13% adenines, then the DNA would contain approximately 13% thymines. Combined, these bases account for approximately 26% of the DNA in the bacterium; therefore, the rest of the DNA (74%) consists of equal amounts of guanines and cytosines, so guanine would comprise approximately 37% of the genome, and cytosine, approximately 37%.

5. DNA is a double-stranded molecule located in the nucleus of cells. Each strand of DNA is made of building blocks called *nucleotides* that consist of a pentose sugar, phosphate group, and a base. DNA is the inherited genetic material of cells because its genes contain instructions for the synthesis of proteins. RNA is copied from DNA through a process called *transcription*. RNA is a single-stranded molecule, which is an important structural difference compared to DNA. RNA not only contains nucleotides including the base uracil, which replaces thymine present in DNA, but also a different pentose sugar (ribose) than DNA (deoxyribose). Three major types of RNA are transcribed: mRNA, tRNA, and rRNA, and other forms of RNA (snRNA and siRNA) are involved in mRNA splicing and gene expression regulation, respectively. After transcription, RNA molecules move into the cytoplasm where they are required for protein synthesis (translation).

6. Seven codons, six amino acids coded.

    Start   Stop

5'-AGCACC<u>AUG</u>CCC<u>CGA</u>ACC<u>UCA</u>AAG<u>UGA</u>AA CAAAAA-3'

The amino acid sequence is methionine-proline-arginine-threonine-serine-lysine. Remember that actual mRNAs and their encoded proteins are much longer than this example sequence and that stop codons do not code for an amino acid.

7. a. Only the bottom strand produces a functional mRNA with a start codon. This sequence is 5'-UUU<u>AUG</u>GG<u>UUGG</u>CCC<u>GGG</u>UC<u>AUGA</u>UU-3'. The amino sequence of a polypeptide produced from this sequence is:

Methionine – glycine – tryptophan – proline – glycine – serine

b. Messenger RNA copied from the bottom strand of DNA with a T inserted between bases 10 and 11 is: 5'-UUUAUGGGAU**A**GGCCCGGGUCAUGAUU-3'. This will result in an A (bold) inserted in the mRNA sequence transcribed, which now creates a stop codon (UAG) in the mRNA, which will produce a truncated (shortened) peptide of only two amino acids, methionine–glycine.

8. Messenger RNA (mRNA) is an exact copy of a gene. It acts like a "messenger" of sorts by carrying the genetic code in the form of codons, encoded by DNA, from the nucleus to the cytoplasm where this information can be interpreted to produce a protein. Ribosomal RNA (rRNA) molecules are important components of ribosomes. Ribosomes recognize and bind to mRNA and "read" along the mRNA during translation. Transfer RNA (tRNA) molecules transport amino acids to the ribosome during protein synthesis. Each tRNA contains an amino acid and an anticodon sequence that base pairs with codon sequences during translation.

9. *Gene regulation* is a broad phrase used to describe ways that cells can control gene expression. Gene regulation is an essential aspect of a cell's functions. Cells use many complex mechanisms to regulate gene expression. In this chapter, we highlighted transcriptional control as a regulatory process. Gene regulation allows cells to tightly control the amount of RNA and protein they produce in response to particular needs of the cell. In the case of the *lac* operon, gene regulation allows cells to respond to environmental changes.

10. Refer to Figure 2.9.

11. Mutations that affect protein structure and function include nonsense mutations (affect a codon by creating a stop codon), missense mutations (change a codon to code for a different amino acid with different properties than the original amino acid coded for), and frameshift created by insertions or deletions. Silent mutations have no effect on protein structure and function because these mutations of a codon do not affect the amino acid coded for by the mutated codon.

12. 1'carbon

Chapter 3

1. Gene cloning is the copying (cloning) of a gene or a piece of gene. The terms *recombinant DNA technology* and *genetic*

engineering are often used interchangeably. Technically, recombinant DNA technology involves combining DNA from different sources, whereas genetic engineering involves manipulating or altering the genetic composition of an organism. For instance, ligating a piece of human DNA into a bacterial chromosome is an example of recombinant DNA technology, and placing this piece of recombinant DNA into a bacterial cell to create a bacterium is considered genetic engineering.

2. DNA ligase is used to form phosphodiester bonds between DNA fragments during recombinant DNA experiments. This is an important step in a cloning experiment because hydrogen bonds between sticky ends of DNA fragments are not strong enough to hold a recombinant DNA molecule together permanently. Restriction enzymes cut DNA at specific nucleotide sequences (recognition sequences). Frequently, DNA is digested into fragments with restriction enzymes as an important step in many cloning experiments prior to using DNA ligase to join fragments together.

3. The gene sequence cloned from rats could be used to create primers that might be used to amplify DNA from human cells in an effort to amplify the complementary gene in humans. If one were successful in obtaining PCR products from this experiment, the PCR products could be sequenced and compared to the rat gene to search for similar nucleotide sequences (suggesting these genes are related). Also, PCR products could be used as probes in library screening experiments to find the full-length human gene or as probes for Northern blot analysis to determine if mRNA for this gene is expressed in human tissues.

4. Refer to Section 3.2.

5. Approximately 32,768 (2^{n-1}, where n = number of cycles)

6. Genomic libraries contain both introns and exons, whereas cDNA libraries contain DNA reverse transcribed copies of mRNA expressed in a given tissue. cDNA libraries are the libraries of choice when cloning expressed genes. When searching for a gene involved in obesity, one could make and screen a cDNA library from adipocytes. Cloning gene regulatory regions, for instance, promoter and enhancer sequences described in Chapter 2, can be accomplished with genomic DNA libraries because they contain both exons and introns.

7. Searching for "diabetes" reveals a list of sequences for genes related to insulin-dependent and non-insulin-dependent diabetes mellitus. Clicking on any of the highlighted numbered links will take you to informative pages providing great detail about each gene. Searching with the accession number 114480 reveals information on genes for familial breast cancer.

8. This piece of DNA can be cut once by *Bam*HI at the beginning of the sequence and once by *Eco*RI at the end of the sequence. There are no cutting sites for *Sma*I in this sequence. Searching for all enzymes in the database

reveals approximately 50 restriction enzymes that can cut this sequence. This demonstration should provide you with an appreciation of how powerful computer programs can be for helping molecular biologists analyze DNA sequences.

9. Corrected answers are available on the website as students work on the questions.

10. Open ended; answers variable.

11. The ddNTPs are missing an oxygen at the 3′ carbon.

12. The human genome consists of approximately 3.1 billion base pairs. The genome is approximately 99.9% the same between individuals of all nationalities and backgrounds. Less than 2% of the human genome codes for genes. The vast majority of our DNA is non-protein coding. The genome contains approximately 20,000 to 25,000 protein-coding genes. Many human genes are capable of making more than one protein. Chromosome 1 contains the highest number of genes. The Y chromosome contains the fewest genes.

Chapter 4

1. Genomics is the complete sequence of DNA nucleotides in an organism; proteomics is a complete set of proteins in an organism (including domains of proteins).

2. Directed molecular evolution technology focuses on the mutations of a specific gene, selecting the best functioning protein from that gene, irrespective of the benefits or hazards it may have for the organism.

3. Only about 25,000 to 35,000 human genes make proteins, narrowing the search to a smaller portion of the total human genome.

4. Only the active genes in a cell make mRNA, the source for cDNA.

5. The process for protein production is being patented, and it must be repeatable. The product must be efficacious if it is to receive a patent.

6. The earliest fractions

7. Last from the column

8. More selective

9. MS, HPLC, X-ray crystallography

10. The procedure lost your protein.

Chapter 5

1. There are several important structural differences between prokaryotic and eukaryotic cells. Bacteria and Archaea are prokaryotes; eukaryotic cells include plant cells, animal cells, fungi, and protozoa, which are structurally more complex than bacteria. Prokaryotic cells are smaller than eukaryotic cells, do not contain a nucleus, have relatively few organelles, and contain a cell wall.

Prokaryotes also have smaller genomes than eukaryotic cells; the prokaryotic genome usually consists of a single circular chromosome. Prokaryotic cells have served many important roles in biotechnology. Bacteria are used as hosts in gene cloning experiments, recombinant proteins produced by bacteria are important for research and some proteins are used in medical applications, microbes are used for manufacturing many foods and beverages, and bacteria are used as sources of antibiotics and as hosts for the production of subunit vaccines.

2. Yeast are unicellular fungal eukaryotes, and bacteria are prokaryotic cells. Most yeast have larger genomes than bacteria. Fermenting yeast is used for making breads and dough, as well as for brewing beers and wines. Yeast serve many important roles in research. The yeast two-hybrid system is a valuable technique for studying protein interactions. Because the yeast genome contains many genes similar to human genes, geneticists study the yeast genome for clues about the functions of many human genes.

3. All vaccines are designed to stimulate the immune system of the recipient to make antibodies or activated T cells to a pathogen, for example, a bacterium or a virus. During antibody production, immune memory cells such as antibody-producing B cells and memory B cells are also produced. By creating antibodies and memory cells in the recipient, the goal of vaccination is to provide the recipient with protection against a pathogen should he or she be exposed to the actual pathogen. Three major types of vaccines are subunit vaccines, which consist of molecules from the pathogen (usually proteins expressing by recombinant DNA technology); attenuated vaccines, which are made from live pathogens that have been weakened to prevent replication; and inactivated vaccines, which are prepared by killing the pathogen and injecting dead or inactivated microbes into the vaccine recipient.

4. Anaerobic microbes are those microorganisms that do not require oxygen to convert sugars into energy (in the form of ATP). Under anaerobic conditions, some microbes use lactic acid fermentation; others use alcohol (ethanol) fermentation to derive energy. Lactic acid and ethanol are waste products of anaerobic fermentation that are important components of many foods. Lactic acid is found in cheeses and yogurts; ethanol is found in alcoholic beverages such as wine and beer.

5. Open-ended questions; answers variable.

6. Studying microbial genomes can help scientists in many ways. Sequencing genomes can reveal new genes, including genes that may be used in biotechnology applications (such as genes that encode novel enzymes with important properties). Scientists can study genes of disease-causing pathogens (including potential agents of bioterrorism) and then use this knowledge to help fight disease. Studying genomes can help scientists learn more about bacterial metabolism, the relatedness of bacterial strains, and lateral gene transfer (the sharing of genes between species).

7. Open-ended activity; answers variable.

8. Open-ended questions, but some people believe that requiring Gardasil vaccination for teens presumes sexual activity in teens, which may be common for some teens, but it should not be assumed that all teens are sexually active.

9. Fusion proteins are generated by inserting a cloned gene for a protein of interest into an expression vector that allows production of a protein fused to a tag protein such as green fluorescent protein or maltose-binding protein. This fused protein can then be passed over an affinity column, which will bind the tag portion of the fusion protein and allow the fusion protein to be isolated from a bacterial extract of proteins.

10. Refer to Table 5.1.

Chapter 6

1. Approval of food crops produced in this manner must be obtained separately from regulatory agencies. GM crops that are destined for animal feeds or are not ingested have been approved, but no human consumption GM crops can be approved without extensive testing to demonstrate lack of allergic response in humans.

2. Agricultural biotechnology innovations can reduce the need for pesticide application because plants have the ability to protect themselves from certain pests and diseases. It not only decreases water usage, soil erosion, and greenhouse gas emissions through more sustainable farming practices but also improves productivity of marginal cropland, especially where acres for planting are decreasing around the world.

3. The benefits of agricultural biotechnology include reduction of undesirable qualities such as saturated fats in cooking oils, elimination of allergens, an increase in nutrients that help reduce the risk of chronic disease, and better delivery of proper nutrients such as vitamin A in commonly consumed crops.

4. With biotechnology, researchers can identify specific genetic characteristics, isolate them, and then transfer them to valuable crop plants. This technique is more precise and efficient than traditional crossbreeding and can increase food production through higher yields.

5. Foods produced from our current biotechnology methods do not require special labeling in the United States (at the time of this printing).

6. Although some evidence of transfer to other plants and insects has been recognized, no long-term damage to beneficial insects has been demonstrated. In fact, genetic engineering has produced more examples of "integrated pest management" in a shorter time than ever before.

7. Biotechnology-enhanced products have been available for only about 5 years. Long-term studies are not required if the gene product already exists in the

environment in another organism. If found, adverse conditions will be appropriately regulated.

8. Monocot plants do not accept the Ti plasmid, requiring other methods to be used to transfect genes.

9. The germplasm was developed and donated for free, although intellectual property rights legal battles have held up its introduction.

10. Regulating multiple genes will make the development of insect resistance much more difficult.

Chapter 7

1. They provide the continuous dividing ability.

2. The HAMA response occurs because mice and humans often construct different antibody structures to the same antigen, and this can cause a human rejection response.

3. They graft human antibody-producing spleen cells into mice that have no functioning immune system (like SCID in humans) and they begin to produce human-like antibodies.

4. Complementary pairing inactivates the naturally produced mRNA.

5. Human genes for specific conditions replace the knockouts, and potential drugs target these human inserted genes, mimicking the human condition.

6. The heat acts as an inducer to the transferred genes that localize with the cancer.

7. It takes only 5 days to determine toxicity and effectiveness against human transgenes because of their rapid development rates.

8. MAbs are specific against one antigen and effective in small quantities.

9. Researchers must prove which animal and the numbers used before research, use alternatives where available, and ensure that the most humane treatments are used.

10. DNA is linked to the sperm cell surface so that it will be carried to the egg.

Chapter 8

1. A polymorphic DNA locus is a sequence of DNA that has many possible alternatives (for example, different numbers of multiple repeats of tandem nucleotide sets, such as GC, GC,GC . . .).

2. The polymorphic DNA sequences tested in fingerprinting have no known effect. They usually occur in regions of DNA that are not translated into protein.

3. Half should occur in the father; half should occur in the mother.

4. Blood typing is cheaper and easier to do than DNA fingerprinting.

5. Heparin inhibits certain enzymes used in a RFLP and could create an incorrect result, producing an artificially similar pattern.

6. DNase is an important enzyme to remove so it will not destroy the DNA that is to be profiled.

7. The DNA fingerprint of the cabernet sauvignon plant contained bands from both parents.

8. The DNA in dot/slot blots only needs to be hybridized with the probe; no DNA digestion, electrophoresis, or Southern blotting is required.

9. It can spill over before electrophoresis and give a false positive to a defendant.

10. Every fifth lane contains reference DNA to allow comparison with the previous (five back) lanes to determine if shifting has occurred.

Chapter 9

1. Open-ended activity; answers depend on the processes described.

2. The addition of fertilizers such as carbon, nitrogen, phosphorus, and potassium is often an important step in many bioremediation processes. Fertilization, also called nutrient enrichment, stimulates the growth and activity (metabolism) of microorganisms in the environment. These microorganisms, usually bacteria, also divide more rapidly to create more bacteria. By stimulating the metabolism of bacteria and increasing their number at a contaminated site, pollutants are usually degraded more rapidly. Adding oxygen to a contaminated site is effective when the microorganisms involved in the cleanup are relying on aerobic biodegradation.

3. Many formerly polluted sites throughout the United States are being redeveloped for industrial and residential use. Sometimes houses developed in these areas are sold at lower prices than other comparable homes; however, not all states require builders to disclose the history of the land to potential homeowners. One obvious problem is the concern that the site may not be fully cleaned up. If the housing development relies on groundwater for its drinking water supply, the water is typically tested regularly to check for chemical pollutants. However, this can be problematic because even trace amounts of some chemicals may go undetected, and the health effects of these chemicals may not be fully known. Similarly, remaining chemicals in the soil can also affect residents' activities, such as playing in the yards and growing gardens in the soil. A major problem with redeveloping bioremediation sites for residential use is that it may take many years (or generations of families) to determine if residents are experiencing health effects from chemicals remaining at the site. Even if residents experience some health problems, it is often very difficult to determine if these effects are caused by pollutants at the site.

4. One approach might involve studying lead-containing structures to identify if bacteria are growing on them. For instance, one could study lead pipes left in the environment and, over time, determine if bacteria are growing on these pipes. Bacterial growth on a lead surface could be an indication that these bacteria have developed a way to avoid the toxic effects of lead. This could be a sign that these cells may be capable of degrading lead. These bacteria could then be isolated, and experiments using microcosms could be designed to test whether these cells can degrade lead.

5. Open-ended activity; answers variable.

6. Open-ended activity; answers variable.

7. Open-ended activity; answers variable.

8. Open-ended activity; answers variable.

9. Open-ended activity; answers variable, although to date no significant bioremediation activities other than bioremediation by indigenous microbes has been reported.

10. Open-ended activity; answers variable.

Chapter 10

1. As discussed in Chapter 8, transgenic animals contain genes from another source. The transgenes may be from related species (for example, introducing the salmon gene for growth hormone into trout) or from very different species (for example, introducing the *luciferase* [*lux*] gene from bioluminescent marine bacteria or fireflies into salmon). Transgenic fish can be created using a number of different techniques, but one prominent technique includes microinjecting the transgene into blastocyst-stage embryos to allow incorporation of the transgene into embryonic tissues. Polyploid fish such as triploids, which contain three complete sets of chromosomes, are usually created by either chemical treatment or electrical treatment of sperm or egg cells to produce diploid gametes that can be used to fertilize haploid gametes.

2. Open-ended activity; answers variable.

3. Examples of benefits include providing food sources, improving populations of fish and shellfish, creating farming industries in noncoastal areas of the world that do not have fishing or seafood industries, and providing fish for recreational fishing. Examples of problems include some species not being suitable for aquaculture or too expensive to raise through farming, disease and illness being able to decimate fish stocks because most farmed fish are genetically similar, wastes from fish farms creating pollution problems, and the genetic potential of native species decreasing when escaped farmed fish enter the population.

4. Open-ended activity; answers variable.

5. Open-ended activity; answers variable. Answers should, however, mention that organisms that grow under

unique or extreme environmental conditions (such as pressure, heat, and ocean depths) are among the most highly studied for the identification of unique molecules that might be potentially valuable. For instance, mussels, which adhere tightly to structures to withstand constant pounding of waves, produce unique molecules in their byssal fibers that provide adhesive strength.

6. Refer to Table 10.1.

7. Many aquatic biotechnologists believe that biotechnology will play important roles in improving finfish and shellfish populations in the future. One obvious way this may be accomplished is to use bioremediation approaches to detect and clean up environmental pollution. Another approach may involve using biotechnology to learn about the pathogens that cause disease in aquatic organisms and develop ways to prevent or treat such diseases. Transgenic and polyploidy approaches may be used to continue to produce aquatic organisms with improved resistance to disease. In addition, aquaculture may be used to grow numbers of species that can be stocked to increase dwindling populations of aquatic organisms.

8. Federal agencies such as the FDA and USDA believed there was no need to regulate the GloFish because it posed no threat to public health since tropical fish are not used for food purposes. The U.S. Fish and Wildlife service determined that the fish posed little threat to the environment because unmodified zebrafish have been sold in the United States and released into the wild with no known consequences. Many animal rights groups and other groups have strongly protested development of the GloFish primarily because they perceive it to be an abuse of genetic engineering to create a transgenic species solely for the purpose of enjoyment as a pet. Concerns have also been raised about the potential release of this GM pet into the wild. Some people have also raised the point that the GloFish has provided a precedent for the unregulated development of other GM pets and GM animals that may pose environmental threats. Anti-biotech groups also cite this as an example to raise public skepticism about biotechnology because if the GloFish can go unregulated, what other areas of biotechnology may go unregulated?

9. Biofilms are accumulations of livings organisms such as bacteria that coat a surface. For example, biofilms occur on teeth, heart implants, and intravenous lines. In marine environments, algae and mollusks growing on the hull of a ship is an example of biofilming. Many aquatic organisms combat biofilming by producing compounds that kill or inhibit growth of biofilming microbes. As a result scientists are interested in identifying these compounds so they can be used for commercial applications to combat biofilming.

10. The LAL test relies on enzymes from the blood of horseshoe crabs (*Limulus polyphemus*) to detect endotoxins produced by toxic microbes. The LAL test is used to check the sterility of instruments such as medical devices.

Chapter 11

1. Amniocentesis and chorionic villus sampling are ways to sample tissue from developing fetuses. Fetal cells or adult tissue (usually white blood cells, skin, cheek, or hair cells) can be tested by karyotype analysis to check for abnormalities in chromosome number and chromosome structure. Molecular techniques such as RFLP analysis and ASO tests can be used to detect defective genes in human cells. In the future, DNA microarrays (gene chips) will play a greater role in genetic testing, allowing for the analysis of thousands of genes at the same time.

2. The purpose of gene therapy is to deliver therapeutic genes into humans to treat or cure disease conditions. *Ex vivo* gene therapy involves removing cells from a patient, inserting a gene or genes into these cells, and then injecting or implanting them into the patient. Treatment of SCID is an example of *ex vivo* gene therapy. *In vivo* gene therapy occurs within the body by delivering genes directly into the body (for example, treating cystic fibrosis by nasal sprays). Some techniques for delivering therapeutic genes include using viruses as vectors for gene therapy, injecting naked DNA, and using liposomes to deliver genes. Gene therapy scientists face many challenges, including ensuring the safe delivery of genes (especially when viral vectors are used), targeting the therapeutic gene to the correct cells and tissues, and finding ways to get sufficient expression of the therapeutic gene to cure the condition.

3. Embryonic stem cells (ESCs) are isolated from early embryos at the blastocyst stage. Blastocysts are usually derived from embryos left over from *in vitro* fertilization procedures. To isolate and culture ESCs, the inner cell mass is dissected out of the blastocyst, and these cells are grown in a tissue culture dish. In contrast, adult-derived stem cells (ASCs) are isolated from mature adult tissues. These cells are found in small numbers in many tissues of the body such as muscle and bone tissue. A small sample (biopsy) of adult tissue is removed (for instance, a needle can be used to biopsy a sample of adult bone marrow stem cells), and ASCs can then be grown in culture. By treating ESCs or ASCs with different growth factors, scientists can stimulate stem cells to differentiate into different types of body cells. Stem cell biologists envision many ways in which stem cells can be used to treat disease. Stem cells could be implanted in the body to replace damaged tissue, stem cells might be good vectors for the delivery of therapeutic genes, and stem cells could be used to grow tissues in organs in culture for subsequent transplantation into humans.

4. Refer to Table 11.3 for a comparison of therapeutic and reproductive cloning.

5. Pharmacogenomics is customized or personalized medicine created by analyzing a person's genetics and designing a drug or treatment strategy that is specific for a particular person based on the genes involved in that person's medical condition.

6. The Human Genome Project will reveal the location of all human genes including those involved in normal and disease processes. Identifying these genes is an important first step toward understanding how genes function and how certain diseases develop. Identifying disease genes will enable scientists to develop tests that can be used to screen individuals for the likelihood that they will inherit a particular genetic condition. As we learn more about different genes and the proteins they encode, it is expected that more specific forms of medical treatment will be available by designing drugs that will affect specific genes and proteins involved in genetic diseases and conditions that have a genetic basis. Such treatments may also include gene therapy strategies. Understanding how human genes are regulated and affected by conditions such as stress and environmental factors will lead to a greater understanding of preventive measures that may be used to minimize genetic disease.

7. Open ended; answers variable.

8. HR 810 was designed to loosen current restrictions on using federal funds to produce embryos for deriving embryonic stem cells. The bill was intended to make eligible for federal funding human embryonic stem cells that were derived from human embryos donated from *in vitro* fertilization clinics, embryos created for the purposes of fertility treatment, and excess embryos that would otherwise be discarded.

9. Antisense RNA technologies and RNAi are two common gene silencing techniques that could be used to inhibit a gene involved in a disease process. Refer to the chapter for a description of each technique.

10. Regenerative medicine involves creating cells, tissues, and organs that can be used to repair or replace dead or damaged tissues in a person.

11. SNPs, or single-nucleotide polymorphisms, are single base changes in DNA sequence that are responsible for the subtle genetic differences between individual humans. When they occur in protein-coding gene sequences, SNPs can represent mutations that affect gene function and cause disease. Thus SNPs can be detected as a way to identify genetic changes involved in disease.

Chapter 12

1. FDA

2. EPA

3. EPA

4. FDA, EPA, USDA

5. For the U.S. PTO to issue a patent, the product must have utility, be nonobvious, and be functional.

6. Approvals that took four years can now occur in one.

7. Europe, where every step of the purification process is usually patented.

8. Firms should evaluate every aspect of production, including people, procedures, equipment, materials, record keeping, audits, and training.

9. Patented products must be nonobvious and have utility. ESTs are not products and only function to identify a location. They do not qualify as patentable.

10. The EMEA has no knowledge of a drug product until it is submitted for approval; therefore a drug will take a shorter time for approval.

Chapter 13

1. Utilitarian approach and deontological (Kantian) approach. Utilitarianism tries to weigh all possible outcomes and produce the greatest good for the greatest number, and the end result of a consequence justifies the means. The deontological approach starts with certain absolutes that cannot be crossed, no matter how much good might be achieved, to maintain an ethically correct outcome.

2. Most bioethicists concluded that this was ethically appropriate because Adam was not physically harmed by providing cord blood for his sister Molly. In addition, no evidence indicated that Adam's parents treated Adam poorly or as if he was conceived only to save his sister. However, what do you think about parents selecting the offspring they want based on genetics (a process that is not uncommon when doing *in vitro* fertilization)?

3. Open-ended activity; answers variable.

4. Open-ended question; answers variable.

5. Open-ended question; answers variable.

6. Open-ended question; answers variable.

7. A bioethicist following a deontological approach would propose a "greatest good for the greatest number" answer. If this form of gene therapy can help lower cholesterol and saturated fats levels in humans and allow people to enjoy fatty foods for a majority of individuals, then it is ethically all right to assume the risk that some smaller number of people may die prematurely because of this therapy.

8. From an ethics perspective one could argue that for the overall benefit of improving human health, drug companies should share such information to help facilitate the development of drugs that cure. However, there is no legal requirement for companies to do so.

9. Open-ended question; answers variable.

10. Open-ended question; answers variable.

Appendix 2

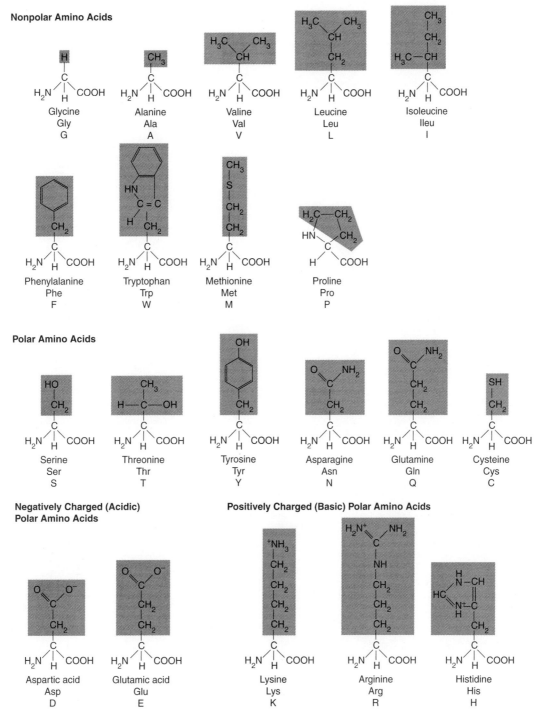

Nonpolar Amino Acids

Glycine
Gly
G

Alanine
Ala
A

Valine
Val
V

Leucine
Leu
L

Isoleucine
Ileu
I

Phenylalanine
Phe
F

Tryptophan
Trp
W

Methionine
Met
M

Proline
Pro
P

Polar Amino Acids

Serine
Ser
S

Threonine
Thr
T

Tyrosine
Tyr
Y

Asparagine
Asn
N

Glutamine
Gln
Q

Cysteine
Cys
C

Negatively Charged (Acidic) Polar Amino Acids

Positively Charged (Basic) Polar Amino Acids

Aspartic acid
Asp
D

Glutamic acid
Glu
E

Lysine
Lys
K

Arginine
Arg
R

Histidine
His
H

The 20 Amino Acids of Proteins

Up to 20 different amino acids can be used to synthesize a protein. Each amino acid has a common structure but side chains (shaded) vary from amino acid to amino acid. These side chains determine the chemical properties of amino acids. Notice that some amino acids are nonpolar while others are polar and there are also negatively charged and positively charged amino acids. Both the three letter code and single letter code of each amino acid is also shown.

Credits

Photo Credits

Chapter 1: Opener: ©Yorgas Nikas/Photo Researchers. **1.1(a):** Doebley, *J Plant Cell*, 2005 Nov; 17(11): 2859–72. Courtesy of John Doebley, University of Wisconsin. **1.1(b):** Richard White, Children's Hospital Boston.**1.4(b):** SIU BioMed/Custom Medical Stock Photo. **1.5:** Noah Berger/AP Images. **1.6:** Roslin Institutes/PA Photos Ltd. **1.7:** Orchid Cellmark. **1.8:** Ken Graham/Accent Alaska.com.**1.9:** AP Images. **1.12:** Anthony Canamucio. **1.15:** Jim Reme, Monmouth University. **UN1:** ScienceCartoonsPlus.com.

Chapter 2: Opener: Robert Silvers, www.photomosaic.com. **2.7:** David Adler, University of Washington Department of Pathology. **2.20(a):** Andrew Syred/Stone/Getty Images. **2.20(b):** National Institutes of Health. **2.21:** David Cooney

Chapter 3: Opener: Michael Palladino. **3.3(a):** Gopal Murti/Photo Researchers. **3.3(b):** Michael Palladino. **3.4(b):** Robley Williams. **3.11(b):** Michael Palladino. **3.13(c):** Registered Trademark of Pfizer Inc. Reproduced with permission. **3.14 (part 2):** P.M. Lansdorp, Terry Fox Laboratory, B.C. Cancer Research Center, U.B.C, Vancouver, Canada. **3.16(b):** Michael Palladino. **3.18 (part 1):** Alfred Pasieka/SPL/Photo Researchers. **3.18 (part 2):** GeneChip® Human Genome U133 Plus 2.0 Array. Courtesy of Affymetrix.

Chapter 4: Opener: Michael Palladino. **4.4:** Garman, S.C., Sechi, S., Kinet, J.P., Jardetzky, T.S.: The Analysis of the Human High Affinity Ige Receptor Fc(Epsilon) Ri(Alpha) from Multiple Crystal Forms *J.Mol.Biol.* 311 pp. 1049 (2001). **4.15 (parts 1 and 2):** Amersham Biosciences.

Chapter 5: Opener: Gopal Murti/Photo Researchers. **5.1(a):** Scimat/Photo Researchers. **5.1(b):** Manfred Kage/Peter Arnold. **5.1(c):** CNRI/Photo Researchers. **5.2(a):** John Durham/Photo Researchers. **5.2(b):** Kevin Schafer/Corbis. **5.3(b):** Catherine Pongratz and Charlotte K. Mulvihill, Biotechnology Program, Oklahoma City Community College. ASM Microbelibrary.org. **5.6(a) and (b):** Ken Lucas/Biological Photo Services. **5.8(a):** Michael Palladino. **5.18(a):** A. Dowsett/Photo Researchers. **5.18(b):** T.H. Chen and S. S. Elberg, *Inf. Imm* 15:972–977, Figure 4, with permission from ASM. **5.20:** AP Images.

Chapter 6: Opener: Bill Thieman. **6.1:** U.S. Department of Agriculture. **6.5(a):** Bio-Rad Laboratories. **6.9(a):** George Chapman/Visuals Unlimited. **6.9(b):** E. R. Degginger/Photo Researchers. **6.11:** Yann Layma/Stone/Getty Images. **6.13:** Reuters NewMedia Inc./Corbis.

Chapter 7: Opener: Keith Weller/ARS/USDA. **7.3(a):** Solnica-Krezel Lab, Vanderbilt University. **7.3(b):** Dorling Kindersley Media Library. **7.6(a):** OptiCell, manufactured by BioCrystal, Ltd. **7.8:** Ben Margot/AP Images. **7.11:** Inga Spence/Visuals Unlimited.

Chapter 8: Opener: Steve Miller/AP Images. **8.6:** Oxford University Press/United Kingdom. **8.8:** Orchid Cellmark. **8.9:** Oxford University Press/United Kingdom. **8.12:** Seminis. **8.13:** Guglich, E.A., Wilson, P.J., White, B.N., 1993, Application of DNA Fingerprinting to Enforcement of Hunting Regulations in Ontario, *Journal of Forensic Sciences*, figure 5, p.56. © ASTM International.

Chapter 9: Opener: Owen Franken/Corbis. **9.4:** Jennifer Bower and Ralph Mitchell, Laboratory of Applied Microbiology, Division of Applied Sciences, Harvard University. **9.9:** Ashbury Park Press. **9.10:** Michael Palladino. **9.12:** Kunen, J. G. and Jetten, M.S.M (2001) Extra ordinary anaerobic ammonium oxidizing bacteria. *ASM News* 67: Figure, p 458. **9.15:** Laboratory of Derek Lovley, University of Massachusetts Amherst. **9.17(a):** Jack Smith/AP Images. **9.17(b):** Vince Streano/Corbis. **9.18:** Exxon Mobil. **9.19:** Science VU/VisualsUnlimited.

Chapter 10: Opener: Rick Price/Corbis. **10.1:** Peggy Greb/ARS/USDA. **10.2 (part 4):** Michael Palladino. **10.3(a):** Kevin Fitzsimmons, University of Arizona. **10.3(b):** China Tourism Press/The Image Bank. **10.3(c):** Paul A. Souders/Corbis. **10.3(d):** Earl & Nazima Kowall/Corbis. **10.4(a):** Kevin Fitzsimmons, University of Arizona. **10.4(b):** G. C. Mair/www.fishgen.com. **10.7(a):** Dwight Smith/Shutterstock. **10.7(b):** Bence NF, Sampat RM, Kopito RR. Impairment of the ubiquitin-proteasome system by protein aggregation. *Science*. 2001 May 25;292(5521):1552–5. Reprinted with permission from the American Association for the Advancement of Science. **10.7(c):** R. E. Mandrell and A. H. Bates; USDA/Agricultural Research Service, Albany, CA. **10.8:** M. Palladino and G. Sarinsky. **10.9:** Choy Hew, National University of Singapore, and Garth Fletcher, Memorial University of Newfoundland, St. John's, Newfoundland. **10.12(a and b):** A.C. Duxbury, and A.B. Duxbury, 1997, *Introduction to the*

World's Oceans, Fifth Edition, William C. Brown Publishing Co., Fig. 17.1, p. 475. Reproduced with permission of The McGraw-Hill Companies. **10.13:** Herbert Waite, Department of Molecular, Cellular, and Developmental Biology, University of California, Santa Barbara. **10.14:** Anne Komarisky. **10.15(a):** Center for Great Lakes and Aquatic Sciences. **10.15(b):** Betty Sederquist/Visuals Unlimited. **10.16:** Deborah L. Santavy, Gulf Ecology Division, US Environmental Protection Agency.

Chapter 11: Opener: Spencer Jones/Taxi (Text overlay: Marlene J. Viola). **11.1:** John Sholtis, Amgen Inc. **11.2 (part 2):** March of Dimes Birth Defects Foundation. **11.3(a,b):** American Cancer Society. **11.3(c):** Lisa G. Shaffer, Health Research and Education Center, Washington State University, Spokane. **11.9 (part 2):** Robert S. Langer and Kenneth J. Germeshausen. **11.13:** Zhang, G., Budker V., Williams P., Subbotin V., Wolff J. A. Efficient Expression of Naked DNA Delivered Interaarterially to Limb Muscles of Nonhuman Primates, *Hum Gene Ther.* 2001 Mar I; 12(4): 42738. **11.18:** Bill Ling/Dorling Kindersley Media Library. **11.20(a):** Becton, Dickinson and Company. **11.20(b):** Copyright © BBC. **11.22 (part 2):** Liz Sanders, Mississippi Fertility Institute. **11.23:** Shinya Yamanaka, HO/AP Images. **11.25 (part 2):** Photography by Anne Bower and Manfred Baetscher, Transgenic Core Laboratory, Oregon Health and Science University, Portland, Oregon.

Chapter 12: Opener: Bettmann/Corbis. **12.1:** Bettmann/Corbis. **12.2:** Steve McCutcheon/Visuals Unlimited. **12.3(a):** ASM Publications. **12.3(b):** Clinton R. Fuller. **12.4:** Steven Lindow, Lindow Lab/The Ice Lab, University of California, Berkeley. **12.5:** Photodisc/Getty Images. **12.6:** Jeffrey L. Rotman/Corbis.

Chapter 13: Opener: Manny Ceneta/AFP/Getty Images. **13.1:** Rooster photo: Eric Carlson, invisiblerobot.com; Digital dental implants: Seelevel.com. **13.2 (part 1):** Clemson University/USDA Cooperative Extension Slide Series, United States. **13.2 (part 2):** Steven Katovich, USDA Forest Service, Bugwood.org. **13.6:** Richard Olsenius/National Geographic. **13.8:** Sidney Harris.

Illustration Credits

Figure 1.3: from *Biotechnology,* Third Edition by John E. Smith, figure 1.3, p. 15, figure 11.1, p. 191 and figure 11.2, p. 193. Reprinted with the permission of Cambridge University Press.

Figure 1.16: from *Focus on Fundamentals*: *The Biotechnology Report, 15th Annual Review,* graph "Private and Public Biotech Companies by Region." Copyright © 2002 Ernst & Young LLP. Permission to use copyrighted material granted by Ernst & Young LLP.

Figure 1.17: from website: http://massbio.org.

Figures 2.1, 2.3, 2.9, 2.12, 2.15, 2.19, 3.3, 3.7, 3.11(a), & 3.15: from *Biology,* Fifth Edition by Neil A. Campbell, Jane B. Reece & Lawrence G. Mitchell.

Figures 2.16, 2.18, 3.19 (Becker): from *The World of the Cell,* Seventh Edition by Wayne M. Becker, Lewis J. Kleinsmith, Jeff Hardin & Gregory Paul Bertoni.

Figure 3.13: from *Biology,* Eighth Edition, by Neil A. Campbell & Jane B. Reece.

Figures 3.21, 3.22: from *Understanding the Human Genome Project,* Second Edition by Michael A. Palladino.

Figure 4.1: from *Protein Biotechnology* by Gary Walsh and Dennis Headon, figure 3.1, p. 42. Copyright © 1994, John Wiley & Sons, Limited. Reproduced with permission.

Figure 4.2: from *Biotechnology,* Third Edition by John E. Smith, figure 1.3, p. 15, figure 11.1, p. 191 and figure 11.2, p. 193. Reprinted with the permission of Cambridge University Press.

Figure 4.3: adapted from illustration by Darryl Leja for National Human Genome Research Institute (NHGRI).

Figure 4.10, 4.11, & 4.13: from *Proteins to PCR* by David W. Burden, figure 5.4, p. 99, figure 5.5, p. 102, figure 5.6, p. 104. Copyright © 1995 by Springer-Verlag. Reprinted with permission.

Figure 4.16: from "Antibodies Detect Protein Building" figure 4.17, p. 44 from *Genomics & Proteomics* November/December 2001. Copyright © 2003 *Genomics & Proteomic,* a publication of Reed Business Information, a division of Reed Elsevier Inc. All rights reserved. Reprinted by permission.

Figures 5.12 & 5.13: from *Microbiology,* Seventh Edition by Gerard J. Tortora, Berdell R. Funke & Christine L. Case.

Figure 5.17: from "Application of High-Density Microarrays for Expression and Genotype Analysis in Infectious Disease" by Michael Lorence, figure 3, p. 11 from *American Biotechnology Laboratory,* January 2006. Reprinted with permission, International Scientific Communications.

Figure 5.19: from "The Magic of Microarrays" by S.H. Friend and R.B Stoughton, from *Scientific American* 286(2) 44–53, February 2002.

Figure 6.12: from "Bonkers about Biofuels" by S. Herrerra. From *Nature Biotechnology,* v. 24 n. 7, July 2007, p. 755–160.

Figures 6.2, 6.3, 6.4, 6.6, 6.10, & 7.13: from *Recombinant DNA,* Second Edition by James D. Watson, et al. © 1992 by James D. Watson, Michael Gilman, Jan Witkowski and Mark Zollar. Used with permission of W. H. Freeman and Company.

Figure 7.2: from "Zebrafish Impacts Bioinformatics Programs" by Henry McCoy, figure 1, p. 25, *Genetic Engineering News,* v. 21, n. 11, 6/1/01 ©2001. Reprinted by permission.

Figure 7.4: from "A Mightier Mouse with Human Adaptive Immunity" by M. Kosko-Vilbios. From *Nature Biotechnology,* v. 22 n. 6, June 2004, p. 684–685, 2004. ©2004 Nature Publishing Group. Used with permission..

Figure 8.5: Courtesy of Dr. Donald E. Riley, University of Washington. ©1997.

Figure 9.5: reprinted with permission from "Anaerobes to the Rescue" by Derek R. Lovley, figure on p. 1445, *Science* 293: 1444–1446 (2001) August 24, 2001. Copyright © 2001 by the American Association for the Advancement of Science.

Figure 9.16: from "Plants tackle explosive contamination" by R.B. Meagher, from *Nature Biotechnology,* v. 24, n.2, February 2006, pp. 161–163. Copyright © 2006 Nature Publishing Group. Used with permission.

Figures 10.5 & 10.6: reprinted with permission from "Antifreeze proteins and their genes" by Garth L. Fletcher, Sally V. Goddary and Yaling Wu, figures 3 and 10, *Chemtech* 1999, 30(6), 17–28. Copyright © 1999 by the American Chemical Society. Used with permission.

Figure 11.4: from *Principles of Cell & Molecular Biology,* Second Edition by Lewis J. Kleinsmith and Valerie M. Kish, figure 3-53, p. 110. Copyright © 1995 by HarperCollins College Publishers. Reprinted by permission of Pearson Education, Inc.

Figure 11.5: from *Concepts of Genetics,* Seventh Edition by William S. Klug and Michael R. Cummings, figure 20-9, p. 530. Copyright © 2003, 2000, 1997 by William S. Klug and Michael R Cummings.

Figure 11.7: from "Pharmacogenomicists dream of customized drugs, but significant technical challenges" by Aileen Constans, figure "Pharmacogenetics: A Case Study," *The Scientist* Volume 16, Issue 19, 44, September 30, 2002.

Reprinted by permission of the National Institute of General Medical Sciences.

Figure 11.8: From J. Derisi "Use of a cDNA microarray to analyse gene expression patterns in human cancer." *Nature Genetics* v. 14, n. 4, pp. 457–460. Copyright © 1996 Nature Publishing Group. Used with permission.

Figure 11.10: from "Nanotechnology takes aim at cancer" by R.F. Service. From *Science* 310: 1132–1134, November 2005.

Figure 11.14: from "Therapeutic Interference" by C.W. Schmidt. From *Modern Drug Discovery,* July 2003, pp. 37–42.

Figure 11.23: from "Reprogramming by the numbers" by S-I Nishikawa. From *Nature Biotechnology* v. 25, n. 8, pp. 877–878, 2007. Copyright © 2007 Nature Publishing Group. Used with permission.

Figure 11.24: from "Cultivating Policy from Cell Types" by Eugene Russo, figure "Repairing Heart Attacks with Adult Stem Cells ," *The Scientist* 15(11):6, May 28, 2001. Reprinted by permission of the National Human Genome Research Institute.

Figure 12.8: by Jensen and Murray. From *Science,* 310, pp. 239–240.

Glossary

A (aminoacyl) site (of a ribosome): Portion of a ribosome into which aminoacyl tRNA molecules bind during translation.

accession number: Unique identifying letter and number code assigned to every cloned DNA sequence that is catalogued in databases such as GenBank (for example, BC009971 is the accession number for one of the human keratin genes, which produces a protein that is a major component of skin and hair cells). The accession number for any sequence can be used by scientists around the world to retrieve database information on that particular sequence.

acetylase: Enzyme produced by the *lacA* gene of the lactose (lac) operon in bacteria.

acquired mutations: These occur in the genome of somatic cells and are not passed along to offspring; can cause abnormalities in cell growth leading to cancerous tumor formation, metabolic disorders, and other conditions.

adenine: Abbreviated A; purine base present in DNA and RNA nucleotides.

adenosine triphosphate (ATP): A nucleoside-triphosphate that contains the nitrogenous base adenine. ATP is the primary form of energy used by living cells.

adenovirus: Virus that causes the common cold.

adult-derived stem cells (ASCs): Stem cells derived from tissues of an adult, as opposed to embryonic stem cells, which are derived from a blastocyst; can differentiate to produce other cell types.

aerobes: Organisms that use oxygen for their metabolism.

aerobic conditions: Conditions in which oxygen is present.

aerobic metabolism: Metabolism that requires oxygen.

affinity chromatography: A separation technique, based on the unique match between a molecule and its column-bound antibody, that involves passing proteins or other substances in solution over a medium that will bind to ("have an affinity for") specific components of the solution. It is used to isolate fusion proteins from a mixture of bacterial cell proteins.

agarose gel electrophoresis: See gel electrophoresis.

agricultural biotechnology: A discipline of biotechnology that involves plants and their applications, including genetic engineering of plants for agricultural purposes.

Agrobacter: Soil bacteria that invade injured plant tissue.

alcohol (ethanol) fermentation: Enzymatic breakdown of carbohydrates (sugars) in the absence of oxygen; products include ATP, CO_2, and ethanol (alcohol) as waste products; important type of microbial metabolism used for the production of certain types of alcohol-containing beverages.

allele-specific oligonucleotide (ASO) analysis: Genetic testing technique that involves using PCR with oligonucleotides specific to a disease gene to analyze a person's DNA.

alternative splicing: Splicing sometimes can join certain exons and *cut out* other exons, essentially treating them as introns. This process creates multiple mRNAs of different sizes from the same gene. Each mRNA can then be used to produce different proteins with different, sometimes unique, functions. Alternative splicing allows several different protein products to be produced from the same gene sequence.

amino acids: Building blocks of protein structure; combinations of 20 different amino acids can join together by covalent bonds in varying order and length to make a polypeptide.

aminoacyl transfer RNA (tRNA): Transfer RNA (tRNA) molecule with an amino acid attached.

amniocentesis: A technique for obtaining fetal cells from a pregnant woman to analyze fetal cells to determine the genetic composition of the cells, such as the number of chromosomes or the sex of the fetus.

amniotic fluid–derived stem cell (AFS): Stem cells in the amniotic fluid of pregnant women that have been found to have many of the same traits as embryonic stem cells.

amylase: An enzyme that digests starch.

anaerobes: Organisms that do not require oxygen for their metabolism.

anaerobic metabolism: Lacking oxygen; an organism, environment, or cellular process that does not require oxygen.

angiogenesis: The growth of new blood vessels.

Animal and Plant Health Inspection Service (APHIS): The branch of the USDA responsible for protecting U.S. agriculture from pests and diseases.

animal biotechnology: A diverse discipline of biotechnology that involves the use of animals to make valuable products such as recombinant proteins and organs for human transplantation; also includes organism cloning.

annotation: Bioinformatics approach that involves searching databases to determine whether the sequence and function of a gene sequence has already been determined.

antibiotic: Substance produced by microorganisms that inhibit the growth of other microorganisms; commonly used to treat bacterial infections in humans, pets, and farm animals.

antibiotic selection: See selection.

antibodies: Proteins produced in response to a non-self molecule by the immune system; antigen-binding immunoglobulins, produced by B cells, that function as the effector in an immune response.

antibody-mediated immunity: Portion of the immune system dedicated to producing antibodies that combat foreign materials; also known as humoral immunity.

anticodon: Three-nucleotide sequence at the end of a tRNA molecule. During translation, the anticodon binds to a specific codon in an mRNA molecule by complementary base pairing.

antifreeze proteins (AFPs): Category of proteins isolated from aquatic organisms that live in cold environments; these proteins have the unique property of lowering the freezing temperature of body fluids and tissues.

antigens: Molecules unique to specific surfaces that can stimulate antibody response; substances that trigger antibody production when introduced into the body.

antimicrobial drugs: Chemicals that inhibit or destroy microorganisms.

antiparallel: Refers to the 5' and 3' directionality of strands of DNA in which the two strands joined together by hydrogen bonds between complementary base pairs are oriented in opposing directions with respect to their polarity.

antisense RNA: An RNA molecule that is complementary to a native RNA.

apoptosis: Involves a complex cascade of cellular proteins that cause cell death; controlled apoptosis is an important biological process that remodels tissues during development; uncontrolled apoptosis is involved in degenerative disease conditions such as arthritis; also known as "programmed cell death."

aquaculture: Farming finfish, shellfish, or plants for commercial or recreational uses.

aquatic biotechnology: The use of aquatic organisms such as finfish, shellfish, marine bacteria, and aquatic plants for biotechnology applications.

Archaea: See domain Archaea.

astaxanthin: A pigment that gives shrimp their pink color. Recombinant astaxanthin is used to change the color of aquaculture species such as salmon.

attenuated vaccine: Vaccine consisting of weakened, live microorganisms.

autograft: Transplanting a patient's own tissues from one region of the body to another; for example, skin or hair grafting, coronary artery bypass surgery.

autoradiograph: An image created by exposing photographic film to a radioactively labeled compound.

autoradiography: Technique that uses film to detect radioactive or light-releasing compounds (such as DNA probe) in cells, tissues, or blots; produces photographic film image called an autoradiogram or autoradiograph.

autosomes: Chromosomes whose genes are not primarily involved in determining an organism's sex; chromosomes 1–22 in humans.

avian flu: "Avian influenza virus" refers to influenza A viruses found chiefly in birds, but infections with these viruses can occur in humans.

avidin: A protein in egg whites that binds biotin and inhibits bacterial growth.

B lymphocytes (B cells): White blood cells (leukocytes) that develop in the bone marrow and can mature into antibody-producing cells called plasma cells.

bacilli: Rod-shaped bacteria.

***Bacillus thuringiensis* (Bt):** A bacterium that produces a crystalline protein that dissolves the cementing substance between certain insect midgut cells.

bacteria: See domain Bacteria.

bacterial artificial chromosomes (BACs): Large circular vectors that can replicate very large pieces of DNA; used to clone pieces of human chromosomes for the Human Genome Project.

bacteriophage: Virus that infects bacteria; often simply called phage.

baculoviruses: Viruses that attack insect cells.

band shifting: Equal molecular-weight fragments of DNA that do not migrate uniformly because of variations in the gel matrix.

Basic Local Alignment Search Tool (BLAST): Internet-based program from the U.S. National Center for Biotechnology Information (NCBI) that can be used for DNA nucleotide-sequence-comparison (alignment) studies and for sequence searches in GenBank and other databases.

basic sciences: Research disciplines that examine fundamental aspects of biological processes, often without obvious direct applications (for curing disease or making a product).

batch (large-scale or scale-up) processes: Growing microorganisms such as bacteria or yeast and other living cells such as mammalian cells in large quantities for the purpose of isolating useful products in a batch.

β-galactosidase: Enzyme produced by the *lacZ* gene of the lactose (lac) operon in bacteria; β-galactosidase degrades the disaccharide lactose.

β-sheet: One of two secondary structures of proteins.

bioaccumulation: Progressive concentration or accumulation of a substance, such as a chemical pollutant, as it moves up the "food chain" from organism to organism.

bioaugmentation (seeding): Adding bacteria or other microorganisms to a contaminated environment to assist native (indigenous) microbes in the bioremediation processes.

biocapsules: Tiny spheres or tubes filled with therapeutic cells or a chemical substance such as a drug; can be implanted for therapeutic purposes; also known as microcapsules.

biodegradation: The use of microorganisms to break down chemicals (not necessarily pollutants).

biodiversity: The range of different species present in an ecosystem.

bioethicist: A person who studies bioethics.

bioethics: The area of ethics (a code of values for our actions) concerning the implications of biologic and biomedical research and biotechnological applications (particularly with respect to medicine).

biofilming (biofouling): The attachment of organisms to surfaces; examples include the attachment of shellfish to the hull of a ship and the attachment of microorganisms to teeth.

bioimpedance: A technique involving the use of low-voltage electricity to measure muscle mass and the fat content of tissues; used, for example, to measure the fat content of fish to be sold for consumption.

bioinformaticists: Scientists specializing in the bioinformatics.

bioinformatics: Interdisciplinary science that involves developing and applying information technology (computer hardware and software) for analyzing biological data such as DNA and protein sequences; also includes the use of computers for the analysis of molecular structures and creating databases for storing and sharing biological data.

biologic: Pertaining to biology.

Biological License Agreement (BLA): Filed by a biotech company seeking approval of a biologically derived product such as a viral therapy, blood compound, vaccine, or protein derived from animals.

bioluminescence: The release of light by living organisms.

biomarkers: Substances used as indicators of a biologic state. A characteristic that is objectively measured and evaluated as an indicator of normal biologic processes, pathogenic processes, or pharmacologic responses to a therapeutic intervention.

biomass: The dry weight of living material in part of an organism, a whole organism, or a population of organisms.

biopiles: Large piles of contaminated soil that have been removed from the original site. Air and vacuum systems are used to help dry these piles and release chemicals by evaporation. Vacuums are sometimes used to collect chemical vapors and trap them in filters.

bioprocessing engineers: Work at the frontiers of biologic and engineering sciences to "bring engineering to life" through the conversion of biologic materials into other forms needed by mankind.

bioprocessing: The use of biologic systems to manufacture (process) a product.

bioprospecting: Endeavors to capitalize on indigenous knowledge of natural resources. However, bioprospecting may also describe the search for previously unknown compounds in organisms that have never been used in traditional medicine.

bioreactors: Cell systems that produce biologic molecules (which may include fermenters).

bioremediation: The use of living organisms to process, degrade, and clean up naturally occurring or man-made pollutants in the environment.

biosensors: Living organisms used to detect or measure biological effects of some factor, such as a chemical pollutant, or condition.

biotechnology: A broad area of science involving many different disciplines designed to use living organisms or their products to perform valuable industrial or manufacturing processes or applications that will solve problems.

bioterrorism: The use of biological materials (live organisms or their toxins) as weapons to cause fear in or harm against civilians.

bioventing: Pumping air or oxygen-releasing chemicals such as hydrogen peroxide into contaminated soil or water to stimulate aerobic degradation by microorganisms.

bioweapons: Biological materials used as weapons.

blastocyst: Hollow cluster of approximately 100 cells formed about 1 week after fertilization of an egg.

blunt ends: Double-stranded ends of a DNA molecule created by the action of certain restriction enzymes.

***BRCA1* gene:** A gene found in about 65% of human breast tumors. Commonly part of breast cancer diagnosis for treatment.

byssal fibers: Ultrastrong, protein-rich threads created by mussels and other shellfish. Byssal fibers have unique adhesive properties and can withstand great stress from stretching and shearing forces; attach shellfish to substrates such as rocks.

CAAT Box: Short nucleotide sequence (CAAT) usually located approximately 80 to 90 base pairs "upstream" (in the 5' direction) of the start site of many eukaryotic genes; part of promoter sequence bound by transcription factors used to stimulate RNA polymerase.

calcitonin: A thyroid hormone that stimulates calcium uptake by digestive organs and promotes bone hardening (calcification).

callus: A loose collection of de-differentiated plant tissue.

Cancer Genome Atlas Project: NIH is working this cancer genome project to map important genes and genetic changes involved in cancer.

carbohydrases: Enzymes that digest carbohydrates.

carcinogen: A cancer-causing chemical.

carrageenan: Polysaccharide (sugar) derived from seaweeds; wide range of uses in everyday products such as syrups, sauces, and adhesives as a "thickening" or bulking agent.

cDNA: DNA synthesized as an exact copy of mRNA called complementary DNA (cDNA). The mRNA is degraded by treatment with an alkaline solution or enzymatically digested; then DNA polymerase is used to synthesize a second strand to create double-stranded cDNA.

cell culture: Growing cells in laboratory conditions outside of a whole organism (*in vitro*); usually a term applied to growing mammalian cells.

cell line: An established or immortalized cell line has acquired the ability to proliferate indefinitely, either through random mutation or deliberate modification. Numerous well-established cell lines are representative of particular cell types.

cell lysis: See lysis.

cellulase: Bacterial enzyme that degrades the polysaccharide cellulose (a primary component of the plant cell wall).

centrifugation: Involves using an instrument called a centrifuge to apply a rotational force to samples to separate components based on their weight. Rotational forces measured in revolutions per minute (rpm) or times gravity (*g*). Centrifugation has a number of important applications related to separating components of a mixture (for instance, separating proteins from DNA, separating different cell types).

centromere: Constricted region of a chromosome formed by intertwined DNA and proteins that hold two sister chromatids together.

chemotherapy: The treatment of cancer and other diseases with specific chemical agents or drugs that have a toxic effect on diseased cells or disease-causing microorganisms.

chimera: An organism with more than one type of cell; some cells are normal and some are genetically modified.

chitin: A complex polysaccharide polymer composed of repeating units of a sugar called *N*-acetylglucosamine. Chitin forms the hard outer shell (exoskeleton) of crabs, lobsters, crawfish, shrimp, and other crustaceans. Also found in insects and other organisms.

chitosan: Polysaccharide polymer derived from chitin. Used in many applications from health care to agriculture to dyes for fabrics and in dietary supplements for weight loss.

chloroplast: The photosynthetic organelle of complex plants.

cholera: Acute intestinal infection of humans caused by the bacterium *Vibrio cholerae* that causes severe diarrhea, which can result in dehydration and death, particularly in young children, if not treated. Spreads by contaminated water and food; bacterium often found in water supplies of developing countries with poor sanitary conditions.

chorionic villus sampling: Technique for sampling fetal cells from the placenta to determine the genetic composition of these cells, such as the number of chromosomes or the sex of the fetus.

chromatin: Strings of DNA wrapped around proteins; present in the nucleus of eukaryotes and the cytoplasm of prokaryotes; coils tightly together to form chromosomes during cell division.

chromatography: A columnar separation method that uses resin beads with special separation capabilities.

chromosomes: Highly folded arrangements of DNA and proteins; they package DNA to allow even separation of genetic material during cell division.

chymosin: See renin.

clinical trials: Experimental process of testing products before approval of a drug, or treatment plan for widespread use in humans; clinical trials involve several "phases" of carefully planned experiments in different numbers of human participants to test drug effectiveness and safety.

clone: A genetically identical copy of a cell or whole organism; also describes the process of making copies of a gene, cell, or organism.

cocci: Bacteria with a spherical shape.

CODIS, Combined DNA Index System: CODIS enables federal, state, and local crime labs to exchange and compare DNA profiles electronically, thereby linking crimes to each other and to convicted offenders.

codon: Combinations of three nucleotide sequences in an mRNA molecule; with few exceptions, each codon codes for a single amino acid; 64 possible codons compose the genetic code of mRNA molecules.

cohesive (sticky) ends: Overhanging single-stranded ends of a DNA molecule created by the action of certain restriction enzymes.

colchicine: Substance derived from crocus flowers that blocks microtubule formation; used to stop cell division (for instance, when creating polyploid organisms).

collagenase: Protease (protein-digesting enzyme) used in a number of biotechnology applications.

colony hybridization: A procedure for binding a single-stranded DNA probe to DNA molecules from bacterial colonies; used for library screening to identify a gene of interest.

comparative genomics: Studies that allow researchers to investigate gene structure and function in organisms in ways designed to understand gene structure and function in other species including humans.

competent cells: Bacterial cells that have been chemically treated to be able to take in DNA (transformation) from their surrounding environment; to be "competent" to accept DNA.

complementary base pairs: Refers to the nucleotides adenine (A) and thymine (T) [or uracil (U) when referring to RNA], and guanine (G) and cytosine (C); in a DNA molecule, A&T and G&C nucleotides join by hydrogen bonds to "complement" each other.

complementary DNA (cDNA): DNA copy of an mRNA molecule; mRNA can be copied into cDNA by the enzyme reverse transcriptase (for instance, when preparing a cDNA library).

complementary DNA (cDNA) library: DNA copies of all mRNA molecules expressed in an organism's cells; can be "screened" to isolate genes of interest.

composting: Mixing soil with hay, straw, grass clippings, wood chips, or other similar "bulking" materials to stimulate biodegradation by soil microbes; also used to degrade everyday materials such as vegetable scraps, grass clippings, weeds, and leaves.

contigs (contiguous sequences): Restriction enzyme–digested pieces of entire chromosomes are sequenced separately, and then computer programs are used to align the fragments based on overlapping sequence pieces called contigs (contiguous sequences).

Coomassie stain: A sensitive stain that reacts with proteins.

Coordinated Framework for Regulation of Biotechnology: A 1984 governmental report that proposed that biotechnology products be regulated much like traditional products.

copy number: The number of copies of a particular DNA molecule or gene sequence such as a plasmid in a bacterial cell.

cosmid vector: Large circular double-stranded DNA vector that is used for gene-cloning experiment in bacteriophages.

credible utility: Requirement of the USPTO that the researcher must convince the patent office that the application is backed by sound science.

crown gall: A hardened mass of protruding dedifferentiated plant tissue usually resulting from an attack from *Agrobacter*.

cryopreservation: Storage of tissue samples at ultra low temperatures (–150 to –20°C).

customer relations specialists: Sometimes work in QA divisions of a company. One function of customer relations is to investigate consumer complaints about a problem with a product and follow up with the consumer to provide an appropriate response or solution to the problem encountered.

cystic fibrosis transmembrane conductance regulator (CFTR): Protein that serves as a "pump" to regulate the amount of chloride ions in cells; mutations in the CFTR are the cause of the genetic disorder cystic fibrosis.

cytokines: Stimulating factors that enhance an animal's immune system function.

cytoplasm: The inner contents of a cell, consisting of fluid (cytosol) and organelles; its boundaries are defined by the plasma membrane.

cytosine: Abbreviated C; pyrimidine base present in DNA and RNA nucleotides.

cytosol: Gel-like, nutrient-rich aqueous (water-based) fluid of the cytoplasm.

dideoxyribonucleotide (ddNTP): modified nucleotide. A ddNTP differs from a normal deoxyribonucleotide (dNTP) because it has a hydrogen group attached to the 3' carbon of the deoxyribose sugar instead of a hydroxyl group-OH.

denaturation: A process that involves using high heat or chemical treatment to break hydrogen bonds in DNA or RNA molecules to separate complementary base pairs and create single-stranded molecules; also refers to the act of changing protein structure (through heat or chemical treatment).

deontological (Kantian) approach: Developed by German philosopher Immanuel Kant, this approach suggests that absolute principles (of moral or ethical behavior) should be followed to dictate our actions.

deoxyribonucleic acid (DNA): Double-stranded nucleic acid consisting of bases (adenines, guanines, cytosines, thymines), sugars (deoxyribose), and phosphate groups arranged in a helical molecule; inherited genetic material that contains "genes," which direct the production of proteins in a cell.

depolymerization: A process of reducing multiunit structures to single units.

deregulated status: First step in the APHIS (USDA) review of a genetically altered product in which field-test reports, scientific literature, and any other pertinent records are checked before APHIS determines whether the plant is as safe to grow as traditional varieties.

diabetes mellitus: A condition of elevated blood sugar. There are several different forms of diabetes. In some people (type 1 or insulin-dependent diabetes mellitus), the hormone insulin is missing.

diafiltration: See dialysis.

dialysis: A separation of water and a solute involving a semipermeable membrane.

Dicer: Enzyme that cuts microRNA transcripts into short single-stranded pieces of 21 to 22 nucleotides.

dicotyledonous: A seed with two embryonic seed leaves (like peanuts).

differential display PCR: A PCR technique that uses random primers to amplify DNA in two or more different samples with the purpose of identifying randomly amplified genes that are different (differentially expressed) in one sample compared with another sample.

differentiation: Cellular maturation process; involves changes in patterns of gene expression that affect the structure and functions of cells. For instance, embryonic stem cells differentiate to produce mature cells such as neurons, liver cells, skin cells, and all other cell types in the body.

diploid: Refers to a cell or an organism consisting of two sets of chromosomes: usually, one set from the mother and another set from the father. In a diploid state, the haploid number is doubled; thus, this condition is also known as 2*n*.

diploid number: Two sets of chromosomes (2*n*); often used to describe cells with two sets of chromosomes, typically one set from each parent.

directed molecular evolution: A process that selects for proteins after extensive mutations.

DNA-binding domains: Regions with folded structural arrangements of amino acids called motifs that interact directly with DNA.

DNA fingerprinting: An analysis of an organism's unique DNA composition as a characteristic marker or fingerprint for identification purposes, such as forensic analysis, remains identification, and paternity. DNA fingerprinting is also used in biologic research (for example, to compare related species based on their DNA sequences).

DNA helicase: Enzyme that separates two strands of a DNA molecule during DNA replication.

DNA library: See genomic DNA library or complementary (cDNA) library.

DNA ligase: Enzyme that forms covalent bonds between incomplete (Okazaki) fragments of DNA during DNA replication; routinely used for joining DNA fragments in recombinant DNA experiments.

DNA microarray (gene chip): A chip consisting of a glass microscope slide containing thousands of pieces of single-stranded DNA molecules attached to specific spots on the slide; each spot of DNA is a unique sequence.

DNA polymerase: Key enzyme that copies strands of DNA during DNA replication to create new strands of nucleotides; has important applications for synthesizing DNA in molecular biology experiments.

DNA sequencing: Laboratory technique for determining the nucleotide "sequence" or arrangement of A, G, T, and C nucleotides in a segment of DNA.

doctrine of equivalents: If a patent holder claims patent infringement on two similar DNA sequences, a court must decide whether the sequences have essentially the same function.

domain Archaea: A classification (domain) of prokaryotic cells; a common feature of these microbes is that they live in extreme environments, for example, boiling hot springs.

domain Bacteria: Taxonomic category of living organisms that includes bacteria; one of three domains (Bacteria, Archaea, Eukarya), domain Bacteria is the most diverse domain consisting of primarily unicellular prokaryotes.

domains: Taxonomic categories above the kingdom level: Archaea, Bacteria, and Eukarya.

dot blot: A detection strip coated with DNA sequences that reflect sites in human DNA where antibody-forming alleles (gene) differ.

double-blind trial: Study, such as a clinical trial, in which neither the patients nor the researchers administering the treatment know which patients receive a drug treatment and which patients receive a placebo; researchers learn the treatment of each patient after the study is completed.

Down syndrome: Human genetic disorder first described by John Langdon Down. Most commonly results from individuals having three copies of chromosome 21 (trisomy 21). Characteristics of affected individuals include a flat face; shortness; short, broad hands and fingers; protruding tongue; and impaired physical, motor, and mental development.

downstream processing: Purification processes involved in producing a pure product such as a recombinant protein.

DPT vaccine: Childhood vaccine designed to provide immune protection against bacterial toxins diphtheria, pertussis, and tetanus (contains killed *Bordetella pertussis* cells). Microbes producing these toxins cause upper respiratory infections, whooping cough, and tetanus (muscle spasm and paralysis), respectively.

E site: A groove through which tRNA molecules leave the ribosome.

efficacy: The effectiveness of a particular product such as a drug or medical procedure.

effluent: Water that has been treated to remove chemical pollutants or sewage; treated water as it leaves a treatment facility, such as water leaving a sewage-treatment plant.

electroporation: A process for transforming bacteria with DNA that uses electrical shock to move DNA into cells; can also be used to introduce DNA into animal and plant cells.

embryo twinning: Splitting embryos in half at the two-cell stage of development to produce two embryos.

embryonic stem (ES) cells: Cells typically derived from the inner cell mass of a blastocyst; cells can undergo differentiation to form all cell types in the body.

endotoxins (lipopolysaccharides): Molecules that are toxic to cells; endotoxins are part of the cell wall of certain bacteria; endotoxins cause cell death.

enhancers: Specific DNA sequences that bind proteins called transcription factors to stimulate ("enhance") transcription of a gene.

enucleation: Removal of the DNA from the nucleus of a cell.

Environmental Genome Project: NIH program designed to study the impacts of environmental chemicals on human genetics and disease.

Environmental Protection Agency: See U.S. Environmental Protection Agency.

environmental genomics (metagenomics): Involves sequencing genomes for entire communities of microbes in environmental samples of water, air, and soils from oceans throughout the world, glaciers, mines—virtually every corner of the globe.

EnviroPig: A transgenic pig expressing the enzyme phytase in its saliva. It is deemed environmentally friendly because it produces substantially less phosphorus in its urine and feces than do nontransgenic pigs.

enzymes: Catalytic proteins.

ES (embryonic stem) cell transfer: Embryonic stem cells (ES cells) are collected from the inner cell mass of blastocysts. Transformed ES cells are usually injected into the inner cell mass of a host blastocyst.

ethidium bromide: Tracking dye that penetrates (intercalates) between the base pairs of DNA. Commonly used to stain DNA in gels because ethidium bromide fluoresces when exposed to ultraviolet light.

eugenics: The creation of a new superior species of human.

eukaryotic cells: Cells that contain a nucleus, including plant and animal cells, protists, and fungi.

European Agency for the Evaluation of Medicinal Products (EMEA): Agency that authorizes medicinal products for human and veterinary use. It also helps spread scientific knowledge by providing information on new products in all 11 official EU languages.

exons: Protein-coding sequences in eukaryotic genes and mRNA molecules.

ex situ **bioremediation:** Removing contaminated soils or water from the site of contamination for clean-up at another location (as opposed to cleaning up pollutants at the contaminated site—a process called *in situ* bioremediation).

ex vivo **gene therapy:** Gene-therapy procedure that involves removing a person's cells (such as blood cells) from the body, introducing therapeutic genes into these cells, and then reintroducing the cells back into a person.

experimental use permit (EUP): Permit issued by the EPA for field experiments with new plant varieties with pesticidal capabilities that involve at least 10 acres of land or 1 acre of water; these experiments cannot be conducted without an experimental use permit.

expressed sequence tags (ESTs): Small incomplete pieces (tags) of gene sequence such as those derived from a cDNA library; represent mRNAs expressed in a given tissue.

expression vector: DNA vector such as a plasmid that can be used to produce (express) proteins in a cell.

extrachromosomal DNA: DNA that is not part of an organism's nuclear (chromosomal) DNA; examples include plasmid and mitochondrial DNA.

fermentation: A metabolic process that produces small amounts of ATP from glucose in the absence of oxygen and also creates byproducts such as ethyl alcohol (ethanol) or lactic acid (lactate). Fermenting microbes (bacteria and yeast) are important for producing a variety of beverages and foods, including beer, wine, breads, yogurts, and cheeses.

fermenters (bioreactors): Containers for growing cultures of microorganisms or mammalian cells in a batch process. Fermenting vessels allow scientists to control and monitor growth conditions such as temperature, pH, nutrient concentration, and cell density.

fertilization (nutrient enrichment): Addition of nutrients (fertilizer such as nitrogen and phosphorus) to a contaminated environment to stimulate growth and activity of naturally occurring soil microorganisms that will aid bioremediation.

field trials: Tests outside of the laboratory required by APHIS (USDA) after a new plant has been engineered in a laboratory; may require several years to investigate everything about the plant, including disease resistance, drought tolerance, and reproductive rates.

5' end: The end of a strand of DNA or RNA in which the last nucleotide is not attached to another nucleotide by a phosphodiester bond involving the 5' carbon of the pentose sugar.

fluorescence *in situ* **hybridization (FISH):** Laboratory technique that uses single-stranded DNA or RNA probes labeled with fluorescent nucleotides to identify gene sequences in a chromosome or cell *in situ* (Latin for "in its original place").

Food and Drug Administration (FDA): An agency of the federal government that regulates food and drug safety.

forensic biotechnology: The analysis and application of biologic evidence such as DNA and protein-sequence data to help solve or recreate crimes.

frameshift mutation: Mutation resulting from the addition or deletion of nucleotides that causes a shift in the genetic code-reading frame of a gene.

functional genomics: Genetic-sequencing analysis of genes active during a cellular event.

fungi: Eukaryotic organisms that belong to Kingdom Fungi.

fusion proteins: A "hybrid" recombinant protein consisting of a protein from a gene of interest connected (fused) to another, well-known protein that serves as a tag for isolating recombinant proteins.

gametes: Haploid cells often called "sex" cells include human sperm and egg cells (ova). Gametes join together during fertilization to form a zygote (see meiosis).

gel electrophoresis: Laboratory procedure that involves using an electrical charge to move and separate biomolecules of different sizes, such as DNA, RNA, and proteins, through a semisolid separating gel matrix. Examples include agarose gel electrophoresis and polyacrylamide gel electrophoresis (PAGE).

GenBank: Renowned public database of DNA sequences provided by researchers throughout the world; resources for sharing and analyzing DNA sequence information.

gene: A specific sequence of DNA nucleotides that serves as a unit of inheritance. Genes govern visible and invisible characteristics (traits) of living organisms in large part by directing the synthesis of proteins in a cell.

gene chip: See DNA microarray.

gene cloning: The process of producing multiple copies of a gene.

gene expression: Term generally used to describe the synthesis ("expression") of RNA by a particular cell or tissue. For instance, if mRNA for the fictitious gene *korn* is detected in a tissue by PCR or Northern blot analysis, that tissue is said to express the *korn* gene.

gene gun: A microprojectile device used to propel genes on the surface of metal particles into cells.

Genentech: California company. Name derived from genetic-engineering technology. Founded in 1976 and widely recognized as the world's first biotechnology company.

gene regulation: General term used to describe processes by which cells can control gene activity or "expression" (RNA and protein synthesis).

gene therapy: The use of therapeutic genes to treat or cure a disease process; also refers to the delivery of genes to improve a person's health.

generally-recognized-as-safe (GRAS) status: Status granted by the FDA if a food product or additive poses no foreseeable threat.

generic drugs: Copies of brand-name products that generally have the same effectiveness, safety, and quality but are produced at a cheaper cost to the consumer than the brand-name drugs.

genetic code: Information contained in bases of DNA and RNA that provide cells with instructions for synthesizing proteins; consist of three-nucleotide combinations called codons.

genetic engineering: The process of altering an organism's DNA. This is usually by design.

Genetic Information Nondiscrimination Act: This act prohibits discrimination based on genetics and the improper use of genetic information in health insurance and employment.

genetically modified foods: Foods that have been genetically altered to produce desired characteristics.

genetically modified (GM) organisms: GM organisms included selectively bred and transgenic organisms. For example, GM crops have been produced to plants that can resist pests, disease, or harsh climates, allowing better production of crops.

genome: All of the genes in an organism's DNA.

genomic DNA library: Collection of DNA fragments containing all DNA sequences in an organism's genome; can be "screened" to isolate genes of interest.

genomics: The study of genomes.

germline genetic engineering: Genetic alteration of sperm, eggs, or embryos to create inheritable genetic changes in offspring.

glycosylation: A natural process of adding sugar units to proteins by complex cells.

glyphosphate: A small molecule that interacts and blocks 5-enolpyruvylshikimate-3-phosphate synthase (EPSPS), a key enzyme in plant photosynthesis. Resistant plants have a substitute pathway obtained by gene transfer.

Golden Rice: Genetically engineered rice with high vitamin A content.

good clinical practice (GCP): An FDA requirement governing clinical trials that protects the rights and safety of human subjects and ensures the scientific quality of experiments.

good laboratory practice (GLP): An FDA requirement that testing laboratories follow written protocols, have adequate facilities, provide proper animal care, record data properly, and conduct valid toxicity tests.

good manufacturing practice (GMP): Regulations instituted by the FDA to govern animal studies of pharmaceutical products.

green fluorescent protein (GFP): Protein produced by the bioluminescent jellyfish *Aequorea victoria*. Protein fluoresces when exposed to ultraviolet light; the *GFP* gene is used as a reporter gene.

Gram stain: Technique for staining the bacterial cell wall that can be used to divide bacteria into different categories, gram-positive or gram-negative bacteria.

growth hormone (GH): A peptide hormone produced by the pituitary gland; accelerates the growth of bone, muscle, and other tissues.

guanine: Abbreviated G; purine base present in DNA and RNA nucleotides.

HAMA response: When mouse hybridoma-produced human antibodies produce enough "mouse" antigens, they evoke this undesirable immune response in humans.

haploid: A single set of 23 chromosomes.

haploid number: A single set of chromosomes (*n*); often used to describe cells with a single set of chromosomes.

helix: One of two secondary structures of proteins.

hemoglobin: Oxygen-binding and oxygen-transporting protein of red blood cells in vertebrates. Mutation of globin genes that encode hemoglobin results in a number of different human blood disorders with a genetic basis, including sickle-cell disease.

high-performance liquid chromatography (HPLC): High-pressure separation of similar proteins by using incompressible beads with special properties.

Hippocratic Oath: An oath that embodies the obligations and duties of medical doctors; one expectation of the oath is that physicians treat patients with the intent of not worsening human health ("first, do no harm").

Histone: DNA-binding proteins that are important for chromosome structure.

homologous recombination: A gene recombines with its corresponding existing gene on a chromosome (with which it has a high degree of sequence similarity), replacing part or all of the gene.

homologues: Related genes in different species; genes share sequence similarity because of a common evolutionary origin.

hormone: Molecules involved in chemical signaling between cells and organs; transported in body fluids, hormones interact with and control the activity of many cells.

human antimouse antibody (HAMA) response: Mouse hybridomas produce enough "mouse" antigens that they still evoke an undesirable immune response in humans.

human embryonic stem cells: Stem cells are immature cells that can grow and divide to produce different types of cells such as skin, muscle, liver, kidney, and blood cells. Most stem cells are obtained from embryos (embryonic stem cells or ESCs). Some ESCs can be isolated from the cord blood of newborn infants.

humane treatment: Compassionate and sympathetic approach, as applied to treating humans or animals.

Human Genome Project: An international effort with overall scientific goals of identifying all human genes and determining (mapping) their locations to each human chromosome.

human papillomavirus (HPV): Causes about 70% of cervical cancers (HPV strains 16 and 18) and a large percentage of genital warts (caused by HPV strains 6 and 11). Cervical cancer affects 1 in 130 women, nearly half a million women worldwide, and approximately 70% of sexually active women will become infected with HPV.

Human Proteome Project: A potential project designed to study the structures and functions of all human proteins (the proteome).

humulin: The recombinant form of human insulin, became the first recombinant DNA product to be approved for human applications by the U.S. Food and Drug Administration.

hybridization: The joining of two DNA strands by complementary base pairing; for instance, binding of a single-stranded DNA probe to another DNA molecule.

hybridomas: Hybrid cells used to create monoclonal antibodies; created by fusing B cells with cancerous cells called myeloma cells that no longer produce an antibody of their own.

hydrophilic: Water-loving (portion of protein molecule).

hydrophobic interaction chromatography (HIC): Chromatographic separation based on binding of hydrophobic parts of proteins to the column.

hydrophobic: Water-hating (portion of protein molecule).

hydroponic systems: A form of aquaculture in which tanks of flowing water are used to grow plants; in some cases, water from fish aquaculture tanks is used to provide nutrients for plant growth in a polyculture approach.

hydroxyapatite (HA): Structural component of bone and cartilage.

***in situ* bioremediation:** Cleaning up pollutants at the actual site of contamination; "in place" clean-up as opposed to removing contaminated soils or water from the clean-up site for remediation at another location—a process called *ex situ* bioremediation.

***in situ* hybridization:** Laboratory technique that uses single-stranded DNA or RNA probes, usually labeled with radioactive or color-producing nucleotides, to identify gene sequences in a chromosome or cell *in situ* (Latin for "in its original place").

in vitro: Occurring within an artificial environment such as a test tube; from Latin "in glass."

***in vitro* fertilization:** Assisted-reproduction technology in which sperm and egg cells are removed from patients and cultured in a dish (*in vitro*) to achieve fertilization.

in vivo: Occurring within a living organism; from Latin "in something alive."

***in vivo* gene therapy:** Gene therapy procedure that involves introducing therapeutic genes directly into a person's tissue or organs without removing them from the body.

inactivated vaccines: Vaccine consisting of killed microorganisms.

inclusion bodies: Foreign proteins that concentrate in transformed cell.

indigenous microbes: Naturally occurring microorganisms living in the environment.

induced pluripotent stem cells (iPSCs): Nuclear reprogramming of mouse and human cells, heralded as a revolution in stem-cell biology research. One approach has involved using retroviruses to deliver four transgenes *Oct3/4, Sox2, c-myc,* and *Klf4* into fibroblasts. Expression of these four genes, which encode transcription factors involved in cell development, "reprograms" the fibroblasts back to an earlier stage of differentiation. iPS cells demonstrate many properties of hESCs, such as self-renewal and pluripotency, and appear to be indistinguishable from hESCs.

influenza: Caused by a large number of viruses that belong to the influenza family of viruses. Influenza kills approximately 500,000 to 1 million people worldwide each year. Because flu viruses mutate so rapidly, no one-size-fits-all vaccine protects against all strains.

informed consent: Applies to clinical trials in humans. Patients are made aware of potential beneficial and harmful effects of a particular treatment so that they are informed of the risks of a procedure before they agree (consent) to participate.

inherited mutations: A change in DNA structure or sequence of a gene passed to offspring through gametes; can be a cause of birth defects and genetic disease.

inner cell mass: Layer of cells in the blastocyst that develop to form body tissues; a source of embryonic stem cells.

Innocence Protection Act: A law that gives the convicted access to DNA testing, prohibits states from destroying biological evidence as long as a convicted offender is imprisoned, prohibits denial of DNA tests to death-row inmates, and encourages compensation for convicted innocents.

inorganic compounds: See inorganic molecules

inorganic molecules: Molecules that do not contain carbon.

insulin: Protein hormone produced by cells of the pancreas; involved in glucose metabolism by cells; deficiencies in insulin production or insulin-receptor production can cause different forms of diabetes.

insulin-dependent (type I) diabetes mellitus: Caused by an inadequate production of insulin by beta cells. The decreased production of insulin results in an elevated blood glucose concentration that can cause a number of health problems such as high blood pressure, poor circulation, cataracts, and nerve damage. People with type I diabetes require regular injections of insulin to control their blood sugar levels.

integration: Inserting DNA into a genome. A process by which retroviruses such as lentivirus and HIV can enter host cells, copy their RNA genome into DNA and then randomly insert their DNA into the genome of the host cell, where it remains permanently.

International Laboratory for Tropical Agricultural Biotechnology (ITLAB): A not-for-profit research laboratory funded by Monsanto and others to research plant varieties that could benefit countries by improving their food quality.

introns: Non–protein-coding sequences in eukaryotic genes and primary transcripts that are removed during RNA splicing.

investigational new drug (IND) application: A formal request to the FDA to consider the results of previous animal experiments, the nature of the substance itself, and the plans for further testing.

ion-exchange chromatography (IonX): The attachment of protein to a column based on its charged side groups.

isoelectric focusing: Migration of a protein until its charge matches the pH of the medium.

isoelectric point (IEP): pH at which the charge on proteins matches that of the surrounding medium.

Kantian approach: See deontological (Kantian) approach.

karyotype: A laboratory procedure for analyzing the number and structure of chromosomes in a cell.

knock-in animals: Animals can have a human gene inserted to replace their own counterpart by homologous recombination to become knock-ins.

knock-outs: An active gene is replaced with DNA that has no functional information.

laboratory technician: Entry-level laboratory job with a range of responsibilities such as preparing solutions and mediums, ordering laboratory supplies, cleaning and maintaining equipment; may sometimes involve bench research.

lac **operon:** A well-characterized bacterial operon that contains a series of genes (*lacZ, Y, Z*) responsible for metabolism of the sugar lactose.

lac **repressor:** Inhibitory protein encoded by the *lacI* gene of the lactose (lac) operon in bacteria; in the absence of lactose, repressor blocks transcription of the lac operon.

lactic acid fermentation: See fermentation.

lagging strand: During DNA replication, the strand of newly synthesized DNA that is copied by DNA polymerase in a discontinuous (interrupted) fashion, 5' to 3' away from the replication fork as a series of short DNA pieces called Okazaki fragments.

landfarming: Process by which contaminated soil is removed from a site and spread into thin layers on a pad that allows polluted liquids (usually water) to leach from the soil. Chemicals also vaporize from spread-out soil as the soil dries.

leachate: Water or other liquids that move (leach) through the ground from the surface or near the surface to deeper layers.

leading strand: During DNA replication, the strand of newly synthesized DNA that is copied by DNA polymerase in a continuous fashion, 5' to 3' into the replication fork.

leaf-fragment technique: A method of plant cloning from asexual plant tissue.

legal specialists: In biotechnology companies, typically work on legal issues associated with product development and marketing, such as copyrights, naming rights, and obtaining patents. Staff in this area also address legal circumstances that may arise if problems are found with a product or litigation from a user of a product.

leukocytes: White blood cells; important cells of the immune system; include B and T lymphocytes and macrophages.

ligands: Binding components.

limulus amoebocyte lysate (LAL) test: A procedure involving blood cells (amoebocytes) from the horseshoe crab (*Limulus polyphemus*); an important test used to detect endotoxin and bacterial contamination of foods, medical instruments, and other applications.

lipases: Fat-digesting enzymes.

liposomes: Small hollow spheres or particles made of lipids; can be packaged to contain molecules such as DNA and medicines for use in therapeutic procedures (for example, gene therapy).

luciferase: A light-releasing enzyme present in bioluminescent organisms.

lyophilization: The process of freeze-drying.

lysis: The dissolution or destruction of cells.

lytic cycle: A process of bacteriophage replication that involves phages infecting bacterial cells and then replicating and rupturing (lysing) the bacterial host cells.

macrophage: Term literally means big eaters; macrophages are phagocytic white blood cells that engulf and destroy dead cells and foreign materials such as bacteria.

major histocompatibility complex (MHC): Tissue-typing proteins present on all cells and tissues; recognized by a person's immune system to determine whether cells are normal body cells or foreign; MHCs must be "matched" for successful organ transplantation.

malaria: Caused by the protozoan parasite *Plasmodium falciparum* and transmitted by insects. Worldwide, *Plasmodium* strains are developing resistance to the most commonly used antimalarial drugs.

manufacturing assistant: Entry-level job includes material handlers, manufacturing assistants, and manufacturing associates. Supervisory and management-level jobs usually require a bachelor's or master's degree in biology or chemistry and several years of experience in manufacturing the products or type of product being produced by a company.

manufacturing technicians: Must strictly comply with Food and Drug Administration (FDA) regulations at all stages of producing biotech drugs.

mariculture: The cultivation or "farming" of aquatic organisms (animals or plants).

marker gene: See reporter gene.

marketing specialist: Marketing and sales position in biotechnology companies; marketing specialists are often involved in designing ad campaigns and promotional materials to market effectively a company's products.

Mass Fatality Identification System (M-FISys): M-FISys was essentially built in response to the 9/11 tragedy. Gene Codes did not have to write entirely new software, and they were able to customize M-FISys as necessary. In addition to analyzing mitochondrial DNA, M-FISys incorporated male-specific variations in the Y chromosome called Y-STRs to aid in the identification of individuals.

mass spectrometry (mass spec): Separation or identification based on charge-to-mass ratio.

maternal chromosomes: The 23 chromosomes inherited from your mother.

medical biotechnology: A diverse discipline of biotechnology dedicated to improving human health; includes a spectrum of topics in human medicine from disease diagnosis to drug discovery, disease treatment, and tissue engineering.

meiosis: A nuclear division process that occurs during formation of gametes. Meiosis involves a series of steps that reduce the amount of DNA in newly divided cells by half compared with those in the original cell.

membrane filtration: Using a thin membrane sheet with small holes (pores) to filter materials (such as large proteins or whole cells) out of a solution.

messenger RNA (mRNA): mRNA, a template for protein synthesis, is an exact copy of a gene, contains nucleotide sequences copied from DNA that serve as a genetic code for synthesizing a protein, and is then bound and "read" by ribosomes to produce proteins.

metabolic engineering: Modifying an energy-generating or energy-requiring chemical process (usually to improve energy generation or to reduce energy use) through a biotechnical or chemical process.

metabolomics: A biochemical snapshot of the small molecules produced during cellular metabolism, such as glucose, cholesterol, ATP, and signaling molecules that result from a cellular change.

metagenomics: Sequencing of genomes for entire communities of microbes in environmental samples of water, air, and soils from oceans throughout the world, glaciers, mines—virtually every corner of the globe.

metallothioneins: Metal-binding proteins.

microbes: Tiny organisms that are too small to be seen individually by the naked eye and must be viewed with the help of a microscope. Although the most abundant microorganisms are bacteria, microbes also include viruses, fungi such as yeast and mold, algae, and single-celled organisms called protozoa.

microbial biotechnology: Discipline of biotechnology that involves the use of organisms (microorganisms such as bacteria and yeast) that cannot be seen individually by the naked eye to make valuable products and applications.

Microbial Genome Program (MGP): U.S. Department of Energy program to map and sequence genomes of a broad range of microorganisms.

microcosms: Test environments designed to mimic polluted environmental conditions; may consist of bioreactors or simply a bucket of polluted soil; allows scientists to test clean-up strategies on a small scale under controlled conditions before trying a particular clean-up approach in the environment.

microfiltration: Removal of particles 40 μm or smaller.

microorganisms (microbes): Organisms that cannot individually be seen with the naked eye; include bacteria, yeast, fungi, protozoans, and viruses.

microRNAs: A new class of non-protein coding RNA molecules (**miRNAs**). MicroRNAs are part of a rapidly growing family of small RNA molecules about 20 to 25 nucleotides in size that play novel roles in regulating gene expression.

microsatellites: Also known as short tandem repeats (STRs), microsatellites are short repeating sequences of DNA usually consisting of one-, two-, or three-base sequences (for example, CACACACACA); microsatellites are important markers for forensic DNA analysis.

microspheres: Tiny particles that can be filled with drugs or other substances and used in therapeutic applications.

missense mutation: A mutation changing a codon to another codon that codes for a different amino acid.

mitochondrial DNA: Small circular DNA found in mitochondria responsible for proteins unique to mitochondria. Mutations in this DNA can be followed from mother to offspring, because the egg is the predominant source of mitochondria.

mitosis: A nuclear-division process that occurs during cell division in eukaryotic somatic cells and prokaryotes; divided into four major stages (prophase, metaphase, anaphase, and telophase). Mitosis results in the even separation of DNA into dividing cells.

MMR vaccine: Measles, mumps, and rubella (German measles) vaccine designed to provide immune protection against common childhood diseases.

model organisms: Nonhuman organisms that scientists use to study biologic processes in experimental laboratory conditions; common examples include mice, rats, fruit flies, worms, and bacteria.

molecular pharming: The use of plants as sources of pharmaceutical products.

monoclonal antibodies (MABs): Antibody proteins produced from clones of a single ("mono") cell; these proteins are highly specific for a particular antigen.

monocotyledonous: Describes a plant with a single embryonic seed leaf (like corn).

moratorium: As it relates to science, a temporary but complete stoppage of any research.

morula: Latin term meaning "little mulberry;" solid ball of cells that forms during embryonic development in animals, created by repeated cell division of the zygote.

motifs: These regions, called DNA-binding domains, have folded structural arrangements of amino acids called motifs that interact directly with DNA.

multilocus probes: The probe will bind only to complementary sequences of DNA, located at more than one site in the genome.

mutagen: A physical or chemical agent that causes a mutation.

mutation: A change in the DNA structure or sequence of a gene.

myelomas: Antibody-secreting tumors.

nanomedicine: Applying nanotechnology to improve health; for instance, microsensors that can be implanted into humans to monitor vital values such as blood pressure.

nanotechnology: Engineering and structures and technologies at the nanometer scale.

National Institutes of Health (NIH): Government agency that is the focal point for medical research in the United States; houses world-renowned research centers and agencies that are an essential source of funding for biomedical research in the United States.

New Drug Authorization (NDA): Approval of a new drug by the Food and Drug Administration (FDA) after the drug has successfully passed through Phase III clinical trials.

nitrogenous base: Important component of DNA and RNA nucleotides; often simply called a "base." Nitrogen-containing structures include double-ring purines (which include the bases adenine and guanine) and single-ring pyrimidines (which include the bases cytosine, thymine, and uracil).

nonsense mutation: A mutation changing a codon into a stop codon; usually produces a shortened, poorly functioning, or nonfunctional protein.

Northern blot analysis: Laboratory technique for separating RNA molecules by gel electrophoresis and transferring (blotting) RNA onto a filter paper blot for use in hybridization studies.

Northern blotting: See Northern blot analysis.

notification: A declaration to the USDA that can be used to "fast-track" some new agricultural products, in which the plant species are well characterized, introduce no new disease, are contained within the nucleus, do not produce a toxin, and are not produced from a plant, animal, or human virus.

nuclear envelope: A double-layered membrane, is typically the largest structure in an animal cell.

nuclear reprogramming of somatic cells: Used for isolating stem cells without creating an embryo. The basic concept of this approach is to use genes involved in cell development to push a somatic cell back to an earlier stage of development and affect gene expression, to reprogram the somatic cell genetically to return to a pluripotent state characteristic of the stem cells from which it was derived.

nucleic acids: Molecules composed of nucleotide building blocks; two major types are DNA and RNA.

nucleotide: Building block of nucleic acids; consists of a five-carbon (pentose) sugar molecule (ribose or deoxyribose), a phosphate group, and the bases adenine (A), guanine (G), cytosine (C), thymine (T), or uracil (U).

nucleus: Membrane-enclosed organelle that contains the DNA of a eukaryotic cell.

nutrient enrichment: See fertilization.

nutrigenomics: A new field of nutritional science focused on understanding interactions between diet and genes.

oligonucleotides (oligos): Short, single-stranded synthetic DNA sequences; used in PCR reactions and as DNA probes.

oncogenes: Cancer-causing genes; they are usually part of the genome and have normal functions; when mutated, oncogenes contribute to the development of cancer.

operator: Region of DNA, such as that located in an operon, that binds to a specific repressor protein to control expression of a gene or group of genes such as an operon.

operon: Gene unit common in bacteria; typically consists of a series of genes, located on adjacent regions of a chromosome, that are regulated in a coordinated fashion. Many operons are involved in bacterial cell metabolism of nutrients such as sugars. Refer to the *lac* operon as a well-characterized bacterial operon.

organelles: Small structures in the cytoplasm of eukaryotic cells that perform specific functions.

organic molecules: Molecules that contain carbon and hydrogen.

origins of replication: Specific locations in a DNA molecule where DNA replication begins.

osteoporosis: Category of bone disorders that generally involve a progressive loss of bone mass.

oxidation: The removal of one or more electrons from an atom or molecule.

oxidizing agents: Atoms or molecules that accept electrons during a redox reaction and cause the oxidation of other atoms or molecules; known as electron acceptors, oxidizing agents become reduced when they accept an electron.

P (peptidyl) site (of a ribosome): Portion of a ribosome into which peptidyl tRNA molecules bind during translation.

p arm: "Petit" or small/short arm of a chromosome.

palindrome: A word or phrase that reads the same forward and backward (for example, "a toyota"). In the context of biotechnology, a DNA sequence with complementary strands that reads the same forward and backward. Most restriction enzyme recognition sequences are palindromes.

papain: Protein-digesting enzyme.

parthenogenesis: Creating an embryo without fertilization by pausing DNA division in an egg and allowing the egg to develop with an increased number of chromosomes; for instance, preventing chromosomes from separating in a human egg creates a diploid cell that may develop to form an embryo.

patent: Legal recognition that gives an inventor or researcher exclusive rights to a product and prohibits others from making, using, or selling the product for a certain number of years (20 years from the date of filing).

pathogens: Disease-causing organisms.

pentose sugar: A five-carbon sugar; important components of DNA and RNA nucleotide structure. The pentose sugar deoxyribose is contained in DNA nucleotides; the pentose sugar ribose is contained in RNA nucleotides.

peptidoglycan: Structure in bacterial cell walls consisting of specialized sugars and short, interconnected polypeptides.

peptidyl transferase: rRNA enzyme that is part of the large ribosomal subunit; catalyzes the formation of peptide bonds between amino acids during translation.

peptidyl tRNA: tRNA molecule attached to a growing polypeptide chain; located in the P site of a ribosome during translation.

permease: Enzyme produced by the *lacY* gene of the lactose (lac) operon in bacteria; permease transports the disaccharide lactose into bacterial cells.

personalized genomes: Sequencing a genome for an individual.

personhood: A term popular in bioethics that defines an entity that qualifies for protection based on certain attributes not intrinsic (built-in) or automatic values.

pharmaceutical companies: Companies creating drugs for the treatment of human health conditions.

pharmacogenomics: A form of customized medicine in which disease-treatment strategies are designed based on a person's genetic information (for a particular health condition).

phase I: The first clinical phase of FDA testing in which 20 to 80 healthy volunteers take the medicine to see whether any unexpected side effects are present and to establish the dosage levels.

phase II: The second clinical phase of FDA testing in which the testing of the new treatment on 100 to 300 patients who actually have the illness occurs.

phase III: The third clinical phase of FDA testing in which between 1,000 and 3,000 patients in double-blinded tests are tested after phase II has shown no adverse side effects; usually lasts for 3.5 years.

phase testing: A statistically significant number of trials are required on cell cultures, in live animals, and on human subjects in the three-phase testing processes specified by the Food and Drug Administration.

phosphodiester bond: Covalent bond between the sugar of one nucleotide and the phosphate group of an adjacent nucleotide; joins nucleotides within strands of DNA and RNA.

phytoremediation: Using plants for bioremediation.

placebo: A blank or ineffective treatment. Used in the scientific practice of having a control group in an experiment, such as a drug trial, that receives an ineffective pill or treatment such as a sugar pill or injection of water instead of the actual medication being tested.

plant transgenesis: Gene transfer to a plant from another species.

plaques: Small clear spots of dead bacteria appearing on a culture plate, caused by bacterial cell lysis by bacteriophage.

plasma (cell) membrane: A double-layered structure, consisting of lipids, proteins, and carbohydrates, that defines the boundaries of a cell; performs important roles in cell shape and regulating transport of molecules in and out of a cell.

plasma cells: Antibody-producing cells that develop from B lymphocytes after B cells are exposed to foreign materials (antigens).

plasmid: Small, circular, self-replicating double-stranded DNA molecules found primarily in bacterial cells. Plasmids often contain genes coding for antibiotic resistance proteins and are routinely used for DNA-cloning experiments.

pluripotent: Term used to describe cells, such as embryonic stem cells, with the potential to develop into other cell types.

point mutations: A single base change in DNA sequence.

polarity: Refers to the 5' and 3' ends of DNA and RNA molecules.

polyadenylation: Addition of a short sequence or "tail" of adenine (A) nucleotides to the 3' end of an mRNA molecule; occurs during RNA splicing in eukaryotes; poly(A) tail is important for mRNA stability in the cytoplasm.

polyculture (integrated aquaculture): Raising more than one aquatic species in the same environment. For instance, cultivating fish together with aquatic vegetation.

polygalacturonase: An enzyme naturally produced by plants that digests tissue, causing rotting.

polymerase chain reaction (PCR): Laboratory technique for amplifying and cloning DNA; involves multiple cycles of denaturation, primer hybridization, and DNA polymerase synthesis of new strands.

polypeptide: A chain of amino acids joined by covalent (peptide) bonds; usually greater than 50 amino acids in length.

polyploids: Organisms with an increased number of complete sets of chromosomes.

posttranslational modification: Protein modification that naturally occurs after initial synthesis.

precipitates: Combines because of mutual attraction.

preimplantation genetic testing: PCR and ASO analysis as well as FISH are being used to screen for gene defects in single cells from eight- to 32-cell– stage embryos created by *in vitro* fertilization. Allows individuals to select a healthy embryo before implantation.

primary sequence: Amino acid sequence of a protein.

primary transcript (pre-mRNA): Initial mRNA molecule copied from a gene in the nucleus of eukaryotic cells; undergoes modifications (processing) to produce mature mRNA molecules that enter the cytoplasm.

primase: Enzyme that adds small RNA segments to a single strand of DNA as an early and necessary step for DNA replication.

primers: Oligonucleotides complementary to specific sequences of interest; used in PCR reactions to amplify DNA and DNA-sequencing reactions.

principal/senior scientists: Science leadership position in biotechnology companies; senior scientists are usually Ph.D. or M.D.-trained individuals who plan and direct the research priorities of a company.

prions: Infections protein particles. They attract normal cell proteins and induce changes in their structure, usually leading to the accumulation of useless proteins that damage cells. Prion diseases can occur in sheep and goats (scrapie) and cows (bovine spongiform encephalitis, or "mad cow" disease). Human forms of these brain-destroying diseases include kuru and transformable spongiform encephalitis (TSE). All these diseases involve changes in the conformation of prion precursor protein, a protein normally found in mammalian neurons as a membrane glycoprotein.

probe: Single-stranded DNA or RNA molecule (labeled, for instance, with radioactive or fluorescing nucleotides) that can bind other DNA or RNA sequences by complementary base pairing and be detected by a process such as autoradiography; important laboratory technique for such applications as identifying genes and studying gene activity.

prokaryotic cells: Cells that lack a nucleus and membrane-enclosed organelles; only examples are bacteria and Archae.

promoter: Specific DNA sequences adjacent to a gene that direct transcription (RNA synthesis); binding site of RNA polymerase to begin transcription.

pronuclear microinjection: This method introduces the transgene DNA at the earliest possible stage of development of the zygote (fertilized egg).

prostate-specific antigen (PSA): A protein released into the bloodstream when the prostate gland is inflamed, and elevated levels can be a marker for prostate inflammation and even prostate cancer.

proteases: Protein-digesting enzymes.

protein: Macromolecule consisting of amino acids joined by peptide bonds; major structural and functional molecules of cells.

protein chip: See protein microarray.

protein microarray: A "chip" similar to a DNA microarray; consists of a glass slide containing thousands of individual proteins attached to specific spots on the slide; each "spot" contains a unique protein.

protein sequencing: Identification of amino acid sequence by cleavage of each amino acid, one at a time.

proteolytic: Protein–lysing characteristic.

proteome: The entire complement of proteins in an organism.

proteomes: Families of proteins.

proteomics: Study of protein families.

protoplast fusion: Fusing of a plant cell with another cell.

protoplast: A naked plant cell (without cell wall).

PulseNet: Partnership of bacterial DNA fingerprinting laboratories, developed by the U.S. Centers for Disease Control and Prevention (CDC) and the U.S. Department of Agriculture, designed to provide rapid analysis of contaminated food, with the purpose of identifying contaminating microbes and preventing outbreaks of foodborne disease.

q arm: Long arm of a chromosome.

quality assurance (QA): All activities involved in regulating the final quality of a product, including quality-control measures.

quality control (QC): Procedures that are part of the QA process involving laboratory testing and monitoring of production processes to ensure consistent product standards (of purity, performance, and the like).

quasi-automated x-ray crystallography: Fast x-ray analysis of protein structure by using algorithms and data analysis.

quaternary structure: Proteins containing more than one amino acid chain that is folded into a tertiary structure.

real-time or quantitative PCR (qPCR): New applications in PCR technology make it possible to determine the amount of PCR product made during an experiment. Uses primers made with fluorescent dyes and specialized thermal cyclers that enable researchers to quantify amplification reactions as they occur.

recognition sequence: See restriction site.

Recombinant DNA Advisory Committee (RAC): NIH panel responsible for establishing and overseeing guidelines for recombinant DNA research and related topics.

recombinant DNA technology: Technique that allows DNA to be combined from different sources; also called gene or DNA splicing. Recombinant DNA is an important technique for many gene-cloning applications.

recombinant proteins: Commercially valuable proteins created by recombinant DNA technology and gene-cloning techniques; examples include insulin and growth hormone.

redox reaction: Combination of oxidation and reduction reactions.

reduction: The addition of one or more electrons to a molecule.

regenerative medicine: A discipline of medical biotechnology that involves repairing or replacing damaged tissues and organs by using tissues and organs grown through biotechnological approaches.

rennin (chymosin): Protein-degrading enzyme derived from the stomach of milk-producing animals such as cows and goats; used in cheese production; recombinant form called chymosin.

replica plating: Technique in which bacteria cells from one plate can be transferred to another plate to produce a replica plate with bacterial cells growing in the same location as on the original plate; can also be carried out with other cells such as yeast, or viruses.

reporter genes: Genes (such as the *lux* genes) that can be used to track or monitor (report on) expression of other genes.

reproductive cloning: Cloning process that creates a new individual; process typically involves using the genetic material of a single cell to create an individual with the single genetic composition of its creator cell.

research assistants/associates: Laboratory positions in which individuals are primarily involved in carrying out experiments under the supervision of other scientists such as principal or senior scientists.

restriction enzymes (endonucleases): DNA-cutting proteins found primarily in bacteria. Enzymes cleave (cut) the phosphodiester backbone of double-stranded DNA at specific nucleotide sequences (restriction sites). Commercially available restriction enzymes are essential for molecular biology experiments.

restriction fragment length polymorphism (RFLP): Unique patterns created when DNA with different numbers of restriction enzyme digestion sites (due to DNA variation) is separated on an electrophoresis gel and detected with dyes or other means.

restriction fragment length polymorphism (RFLP) analysis: DNA fingerprinting technique that involves digesting DNA into fragments of different lengths by using restriction enzymes; patterns of DNA fragments are analyzed to create a "fingerprint."

restriction map: An arrangement or "map" of the number, order, and types of restriction-enzyme cutting sites in a DNA molecule.

restriction site: Specific sequence of DNA nucleotides recognized and cut by a restriction enzyme.

retroviruses: Viruses that contain an RNA genome and use reverse transcriptase to copy RNA into DNA during the replication cycle in host cells.

retrovirus-mediated transgenesis: Infecting embryos with a retrovirus (usually genetically engineered) before the embryos are implanted. The retrovirus acts as a vector for the new DNA.

reverse transcriptase: Viral polymerase enzyme that copies RNA into single-stranded DNA. This commercially available enzyme is used for many molecular biology experiments, such as creating cDNA.

reverse transcription PCR (RT-PCR): Laboratory technique that involves using the enzyme reverse transcriptase to copy RNA from a cell into cDNA and then amplifying cDNA by the polymerase chain reaction (PCR); a valuable technique for studying gene expression.

ribonucleic acid (RNA): Single strands of nucleotides produced from DNA. Different types of RNA have important functions in protein synthesis.

ribosomal RNA (rRNA): Small RNA molecules that are essential components of ribosomes.

ribosome: Organelle composed of ribosomal ribonucleic acid (rRNA) and proteins assembled into packages called subunits. Ribosomes bind to mRNA and tRNA molecules and are the site of protein synthesis in prokaryotes and eukaryotes.

RNA-induced silencing complex (RISC): RISC unwinds the double-stranded siRNAs, releasing single-stranded siRNAs that bind to complementary sequences in mRNA molecules. Binding of siRNAs to mRNA results in degradation of the mRNA (by the enzyme slicer) or blocks translation by interfering with ribosome binding.

RNA interference (RNAi): RNA-based mechanisms of gene silencing. These siRNAs are bound by a protein–RNA complex called the RNA-induced silencing complex (RISC).

RNA polymerase: Copies RNA from a DNA template; different forms of RNA polymerase synthesize different types of RNA.

RNA splicing: The removal of nonprotein coding sequences (introns) from primary transcript (pre-mRNA) and joining of protein-coding sequences (exons).

sales representatives: Salespersons in biotechnology companies; sales representatives are "people persons" who work closely with medical doctors, hospitals, and health care providers to promote a company's products.

scaling-up: Process manufacturing modification from original research purification.

secondary structure: α-Helix or β-sheet structure of proteins.

seeding: See bioaugmentation.

selection (antibiotic, blue-white screening): Laboratory technique used to identify bacteria containing recombinant DNA of interest; involves growing bacteria on media with antibiotic or other selection molecules.

selective breeding: Mating organisms with desired features to produce offspring with the same characteristics.

semiconservative replication (of DNA): Process by which DNA is copied; one original (parent) DNA molecule gives rise to two molecules, each of which has one original strand and one new strand.

senescence: Cellular aging process.

severe combined immunodeficiency (SCID): Genetic condition created by a mutation in the gene encoding the enzyme adenosine deaminase. Affected individuals have no functional immune system and are prone to death by common, normally minor infections. Commonly known as "boy in the bubble" condition because of need for SCID individuals to live in germ-free environments.

sex chromosomes: Contain majority of genes that determine an organism's sex; the X (female) and Y (male) chromosomes in humans.

short interfering RNA (siRNA): Small (21 or 22 nt) double-stranded pieces of nonprotein coding RNA, so named because they were shown to bind to mRNA and subsequently block or interfere with translation of bound mRNAs.

short tandem repeat (STR): One to six nucleotide repeats that are dispersed throughout chromosomes. Because these repeated regions can occur in many locations within the DNA, the probes used to identify them complement the DNA regions that surround the specific microsatellite being analyzed.

"shotgun" cloning: Random cloning of many fragments at once; no individual gene is specifically targeted for cloning.

silent mutation: Base-pair substitution that has no effect on the amino acid sequence of a protein.

single-locus probe: The probe will bind to only complementary sequences of DNA, found in only one location in the genome.

single-nucleotide polymorphism (SNP): A single-nucleotide variation in the gene sequence; or a type of DNA mutation; the basis of genetic variation among humans.

single-strand binding proteins: Proteins that bind to unraveled single strands of parental DNA during DNA replication to prevent DNA from reforming double strands before being copied.

sister chromatids: Exact copies of double-stranded DNA molecules and proteins (chromatin) joined to form a chromosome.

site-directed mutagenesis: With this technique, mutations can be created in specific nucleotides of a cloned gene contained in a vector. The gene can then be expressed in cells, which results in translation of a mutated protein. This allows researchers to study the effects of particular mutations on protein structure and functions as a way to determine what nucleotides are important for specific functions of the protein.

size-exclusion chromatography (SEC): Separation based on molecular size.

sludge: A semisolid material produced from treated sewage waste; consists largely of small particles of feces, waste papers, and microorganisms.

slurry-phase bioremediation: Process in which contaminated soil is removed from a site and mixed with water and fertilizers (and often oxygen) in large drums or bioreactors to create a mixture (slurry) that stimulates bioremediation by microorganisms in the soil.

Sodium dodecylsulfate polyacrylamide gel electrophoresis (SDS-PAGE): Protein separation based on molecular size and the uniform distribution of charged sulfate groups.

solid-phase bioremediation: *Ex situ* soil clean-up strategies that primarily involve composting, landfarming, or biopiles.

somatic cell nuclear transfer: DNA transfer process that can be used for reproductive or therapeutic cloning purposes; involves removing DNA from one cell and inserting it into an egg cell that has had its DNA removed (enucleated).

somatic cells: All cell types in multicellular organisms except for gametes (sperm and egg cells).

Southern blot analysis: Laboratory technique invented by Ed Southern that involves transferring (blotting) DNA fragments onto a filter-paper blot for use in probe hybridization studies.

Southern blotting: See Southern blot analysis.

species integrity: Generally refers to maintaining the natural functions, abilities, and genetic constitution of a particular species; for example, creating recombinant animals or plants such as transgenic or polyploid organisms may change "species integrity" by making a species less well adapted to life in the wild.

specific utility: Requirement of the USPTO that a researcher must know exactly what the DNA sequence does to patent it.

spectral karyotype: Karyotyping technique involving probes specific for each chromosome that fluoresce different colors to identify chromosomes according to their color pattern.

sperm-mediated transfer: The DNA is injected or attached directly to the nucleus of the sperm before fertilization.

StarLink: A transgenic corn not approved for humans that was suspected of contaminating seed intended for food.

startup company: Formed by a small team of scientists who believe they may have a promising product to make (such as a recombinant protein to treat disease). The team must typically then seek investors to fund their company so they can buy or rent laboratory facilities, buy equipment and supplies, and continue the research and development necessary to make their product.

statistical probability: Using statistical measures to calculate the possibility (probability) that a particular event will happen.

stem cells (embryonic and adult-derived stem cells): Immature (undifferentiated) cells that are capable of forming all mature cell types in animals and that can be derived from embryos at several days of age or from adult tissues.

stone-age genomics: A number of laboratories around the world are involved in analyzing "ancient" DNA. These studies are generating fascinating data from minuscule amounts of ancient DNA from bone and other tissues and fossil samples that are tens of thousands of years old.

substantial utility: Requirement of the USPTO that the product must have a real-world function before it can be patented.

substrate: Molecule or molecules on which an enzyme performs a reaction.

subtilisin: Protease derived from *Bacillus subtilis,* a valuable component of many laundry detergents, where it functions to degrade and remove protein stains from clothing. Several bacterial enzymes are also used to manufacture foods, such as carbohydrate-digesting enzymes called amylases that are used to degrade starches.

subunit vaccine: Vaccine created from components of a pathogen such as viral proteins or lipid molecules.

Superfund Program: Program established by the U.S. Congress in 1980 through the U.S. Environmental Protection Agency, designed to identify and clean up hazardous waste sites and protect citizens from harmful effects of waste sites.

T lymphocytes (T cells): Play essential roles in helping B cells recognize and respond to antigens.

***Taq* DNA polymerase:** DNA-synthesizing enzyme isolated from *Thermus aquaticus,* a thermophilic Archae that lives in hot springs; its ability to withstand high temperatures (thermostable) without denaturation makes it valuable for use in PCR experiments.

TATA box: Short nucleotide sequence (TATA) usually located approximately 20 to 30 base pairs "upstream" (in the 5' direction) of the start site of many eukaryotic genes; part of promoter sequence bound by transcription factors used to stimulate RNA polymerase.

telomerase: Enzyme that fills in gaps of DNA nucleotides at the ends of chromosomes (telomeres) that remain after DNA polymerase has copied DNA.

telomere: The end structures of a eukaryotic chromosome; in humans, consists of specific repeating sequences of DNA (TTAGGG).

template strand: After RNA polymerase binds to a promoter, it unwinds a region of the DNA to separate the two strands. Only one of the strands, called the template strand (the opposite strand is called the *coding strand*), is copied by RNA polymerase.

tertiary structure: The unique three-dimensional shape of a protein.

thalidomide: Compound used initially to combat morning sickness in pregnant women; certain chemical forms of thalidomide caused severe birth defects; thalidomide derivatives are still being investigated for their use in treating cancer, HIV, and other diseases.

therapeutic cloning: Using a patient's DNA to create (clone) an embryo as a source of stem cells that could be used in therapeutic applications to treat the patient.

thermophiles: Organisms with high optimal growth temperatures. For example, bacteria that live in hot springs are thermophiles.

thermostable enzymes: Enzymes that are capable of withstanding high temperatures and that are isolated from theromophiles. For example, *Taq* DNA polymerase is a thermostable enzyme.

3' end: The end of a strand of DNA or RNA in which the last nucleotide is not attached to another nucleotide by a phosphodiester bond involving the 3' carbon of the pentose sugar.

three Rs of animal research: *Reduce* the number of higher species used; *Replace* animals with alternative models whenever possible; and *Refine* tests and experiments to ensure the most humane conditions possible.

thymine: Abbreviated T; pyrimidine base present in DNA nucleotides.

Ti vector: Plasmid DNA vector derived from soil bacterium that can be used to clone genes in plant cells and deliver genes into plants.

T lymphocyte: Type of white blood cell (leukocyte); T lymphocytes, also called "T cells" play an essential role in helping the immune system recognize and respond to foreign materials (antigens).

tissue engineering: Designing and growing tissues for use in regenerative medicine applications.

tobacco mosaic virus (TMV): A virus that naturally invades certain plants.

traits: Inherited features of an organism such as skin color and body shape.

transcription: The synthesis of RNA from DNA, which occurs in the nucleus of eukaryotic cells and the cytoplasm of prokaryotic cells.

transcription factors: DNA-binding proteins that bind promoter regions of a gene and stimulate transcription of a gene by RNA polymerase.

transcriptional regulation: Form of gene-expression regulation that involves controlling the process of transcription by controlling the amount of RNA produced by a cell.

transfection: Introducing DNA into animal or plant cells.

transfer RNA (tRNA): Small RNA molecules that transport amino acids to a ribosome during protein synthesis. tRNA binds to specific codons in mRNA sequences during translation.

transformable spongiform encephalitis (TSE): Prion disease that can occur in sheep and goats (scrapie) and cows (bovine spongiform encephalitis, or "mad cow" disease). Human forms of these brain-destroying diseases include kuru and TSE. All these diseases involve changes in the conformation of prion precursor protein, a protein normally found in mammalian neurons as a membrane glycoprotein.

transformation: The process by which bacteria take in DNA from the surroundings. Term also is used to define changes that cause a normal cell to become a cancer cell.

transgene: Gene from one organism introduced into another organism to create a transgenic; term usually applies to genes used to create transgenic animals and plants.

transgenic: Animals that contain genes from another source. For instance, human genes for clotting proteins can be introduced into cows for the production of these proteins in their milk.

triploid: Three sets of chromosomes ($3n$); used to describe organisms with three sets of chromosomes.

tuberculosis (TB): TB is caused by the bacterium *Mycobacterium tuberculosis,* which grows slowly and can exist in a human for several years before the individual develops TB.

tumor-inducing (Ti) plasmid: A plasmid vector found in *Agrobacter.*

translation: The synthesis of proteins from genetic information in messenger (mRNA) molecules. Translation occurs in the cytoplasm of all cells.

two-dimensional electrophoresis: Separation based on charge in two directions.

type I, insulin-dependent diabetes mellitus: Disease caused by a lack of the pancreatic hormone insulin, which is required for carbohydrate metabolism. Creates elevated blood sugar levels (hyperglycemia).

U. S. Department of Agriculture (USDA): Agency created in 1862 that has many functions related to the advancement and regulation of agriculture. Some of those functions include the regulation of plant pests, plants, and veterinary biologics.

U.S. Environmental Protection Agency (EPA): Agency whose primary purpose is to protect human health and to safeguard the natural environment (air, water, and land) by working with other federal agencies in the United States and state and local governments to develop and enforce regulations under existing environmental laws.

U. S. Patent and Trademark Office (USPTO): The office of the federal government that issues patents (including patents for gene sequences).

ultrafiltration: Separation of particles smaller than 20 μm.

upstream processing: Adjustments in purification process based on changes in the biologic process, making processing more efficient.

uracil: Abbreviated U; pyrimidine base present in RNA nucleotides.

utilitarian approach: Line of ethical thought that states that actions are moral if the result produces the greatest good for the greatest number of humans; also referred to as consequential ethics because it focuses on results or consequences, not on intentions.

vaccination: The process of administering a vaccine to provide an organism with immunity to an infectious microorganism.

vaccines: A preparation of a microorganism or its components that is used to stimulate the production of antibodies (antibody-mediated immunity) in an organism.

variable number tandem repeats (VNTRs): The recognition that variable numbers of repeated nucleotides can be found in DNA and can be used for identification of individuals.

vector: DNA (or viruses) that can be used to carry and replicate other pieces of DNA in molecular biology experiments; for example, plasmid DNA, viruses used for gene therapy; also refers to organisms that carry disease.

Western blot analysis: Laboratory technique for separating protein molecules by gel electrophoresis and transferring (blotting) proteins onto a filter paper blot that is usually probed with antibodies to study protein structure and function.

Western blotting: See Western blot analysis.

xenotransplantation: The transfer of tissues or organs from one species to another; for instance, transplanting a pig organ into a human.

x-ray crystallography: Identification of molecules based on the patterns produced after bombardment of their crystals with x-rays.

yeast: A unicellular fungus.

yeast artificial chromosome (YAC): Plasmid vectors grown in yeast cells that can replicate very large pieces of DNA; used to clone pieces of human chromosomes for the Human Genome Project.

yeast two-hybrid system: Laboratory technique involving the use of yeast to join together different proteins (creating a hybrid protein) as a way to study protein function.

zygote: Diploid cell formed when haploid gametes such as a sperm and egg cell unite.

Index

A site, 44, 45f
ABO blood typing, 273
Accession number, 89
Acetylase, 49
Acquired mutations, 53
Activators, 48
Adenine, 32, 32f, 34f
Adenosine triphosphate (ATP), 28, 129, 129f
Adenoviruses, 276
Adhesives, 101–102, 101t, 252
Adult-derived stem cells. *See* Stem cell(s)
Advanced Cell Technology, 297
Aerobic conditions, 122
    in bioremediation, 211–213, 213f, 214f
    in glucose metabolism, 129
Aerobic metabolism, 211
Affinity chromatography, 111–113, 113f
Agarose gel electrophoresis, 75–78, 77f
Agglutinogens, 273
Aging, telomeres in, 286–287, 287f
Agricultural biotechnology, 5f, 8
    animal, 10, 171–188, 181–184. *See also* Animal biotechnology
    fish, 232–242. *See also* Aquaculture
    plant, 155–169. *See also* Plant biotechnology
    recombinant bacteria in, 134–135
    regulation of, 308–309
    selective breeding in, 3, 157, 181, 238–239
Agrobacterium tumefaciens, 68
AIDS. *See* HIV/AIDS
Alcohol, fermentation of, 2–3, 99, 129–130, 129f, 166
Algae, 255, 259
Allele-specific oligonucleotide analysis, 266, 266f
Allergies, transgenic plants and, 168
Alpha helix, 104
Alternative splicing, 41–42, 42f
Alzheimer's disease, 103, 282
American Association for the Advancement of Science, 204
Amino acids. *See also* Protein(s)
    arrangement of, 104
    in protein synthesis, 43–45, 45f
Aminoacyl tRNA, 44
Ammonium, removal from wastewater, 219, 221f

Ammonium sulfate, in protein precipitation, 110, 110f
Amniocentesis, 263–264, 264f
    in paternity testing, 202, 203f
Amniotic fluid–derived stem cells, 290
Amylase, 99
Anaerobic conditions, 122
    in bioremediation, 211–214, 213f
    in fermentation, 122
Anaerobic metabolism, 211
ANDi, 186
Androgen-responsive element, 48
Angiogenesis, 253
Animal(s), 10, 10f
    cell structure in, 28, 28f, 29t–30t
    DNA fingerprinting for, 204, 205–206, 206f
    humane treatment of, 328
    model, 58, 61, 90, 175–176, 261–263, 328
    transgenic, 10, 10f, 180–186, 187t
Animal and Plant Health Inspection Service (APHIS), 308–309
Animal biotechnology, 10, 171–188
    agricultural applications of, 181–184
    bioreactors in, 183–184, 183f
    careers in, 188
    cloning in, 177–180, 178f, 180f
    computer models in, 174–175
    ethical aspects of, 175–176, 328, 331–332
    gene discovery in, 172–173, 173f
    knock-ins in, 185
    knockouts in, 184–186, 185f
    legal aspects of, 317–321, 331
    medical applications of, 172f, 176–177
    model organisms in, 58, 61, 90, 172–175, 261–263
    monoclonal antibody production in, 186–187, 187f
    overview of, 172
    phase testing in, 172–173
    regulation of, 175–176, 308–309
    3Rs of, 176
    transgenic animals in, 180–186, 187t
    veterinary medicine in, 176
Animal cells, in protein engineering, 108
Animal pathogens, in bioterrorism, 150, 151t
Animal technicians, 188
Animal Welfare Act, 175

Animal-human transplants, 176, 284
Annotation, 142
Anthrax, in bioterrorism, 148, 149t, 150f, 152
Antibiotic(s), 3, 131–134
    animal, 188
    development/production of, 3, 131–134
    discovery of, 131
    mechanism of action of, 132, 134f
    microbial sources for, 131–132, 133t
    resistance to, 132
Antibiotic selection, 63, 64f, 123
Antibodies, 10, 99, 136–137, 137f, 138f
    monoclonal, 108–109, 186–187, 187f, 274, 274f, 275
Antibody test kits, in aquaculture, 239
Anticodons, 44
Antifouling agents, 256–257, 257f
Antifreeze proteins (AFPs), 243–244, 244f
Antigens, 108–109, 136–137, 137f, 138f
Antimicrobials, 131. *See also* Antibiotic(s)
Antiparallel strands, 33, 34f
Antisense RNA technology, 159–161
    in gene therapy, 277–279, 278f
Antithrombin III, 183–184
Antiviral drugs, 145
APHIS (Animal and Plant Health Inspection Service), 308–309
Apoptosis, 262
Aquaculture, 232–242
    barriers and limitations in, 239–240
    bioimpedance in, 238
    careers in, 258
    controversial aspects of, 240–241
    cryopreservation in, 238
    definition of, 232
    disease control in, 239, 240, 245–246
    economics of, 232–235
    enhanced quality and safety in, 239
    environmental impact of, 240–241
    fish-rearing practices in, 235–237, 236f–238f
    future directions in, 242
    hydroponic systems in, 238
    improved strains for, 238–239
    innovations in, 237–238
    integrated, 238
    major organisms in, 235–237, 235t
    polyploids in, 247–250, 250f

Aquaculture (*continued*)
  regulation of, 241–242
  selective breeding in, 238–239
  transgenic/polyploid species in, 241,
    243, 247–250
Aquatic biotechnology, 12, 12f, 231–259
  antifreeze proteins in, 243–244, 244f
  aquaculture in, 232–242
  biomass in, 255
  bioprocessing in, 255
  in bioremediation, 257–259
  careers in, 258
  differential display polymerase chain
    reaction in, 234
  environmental applications of,
    256–259
  gene cloning in, 245–246
  gene discovery in, 242–246
  genomics in, 245–246
  industrial applications of, 254–255
  medical applications of, 251–254
  molecular genetics in, 242–250
  overview of, 232, 236f
  regulation of, 241–242
Arabidopsis thaliana, 68t, 224
Arber, Werner, 59
Archaea, 120. *See also* Bacteria
Arthritis, 15, 16
Arthrobacter luteus, 60t
Artificial blood, 273
Assisted reproduction, 287–288,
    335–336, 337
Astaxanthin, 239
ATP (adenosine triphosphate), 28, 129,
    129f
Attenuated vaccines, 137, 138
Autografts, 283–284
Automated DNA sequencing, 78–80,
    79f
Autoradiography, 70
  in DNA fingerprinting, 194, 195f
Autosomes, 35
Avery, Oswald, 31–32
Avian flu, 140–141
Avidin, 163

B lymphocytes, 136, 137f
Baby Fae, 284
Bacilli, 121, 121f
Bacillus amyloliquefaciens, 60t
Bacillus anthracis, 148, 149t, 150f, 152
Bacillus subtilis, 68, 126
Bacillus thuringiensis (Bt), 162–163, 163f,
    165t
Bacteria, 27–28, 27f, 119f, 120. *See also*
    Microbes
  antibiotic-resistant, 132
  bioluminescent, 126–127, 224, 257
  in bioremediation, 213–214
  classification of, 120
  cloning of, 58, 61–63, 62f. *See also*
    Microbial biotechnology;
    Recombinant bacteria
  culture of, 121, 122f
  disease-causing. *See* Pathogens
  gene expression in, 48–50, 49f, 50f
  ice-minus, 134–135, 310, 310f
  petroleum-eating, 222–223, 226

recombination protein engineering,
    107, 107t
  size and shape of, 121, 121f
  staining of, 120–121
  structure of, 27–28, 27f, 120–121, 121f
Bacterial artificial chromosomes, 68
Bacterial biobatteries, 221
Bacterial cells, 27–28, 27f, 120–121, 121f
Bacterial chromosomes, 121
Bacterial infections, diagnosis of,
    145–147, 146f
Bacterial plaques, 66
Bacterial plasmids. *See* Plasmid(s)
Bacterial replication, inhibitors of,
    133–134
Bacterial transformation, 31–32, 31f,
    61–63, 123–124, 124f
  calcium chloride technique in,
    123–124, 124f
  electroporation in, 124–125, 125f
Bacteriocins, 130
Bacteriophage vectors, 66–67, 66t
Bacteriophages, 58, 66–67, 66t
Baculoviruses, 109
Bases, 32–33, 32f, 34f
Basic Local Alignment Search Tool
    (BLAST), 89
Basic sciences, 5–6, 5f
Batch centrifugation, 110–111, 110f
Batch processing, 122
Bentham, Jeremy, 325
Berg, Paul, 60–61
Beta sheet, 104
Beta turn, 104
Bioaccumulation, 223
Bioaugmentation, 215–216
  in tissue engineering, 285, 286f
Biocapsules, 284–285, 285f
Biodegradation, 209. *See also*
    Bioremediation
  aerobic vs. anaerobic, 212–213, 213f
  of jet fuel, 225
  of petroleum, 11, 11f, 222–223,
    225–227, 226f, 227f
Biodiversity, 329
Bioethicists, 325, 327
Bioethics, 8, 324–342
  animal biotechnology and, 175–176,
    328, 331–332
  assisted reproduction and, 335–336,
    337
  biodiversity and, 329
  careers in, 325, 327, 342
  cloning and, 296, 328, 336–337
  communication and, 341
  decision making and, 325, 326–327
  definition of, 325
  deontological (Kantian) approach in,
    326
  DNA fingerprinting and, 200
  drug development and, 328
  efficacy and, 328
  gene therapy and, 339–340
  genetic testing and, 338–339
  genetically modified organisms and,
    328–332
  genomics and, 338–339
  germline genetic engineering and,
    339–340

human experimentation and,
    332–335, 339–340
  in vitro fertilization and, 335–336
  patient rights and, 337–338
  personhood and, 334–335
  placebos and, 332–333
  regenerative medicine and, 333–335
  rights to biological material and,
    337–338
  risk assessment and, 330
  species integrity and, 328–329
  stem cells and, 296, 333–335
  utilitarian approach in, 325–326
Biofilms, 256–257, 256f, 257f
Biofuels, 166
Bioimpedance, 238–239
Bioinformatics, 6, 21, 87–89, 88f
Biological License Agreement, 313
Biologics, 308
Bioluminescence, 126–127, 127f, 224,
    257
Biomanufacturing, 21f, 22–23
Biomarkers, 263
Biomass, 213
  marine, 255
Biopiles, 218, 218f
Bioprocessing, 255
Bioprocessing engineer, 117
Bioprospecting, 123
Bioreactors, 7, 107, 108f
  for ammonium degradation, 220, 221f
  transgenic animals as, 183–184, 183f
Bioremediation, 11, 11f, 102, 208–230
  aerobic vs. anaerobic biodegradation
    in, 212–213, 213f
  aquatic biotechnology in, 257–259
  bioaugmentation in, 215–216
  biosensors in, 224
  careers in, 229
  case studies of, 225–227
  chemical types and sources in, 211,
    212t
  contamination zones in, 210–211,
    211f
  definition of, 11, 209
  dredging in, 228
  in energy generation, 220–222
  ex situ, 217, 220, 221f
  fertilization in, 215, 216f, 226, 227f
  future directions for, 227–228
  genomics in, 143–144, 215, 215t
  of heavy metals, 223–224
  importance of, 209–210
  in situ, 217, 220, 221f
  limitations of, 222
  in metal recovery, 228–229
  microbes in, 213–214
  microcosm studies of, 223
  nutrient enhancement in, 215, 216f
  overview of, 209
  oxidation and reduction reactions in,
    211–212, 213f
  of PCBs, 228
  of petroleum, 222–223
  phytoremediation in, 216–217, 216f,
    224, 225f
  plants in, 224, 225f
  of radioactive waste, 214, 229–230
  recombinant bacteria in, 222–224

seeding in, 215–216
slurry-phase, 217
of soil, 217–218, 218f, 225–227
solid-phase, 217
treatment environments in, 210–211, 211f
validation technicians in, 229
of water, 218–220, 219f–221f, 228
Biosensors, 126–127, 224, 257
Biosteel, 183
Biotechnology
agricultural, 5f, 8
animal, 10, 10f, 181–184
plant, 155–169
aquatic, 12, 12f, 231–259
basic sciences in, 5–6, 5f
"big" picture of, 13–14
in bioremediation, 11, 11f, 102, 208–230
career opportunities in. See Careers
controversial aspects of, 8, 314–315
definition of, 2
disciplines in, 5–6, 5f
forensic, 10–11, 11f, 190–204
funding for, 320, 340–341
globalization and, 321–322
historical perspective on, 2–5
medical, 12, 260–303
microbial, 8, 119–153
patents in, 76, 317–321
products of, 6–7
protein engineering in, 97–118
regulation of, 13, 23, 305–322, 338.
See also under Regulatory
biotechnology safety of. See Health
and environmental concerns
scale-up processes in, 22
in 21st century, 14–17
types of, 8–14
Biotechnology century, 1–25
Biotechnology companies
biomanufacturing in, 21f, 22–23
current status of, 19, 19f
customer relations in, 21f, 23
distribution of, 19, 19f
finance division in, 23
hiring trends for, 24–25
legal division in, 23
marketing division in, 23
operations in, 21f, 22–23
organization of, 19–20, 21f
production in, 21f, 22–23
quality assurance/quality control in, 21f, 23
research and development in, 20–22, 21f
salaries in, 23–24
sales division in, 23
startup, 20
top ten, 20t
Biotechnology industry, 18–25
vs. pharmaceutical industry, 19
Bioterrorism, 148–153
definition of, 148
pathogen delivery in, 150
prevention/management of, 151–153
targets of, 150, 151t
Bioventing, 217
Bioweapons, 149

Black Death, 148–149, 149t
BLAST, 89
Blastocyst, 288
Blood, artificial, 273
Blood matching, 273
Blue-white selection, 63, 64f
Blunt ends, 59, 60t
Bonds
hydrogen, 104
phosphodiester, 32f, 33
Bovine growth hormone, 165t, 311
Bovine somatotropin (BST), 165t, 311, 316
Bovine spongiform encephalopathy, 103, 105–106
Boyer, Herbert, 60, 61
Brachydanio rerio, 174, 174f
BRCA genes, 15, 270, 320, 339
Breadmaking, 130
Breast cancer, 15, 270, 320, 339
Breeding, selective, 3, 157, 181
in aquaculture, 238–239
Bt crops, 162–163, 163f, 165t, 329, 329f
Bubonic plague, 148–149, 149t, 150f
Bulls, transgenic, 181–183
Byssal fibers, 252, 252f

CAAT box, 47, 47f
Calcitonin, 251, 253t
Calgene, 159
Callus, 157
Cancer
breast, 15, 270, 320, 339
carcinogens and, 211, 212t
chemotherapy for, 3, 7t, 16, 269–271
cytokines for, 176
hyperthermia for, 176
oncogenes in, 269
telomerase in, 286–287
Cancer Genome Atlas Project, 92, 300–301
Candidatus 'Brocadia anammoxidans,' 219, 221f
Capillary gel, 78
Carcinogens, 211, 212t
Careers, 20–25, 53, 94
in animal biotechnology, 188
in aquatic biotechnology, 258
in bioethics, 325, 327, 342
in bioremediation, 229
in drug development/production, 20–25, 302
education and training for, 23–24
in genomics, 53
hiring trends and, 24–25
for laboratory technicians, 20, 205
in medical biotechnology, 302
in microbial biotechnology, 153
in plant biotechnology, 169
in regulatory biotechnology, 320, 322
salaries and, 23–24
Carrageenan, 255
Casein, 99f, 128–129
Cattle, transgenic, 181–183
cc (cat), 336–337, 337f
cDNA, 69, 107, 298
cDNA libraries, 69–71, 70f, 71f
Celera Genomics, 90, 199

Cell(s)
competent, 123
eukaryotic, 28, 28f, 29t–30t, 120
immortal, 186–287
nucleus of, 28, 28f, 30t
patented, 337–338
prokaryotic, 27–28, 27f, 119f, 120–121
somatic, 35, 290, 294–297, 294t, 336–337
stem, 13, 17, 18f
structure and function of, 27–28, 27f, 28f
Cell culture, 7
bacterial, 121, 122f
equipment for, 175f
limitations of, 173, 174–175
stem cell, 288–289, 288f
vs. computer models, 174–175
yeast, 122
Cell cycle, 37
Cell death
programmed, 262
senescent, 286
Cell lines, 289
Cell lysis, 109, 137
Cell membrane, 28, 28f, 29t
Cellular therapeutics, 284–285
Cellulase, 123
Cellulose, 166
Center for Drug Evaluation and Research, 312
Centers for Disease Control and Prevention (CDC), 147, 175
Central vacuole, 29f, 30t
Centrifugation, 110–111, 110f
Centrioles, 29t
Centromeres, 35f, 36
Chain of custody, 201
Chakrabarty, Ananda, 222–223
Chargaff, Erwin, 32, 33
Cheese-making, 99, 99f, 128–129, 165t
Chemotherapy, 3, 7t, 16, 269–271
Chickens, transgenic, 181
Chimeras, 184–185
China, plant biotechnology in, 320
Chitin, 254
Chitosan, 254
Chloroplast, 29f, 30t, 158
Chloroplast engineering, 159, 161f
Cholera, 143, 239
Chorionic villus sampling, 263–264, 264f
Chromatids, sister, 35f, 36
Chromatin, 29f, 30t, 35, 35f
Chromatography, 111–114, 112f–114f
Chromosomes, 33–37
autosomes, 35
bacterial, 121
bacterial artificial, 68
diploid number of, 35
eukaryotic vs. prokaryotic, 35–36
haploid number of, 35
homologous, 35
Internet resources for, 300
karyotyping of, 36–37, 36f
mapping of, 75–78, 77f, 298, 299f, 300–302, 301f
maternal, 35
number of, 37
paternal, 35

Chromosomes (*continued*)
  sex, 35
  structure of, 35, 35f
  Y, in paternity testing, 203–204, 203f
  yeast artificial, 68
Chymosin, 99, 99f, 129, 165t, 311
Cilia, 29t
Clinical research
  animal models in. *See* Animal
    biotechnology
  regulation of, 23. *See also under*
    Regulatory agencies
Clinical trials, 261–262
  double-blind, 333
  publication of, 332
Clones/cloning, 3–4, 10, 10f, 58–74,
    63–87, 177. *See also* Genetic
    engineering; Recombinant DNA
    technology
  animal, 177–180, 178f, 189f. *See also*
    Animal biotechnology
  applications of, 75–87
  in aquatic biotechnology, 245–246
  autoradiography in, 70
  bacterial, 58, 61–63, 62f
  colony hybridization in, 69–70
  controversial aspects of, 296, 314–315
  definition of, 58, 63, 177
  development of, 58, 60–61
  DNA libraries in, 68–71, 70f, 71f
  ethical aspects of, 336–337
  gene-of-interest identification in,
    68–71, 70f, 71f
  human, 64–65, 296–297, 336–337,
    336f, 337f, 338
  medical applications of, 46
  organism, 10, 10f, 58
  overview of, 63–64
  patents and, 76, 317–321
  plant, 155–169, 157, 161f. *See also*
    Plant biotechnology
  polymerase chain reaction in, 73, 74f
  process of, 63–74
  random, 68
  regulation of, 338
  reproductive, 293–298, 294t, 295f
  in restriction mapping, 298, 299f
  shotgun, 87, 88f, 142
  steps in, 63, 64f
  therapeutic, 293–298, 294t, 295f
  transformation in, 61–63, 62f
Cloning vectors. *See* Vectors
Cocci, 121, 121f
Cocktails, vaccine, 141
Coding strand, 40
CODIS (Combined DNA Index System),
    192
Codons, 43, 44f
Cohen, Stanley, 60, 61
Cohesive ends, 59, 60t
Colchicine, 249
Collagenase, 255
Colony hybridization, 69
Combined DNA Index System (CODIS),
    192
Communication, 341
Comparative genomics, 92–93, 93t
Competent cells, 123
Complementary base pairs, 33, 34f

Complementary DNA, 69, 107, 298
Complementary DNA libraries, 69
Composting, 217–218
Computer-automated DNA sequencing,
    78–80, 79f
Conotoxins, 252, 253t
Consent, 332
Contigs, 88, 88f
Coomassie stain, 115
Coppolino standard, 201
Copy number, 65
Copyrights, 317–321, 318t
Cosmid vectors, 66t, 67
Counterfeit products, DNA fingerprinting
    for, 206
Cows, transgenic, 181–183
Credible utility, 320
Crick, Francis, 33, 58, 261
Criminal investigations. *See* Forensic
    biotechnology
Crops. *See* Plant biotechnology
Crossbreeding. *See* Selective breeding
Crown gall, 158, 159f
Cryopreservation, 238, 310
Cryoprotection, 134–135, 243–244, 244f,
    310
Crystallography, X-ray, 116–117
Culture. *See* Cell culture
Customer relations specialists, 21f, 23
Cystic fibrosis transmembrane
    conductance regulator, 279–281,
    280f
Cytokines, for cancer, 176
Cytoplasm, 28, 29t
Cytosine, 32, 32f, 34f
Cytosol, 28

DaSilva, Ashanti, 279
Daubert standard, 201
Deinococcus radiodurans, 214, 229–230
Deontology, 326
Deoxyribonucleic acid. *See* DNA
Deoxyribose, 32
Depolymerization, 99
Deregulated status, 308–309, 310–311
Desulfovibrio desulfuricans, 229
Desulfuromonas acetoxidans, 221
Detergents, 101, 101t, 123
Diabetes mellitus, 64–65, 100, 131, 132f,
    133t, 272
Diafiltration, 111, 111f
Diagnostics, 263–269
Dialysis, 111, 111f
Dicer, 86
Dicotyledonous plants, 158
Dideoxyribonucleotide (ddNTP), 78
Differential display polymerase chain
    reaction, 234
Differentiation, in stem cells, 288
Diphtheria/pertussis/tetanus (DPT)
    vaccine, 136
Diploid number, 35, 248
Directed molecular evolution technology,
    104–106, 105f
Disease screening
  DNA microarrays in, 268, 268f
  legal/ethical aspects of, 338–340
Disease-causing microbes. *See* Pathogens

Disease-causing mutations, 54–55, 54f
DNA, 27, 30–33
  chromosomal, 34–35
  complementary, 69, 107, 298
  early studies of, 30–32
  extrachromosomal, 61
  as inherited genetic material, 30–32,
    31f
  insert, 63, 186
  mitochondrial, 202–203, 204
  naked, 277, 277f
  nuclear, 28
  nucleotides in, 32–33, 32f, 34f
  patented, 76, 319–321, 321f
  plasmid, 58, 61–63, 62f, 121. *See also*
    Plasmid(s)
  recombinant. *See under* Recombinant
    DNA technology
  structure of, 32–33, 34f. *See also*
    Double helix
  synthesis of, 37–38, 37f, 38f
  in transformation, 30–32, 31f
  variations in, 55
DNA amplification, 196, 196f. *See also*
    Polymerase chain reaction (PCR)
DNA cloning. *See* Clones/cloning
DNA database searching, 89
DNA fingerprinting, 10–11, 11f,
    190–204
  for animals, 204, 205–206, 206f
  applications of, 196–200
  autoradiography in, 194, 195f
  CODIS and, 192
  in criminal investigations, 199–202
  definition of, 191
  DNA extraction in, 193
  dot blot analysis in, 196, 196f
  ethical aspects of, 200
  as evidence, 200–201
  of identical twins, 197f
  known samples in, 197, 198f
  limitations of, 197, 198f
  in mass casualty events, 199–200
  methodology of, 191–196, 194f–196f
  in microbe tracking, 147
  microsatellites in, 192
  mitochondrial DNA analysis in,
    202–203, 203f
  in paternity testing, 202–204, 203f
  for plants, 204–205, 205f
  polymerase chain reaction in, 191
  privacy concerns and, 200
  in product authentication, 206
  restriction fragment length polymor-
    phisms in, 191, 193–195, 194f,
    195f
  short tandem repeats in, 192, 196
  Southern blotting in, 193–195, 195f
  specimen collection for, 192–193,
    201
  variable number tandem repeats in,
    192, 192f
  Y-chromosome analysis in, 203–204,
    203f
DNA helicase, 38, 40
DNA hybridization, 194
DNA libraries, 68–71, 70f, 71f, 142
DNA ligase, 38, 40
DNA microarrays, 15, 84–86, 85f, 118

in bioterrorism, 151
in disease screening, 268, 268f
in disease tracking, 147–148
field-based, 151
DNA polymerases, 38, 40, 41f
DNA probes. *See* Probes
DNA replication, 37–38, 37f, 38f
DNA sequencing, 78–80, 79f
DNA Shoah Project, 200
DNA tagging, 206
DNA-binding domains, 48
DNA-binding motifs, 48, 49f
DNA-profiling databases, 200
DNase, 31–32, 133t
Doctrine of equivalents, 320
Dolly (sheep), 10, 10f, 172, 177–178, 293, 295
Domains, 120
Dopamine, 282
Dot blot analysis, 196, 196f
Double helix, 32–33, 34f. *See also* DNA
unwinding of, 37–38, 38f, 40
Double-blind trials, 333
Down syndrome, 263–264, 264f
Downstream processing, 107
DPT vaccine, 136, 137
Dredging, in bioremediation, 228
Drososphilia melanogaster, genome of, 68t
Drug(s)
cancer, 3, 7t, 16, 269–271
generic, 22
teratogenic, 173, 328
toxicity of, 173
Drug approval process, 311–314
Drug delivery, 271–273, 272f
biocapsules in, 284–285, 285f
Drug development/production, 97–100, 100t, 268–275. *See also* Medical biotechnology
animal testing in, 172–175. *See also* Animal biotechnology
of antibiotics, 3
aquatic biotechnology in, 251–254
of cancer drugs, 3, 7t
careers in, 20–25, 302
clinical trials in, 261–262
ethical aspects of, 328
FDA approval in, 311–314
financing of, 102
generally-recognized-as-safe status and, 311
of generic drugs, 22
genomics in, 269–271
medical applications of, 100, 100t. *See also* Medical biotechnology
metabolomics in, 16
methods of, 99–100
nanotechnology in, 16–17, 17f, 272–273, 272f
overview of, 99–100
patents in, 76, 317–321
pharmacogenomics in, 15–16
phase testing in, 172–173, 312–313
plant biotechnology in, 164–166
regulation of, 311–314
top-ten drugs in, 7t
Drug-resistant bacteria, 132
Duty ethics, 326
Dwarfism, 46

E site, 44, 45f
Economic issues, 320, 340–341
Education and training. *See* Careers
Efficacy, 328
Effluent, 219
Eggs
from bioreactor hens, 184
genetically engineered, 181
protein harvesting from, 184, 188
Eight-base pair cutters, 59
Electicigens, 221
Electrophoresis, two-dimensional, 113, 113f
Electroporation, 62–63, 124–125, 125f
ELISA, 115
Embryo(s)
cloning of, 294–297, 336–337, 336f. *See also* Clones/cloning
moral/ethical status of, 333–336
Embryo twinning, 177
Embryonic development, 288–289, 288f
Embryonic stem cell method, 181. *See also* Stem cell(s)
Employment opportunities. *See* Careers
Endotoxins, 254
Energy sources
biomass as, 213, 255
microbes as, 222
plants as, 166, 255
sediment as, 221, 222f
Enhancers, 40, 47–48, 47f
Enucleation, 177
Environmental Genome Project, 93, 210
Environmental genomics, 143–144, 215–216, 215f
Environmental Protection Agency (EPA), 169, 209, 307, 307t, 309–311, 316t
Environmental toxins. *See also* Health and environmental concerns
bioremediation for, 208–230. *See also* Bioremediation
EnviroPig, 182
Enzyme(s). *See also* Protein(s)
in alcohol production, 99
definition of, 99
in DNA replication, 38, 40
in fermentation, 99
in food processing, 100–101
functions of, 99
industrial applications of, 99–100, 123
from marine organisms, 254–255
medical applications of, 100–101, 101t
microbial, 123. *See also* Microbial biotechnology
proteolytic, 108
restriction. *See* Restriction enzymes
substrates for, 59
thermostable, 123
Enzyme-linked immunosorbent assay (ELISA), 115
Erythropoietin, 133t
Escherichia coli
in bioremediation, 214, 223–224
genome of, 93t
as model organism, 58
protein production in, 107, 107t
Escherichia coli restriction enzyme, 60–61, 60t

Ethanol, fermentation of, 2–3, 99, 129–130, 129f, 166
Ethics. *See* Bioethics
Ethidium bromide, 76
Eugenics, 340
Eukaryotic cells, 28, 28f, 29t–30t, 120
European Agency for the Evaluation of Medicinal Products (EMEA), 321–322
European Union, 321–322, 322t
Eve hypothesis, 204
Evidence rules, 201–202
Ex situ bioremediation, 217, 220, 221f
Ex vivo gene therapy, 275
Exons, 41, 42f
Experimental use permits (EUPs), 309–310
Explosive residue, biodegradation of, 224, 225f
Expressed sequence tags, 76, 298–300
patented, 321
Expression vectors, 66t, 67–68, 125, 126f
Extrachromosomal DNA, 61
Exubera, 272
Exxon *Valdez* oil spill, 11, 11f, 225–226, 227

Factor VIII, 133t
Fermentation, 2–3, 99, 129–130, 129f
Fermenters. *See* Bioreactors
Fertilization
in bioremediation, 215, 216f, 226, 227f
in vitro, 287–288
Fetal testing, 263–265, 264f
Fetal tissue grafts, 282–283
Field trials, 308
Filtration separation, in protein purification, 110–111, 110f
Finance divisions, 23
Finasteride (Propecia), 173
Fingerprint analysis, 191
Fire, Andrew, 50, 87
Fish. *See also* Aquaculture
genetically engineered, 12, 12f
genome sequencing for, 246
marker genes in, 242–243
transgenic, 241
5' cap, 42, 42f
5' end, 33
Fixed-angle centrifugation, 110–111, 110f
Flagella, 30t
Flavr Savr tomato, 159, 161f, 277, 311
Fleming, Alexander, 131
Fletcher, Joseph, 326
Fluorescence in situ hybridization (FISH), 80, 81f, 264–265, 265f
Fluorescent bioassays, 127
Fluvirin, 316–317
Folded proteins, 103
Food
as bioterrorism target, 150, 151t
genetically modified. *See* Genetically modified foods; Plant biotechnology
regulation of, 311
vaccine delivery via, 165
Food additives, regulation of, 311

Food and Agriculture Organization, 321
Food and Drug Administration (FDA), 169–170, 175, 307, 307t, 311–314, 316, 316t, 321–322
Food processing, 100–101, 127–129
Foodborne diseases, pathogen tracking in, 147
Forensic biotechnology, 10–11, 11f, 190–204
    DNA fingerprinting in, 191. *See also* DNA fingerprinting
    evidence contamination in, 201–202
    jury characteristics and, 202
    mitochondrial DNA analysis in, 202–203
    in paternity testing, 202–204, 203f
    rules of evidence and, 201–202
    sources of error in, 201–202
    specimen collection in, 192–193, 201
    Y-chromosome analysis in, 203–204, 203f
Forest Hills rapist, 197–198
454 Life Sciences, 92
Four-base pair cutters, 59, 60t
F-plasmids, 63
Frameshift mutations, 52f, 53
Freeze-dried proteins, 115–116
Frye standard, 200–201
Fuel cells
    microbe-powered, 222
    underwater, 255
Functional genomics, 166
Funding sources, 320, 340–341
Fungi, 121–122. *See also* Microbial biotechnology; Yeast
    in bioremediation, 214
    in protein engineering, 107–108, 108t
Fusarium graminearum, 142
Fusarium oxysporum, 214
Fusion proteins, 107, 125–126

Gametes, 35
Gardasil, 139
G-banding, 36
Gel electrophoresis, 75–78, 77f
Gelsinger, Jesse, 281, 340
GenBank, 89
Gene(s)
    definition of, 33
    disease-causing, 263
    functions of, 91, 91f
    homolog, 262
    Internet resources for, 300
    knock-in, 185
    knockout, 10, 184–186, 185f, 284
    mapping of, 75–78, 77f, 298, 299f, 300–302, 301f
    marker. *See* Marker genes
    patented, 76, 319–321, 321f
    reporter, 244–245
    structure of, 33
Gene chips. *See* DNA microarrays
Gene cloning. *See* Clones/cloning
Gene Codes Forensics, 199, 200
Gene expression, 46–50
    in bacteria, 48–50, 49f, 50f
    experimental methods for, 81–87
    in pharmacogenomics, 271, 271f

Gene guns
    in animal transgenics, 181
    in plant transgenics, 158–159, 160f
Gene mapping, 4, 4f
    restriction, 75–78, 77f, 298, 299f, 300–302, 301f
Gene microarrays, 84–86, 85f
Gene probes. *See* Probes
Gene regulation, 46–50
Gene silencing, 50–51, 51f
    in gene therapy, 277–279, 278f
    in plants, 160–161
Gene therapy, 7, 13, 16–17, 275–282
    antisense technology in, 277–279, 278f
    challenges facing, 281–282
    for cystic fibrosis, 280f, 281
    early experiences with, 279, 280f
    ethical aspects of, 339–340
    ex vivo, 275, 276f
    gene silencing in, 277–279, 278f
    in vivo, 275–276, 276f
    patents for, 318t
    procedures in, 275–276
    regulation of, 318t
    RNA interference in, 278–279, 278f
    safety of, 281–282, 340
    for severe combined immunodeficiency, 279, 281
    targets for, 279–281
    vectors in, 276–277
Genentech, 19, 65, 131, 270
Generally-recognized-as-safe (GRAS) status, 311
Generic drugs, 22
Genetic code, 43–44, 44t
Genetic diseases
    diagnosis of, 263–269
    treatment of. *See* Gene therapy
Genetic engineering, 3–4, 58–87. *See also* Protein engineering; Recombinant DNA technology
    of animals, 171–188. *See also* Animal biotechnology
    definition of, 58
    germline, 339–340
    of plants, 155–169. *See also* Plant biotechnology
Genetic Information Nondiscrimination Act, 339
Genetic mutations. *See* Mutations
Genetic pesticides, 162–163, 165t
Genetic recombination, homologous, 184, 185f
Genetic screening
    DNA microarrays, 268, 268f
    legal/ethical aspects of, 338–340
Genetic testing, ethical aspects of, 338–339
Genetic variation, 55
Genetically modified foods, 2, 108, 155–170. *See also* Plant biotechnology
    benefits of, 9
    in China, 320
    ethical aspects of, 328–331
    health benefits of, 167t
    labeling of, 316, 331
    legal aspects of, 330, 331

opposition to, 9
    regulation of, 169–170
    safety of, 9, 61, 162–163, 167–170
Genetically modified organisms. *See under* Transgenic organisms
Genome(s), 38–39
    definition of, 4, 38
    interspecies similarity in, 55, 262–263
    microbial, 141–145, 143f, 144t, 145t
    personalized, 92
    relative size of, 39, 42
Genome sequencing, 4, 4f, 13, 39, 76, 87–92. *See also* Genomics; Human Genome Project
    annotation in, 142
    applications of, 142–143
    bioinformatics in, 87–89
    BLAST search in, 89
    DNA libraries in, 68–71, 70f, 71f, 142
    methods of, 87, 142
    microbial, 141–145
    for model organisms, 92
    shotgun, 87, 142
    whole-genome, 87, 142
Genomic DNA libraries, 69–71, 70f
Genomics, 39, 87–95. *See also* Genome sequencing
    applications of, 141–142
    in aquatic biotechnology, 245–246
    in bioremediation, 215, 215t
    careers in, 53
    comparative, 92–93, 93t
    definition of, 87
    in drug development, 269–271
    environmental, 143–144, 215–216, 215t
    ethical aspects of, 338–339
    functional, 166
    microbial, 141–145
    Stone Age, 93–95
    viral, 144–145
Germline genetic engineering, 339–340
Gilbert, Walter, 61
Gill, Peter, 197
Ginseng, 204
Gleevec, 270–271
Globalization, 321–322
Glucose metabolism, 129
Glues, 101–102, 101t
Glycomics, 92
Glycosylation, 104, 104f
Glyphosphate, 163, 164f
Golden Rice, 164, 164f
Goldman, Ronald, 201
Golgi apparatus, 29t
Good clinical practices, 312
Good laboratory practices, 312
Good manufacturing practices, 312
Government funding, for biotechnology, 320, 340–341
Government regulation, 23, 305–317. *See also under* Regulatory agencies
Grafts, 283–284
Gram stain, 120–121
Granulocyte colony-stimulating factor, 133t
Green fluorescent protein (GFP), 124, 124f, 244–245, 245f
Griffith, Frederick, 30–31, 31f

Groundwater pollution, bioremediation of, 219–220, 221f
Growth factors, 289
Growth hormone
    bovine, 165t, 311
    human, 46, 133t, 243, 246
Guanine, 32, 32f, 34f

Haemophilus aegypticus, 60t
Haemophilus influenzae, genome sequencing for, 87, 90, 144t
Haemophilus influenzae restriction enzyme, 58, 60t
HAMA response, 186
Haploid number, 35, 248–249
HapMap, 15, 267–268
Health and environmental concerns
    in aquaculture, 239, 240–241
    bioremediation and, 208–230. *See also* Bioremediation
    drug approval process and, 313–314
    efficacy and, 328
    ethical aspects of, 327–338
    in gene therapy, 281–282, 340
    for genetically modified foods, 9, 61, 328–331
    for genetically modified organisms, 9, 61
    government regulation and, 305–317
    in plant biotechnology, 162–163, 167–170
Heavy metals, bioremediation for, 102, 223–224, 257–259
Helix
    alpha, 104
    double, 32–33, 34f. *See also* DNA unwinding of, 37–38, 38f, 40
Hemoglobin, in sickle-cell disease, 54, 54f
Hemophilia, 183–184
Hepatitis B vaccine, 138–139
Herbicides
    commercialization of, 310
    deregulated status of, 308–309, 310–311
    experimental use permits for, 309–310
    resistance to, 163
Herceptin, 270, 275
hGH, 46
High-performance liquid chromatography, 114, 114f
Hippocratic Oath, 325
Hiring trends, 24–25
Histones, 35, 35f
HIV/AIDS, 69, 263
    vaccine for, 139, 140f, 141
Homo neanderthalensis, 95
Homologous recombination, 184, 185f
Homologs, 262
Homologues, 35
Hormones, 48, 99
HPV vaccine, 139
Hudson River, PCBs in, 228
Human antimouse antibody (HAMA) response, 186
Human cloning, 64–65, 296–297, 336–337, 336f, 337f, 338. *See also* Clones/cloning

Human embryonic stem cells. *See* Stem cell therapy
Human experimentation, 332–335, 339–340
Human Genome Nomenclature Commission, 89
Human Genome Project, 4, 4f, 13, 39, 76, 89–92, 262, 303
    ethical aspects of, 338–339
    future applications of, 14–17
    history of, 90
    Internet resources for, 300
    medical applications of, 298–303
    "omics" revolution and, 92
    results of, 90–92
Human growth hormone (hGH), 46
Human immunodeficiency virus infection. *See* HIV/AIDS
Human leukocyte antigens (HLAs), 284
Human Microbiome Project, 141
Human papillomavirus vaccine, 139
Humane treatment, 328
Humulin (recombinant insulin), 64–65, 100, 131, 132f, 133t, 272
Hwang, Woo Suk, 297, 334
Hybridization, 69–70
    colony, 69, 71f
    in situ, 84
    in library screening, 69–70
    plant, 157
    in polymerase chain reaction, 72
    in Southern blotting, 194
Hybridomas, 274–275
Hydrogen, 104
Hydrogen bonds, 104
Hydrophilic molecules, 102
Hydrophobic interaction chromatography, 113
Hydrophobic molecules, 102
Hydroponic systems, 238
Hydroxyapatite, 251–252
Hyperthermia, for cancer, 176

Ice-minus bacteria, 134–135, 310, 310f
Immortality, cellular, 186–287
Immune response, 136–137, 137f, 138f
Immunity, antibody-mediated, 136–137, 137f, 138f
Immunization. *See* Vaccines
In situ bioremediation, 217, 220, 221f
In situ hybridization, 84
In vitro fertilization, 287–288, 335–336, 337
In vivo gene therapy, 275
Inactivated vaccines, 137–138
Inclusion bodies, 107
Indigenous microbes, 214
Induced pluripotent stem cells, 290
Industrial proteins, 100–102, 101t
    in adhesives, 101–102
    in detergents, 101, 101t
    in leather goods, 101, 101t
    in paper manufacturing/recycling, 101
    in textiles, 101
Influenza vaccine, 139–140
Informed consent, 332
Inherited mutations, 53
Inner cell mass, 288, 288f

The Innocence Project, 202
Innocence Protection Act, 202
Inorganic compounds, 213
Insects, in protein engineering, 109
Insert DNA, 63, 186
Institute for Genomic Research, 141
Insulin
    in biocapsules, 285
    recombinant, 64–65, 100, 131, 132f, 133t, 272
Integrated aquaculture, 238
Integration, 276
Intellectual property, 340–341
Interactomics, 92
Interferons, 133t
Interleukins, 133t
    for cancer, 176
Intermediate filaments, 29t
Introns, 41, 42f
Invasive species, in aquaculture, 241
Investigational new drug applications, 312
Ion exchange chromatography, 111, 112f
Iso-electric focusing, 113
Iso-electric point, 113

James, Jesse, 204
Jeffries, Alex, 197
Jenner, Edward, 135–136
Job opportunities. *See* Careers
Juries, 202
Justice for All Act, 202

Kant, Immanuel, 326
Karyotype analysis, 36–37, 36f, 264
Killed vaccines, 137–138
Knock-ins, 185
Knockout, 10, 184–186, 185f, 284
Kuru, 105–106
Kuwait oil fields, bioremediation in, 226–227

Labeling, of genetically modified food, 316, 331
Labels, DNA, 206
Laboratory technicians, 20, 205
lac operon, 48–50, 50f
lac repressor, 48, 49–50, 50f
lacZ gene, in insulin production, 131
Lactase, 129, 130
Lactic acid fermentation, 129, 130
Lactic-acid bacteria, genome sequencing for, 142
Lactobacilli, 130
Lactococcus lactis, 143
Lagging strand, 38
Landfarming, 218
Large-scale (batch) processing, 3
Leachate, 210
Leading strand, 38
Leaf fragment technique, 158, 159f
Leather industry, 101
Legal issues, 23
    forensics. *See* Forensic biotechnology genetically modified foods, 330, 331
    intellectual property, 340–341

Legal issues (*continued*)
  labeling, 316
  patents, 76, 317–321, 331, 337–338, 340–341
  rights to biological material, 337–338
Legal specialists, 23
Leukemia, 270–271, 281, 291
Leukocytes, 136
Ligands, 111–113, 113f
Lignocellulose, 166
Limulus amoebocyte lysate (LAL) test, 254
Lindow, Steven, 134–135
Linker sequences, 69
Lipases, 99
Liposomes, in gene therapy, 277, 378
Locus, 193
Lopez, Victor, 198
Luciferase, 127
lux genes, 127, 127f, 224
Lymphocytes, 136, 137f
Lyophilization, 115–116
Lysosomes, 29t, 137
Lytic cycle, 66, 67f

MacLeod, Colin, 31–32
Macrophages, 137
Macular degeneration, 278–279
Mad cow disease, 103, 105–106
Major histocompatibility complex, 283–284
Malaria, 141, 177
Mammalian cells, in protein engineering, 108
Manoalide, 252
Manufacturing assistants, 117
Manufacturing technicians, 117
Mapping, chromosome, 75–78, 77f, 298, 299f, 300–302, 301f
Mariculture, 232. *See also* Aquaculture
Marker genes, 127, 158–159, 160f
  excision of, 168
  in fish, 242–243
  in gene guns, 158–159, 160f
  in vectors, 65
Marketing specialists, 23
Marx standard, 201
Mass Fatality Identification System (M-FISys), 199–200
Mass spectrometry, 114, 114f
Maternal chromosomes, 35
McCarty, Maclyn, 31–32
Measles/mumps/rubella (MMR) vaccine, 136
Medical biotechnology, 12, 260–303
  artificial blood and, 273
  biocapsules in, 284–285, 285f
  careers in, 302
  clinical trials in, 261–263
  definition of, 261
  in disease diagnosis, 263–269
  in drug delivery, 271–273, 272f
  in drug development/production, 97–100, 100t, 268–275
  gene expression data in, 271, 271f
  gene therapy in, 7, 13, 275–282
  Human Genome Project in, 298–303
  model organisms in, 261–263
  overview of, 261

proteomics in, 92, 117–118
regenerative medicine in, 272–298
regulation of, 297–298
stem cell therapy in, 13
tissue engineering in, 285–286, 286f
transplantation in, 282–284
vaccines and, 133t, 135–141, 273–275. *See also* Vaccines
Meiosis, 37
Meischer, Friedrich, 30
Mello, Craig, 50, 87
Membrane, cell, 28, 28f, 29t
Membrane filtration, 111
Memory B cells, 136, 137f
Mercury, bioremediation for, 102, 223–224, 257–259
Messenger RNA, 39
Metabolic engineering, 166
Metabolomics, 16, 92
Metagenomics, 143–144
Metal recovery, bioremediation in, 102, 228–229, 258–259
Metallothioneins, 102
Microarrays. *See* DNA microarrays; Protein microarrays
Microbes
  antibiotic-producing, 131–132
  in bioremediation, 213–214
  classification of, 120
  definition of, 120
  disease-causing. *See* Pathogens
  indigenous, 214
  structure of, 120–121
Microbial biotechnology, 8, 119–153
  agricultural applications of, 134–135
  bacteria in, 120–124. *See also* Recombinant bacteria
  bioluminescence in, 126–127, 127f
  in bioterrorism, 148–153
  career opportunities in, 153
  competent cells in, 123
  diagnostic applications of, 145–148, 146f
  in drug development, 131–132, 133t
  enzymes in, 123
  fermentation in, 129–130, 129f
  in food production, 100–101, 127–130
  fusion proteins in, 125–126
  genomics in, 141–145, 143f, 144t, 145t
  in insulin production, 64–65, 100, 131
  medical applications of, 131–132, 133t, 135–148
  protein engineering in, 107
  safety concerns in, 61, 135
  in vaccine development, 133t, 137–141
  yeasts in, 121–122
Microbial diagnostics, 145–148
Microbial enzymes, 123
Microbial Genome Program, 215
Microcapsules, 284–285, 285f
Microcosm studies, of bioremediation, 223
Microfilaments, 29t
Microfiltration, 111
Microorganisms. *See* Microbes
MicroRNAs, 41, 50, 51f
Microsatellites, 192
Microspheres, 271–272, 272f

Microtubules, 29t
Milk, bovine somatotropin and, 165t, 311, 316
Mill, John Stuart, 325–326
miRNAs, 41
Missense mutations, 52, 52f
Mitochondria, 29t, 202
Mitochondrial DNA analysis, 202–203, 203f, 204
Mitosis, 37
MMR vaccine, 136
Model organisms, 58, 61, 90, 92, 261–263
Molecular pharming, 9
Molt-inhibiting hormone, 243
Monoclonal antibodies, 108–109, 274, 274f, 275
  production of, 186–187, 187f
Monocotyledonous plants, 158
Monsanto, 331
Moratorium, 327
Mortierella hyaline, in bioremediation, 214
Morula, 288
Mosquitoes, bioengineered, 177
Mothers of Placo de Mayo, 203–204
Motifs, 48, 49f
Mouse hybridoma cells, 186, 187f
mRNA, 39
mtDNA analysis, 202–203, 203f, 204
Mullis, Kary, 72
Multilocus probe, 194
Multiple cloning sites, 65
Murder cases. *See* Forensic biotechnology
Mutagenesis studies, 86
Mutagens, 51
Mutations, 15, 51–55
  acquired, 53
  causes of, 51
  definition of, 51
  disease-causing, 54–55, 54f
  frameshift, 52f, 53
  inherited, 53
  missense, 52, 52f
  nonsense, 52f, 53
  point, 52, 52f
  silent, 52, 52f
Mycobacterium tuberculosis, 127, 141
Myelomas, 186, 187f

NAD+/NADH, in fermentation, 129–130, 129f
Naked DNA, 277, 277f
Nanomedicine, 272–273, 272f
Nanotechnology, 16, 17f, 272–273, 272f
Narborough Village murders, 197, 200
Nathans, David, 59
National Academy of Sciences, 176
National Center for Biotechnology Information (NCBI), 89
National Institute of Environmental Health Sciences, 210
National Institutes of Health (NIH), 61, 89, 90, 93, 98, 175, 210, 321
Natural protein adhesives, 102
Neanderthals, genome of, 95
New drug authorization (NDA), 312–313
New York v. Castro, 201
Nitrogenous bases, 32

Nonsense mutations, 52f, 53
Northern blotting, 81, 82, 83f
Notification, 309
Nuclear DNA, 28
Nuclear envelope, 28, 29f, 30t
Nuclear reprogramming, of somatic cells, 290, 291f
Nucleic acids, 30. *See also* DNA; RNA
Nuclein, 30
Nucleoli, 29f, 30t
Nucleotides, 32–33, 32f, 34f
   in genetic code, 43–44, 44t
Nucleus, cell, 28
Nutrient enrichment, 215, 216f
Nutrigenomics, 92

Objectivism, 326
Occupational opportunities. *See* Careers
Oil spills, 11, 11f, 222–223, 225–227, 226f, 227f
Okazaki fragments, 38, 38f
Oligonucleotides, synthesis of, 71
Oncogenes, 269
Operations, 21f, 22–23
Operator, 50
Operons, 48–50, 50f
Oragenics, 134
Oral polio vaccine (OPV), 136
Organ transplantation
   animal-human, 176
   legal/ethical aspects of, 337–338
Organelles, 28, 29t
Organic molecules, 213
Origins of replication, 38, 65
Orphan Drug Act, 313
Osteoporosis, 251
Oxidation and reduction reactions, in bioremediation, 211–212, 213f
Oxidizing agents, 212
Oxygen, in aerobic metabolism, 213

p arm, 35f, 36
P site, 44, 45f
Pääbo, Svante, 95
Paleogenomics, 93–95
Palindromes, 59
Pandemic influenza, 139–141
Panitumab, 186–187
Papain, 108
Paper manufacturing/recycling, 101–102
Parkinson's disease, 282–283
Parthenogenesis, 289
Patents, 76, 317–321, 331, 337–338, 340–341
Paternal chromosomes, 35
Paternity testing, 202–204, 203f
Pathogens, 135. *See also* Microbes
   in bioterrorism, 148–153, 149t, 151t
   identification of, 145–147, 146f, 148f
   tracking of, 146f, 147–148
   in transgenic plants, 168
Patient rights, 337–338
PCBs, bioremediation of, 228
Penicillin, 3, 131, 133t. *See also* Antibiotic(s)
Pentose sugars, in nucleotides, 32, 32f, 34f
Peptidoglycan, 120

Peptidyl transferase, 45
Peptidyl tRNA, 45, 45f
Permitting processes, 308
Peroxisomes, 29t
Personalized genomes, 92
Personhood, 334–335
Pesticides
   commercialization of, 310
   deregulated status of, 308–309, 310–311
   experimental use permits for, 309–310
   genetic, 162–163, 165t
Petroleum products, bioremediation for, 11, 11f, 222–223, 225–227, 226f, 227f
Pfiesteria piscicida, 246
Phage. *See under* Bacteriophage
Phanerochaete chrysosporium, 214
Phanerochaete sordida, 214
Pharmaceutical companies. *See also* Drug development/production
   top ten, 20t
   vs. biotechnology companies, 19
Pharmacogenomics, 15–16, 269–271
Phase testing, 172–173, 312–313
Phosphodiester bonds, 32f, 33
Photoreceptors, 144
Phytoremediation, 216–217, 216f
Pichia pastoris, 122
Pigs, in xenotransplantation, 284, 284f
Pirating, DNA fingerprinting and, 206
Placebos, 332–333
Placo de Mayo mothers, 203–204
Plague, 148–149, 149t, 150f
Plankton, 255
Plant(s). *See also* Food
   in bioremediation, 216–217, 216f, 224, 225f
   cell structure in, 28, 28f, 29t–30t
   dicotyledonous, 158
   DNA fingerprinting for, 204–205, 205f
   genetically modified, 155–170, 320, 328–331. *See also* Transgenic organisms
   in integrated aquaculture, 238
   in molecular pharming, 9
   monocotyledonous, 158
   polyploid, 157
   in protein engineering, 108
Plant biotechnology, 155–169. *See also* Genetically modified foods
   antisense technology in, 159–161
   applications of, 161–166, 165t, 172f
   in biofuel development, 166
   careers in, 169
   in China, 320
   chloroplast engineering in, 159, 161f
   cloning in, 157, 161f
   for disease resistance, 161–162, 165t
   in drug development, 164–166
   for enhanced nutrition, 164, 165t
   functional genomics in, 166
   gene guns in, 158–159, 160f
   gene silencing in, 160–161
   genetic pesticides in, 162–163, 165t
   for herbicide resistance, 163, 165t
   leaf fragment technique in, 158, 159f
   legal aspects of, 317–321, 331
   metabolic engineering in, 166

   patents in, 317–321, 318t, 331
   protoplast fusion in, 157, 158f
   regulation of, 169–170, 308–309
   for safe storage, 163, 165t
   safety of, 162–163, 167–170
   for stronger fibers, 163–164
   transgenesis in, 157–161
   vaccines in, 161–162, 162f, 165
   vs. conventional breeding/hybridization, 157
Plant pathogens, in bioterrorism, 150, 151t
Plant pest risks, 308–309
Plaques, 66
Plasma cells, 136
Plasma membrane, 28, 28f, 29t
Plasmid(s), 61–63, 62f, 121
   as cloning vectors, 65–68, 66t, 67f
   extrachromosomal nature of, 61
   F, 63
   Ti, 68
   tumor-inducing, 158
Plasmid DNA, 58
Plasmodium falciparum, 141
Pluripotent stem cells, 288, 290. *See also* Stem cell(s)
Point mutations, 52, 52f
Polarity, of nucleotides, 33
Polio vaccine, 136
Pollution, bioremediation for, 11, 102, 208–230. *See also* Bioremediation
Polyadenylation, 42–43
Polyculture, 238, 238f
Polycyclic aromatic hydrocarbons, biodegradation of, 224
Polygalacturonase, 159
Polymerase chain reaction (PCR), 71–73, 72f
   applications of, 73, 74f, 146–147
   in bioweapon identification, 151–152
   differential display, 234
   in disease diagnosis, 266, 266f
   in DNA fingerprinting, 191, 196
   in gene cloning, 73, 74f
   process of, 71–73, 72f
   quantitative (real-time), 73, 83–84, 83f
   reverse transcription, 82–83, 83f
Polypeptides, 43–45, 45f
Polyploids
   in aquaculture, 247–250, 250f
   in plant biotechnology, 157
Poly(A) tail, 42f, 43
Posttranslational modifications, 104, 104f
Precipitate, 110
Preimplantation genetic testing, 266
Pre-mRNA, 41
Primary protein structure, 103, 103f
Primary transcript, 41
Primase, 38
Primers, 66, 72, 72f, 73
Principal scientists, 20–21
Prions, 103, 105–106
Privacy concerns, for DNA fingerprinting, 200
Probability, 329–330
Probes, 69–70
   in aquaculture, 239
   multilocus, 193–194
   single-locus, 193–194

Production, 21f, 22–23
Programmed cell death, 262
Project BioShield, 151
Prokaryotic cells, 27–28, 27f, 119f, 120–121. See also Bacteria
Promoters, 40, 47, 47f
Pronuclear microinjection, 181
Propecia (finasteride), 173
Prostate-specific antigen, 263
Proteases, 99
Protein(s). See also Enzyme(s)
    biosynthesis of, 39, 43–45
    as biotechnology products, 97–118.
        See also Protein engineering
    definition of, 98
    freeze-dried, 115–116
    functions of, 28, 39
    fusion, 107, 125–126
    glycosylated, 104, 104f
    hydrogen bonds in, 104, 104f
    industrial. See Industrial proteins
    misfolded, 105–106, 105f
    recombinant, 6–7, 7t. See also Protein engineering
    single-strand binding, 38
Protein engineering, 6–7, 6f, 7t, 98, 98f, 100–102. See also Genetic engineering; Recombinant DNA technology
    in adhesive production, 101–102
    bioreactors in, 107, 108f
    in bioremediation, 11, 102
    in detergent production, 101
    directed molecular evolution technology in, 104–106, 105f
    in drug development/production, 97–100, 100t, 101t
    in E. coli, 107, 107t
    ELISA in, 115
    in food processing, 100–101, 101t
    insects in, 109
    in leather processing, 101
    methodology of, 98–100, 98f, 99f, 107–117
    microarrays in, 106, 117–118, 117f, 118f
    overview of, 98
    in paper manufacturing/recycling, 101
    protein expression in, 107–109
    protein preservation in, 115–116
    protein purification in, 109–114.
        See also Protein purification
    protein sequencing in, 116
    protein sources for, 107–108
    proteomics and, 117–118
    purification of, 98
    recombinant, 6–7, 6f, 7t
    SDS-PAGE in, 114–115, 115f
    steps in, 107–117
    structure of, 102–104, 103f, 104f
        elucidation of, 98
    in textile manufacturing, 101
    verification in, 114–115
    X-ray crystallography in, 116–117
Protein folding, 103, 105–106, 105f
Protein microarrays, 106, 117–118, 117f, 118f, 268
    in bioterrorism, 151, 152f

in pathogen identification, 147–148, 148f, 151, 152f, 268
Protein precipitation, 110
Protein preservation, 115–116
Protein purification, 109–114
    cell lysis in, 109
    chromatography in, 111–114, 112f–114f
    dialysis in, 111, 111f
    ELISA in, 115
    filtration separation in, 110
    iso-electric focusing in, 113
    mass spectrometry in, 114, 114f
    membrane filtration in, 111
    method selection for, 116
    protein harvesting in, 109
    protein precipitation in, 110
    protein stabilization in, 109
    scale-up of, 116
    two-dimensional electrophoresis in, 113, 113f
    verification in, 114–115
Protein sequencing, 116
Protein Structure Initiative, 98
Proteolytic enzymes, 108
Proteomes, 14, 106, 117–118
Proteomics, 92, 117–118
Protoplast, 157
Protoplast fusion, 157, 158f
pSC101 cloning experiments, 61
Pseudomonas aeruginosa, 143
Pseudomonas fluorescens, 135, 224
Pseudomonas putida, 214
Pseudomonas stutzeri, 120
Pseudomonas syringae, 134–135
PulseNet, 147

q arm, 35f, 36
Quality assurance (QA), 13, 322
Quality control (QC), 13, 322
Quality control inspector, 322
Quantitative polymerase chain reaction, 73, 83–84, 83f, 84f
Quaternary protein structure, 103, 103f

Raceways, 236, 236f
Radioactive waste, bioremediation of, 214, 229–230
Ramsey, Paul, 326
Random cloning, 68
Rape cases. See Forensic biotechnology
RDX, biodegradation of, 224, 225f
Real-time polymerase chain reaction, 73, 83–84, 83f, 84f
Recognition sequences, 59, 59f
Recombinant bacteria. See also Microbial biotechnology
    in bioremediation, 222–224, 226
    as biosensors, 224
    in drug development, 133–134
    field applications of, 134–135
    in food production, 127–130, 129f
    petroleum-eating, 222–223, 226
    safety of, 61, 135
    selection of, 63, 64f
    transformation in, 61–63, 123–124, 124f

Recombinant DNA Advisory Committee (RAC), 61
Recombinant DNA technology, 4, 57–87, 61–63, 62f. See also Genetic engineering
    applications of, 75–87
    controversial aspects of, 296
    definition of, 58
    development of, 60–61
    DNA microarrays in, 84–86, 85f
    DNA sequencing in, 76–80, 79f
    fluorescence in situ hybridization in, 80, 81f
    gel electrophoresis in, 75
    in gene cloning, 63–74
    in situ hybridization in, 84
    Northern blotting in, 81, 82, 83f
    plasmids in, 61–63
    polymerase chain reaction in, 71–73, 72f. See also Polymerase chain reaction (PCR)
    protein expression and purification in, 86
    regulation of, 297–298
    restriction enzymes in, 58–59
    RNA interference in, 86–87, 86f
    site-directed mutagenesis in, 86
    Southern blotting in, 81, 82f
    vectors in, 61, 65–68
    Western blotting in, 81
Recombinant proteins, 6–7, 7t. See also Protein(s); Protein engineering
Recombination, homologous, 184, 185f
Recycling, paper, 101–102
Red algae, 255
Red seaweed, 255
Redox reactions, in bioremediation, 211–212, 213f
Regenerative medicine, 17, 272, 333–335
    cellular therapeutics in, 284–285
    cloning in, 293–298
    stem cell therapy in, 13, 17, 287–293
    telomeres and, 286–287
    tissue engineering in, 285–286
    transplantation in, 282–284
Regulatory agencies, 306–317, 307t
    controversial issues and, 314–315
    Environmental Protection Agency (EPA), 169, 209, 309–311, 316t
    Food and Drug Administration, 169–170, 311–314
    labeling and, 316
    U.S. Department of Agriculture, 169, 308–309
    U.S. Department of Energy, 141, 229
Regulatory biotechnology, 13, 305–322
    controversial issues and, 314–315
    ethical aspects of, 338–340
    labeling and, 316
    patents and, 76, 317–321
Rennin, 8, 128
Replication origins, 38, 65
Reporter genes, 244–245
Reproductive cloning, 293–298, 294t, 295f. See also Clones/cloning
Research and development (R&D), 20–22, 21f
Research assistants/associates, 20
Research funding, 320, 340–341

Restriction endonuclease, 193
Restriction enzymes, 58–59, 59f, 60t, 80
    in gene mapping, 75–78, 77f
    methylation and, 61
    types of, 60t
Restriction fragment length
        polymorphism (RFLP) analysis,
        146
    in disease diagnosis, 265–266, 266f
    in DNA fingerprinting, 191, 193–195,
        194f, 195f
Restriction mapping, 75–78, 77f, 298,
        299f, 300–302, 301f
Restriction sites, 59, 59f
Retroviruses, 69, 139
    in transgenics, 180
Reverse transcriptase, 69, 70f
Reverse transcription polymerase chain
        reaction, 82–83, 83f
Ribonucleic acid. *See* RNA
Ribosomal RNA, 40–41
Ribosomes, 29t, 43, 44
Risk assessment, 330
RNA, 30
    in DNA replication, 38, 38f
    messenger, 39
    micro, 41, 50, 51f
    processing of, 39, 41–43, 42f
    in protein synthesis, 39, 43–45
    ribosomal, 40–41
    short interfering, 50
    small interfering, 17, 86, 278
    in transcription, 39–41, 39f, 41f
    transfer, 40–41
    in translation, 39, 39f, 43–45
RNA interference (RNAi), 50, 86–87, 86f,
        186
    in gene therapy, 278–279, 278f
RNA polymerase, 40, 41f
RNA polymerase promoter sequences,
        65–66
RNA primers, 38, 38f, 66, 72, 72f, 73
RNA splicing, 41–42, 42f
RNA-induced silencing complex (RISC),
        86, 278. *See also* Gene silencing
RNase, 31–32
Rough endoplasmic reticulum, 29t
Royal demolition explosive (RDX),
        biodegradation of, 224, 225f
rRNA, 40–41

Saccharomyces cerevisiae, 93t, 122, 128.
        See also Yeast
Safety concerns. *See* Health and
        environmental concerns
Salaries, 23–24
Sales representatives, 23
Salmon farming, 239, 241, 245. *See also*
        Aquaculture
Salmonella infection, 147, 182
Sanger, Frederick, 61
Scale-up processes, 22–23
Scientific communication, 341
Scrapie, 105–106
Screening, disease
    DNA microarrays in, 268, 268f
    legal/ethical aspects of, 338–340
SDS-PAGE, 114–115, 115f

Seaweed, 255
Secondary protein structure, 103, 103f
Sediment, as energy source, 221, 222f
Seeding
    in bioremediation, 215–216
    in tissue engineering, 285, 286f
Selection, 63
    antibiotic, 63, 64f, 123
Selective breeding, 3, 157, 181
    in aquaculture, 238–239
Semiconservative replication, 37–38, 38f
Senescence, 286
Senior scientists, 20–21
Septic tanks, 219, 219f
Serratia marcescens, 60t
Severe acute respiratory syndrome
        (SARS), 147–148, 148f
Severe combined immunodeficiency
        (SCID), 279, 281
Sewage treatment, 219
Sex chromosomes, 35
Sex hormones, 48
Shellfish. *See* Aquaculture
Short tandem repeats (STRs), 192, 196
Shotgun cloning (sequencing), 87, 88f,
        142
Sickle-cell disease, 54–55, 54f, 265, 266,
        292
Silent mutations, 52, 52f
Silk protein production, 183
Simpson, O.J., 201
Single nucleotide polymorphisms (SNPs),
        15, 15f, 55, 267–268
Single-locus probe, 193–194
Single-strand binding proteins, 38
siRNA, 17, 50, 86, 278, 278f
Sister chromatids, 35f, 36
Site-directed mutagenesis, 86
Six-base pair cutters, 59, 60t
Size-exclusion chromatography, 111,
        112f
Slot blot analysis, 196, 196f
Sludge, 219
    as energy source, 221, 222f
Slurry-phase bioremediation, 217
Small interfering RNA (siRNA), 17, 50,
        86, 278, 278f
Smallpox, 149–150, 149t, 150f
Smith, Hamilton, 59
Smooth endoplasmic reticulum, 29t
Soil bioremediation, 217–218, 218f,
        225–227. *See also* Bioremediation
Solid-phase bioremediation, 217
Somatic cell(s), 35
    nuclear reprogramming of, 290, 291f
Somatic cell nuclear transfer, 294–297,
        294t, 336–337, 336f
Sorcerer II Expedition, 143–144
South Asian tsunami, DNA
        fingerprinting in, 199
Southern blotting, 81, 82f
    in DNA fingerprinting, 193–195, 195f
Species integrity, 328–329
Specific utility, 319
Spectral karyotype, 264–265
Sperm DNA, isolation of, 197, 198f
Sperm-mediated transfer, 181
Spinal cord injuries, 283, 283f, 292
Splicing, RNA, 41–42, 42f

Squalamine, 253
Stain
    Coomassie, 115
    Gram, 120–121
Staphylococcus aureus, 182
Staphylococcus simulans, 182
StarLink corn, 169
Startup companies, 20
Statistical probability, 329–330
Stem cell(s), 287–293
    adult-derived, 17, 289–290, 294t, 335
    amniotic fluid–derived, 290
    embryonic, 17, 18f, 288–290, 294t,
        334–336
    from nuclear reprogramming of
        somatic cells, 290, 291f
    pluripotent, 288
    induced, 290
Stem cell therapy, 13, 17, 287–293
    applications of, 290–293, 292t
    cell isolation and culture in, 288–289,
        288f
    controversial aspects of, 296, 314–315
    ethical aspects of, 333–335
    regulation of, 297–298
    in therapeutic cloning, 293–298, 294t
    in tissue engineering, 290–293, 293f,
        294t
    unanswered questions in, 292–293
    in vitro fertilization and, 287–288
Stemagen, 297
Stone Age genomics, 93–95
STR analysis, 192, 196
Streptococcus mutans, 134
Streptococcus pneumoniae, genome
        sequencing for, 141–142, 144t
Substantial utility, 319–320
Substrates, 59
Subtilisin, 123
Subunit vaccines, 137
Superfund Program, 209
Superoxide dismutase, 133t
SYBR Green probes, 83–84, 84f

T lymphocytes, 136, 137f
Tag proteins, 125–126, 126f
Taq DNA polymerase, 72, 196, 254
TaqMan probes, 83–84, 84f
TATA box, 47, 47f
Telomerase, 286–287, 289
Telomeres, 35f, 36, 286–287, 287f
Template strand, 40
Teratogenic drugs, 173, 328
Terrorism. *See* Bioterrorism
Tertiary protein structure, 103, 103f
Testosterone, in gene expression, 48
Tetrodotoxin, 253
Textile industry, 101
Thalidomide, 328
Therapeutic cloning, 293–298, 294t,
        295f. *See also* Clones/cloning
Thermal cycler, 72, 72f, 196
Thermophiles, 123
Thermostable enzymes, 123
Thermus aquaticus, 60t, 72, 254
3' end, 33
Thymine, 32, 32f, 34f
Ti vectors, 68

Tissue donation, legal/ethical aspects of, 337–338
Tissue engineering, 285–286, 286f
  stem cells in, 290–293, 293f, 294t
Tissue plasminogen activator, 133t
Tissue typing, 283–284
TNT, biodegradation of, 224, 225f
Tobacco mosaic virus vaccine, 162, 162f
Trade secrets, 318t
Trademarks, 318t
Traits, inheritance of, 33
Transcription, 39–41, 39f, 41f
Transcription factors, 40, 47, 47f, 48
Transcriptional regulation, 46–50
Transcriptomics, 92
Transfection, 275
Transfer RNA, 44–45, 45f
Transformable spongiform encephalitis, 105–106
Transformation. See Bacterial transformation
Transgenes, 180
Transgenic biomanufacturing, 172–174
Transgenic organisms
  animals, 10, 10f, 180–186. See also Animal biotechnology
  ethical aspects of, 328–332, 331–332
  fish/shellfish, 241, 243, 247–250, 247t–248t, 249f, 250f. See also Aquaculture
  patents for, 76, 317–321, 331, 337–338, 340–341
  plants, 108, 155–170. See also Genetically modified foods; Plant biotechnology
    in China, 320
  ethical aspects of, 328–331
  in phytoremediation, 224, 225f
Translation, 39, 39f, 43–45
  ribosomes in, 44
  stages of, 44–45, 45f
Transplantation
  animal-human, 176, 284
  cell and tissue, 272–273
  legal/ethical aspects of, 337–338
  organ, 283–284
Trinitrotoluene (TNT), biodegradation of, 224, 225f
Triploids, in aquaculture, 249–250, 250f
Trisomy 21, 263–264, 264f
tRNA, 40–41, 44–45, 45f
Tsunami victims, DNA fingerprinting for, 199–200
Tuberculosis, 127, 141
Tumor-inducing plasmids, 158

Two-dimensional electrophoresis, 113, 113f
Type-I diabetes mellitus, 64–65, 100, 131, 132f, 133t, 272

Ultrafiltration, 111
Unjust enrichment, 338
Upstream processing, 107
U.S. Department of Agriculture, 169, 307t, 308–309, 310, 316t
U.S. Department of Energy, 141, 215, 229
U.S. Environmental Protection Agency (EPA), 169, 206, 307, 307t, 309–311, 316t
U.S. Patent and Trademark Office, 317
Utilitarian approach, 325–326
Utility, in patent law, 319–320

Vacanti, Charles, 285
Vaccines, 133t, 135–141, 273–275
  in aquaculture, 239
  attenuated, 137, 138
  bacterial/viral targets for, 139–141
  cocktail, 141
  DNA-based, 138
  DPT, 136, 137
  edible, 165
  HIV, 139, 140f, 141
  HPV, 139
  inactivated (killed), 137–138
  MMR, 136
  plant, 161–162, 162f
  polio, 136
  production of, 137–139
  regulation of, 316–317
  subunit, 137
  tuberculosis, 141
  types of, 137–139
Vaginal DNA, isolation of, 197, 198f
Validation technician, 229
Variable number tandem repeats, 192, 192f
Vectors, 61, 65–68
  bacterial artificial chromosome, 68
  bacteriophage, 66–67, 66t
  cosmid, 66t, 67
  definition of, 61
  development of, 61
  DNA libraries and, 69
  expression, 66t, 67–68, 125, 126f
  in gene therapy, 276–277
  ideal characteristics of, 65–66

Ti, 68
  types of, 66–68, 66t
Venter, J. Craig, 76, 90, 92, 143
Veterinary medicine, 176, 188
Vibrio spp.
  bioluminescent, 126–127, 257
  collagenase from, 255
Vibrio cholerae
  genome sequencing for, 143, 144t
  in shellfish, 239
Vibrio fisheri, 126–127, 127f
Vibrio harveyi, 127
Viral genomics, 144–145
Viral vectors, in gene therapy, 276–277
Viruses, synthetic, 145

Wastewater treatment, 219
Water pollution
  in aquaculture, 240–241
  bioremediation of, 11, 102, 218–220, 219f–221f, 228, 257–259. See also Bioremediation
Watson, James, 33, 58, 92, 261
Websites, for genes and chromosomes, 300
Weed consequences, 309
Werrett, Dave, 197
Western blotting, 81
  SDS-PAGE and, 115
White blood cells, 136
Whole-animal production systems, 108–109
Wildlife. See Animal(s)
Wine, grape lineages and, 204–205
World Trade Center attack, DNA fingerprinting in, 199

Xenotransplantation, 176, 284
X-ray crystallography, 116–117

Y-chromosome analysis, 203–204, 203f
Yeast, 121–122. See also Microbial biotechnology
  culture of, 122
  genome of, 93t
Yeast artificial chromosomes, 68
Yeast two-hybrid system, 128
Yogurt, 130

Zebrafish, 174, 174f
Zygotes, 37, 288, 288f